Lecture Notes in Mechanical Engineering

ABCM Series on Mechanical Sciences and Engineering

This series publishes selected papers as well as full proceedings of events organized and/or promoted by the Brazilian Society for Mechanical Sciences and Engineering (ABCM) on an international level. These include the International Congress of Mechanical Engineering (COBEM) and the International Symposium on Dynamic Problems of Mechanics (DINAME), among others.

More information about this series at http://www.springer.com/series/14172

José Eduardo Souza De Cursi
Editor

Proceedings of the 5th International Symposium on Uncertainty Quantification and Stochastic Modelling

Uncertainties 2020

ABCM
Associação Brasileira de Engenharia e Ciências Mecânicas

Springer

Editor
José Eduardo Souza De Cursi
Department Mechanics/Civil Engineering,
INSA Rouen Normandie
Saint-Etienne du Rouvray, France

ISSN 2195-4356 ISSN 2195-4364 (electronic)
Lecture Notes in Mechanical Engineering
ISSN 2662-3021 ISSN 2662-303X (electronic)
ABCM Series on Mechanical Sciences and Engineering
ISBN 978-3-030-53668-8 ISBN 978-3-030-53669-5 (eBook)
https://doi.org/10.1007/978-3-030-53669-5

This Springer imprint is published by the registered company Springer Nature Switzerland AG
The registered company address is: Gewerbestrasse 11, 6330 Cham, Switzerland

Preface

Uncertainty quantification (UQ) is a field of research that has taken off recently, driven by the ever-increasing need to assess risks, take into account variability and provide quantified estimates of the probabilities of failure, as well as estimates of their consequences and costs in human, social, environmental and economic terms.

The growing interest in UQ can be explained by the understanding that, despite all efforts in terms of procedures, norms and standardization, variabilities that cannot be eliminated remain. For example, the distribution of atoms and impurities in a material cannot yet be controlled to eliminate any variability: While such variations play only a limited role when dealing with a few atoms, the same cannot be said for macroscopic parts, which can reach large dimensions—the behavior can vary significantly from one point to another. Similarly, parts made in the same factory, in the same machine, by the same worker from the same material will not be identical and will exhibit variability. The effects of the environment, loads and the history of their use will then be factors determining the evolution of their behavior, introducing uncertainties about their condition. In addition, observations made on a system integrating them will be affected by measurement and model errors, which will be added to the uncertainties on the physical parameters.

In other fields, such as the life sciences, variability is intrinsic, as individuals possess characteristics that make them unique.

Thus, considering uncertainties, variability and errors is fundamental to producing solutions that are resilient, robust to environmental variations—or simply effective for the greatest number of individuals.

This book brings together contributions from European and Latin American experts in this field and presents applications, methods and models integrating stochastic, uncertain or variability aspects. The main techniques of the field are implemented in the contributions presented. It also presents new methods that will probably be developed in the coming years.

The Congress from which it originated is the product of a fruitful and long-standing Franco-Brazilian collaboration, part of the results of which are presented in this volume. This collaboration relies on many physical and institutional actors in both countries, including scientific societies, funding institutions, universities and schools, research centers and the fundamental actors which are the researchers, productors of knowledge, of which this book provides an overview.

May 2020 Eduardo Souza de Cursi

Organization

Program Chair

Eduardo Souza de Cursi	INSA Rouen, France

Editorial Board

Sondipon Adhikari	Swansea University, UK
Jose Arruda	University of Campinas, Brazil
Anas Batou	Université Paris-Est Marne-la-Vallée, France
André Beck	University of Sao Paulo, Brazil
Michael Beer	Leibniz University Hannover, Germany
Philippe Besnier	INSA Rennes, France
Jean-Marc Bourinet	SIGMA Clermont, France
Haroldo Campos Velho	INPE (National Institute for Space Research), Brazil
Rachid Ellaia	Ecole Mohamadia d'Ingenieurs, Morocco
Adriano Fabro	University of Brasilia, Brazil
Nicolas Gayton	SIGMA Clermont-Ferrand, France
Hanaa Hachimi	ENSA UIT, Morocco
Peter Hagedorn	TU Darmstadt, Germany
Rafael Holdorf Lopez	Universidade Federal de Santa Catarina, Brazil
Hector Jensen	Federico Santa Maria University, Chile
Ioannis Kougioumtzoglou	Columbia University, USA
Carlile Lavor	UNICAMP, Brazil
Jinglai Li	University of Liverpool, UK
Roberta Lima	PUC-Rio, Brazil
Guang Lin	Purdue University, USA
David Moens	KU Leuven, Belgium
He Qing Mu	South China University of Technology, China

Contents

Uncertainty Analysis and Estimation

Optimization

Modeling and Tools

Stick-Slip Oscillations in a Stochastic Multiphysics System

Roberta Lima(✉) and Rubens Sampaio

Department of Mechanical Engineering, PUC-Rio,
Rua Marquês de São Vicente, 225, Gávea, RJ 22451-900, Brazil
{robertalima,rsampaio}@puc-rio.br

Abstract. This work analyzes the stochastic response of a multiphysics system with stick-slip oscillations. The system is composed of two subsystems that interact, a mechanical with Coulomb friction and an electromagnetic (a DC motor). An imposed source voltage in the DC motor stochastically excites the system. This excitation combined with the dry-friction induces in the mechanical subsystem stochastic stick-slip oscillations. The resulting motion of the mechanical subsystem can be characterized by a random sequence of two qualitatively different and alternate modes, the stick- and slip-modes, with a non-smooth transition between them. The onset and the duration of each stick-mode are uncertain and depend on electromagnetic and mechanical parameters and variables, specially the position of the mechanical subsystem during the stick-mode. Duration and position are dependent random variables and must be jointly analyzed. The objective of this paper is to characterize and quantify this dependence, a novelty in the literature. The high amount of data required to perform the analysis and to construct joint histograms puts the problem into the class of big data problems.

Keywords: Multiphysics systems · Electromechanical systems · Nonlinear dynamics · Dry-friction · Uncertainties · Big data

1 Introduction

The presence of dry-friction in mechanical systems may induce the occurrence of stick-slip oscillations, a type of motion with a non-smooth behavior [1,3,11,13,17]. When stick-slip occurs, the system response is characterized by two qualitatively different modes, the stick- and slip-modes [10,15,31,34]. We call stick when the relative velocity between the bodies in contact is null in a time-interval of non-null duration and we call slip if the relative velocity is non-zero, or zero in isolated points. These two modes alternate with a non-smooth transition between them.

The literature dealing with sick-slip oscillations in mechanical systems is vast [2,9,36]. Nevertheless, this paper brings a novelty. It analyzes the sick-slip oscillations in a multiphysics system, an electromechanical system. The analyzed system is composed of two subsystems that interact, a mechanical and an electromagnetic (DC motor). In the mechanical subsystem there is dry-friction.

J. E. S. De Cursi (Ed.): Uncertainties 2020, LNME, pp. 3–17, 2021.
https://doi.org/10.1007/978-3-030-53669-5_1

By coupling between the two subsystems we mean mutual influence, i.e., the dynamics of the mechanical subsystem influences and is influenced by dynamics of the electromagnetic subsystem [19,23,30,32,33]. Consequently, the dry-friction present in the mechanical subsystem affects and is affected by the interaction. Furthermore, the sick-slip oscillations affect and are affected by the electromechanical coupling [25,26,28,29]. The system mode at each instant (stick or slip) depends on the imposed source voltage and on the state of the whole system, i.e., depends on mechanical and electromagnetic variables [27].

Traditionally, stick-slip is analyzed as the interplay of a friction and an elastic force. A stick lasts as long as the these two forces balance themselves. In our case, there is no elastic force, it is replaced by a force generated by the motor. During the stick modes, the DC motor rules the system dynamics. This is described by an initial value problem involving the current and the imposed voltage in the electric circuit of the motor.

To better characterize stick-slip oscillations, it is important to consider the uncertainties that affect the system behavior. These uncertainties influence the instants at which each mode starts, the duration of the modes and consequently the number of time intervals in which stick or slip occur. Due to the uncertainties, the sequence of stick- and slip-modes becomes stochastic and the variables that characterize it become random [18,20,21]. Deterministic predictions of these variables will probably fail. Predictions should be made with a stochastic approach.

A common source of uncertainties in DC motors is the imposed source voltage. Usually, transmission lines are subjected to noises which perturb transmitted signals [12,35]. These uncertainties affect the behavior of the systems powered by these lines and are critical in planning and modeling.

The analysis performed in this paper takes into account two sources of uncertainties. Besides the imposed source voltage in the electromagnetic subsystem, is considered that the initial condition of the position of the mechanical subsystem is random. We aim to quantify the influence of these two sources of uncertainties in the system response. The focus is on the analysis of the stick duration, a random variable that depends on electromagnetic and mechanical parameters and variables, specially the position of the mechanical subsystem during the stick-mode. Duration and position are dependent random variables and should be jointly analyzed. Only marginal distributions are not enough to characterize them. The objective of this paper is to characterize and quantify this dependence [22,24], a novelty in the literature.

With Monte Carlo simulations, scatter plots, marginal and joint histograms of the stick duration and the position of the mechanical system during the stick were construct for stick duration for different values of friction coefficient. To construct the joint histograms with accuracy, several realizations of the system response were required. Since each realization was obtained by a numerical integration of the initial value problem that gives the dynamics of the system, the computational cost to perform the simulations was high. The developed analysis belong to a class of big data problems.

This paper is organized as follows. Section 2 presents the dynamics of an electromechanical system with dry-friction, i.e., the initial value problem (IVP) that describes the dynamics of the coupled system. The dry-friction force model, and the necessary conditions for the occurrence of the stick- and slip-modes are defined in Sect. 3. In Sect. 4, an analytical approximation to the upper bound for the stick duration is presented. The influence of mechanical and electromagnetic variables and parameters values in this upper bound is discussed in Sect. 5. With the analytical approximation, it is possible to determine the influence of the position of the mechanical system during the stick in the stick duration. The upper bound will define the support of the joint distributions of interest in the paper. The construction of the probabilistic model of the uncertain source voltage is given in Sect. 6. Scatter plots, marginal and joint histograms constructed for stick duration and position of the mechanical system during the stick are presented in Sect. 7.

2 Dynamics of the Electromechanical System with Dry-Friction

The system analyzed in this paper is composed by a cart-disk whose motion is driven by a DC motor. The motor is coupled to the cart through a pin that slides into a slot machined on a plate that is part of the cart, as shown in Fig. 1. The pin hole is drilled off-center on a disk fixed in the axis of the motor, so that the motor rotational motion is transformed into horizontal cart motion over a rail. The eccentricity influences heavily the nonlinearity of the system. Even small eccentricities produces high nonlinearities [4,6–8,14]. The initial value problem to the coupled motor-disk-cart system with dry-friction is given Eqs. (1) and (2). Find $(\alpha,\ c)$ such that, for all $t > 0$,

$$l\dot{c}(t) + r\,c(t) + k_e\dot{\alpha}(t) = \nu(t)\,, \tag{1}$$

$$\begin{aligned}\ddot{\alpha}(t)[j_m + m\,d^2\,(\sin(\alpha(t)))^2] + \dot{\alpha}(t)[b_m + m\,d^2\,\sin(\alpha(t))\cos(\alpha(t))]\\ -k_e c(t) = -f_r(t)\,d\,\sin(\alpha(t))\,,\end{aligned} \tag{2}$$

with the initial conditions

$$\dot{\alpha}(0) = \beta, \quad \alpha(0) = \alpha_0, \quad c(0) = c_0, \tag{3}$$

where t is the time, ν is the source voltage, c is the electric current, $\dot{\alpha}$ is the angular speed of the motor, l is the electric inductance, j_m is the motor moment of inertia, b_m is the damping ratio in the transmission of the torque generated by the motor, k_e is the motor electromagnetic force constant, r is the electrical resistance, m is the mass of the cart, d is the eccentricity of the pin and f_r is the dry-friction force between the cart and the rail. The source voltage ν is considered to be

$$\nu(t) = \nu_0 + \nu_1 \sin(\omega_v\, t + \theta)\,, \tag{4}$$

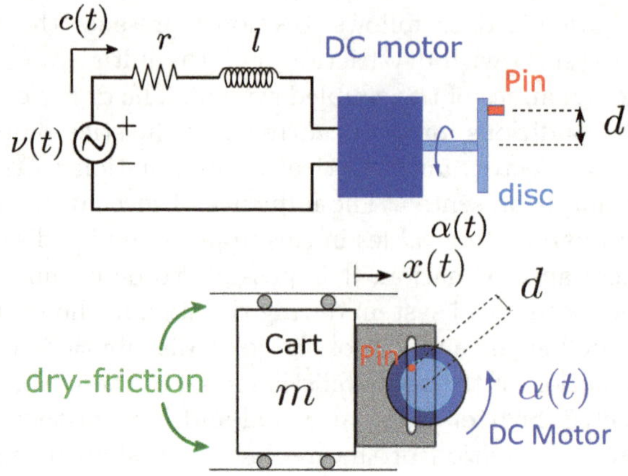

Fig. 1. Electromechanical system with dry-friction between the cart and the rail.

i.e., the source voltage oscillates around ν_0 with amplitude ν_1, frequency ω_v and phase θ. Please note that the system state is given by three variables, two of them mechanical (angular velocity and position of the motor) and one of them electromagnetic (current). The dynamics of the electromechanical system analyzed is given by an initial value problem comprising a set of two coupled differential equations. The coupling between the mechanical and electromagnetic subsystems is not given by a functional relation. It depends on the system state and, consequently, depends on initial conditions.

3 Dry-Friction Model and Necessary Conditions for the Occurrence of the Stick- and Slip-Modes

The friction is modeled as Coulomb's and as simple as possible, Fig. 2.

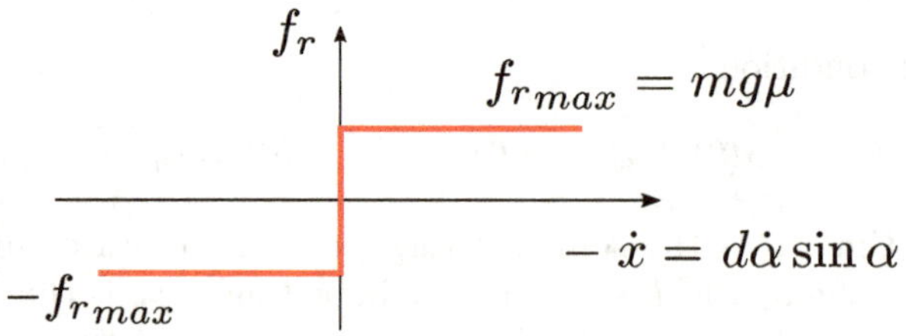

Fig. 2. Coulomb dry-friction.

When the cart velocity is not zero, $\dot{x} \neq 0$, the dry-friction force can only assume the values $-f_{r\,max}$ or $f_{r\,max}$. However, when $\dot{x} = 0$, it can assume any value in the interval $[-f_{r\,max}, f_{r\,max}]$. It is considered $f_{r\,max} = \mu m\, g$, where g is the gravity and μ is the friction coefficient between the cart and the rail. Note that the dry-friction force is not a function of the cart velocity. For one value of cart velocity, $\dot{x} = 0$, the friction can take infinite values, but it is confined in an interval, i.e., its magnitude is bounded.

Regarding Coulomb's friction models, the selected model is considered to be the simplest. It does not present, for example, hysteresis loops or a difference between the kinetic and static friction coefficients. It is a bare minimum to study stick-slip. Furthermore, the friction model does not influence the analysis developed in the paper regarding the upper bound for the stick duration. Please remark that during the stick mode, the important parameter is only the static friction coefficient.

The presence of the dry-friction force can induce in the system stick-slip oscillations. Depending on the values of the system parameters, the response of the system may be composed of a sequence alternating stick- and slip-modes. In Fig. 3 it is shown a possible sequence of stick- and slip-modes for the interval of analysis $[0, t_a]$.

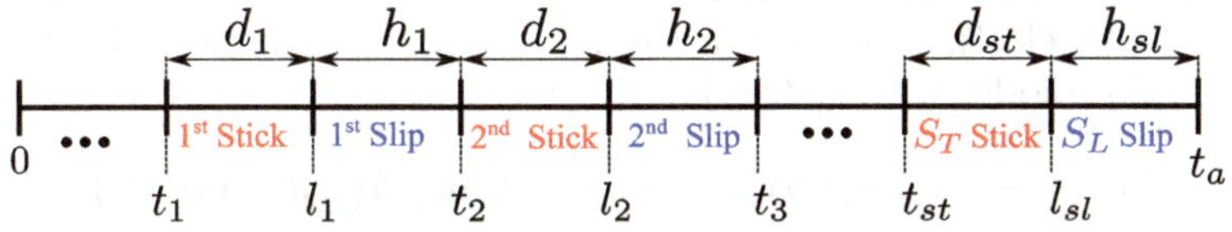

Fig. 3. Sequence of stick- and slip-modes in a electromechanical system response.

During a stick-mode, the disk-cart does not move, so that the angle, describing the angular position of the disk, is constant. The frictional force and the current, however, can vary. Hence, stick means only no motion of the disk-cart, the mechanical subsystem. The electromagnetic subsystem continues to change its state until it gathers enough power to move the disk-cart again. The stick-mode occurs when $\dot{\alpha} = 0$ in an time interval and the frictional force is the interval $[-f_{r\,max}, f_{r\,max}]$. To better understand the effect of the stick-mode in the dynamics of the electromechanical system, let us investigate how it influences the IVP that describes the system dynamics.

Considering $\dot{\alpha} = 0$ and $\ddot{\alpha} = 0$ in Eqs. (1) and (2), the following equations are obtained

$$l\dot{c}(t) + r\, c(t) = \nu_0 + \nu_1 \sin(\omega_v\, t)\,, \tag{5}$$

$$k_e\, c(t) = f_r(t)\, d\, \sin(\alpha(t))\,. \tag{6}$$

Observe that the first differential equation of the IVP becomes a differential equation (Eq. (5)) which just depends on the current, an electromagnetic variable. The second differential equation of the IVP becomes an algebraic equation (Eqs. (6)). During the stick-mode, the dynamics of the system is not given anymore by an IVP with two coupled equations. It is reduced to one differential equation depending on an electromagnetic variable. The initial condition for the current is the value of it in the beginning of the stick-mode.

During the stick-mode, the coupling between the electromagnetic and mechanical subsystems is made by Eq. (6). The value of the friction force, during the stick-mode, can vary and follows Eqs. (6), i.e., it depends on an electromagnetic variable and on the angular position of the motor, a mechanical variable. We consider this a novelty in the literature.

Remark that during the stick-mode, the sum of the forces that act over the cart is zero (it does not move). The horizontal coupling force between the DC motor and the cart, f, is balanced by the dry-friction force, f_r. This balance lasts until the frictional force, given in Eq. (6), reaches its maximum value, $f_{r_{max}}$. Recall that when $\dot{x}(t) = 0$, the magnitude of the frictional force is bounded.

During the stick-mode, the dynamics of the system is governed only by the dynamics of the electrical circuit of the motor, the electromagnetic subsystem. During the slip-mode, the dry-friction force is

$$f_r(t) = -m\,g\mu\,\mathrm{sgn}(\dot{x}(t)) = -m\,g\mu\,\mathrm{sgn}(-\dot{\alpha}(t)\,d\,\sin(\alpha(t)))\,. \tag{7}$$

4 Approximation to the Upper Bound for the Stick Duration

As explained in the introduction, one of the variables of great interest in systems with stick-slip dynamics is the stick duration. To obtain an analytical approximation to the upper bound for this variable for the electromechanical system analyzed, the starting point is the initial-value problem that describes the dynamics of the coupled motor-disk-cart system with dry-friction during the stick-phase given by Eq. (5), i.e., the reduced initial-value problem. Calling

- the instant of beginning of a stick as t_b;
- the instant of end of a stick as t_e;
- the angle of the disk in the beginning of a stick, the angle where the disk is stuck, as $\alpha(t_b) = \alpha^\star$;
- and the current in the beginning of a stick as $c(t_b) = c_0$.

The stick-mode ($t_b \leq t \leq t_e$) is governed by the following linear initial-value problem:

$$\begin{aligned} &l\dot{c}(t) + \ r\,c(t) = \nu_0 + \nu_1 \sin(\omega_v\,t). \\ &c(t_b) = c_0 \end{aligned} \tag{8}$$

The analytical solution of this initial-value problem ($t_b \leq t \leq t_e$) is known and it is given by

$$c(t) = \left[\frac{c_0 - \dfrac{\nu_0}{r} - \dfrac{\nu_1\sqrt{r^2+l^2}}{l^2\omega_v^2+r^2}\sin(\omega_v t_b + \phi)}{e^{-r/l\, t_b}}\right] e^{-r/l\, t} + \frac{\nu_1\sqrt{r^2+l^2}}{l^2\omega_v^2+r^2}\sin(\omega_v t + \phi) + \frac{\nu_0}{r}, \quad (9)$$

where $\phi = \arctan \dfrac{-l}{r}$. Since, the variable of interest is the upper bound for the stick duration, the influence of the initial condition (c_0) can be neglected, and the longer the stick duration the better the approximation. Doing this, the solution of the IVP during the stick-mode ($t_b \leq t \leq t_e$) can be approximated by:

$$c(t) \approx \frac{\nu_1\sqrt{r^2+l^2}}{l^2\omega_v^2+r^2}\sin(\omega_v t + \phi) + \frac{\nu_0}{r}. \quad (10)$$

Recalling that during the stick-mode

- $f_r \in [-f_{r\,max}, f_{r\,max}]$ and
- $k_e\, c(t) = f_r(t)\, d\, \sin\alpha(t)$.

The stick ends when

$$c(t_e) = \frac{f_{r\,max}\, d\, \sin\alpha^\star}{k_e}. \quad (11)$$

The upper bound for the stick duration $t_e - t_b$ is the maximal interval during which the current, given by Eq. (10), is lower than the value $\dfrac{f_{r\,max}\, d\, \sin\alpha^\star}{k_e}$. Please observe that if c is always under this value, the stick can last forever. If c is always above, the stick does not happen. The upper bound for the stick duration depends on mechanical and electrical parameters as well as a characteristic of the stick, its angular position.

5 Influence of Mechanical and Electromagnetic Variables and Parameters in the Approximation to the Upper Bound for the Stick Duration

The upper bound for the stick duration depends on mechanical and electrical variables and geometry of the stick. Among all variables, two of them can be highlighted. They are the angle of the disk in the beginning of a stick, $\alpha(t_1) = \alpha^\star$, and the value of the friction coefficient, μ. To better visualize the influence of these two variables, values to all the others parameters were selected and fixed (they are given in Table 1) and the approximation to the upper bound for the stick duration was computed for different combinations of $\alpha^\star$ and μ. The motor parameters were obtained from the specifications of the motor Maxon DC brushless number 411678. They are listed in Table 1.

Table 1. Parameter values

$l = 1.880 \times 10^{-4}$ H	$k_e = 5.330 \times 10^{-2}$ V/(rad/s)
$j_m = 1.210 \times 10^{-4}$ kg m^2	$r = 0.307\ \Omega$
$b_m = 1.545 \times 10^{-4}$ Nm/(rad/s)	$m = 5.000$ kg
$\nu_0 = 1.000$ V	$\nu_1 = 0.500$ V
$\omega_v = 10.000$ rad/s	$d = 0.010$ m

The results are shown in Figs. 4 and 5. In the first graph, the values of friction coefficient are 0.1, 0.2, 0.3, 0.4, 0.5. It can be observed that with $\mu = 0.1$, the approximation to the upper bound for the stick duration is 0 for all $\alpha^\star$, which means that there is no stick. When $\mu = 0.2$, it is possible to have stick. The longest stick happens when $\alpha^\star = \pi/2$ or $\alpha^\star = 3\pi/2$. In the region that $\alpha^\star$ is around 0 or π, there is no stick. As the friction coefficient increases, the region where there is no stick shrinks.

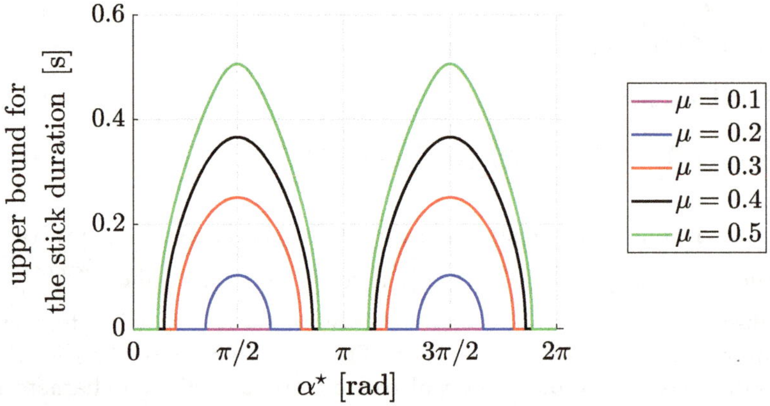

Fig. 4. Approximation to the upper bound for the stick duration as function of the angle of the disk in the beginning of a stick for different values of the friction coefficient.

In the second graph, the values of friction coefficient are 0.6, 0.7, 0.8, 0.9, 1.0. For these values, an interesting behavior occurs. There are regions of $\alpha^\star$ around $\pi/2$ and $3\pi/2$ that the stick can last forever. This means that if a stick happens when $\alpha^\star$ is in a certain region, the stick will last forever. In other words, the motor does not gather enough power to move the mechanical part again.

6 Construction of a Probabilistic Model to the Source Voltage

As it was explained in the introduction, in this paper we analyze the stick-slip oscillator with a stochastic approach. We consider the source voltage and the

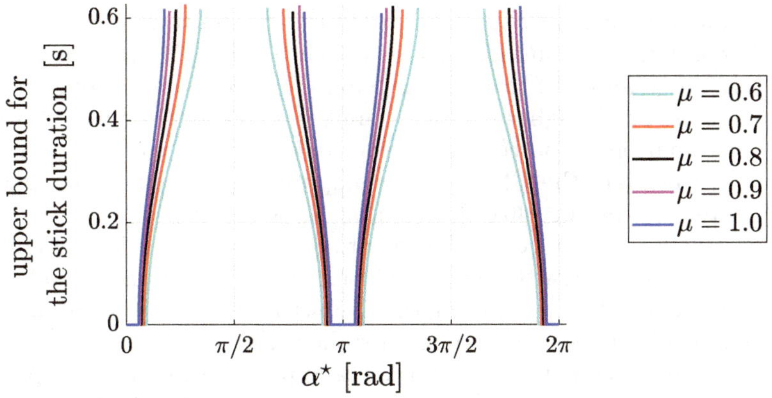

Fig. 5. Approximation to the upper bound for the stick duration as function of the angle of the disk in the beginning of a stick for different values of the friction coefficient.

initial condition of the angular position of the disk as sources of uncertainties in the stick-slip oscillator problem. We model the phase of the source voltage and the initial condition of the angular position of the disk as uniform random variables over $[0, 2\pi]$.

Due to the consideration of uncertainties, the equation of motion of the system, became a stochastic differential equation. Thus, the response of the stochastic stick-slip oscillator is a random process which presents a sequence alternating stick and slip-modes. We are interested in the stochastic characterization of these sequences. Defined a time interval for analysis, the variables of interest are the number of time intervals in which stick or slip occur, the instants at which they start, and their duration. These variables are modeled as stochastic objects in order to allow the stochastic characterization the dynamics of the oscillator.

The analysis of all these variables is important to characterize the stick-slip process. However, in this paper, we focus on the position of the mechanical system (disk) and duration of the first stick. The objective is to determine the stochastic dependence of these variables, i.e., their joint distribution.

7 Scatter Plots, Marginal and Joint Histograms for Stick Duration

With Monte Carlo simulations, scatter plots and joint histograms of the duration and the position of the disk during the first stick were construct for different values of friction coefficient. We considered $\mu = 0.2$, $\mu = 0.3$, $\mu = 0.4$ and $\mu = 0.5$. For each value of μ, the initial value problem that characterizes the system dynamics were numerically integrated $10,000$ times, totalizing $40,000$ numerical integrations. The 4th- and 5th-order Runge-Kutta method is used for the time integration scheme with a time-step equal to 10^{-3} s in a range of $[0.0, 2.0]$ seconds. A small time-step was required in order to predict the instants of change of mode with accuracy.

Besides, a large number of realizations of the system response was necessary in order to construct the joint histograms with accuracy [5,16]. It is important to remark that the number of realizations required to construct a joint histogram with accuracy is much higher than the number of realizations required to construct a marginal one. The computational and temporal costs to perform the simulations were high, the CPU time was approximately 60 h. The amount of data generated was around 40 GB, i.e., a big data problem.

The parameters values used in all simulations are listed in Table 1. Figures 6, 7, 8, 9, 10, 11, 12 and 13, show the marginal histograms of the duration and the position of the mechanical system during the first stick, scatter plots and normalized joint histograms for these two variables for the different values of friction coefficient. In the scatter plots, it is also shown the approximation to the upper bound for the stick duration as function of the position of the disk ($\alpha^\star$) during the stick. Please remark that the upper bounds defines the supports of the joint distributions of the duration and the position of the disk during a stick.

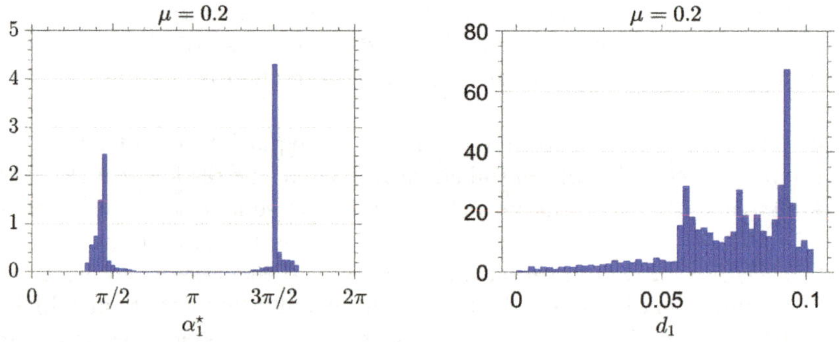

Fig. 6. Normalized marginal histograms of the duration and the position of the mechanical system of the first stick. The graphs were constructed with 10,000 realizations of the system response for $\mu = 0.2$.

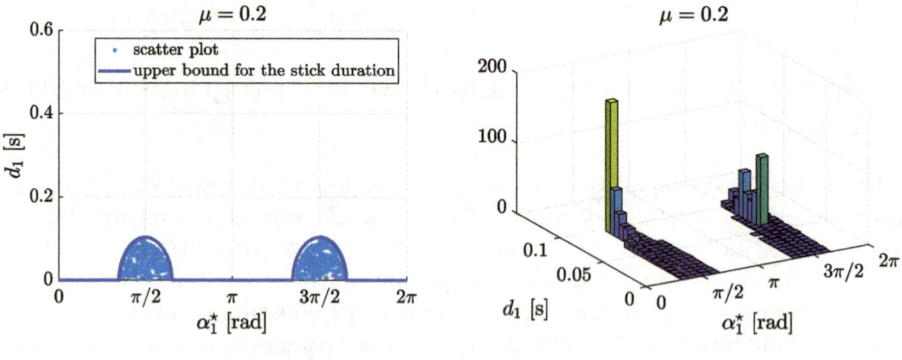

Fig. 7. Scatter plots and normalized joint histograms of the duration and the position of the mechanical system of the first stick. The graphs were constructed with 10,000 realizations of the system response for $\mu = 0.2$.

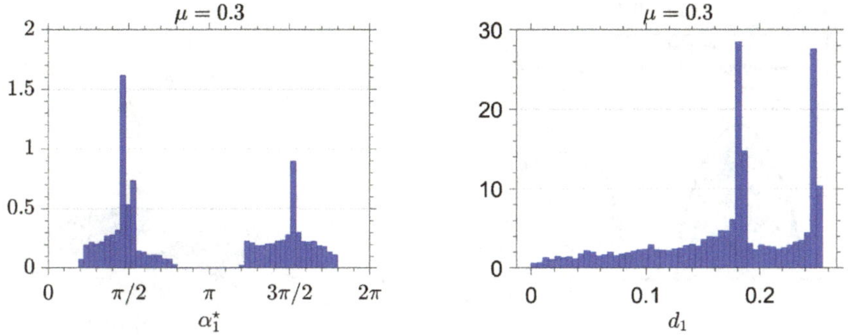

Fig. 8. Normalized marginal histograms of the duration and the position of the mechanical system of the first stick. The graphs were constructed with 10,000 realizations of the system response for $\mu = 0.3$.

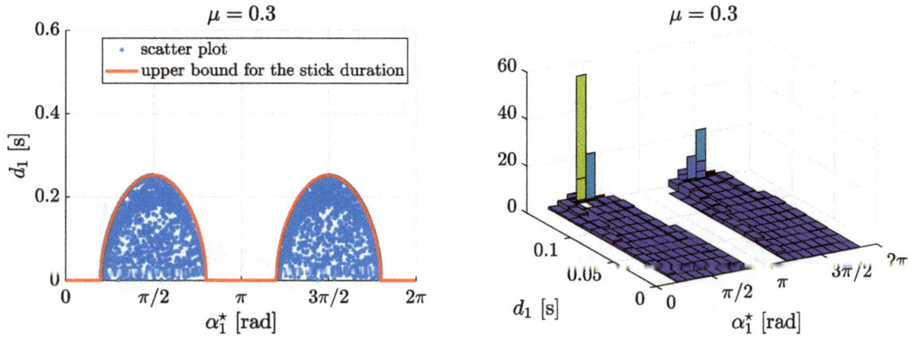

Fig. 9. Scatter plots and normalized joint histograms of the duration and the position of the mechanical system of the first stick. The graphs were constructed with 10,000 realizations of the system response for $\mu = 0.3$.

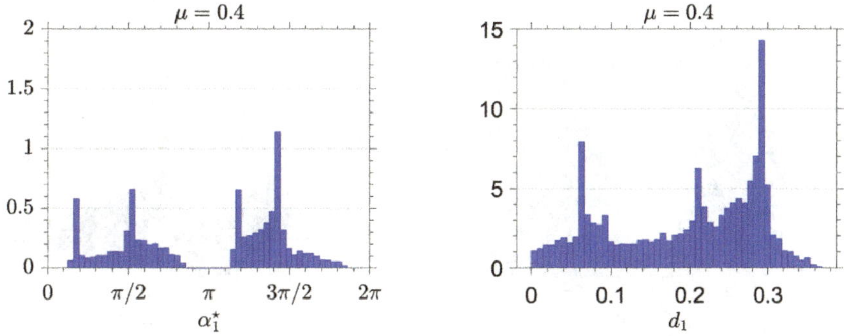

Fig. 10. Normalized marginal histograms of the duration and the position of the mechanical system of the first stick. The graphs were constructed with 10,000 realizations of the system response for $\mu = 0.4$.

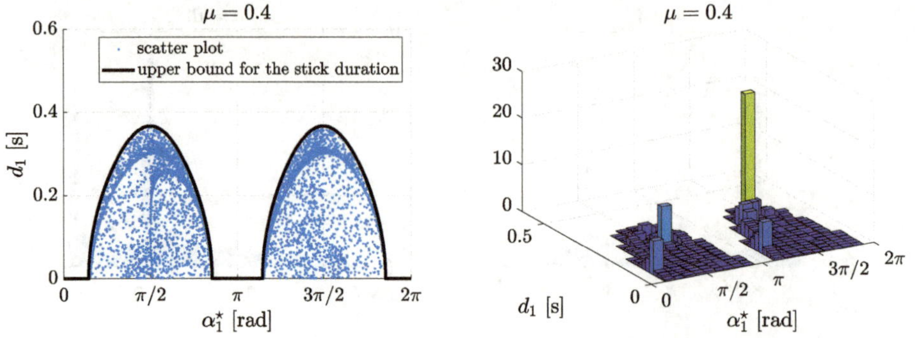

Fig. 11. Scatter plots and normalized joint histograms of the duration and the position of the mechanical system of the first stick. The graphs were constructed with 10, 000 realizations of the system response for $\mu = 0.4$.

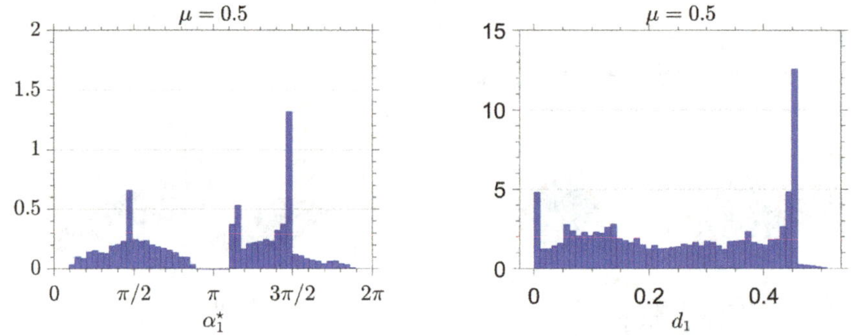

Fig. 12. Normalized marginal histograms of the duration and the position of the mechanical system of the first stick. The graphs were constructed with 10, 000 realizations of the system response for $\mu = 0.5$.

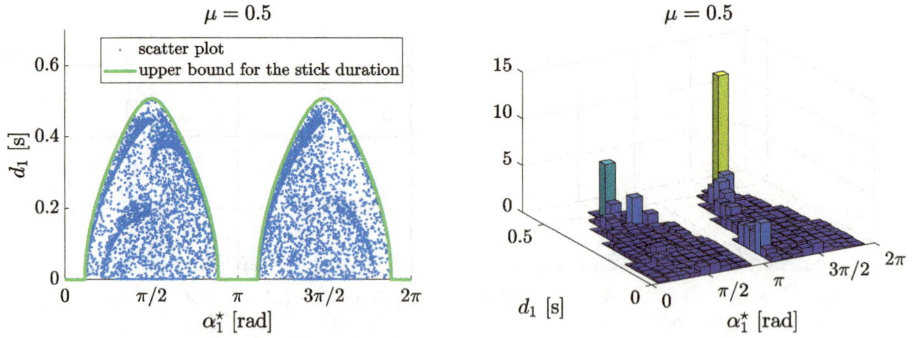

Fig. 13. Scatter plots and normalized joint histograms of the duration and the position of the mechanical system of the first stick. The graphs were constructed with 10, 000 realizations of the system response for $\mu = 0.5$.

8 Conclusions

This article analyzes the dynamics of an electromechanical system composed by a cart and a DC motor. The coupling between the motor and the cart is made by a mechanism called scotch yoke, so that the motor rotational motion is transformed in horizontal cart motion on a rail. There is dry-friction between the cart and the rail. The resulting motion of the cart can be characterized by stick- and slip-modes, with a non-smooth transition between them.

The source of uncertainties of the multiphysics system analyzed in this paper are the imposed source voltage in the electromagnetic subsystem and the initial condition of the position of the mechanical subsystem. The excitation induces in the mechanical subsystem stochastic stick-slip oscillations. Mechanical and electromagnetic parameters influence this stochastic sequence. The focus is on the analysis of the stick duration, a random variable that depends on electromagnetic and mechanical parameters and variables, specially the position of the mechanical subsystem during the stick-mode. Duration and position are dependent random variables and were jointly analyzed. With Monte Carlo simulations, scatter plots, marginal and joint histograms were construct for different values of friction coefficient.

Acknowledgments. The authors acknowledge the support given by FAPERJ, CNPq and CAPES.

References

1. Anh, L.: Dynamics of Mechanical Systems with Coulomb Friction, vol. 1. Springer, Berlin (2002)
2. Berger, E.: Friction modeling for dynamic system simulation. Appl. Mech. Rev. **55**(6), 535–577 (2002)
3. Cao, Q., Léger, A.: A Smooth and Discontinuous Oscillator: Theory, Methodology and Applications, vol. 1. Springer, Berlin (2017)
4. Cartmell, M.: Introduction to Linear, Parametric and Nonlinear Vibrations, vol. 260. Springer, Heidelberg (1990)
5. Souza de Cursi, E., Sampaio, R.: Uncertainty Quantification and Stochastic Modeling with Matlab. Elsevier, ISTE Press (2015)
6. Dantas, M., Sampaio, R., Lima, R.: Asymptotically stable periodic orbits of a coupled electromechanical system. Nonl. Dyn. **78**, 29–35 (2014)
7. Dantas, M., Sampaio, R., Lima, R.: Existence and asymptotic stability of periodic orbits for a class of electromechanical systems: a perturbation theory approach. Zeitschrift für angewandte Mathematik und Physik **67**, 2 (2016)
8. Dantas, M., Sampaio, R., Lima, R.: Phase bifurcations in an electromechanical system. IUTAM Procedia **1**(19), 193–200 (2016)
9. Fidlin, A.: Nonlinear Oscillations in Mechanical Engineering. EUA. Springer, Heidelberg (2006)
10. Galvanetto, U., Bishop, S.: Dynamics of a simple damped oscillator under going stick-slip vibrations. Meccanica **34**, 337–347 (1999)
11. Glendinning, P., Jeffrey, M.: An Introduction to Piecewise Smooth Dynamics, vol. 1. Birkhäuser, Switzerland (2019)

12. Hlalele, T., Du, S.: Analysis of power transmission line uncertainties: status review. J. Electr. Electron. Syst. **5**(3), 1–5 (2016)
13. Jeffrey, M.: Hidden Dynamics: The Mathematics of Switches, Decisions and Other Discontinuous Behaviour, vol. 1. Springer, Cham (2018)
14. Jordan, D., Smith, P.: Nonlinear Ordinary Differential Equations, vol. 560. Oxford University Press, Oxford (2007)
15. Leine, R., Van Campen, D., Kraker, A., Van den Steen, L.: Stick-slip vibrations induced by alternate friction models. Nonlinear Dyn. **16**(1), 45–54 (2019)
16. Lima, R., Sampaio, R.: Modelagem Estocástica e Geração de Amostras de Variáveis e Vetores Aleatórios. Notas de Matemática Aplicada, vol. 70. SBMAC (2012). http://www.sbmac.org.br/arquivos/notas/livro_70.pdf
17. Lima, R., Sampaio, R.: Stick-mode duration of a dry-friction oscillator with an uncertain model. J. Sound Vib. **353**, 259–271 (2015)
18. Lima, R., Sampaio, R.: Analysis of a dry-friction oscillator driven by a stochastic base motion. In: Third International Symposium on Uncertainty Quantification and Stochastic Modeling (Uncertainties 2016). Maresias, Brazil (2016)
19. Lima, R., Sampaio, R.: Two parametric excited nonlinear systems due to electromechanical coupling. J. Brazil. Soc. Mech. Sci. Eng. **38**, 931–943 (2016)
20. Lima, R., Sampaio, R.: Construction of a statistical model for the dynamics of a base-driven stick-slip oscillator. Mech. Syst. Signal Process. **91**, 157–166 (2017)
21. Lima, R., Sampaio, R.: Parametric analysis of the statistical model of the stick-slip process. J. Sound Vib. **397**, 141–151 (2017)
22. Lima, R., Sampaio, R.: Uncertainty quantification and cumulative distribution function: how are they related? In: Polpo, A., Stern, J., Louzada, F., Izbicki, R., Takada, H. (eds.) Bayesian Inference and Maximum Entropy Methods in Science and Engineering. Springer, Cham (2017)
23. Lima, R., Sampaio, R.: Pitfalls in the dynamics of coupled electromechanical systems. In: Proceeding Series of the Brazilian Society of Computational and Applied Mathematics (CNMAC 2018), pp. 1–7. Campinas, Brazil (2018)
24. Lima, R., Sampaio, R.: What is uncertainty quantification? J. Brazil. Soc. Mech. Sci. Eng. **40**, 155 (2018)
25. Lima, R., Sampaio, R.: Dynamics of a system with dry-friction and electromechanical coupling. In: Proceeding of the 25th International Congress of Mechanical Engineering (COBEM 2019). Uberlândia, Brazil (2019)
26. Lima, R., Sampaio, R.: Electromechanical system with a stochastic friction field. Mecánica Computacional **XXXVII**(18), 667–677 (2019)
27. Lima, R., Sampaio, R.: Nonlinear dynamics of a coupled system with dry-friction. In: Proceeding Series of the Brazilian Society of Computational and Applied Mathematics (CNMAC 2019). Uberlândia, Brazil (2019)
28. Lima, R., Sampaio, R.: Stick-slip oscillations or couple-decouple oscillations? In: Proceeding of the XVIII International Symposium on Dynamic Problems of Mechanics (DINAME 2019). Búzios, Brazil (2019)
29. Lima, R., Sampaio, R.: Stick-slip oscillations in a multiphysics system. Nonlinear Dyn. **100**, 2215–2224 (2020)
30. Lima, R., Sampaio, R., Hagedorn, P., Deü, J.F.: Comments on the paper 'on nonlinear dynamics behavior of an electro-mechanical pendulum excited by a nonideal motor and a chaos control taking into account parametric errors' published in this journal. J. Brazil. Soc. Mech. Sci. Eng. **41**, 552 (2019)
31. Luo, A., Gegg, B.: Stick and non-stick periodic motions in periodically forced oscillators with dry friction. J. Sound Vib. **291**(1–2), 132–168 (2006)

32. Manhães, W., Sampaio, R., Lima, R., Hagedorn, P.: Two coupling mechanisms compared by their lagrangians. In: Proceeding of the XVIII International Symposium on Dynamic Problems of Mechanics (DINAME 2019). Búzios, Brazil (2019)
33. Manhães, W., Sampaio, R., Lima, R., Hagedorn, P., Deü, J.F.: Lagrangians for electromechanical systems. Mecánica Comput. XXXVI **42**, 1911–1934 (2018)
34. Olejnik, P., Awrejcewicz, J.: Application of hénon method in numerical estimation of the stick-slip transitions existing in filippov-type discontinuous dynamical systems with dry friction. Nonlinear Dyn. **73**(1), 723–736 (2013)
35. Sivanagaraju, G., Chakrabarti, S., Srivastava, S.: Uncertainty in transmission line parameters: estimation and impact on line current differential protection. IEEE Trans. Instrum. Meas. **63**(6), 1496–1504 (2014)
36. Vande Vandre, B., Van Campen, D., De Kraker, A.: An approximate analysis of dry-friction-induced stick-slip vibrations by a smoothing procedure. Nonlinear Dyn. **19**, 157–169 (1999)

Some Tools to Study Random Fractional Differential Equations and Applications

Clara Burgos, Juan-Carlos Cortés(✉), María-Dolores Roselló, and Rafael-J. Villanueva

Instituto Unversitario de Matemática Multidisciplinar, Universitat Politècnica de València, Camino de vera s/n, 46022 Valencia, Spain
jccortes@imm.upv.es
https://www.imm.upv.es/

Abstract. Random fractional differential equations are useful mathematical tools to model problems involving memory effects and uncertainties. In this contribution, we present some results, which extent their deterministic counterpart, to fractional differential equations whose initial conditions and coefficients are random variables and/or stochastic process. The probabilistic analysis utilizes the random mean square calculus. For the sake of completeness, we study both autonomous and non-autonomous initial value problems. The analysis includes the computation of analytical and numerical solutions, as well as their main probabilistic information such as the mean, the variance and the first probability density function. Several examples illustrating the theoretical results are shown.

Keywords: Fractional differential equations · Mean square calculus · Mean square caputo fractional derivative · Fröbenius method · Random fractional numerical methods · Maximum Entropy Principle · Random Variable Transformation Technique

1 Introduction

Increasingly, fractional calculus is becoming a useful tool in mathematical modeling. This kind of derivatives are able to model some processes that the classical derivatives can not. In particular, this happens in those processes subject to after-effects, i.e., systems where past stage affects significantly to the present state of the system. For example, this memory effect appears when modelling the dynamics of physical properties in viscoelastic materials or the dynamics of some diseases [1]. In these two examples, the present state of the system depends strongly upon the previous evolution of the system on a certain whole interval. This key feature is taken into account when considering adequate fractional derivatives that, as we will see later, consider the behaviour of the function over an interval rather than at a point. In practice, we also must consider the uncertainties often present in the model due to measurement errors, lack of information, complexity of the phenomenon under analysis, etc. These facts motivate

J. E. S. De Cursi (Ed.): Uncertainties 2020, LNME, pp. 18–34, 2021.
https://doi.org/10.1007/978-3-030-53669-5_2

that random fractional differential equations are plausible mathematical tools to model complex problems subject to both after-effects and uncertainties.

This contribution is addressed to present some useful results in the realm of random fractional differential equations using the mean square calculus. It is important to point out that we will deal with random differential equations instead of stochastic fractional differential equations. In the former case uncertainties have regular sample paths (continuous, differentiable, etc.), while in the later case the noise is driven by irregular sample paths, typically by the Wiener process whose trajectories are nowhere differentiable. The mathematical treatment of stochastic differential equations requires a Itô calculus and in this type of equations uncertainties are considered using the Gaussian pattern (via the Wiener process or via its formal derivative, the white noise process) [2]. This fact restricts the applicability of these equations since there are many phenomena whose uncertainties cannot be properly described using Gaussian noises. Complementary, random differential equations can be analysed using the random mean square calculus that is a natural extension of the classical Newton-Leibniz calculus to the random setting. The main advantage of this class of differential equations is that a wide range of probabilistic distributions can be assigned to their random inputs including the Gaussian pattern [3].

This contribution is organized as follows. In Sect. 2, we introduce the extension of the deterministic Riemann-Liouville fractional integral and Caputo fractional derivative to the random context in the mean square sense. In Sect. 3, we apply generalized power series to construct via the Fröbenius method analytical solutions of some classes of autonomous and non-autonomous initial value problems (IVP) with randomness. In Sect. 4 we present a numerical method for solving random fractional differential equations [4]. In all the aforementioned problems, we show how to calculate reliable approximations for the mean and the variance of the solution stochastic process. We take advantage of this parametric information along with the maximum entropy technique to construct approximations of the first probability density function of the solution. Conclusions are summarized in Sect. 5.

2 Random Fractional Differential Operators

In dealing with fractional differential equations different non-integer derivatives can be considered depending upon their real world applications [5]. In this contribution we will consider the Caputo derivative because the initial value of fractional differential equation with Caputo derivative is the same as that of integer differential equation [6,7]. This key property makes it suitable for our goals as it will be apparent later. The definition of the Caputo derivative is based on the Riemann-Liouville integral [6]. So, in order to present their corresponding definitions in the random scenario, we will first introduce the mean square Riemann-Liouville integral and, afterwards, the mean square Caputo derivative.

Throughout this paper we will assume that $(\Omega, \mathcal{F}, \mathbb{P})$ is a complete probability space. We will denote by $(\mathrm{L}^2(\Omega), \|\cdot\|_2)$ the Hilbert space of second-order real random variables $X : \Omega \to \mathbb{R}$, i.e. such that $\mathbb{E}[X^2] < \infty$, so having finite variance. Here $\mathbb{E}[\cdot]$ denotes the expectation operator. Given $X \in \mathrm{L}^2(\Omega)$, its norm is defined by $\|X\cdot\|_2 = (\mathbb{E}[X^2])^{1/2}$. This norm infers a convergence usually termed mean square convergence. A sequence of second order random variables, $\{X_n : n \geq 0\}$, is said to be mean square convergent to $X \in \mathrm{L}^2(\Omega)$ if and only if $\|X_n - X\|_2 \to 0$ as $n \to \infty$. This is denoted by $X_n \xrightarrow[n\to+\infty]{\text{m.s.}} X$. Mean square convergence has the following key property for the mean and the variance: if $X_n \xrightarrow[n\to\infty]{\text{m.s.}} X$ then

$$\mathbb{E}[X_n] \xrightarrow[n\to\infty]{\text{m.s.}} \mathbb{E}[X], \quad \mathbb{V}[X_n] \xrightarrow[n\to\infty]{\text{m.s.}} \mathbb{V}[X]. \tag{1}$$

A stochastic process $X(t)$, $t \in \mathcal{T} \subset \mathbb{R}$, is termed of second-order if $X(t)$ is a second-order real random variable for each $t \in \mathcal{T}$. In our context $t \in \mathcal{T} = [a, b]$ is an interval.

Definition 1. *Let $\mathcal{D} = [a, b]$, $-\infty < a < b < +\infty$, be a finite interval of the real line, $\mathbb{R}$. Let $\{X(t) : t \in \mathcal{D}\}$ be a second-order stochastic process, i.e., having finite variance. The random mean square Riemann-Liouville fractional integral of $X(t)$, $\mathrm{J}^{\alpha}_{a+} X$, of order $\alpha > 0$, is defined by*

$$\left(\mathrm{J}^{\alpha}_{a+} X\right)(t) := \frac{1}{\Gamma(\alpha)} \int_a^t (t-u)^{\alpha-1} X(u)\, \mathrm{d}u, \tag{2}$$

where $\Gamma(\alpha)$ denotes the deterministic gamma function [8].

The following Proposition gives us a sufficient condition in order to guarantee the existence of the random mean square Riemann-Liouville fractional integral, $\left(\mathrm{J}^{\alpha}_{a+} X\right)(t)$. The proof is detailed in [7].

Proposition 1. *Let $\alpha > 0$ and $\{X(t) : t \in \mathcal{D}\}$ be a second-order stochastic process such as*

$$\int_a^t (t-u)^{\alpha-1} \|X(u)\|_2 \, \mathrm{d}u < \infty. \tag{3}$$

Then, the random mean square Riemann-Liouville fractional integral $(\mathrm{J}^{\alpha}_{a+} X)(t)$ exists.

Once the random fractional integral has been introduced, we can define the fractional Caputo derivative in the mean square sense. Intuitively, the concept of this fractional derivative of order $\alpha > 0$ is defined first making the $[\alpha + 1]$-th derivative, being $[\cdot]$ the integer part function, and then integrating the "excess", that is to say, calculating the $([\alpha + 1] - \alpha)$ integral of the $[\alpha + 1]$ classical derivative. Formally, the Caputo mean square derivative is defined as

Definition 2. *Let $\mathcal{D} = [a, b]$, $-\infty < a < b < \infty$, be a finite interval of the real line $\mathbb{R}$. Let $\{X(t) : t \in \mathcal{D}\}$ be a second-order stochastic process. The random mean square Caputo fractional derivative of $X(t)$, $({}^{C}D^{\alpha}_{a+}X)(t)$, of order $\alpha > 0$, is defined by*

$$\left({}^{C}D^{\alpha}_{a+}X\right)(t) := \left(\mathrm{J}^{n-\alpha}_{a+} X^{(n)}\right)(t) = \frac{1}{\Gamma(n-\alpha)} \int_a^t (t-u)^{n-\alpha-1} X^{(n)}(u)\, \mathrm{d}u, \tag{4}$$

where $n = -[-\alpha]$, being $[\cdot]$ the integer part function and, $X^{(n)}(t)$ denotes the n-th mean square derivative of $X(t)$.

Analogously to Proposition 1, the next result provides sufficient conditions in order to guarantee the existence of mean square Caputo derivative [7].

Proposition 2. *Let $\alpha > 0$ and $\{X(t) : t \in \mathcal{D}\}$ be a 2-SP n-times mean square differentiable such that*

$$\int_a^t (t-u)^{n-\alpha-1} \left\| X^{(n)}(u) \right\|_2 \, \mathrm{d}u < \infty.$$

Then, the random Caputo fractional derivative, $\left({}^C D_{a+}^{\alpha} X\right)(t)$, exists.

As we have previuosly pointed out, the Caputo derivative is commonly used in dealing with models formulated via IVP, [5,9]. Despite its physical meaning is not completely well known, this derivative allows us to express IVPs in terms of classical derivatives, where the physical meaning is well-known. From this point on, we will work with the Caputo derivative, $\left({}^C D_{a+}^{\alpha} X\right)(t)$.

3 Solving Random Fractional Differential Equations by Generalized Fröbenius Method: Moments and Density

In this section we will show how to solve two important classes of autonomous and non-autonomous random IVPs. In both cases, the solution will be constructed analytically via a random generalized power series, [8]. We will take advantage of this analytic representation to compute approximations of the mean and the variance. From this key information, we will apply the Maximum Entropy Principle to approximate the first probability density function of the solution [10]. Finally, this approximation will be compared with the one obtained using the Random Variable Transformation technique [3, Section 2.4.2].

3.1 Solving an Autonomous Random Fractional Initial Value Problem

In this subsection we deal with the following random fractional IVP

$$\begin{cases} \left({}^C D_{0+}^{\alpha} Y\right)(t) - \lambda Y(t) = \gamma, & t > 0, \quad 0 < \alpha \leq 1, \\ \qquad Y(0) = \beta_0, \end{cases} \tag{5}$$

with the following assumptions on the random inputs parameters.

H1: Inputs β_0, γ and λ are mutually independent second-order random variables.

H2: The growth of the moments of the random variable λ satisfy

$$\exists \eta, H > 0, \ p \geq 0 : \|\lambda^m\|_2 \leq \eta H^{m-1}((m-1)!)^p, \ \ \forall m \geq 1 \text{ integer}. \tag{6}$$

We remark that hypothesis H2 is satisfied by several important families of second order random variables, like any bounded random variable (taking $p = 0$), Gaussian ($p = 1/2$) or Exponential random variables ($p = 1$), see [11].

We seek for solutions of the form $Y(t) = \sum_{n\geq 0} Y_n t^{\alpha n}$. By applying the Fröbenius method, [8], it can be checked that the solution of the IVP (5) is given by the following random generalized power series

$$Y(t) = \sum_{m\geq 0} \frac{\lambda^m \beta_0}{\Gamma(\alpha m + 1)} t^{\alpha m} + \sum_{m\geq 1} \frac{\lambda^{m-1}\gamma}{\Gamma(\alpha m + 1)} t^{\alpha m}. \tag{7}$$

Under hypotheses H1 and H2, it can be seen that this infinite sum is mean square convergent for all $t > 0$ if $p < \alpha$. If $p = \alpha$, the domain of convergence reduces to $t \in [0, \frac{\alpha}{H^{\frac{1}{\alpha}}}]$, being $H > 0$ the constant involved in hypothesis H2.

Now, we take of the key property (1) of mean square convergence to approximate the mean and the variance of the solution from the approximations obtained via truncation of random generalized series (7). So, let us consider the finite sum

$$Y_M(t) = \sum_{m=0}^{M} \frac{\lambda^m \beta_0}{\Gamma(\alpha m + 1)} t^{\alpha m} + \sum_{m=0}^{M} \frac{\lambda^{m-1}\gamma}{\Gamma(\alpha m + 1)} t^{\alpha m}. \tag{8}$$

Applying the expectation and the variance operators in $Y_M(t)$, we obtain the following approximations for the mean and for the variance, respectively,

$$\mathbb{E}[Y_M(t)] = \mathbb{E}[\beta_0] \sum_{m=0}^{M} \frac{\mathbb{E}[\lambda^m]}{\Gamma(\alpha m + 1)} t^{\alpha m} + \mathbb{E}[\gamma] \sum_{m=1}^{M} \frac{\mathbb{E}[\lambda^{m-1}]}{\Gamma(\alpha m + 1)} t^{\alpha m}, \tag{9}$$

and

$$\begin{aligned}
\mathbb{V}[Y_M(t)] = {} & \mathbb{E}\left[(\beta_0)^2\right] \sum_{m=0}^{M}\sum_{n=0}^{M} \frac{\mathbb{E}\left[\lambda^{m+n}\right]}{\Gamma(\alpha m+1)\Gamma(\alpha n+1)} t^{\alpha(m+n)} \\
& - (\mathbb{E}[\beta_0])^2 \left(\sum_{m=0}^{M} \frac{\mathbb{E}[\lambda^m]}{\Gamma(\alpha m+1)} t^{\alpha m}\right)\left(\sum_{n=0}^{M} \frac{\mathbb{E}[\lambda^n]}{\Gamma(\alpha n+1)} t^{\alpha n}\right) \\
& + \mathbb{E}[\beta_0]\,\mathbb{E}[\gamma] \sum_{m=0}^{M}\sum_{n=1}^{M} \frac{(\mathbb{E}\left[\lambda^{m+n-1}\right])}{\Gamma(\alpha m+1)\Gamma(\alpha n+1)} t^{\alpha(m+n)} \\
& - \mathbb{E}[\beta_0]\,\mathbb{E}[\gamma] \left(\sum_{m=0}^{M} \frac{\mathbb{E}[\lambda^m]}{\Gamma(\alpha m+1)} t^{\alpha m}\right)\left(\sum_{n=1}^{M} \frac{\mathbb{E}[\lambda^{n-1}]}{\Gamma(\alpha n+1)} t^{\alpha n}\right) \\
& + \mathbb{E}[\beta_0]\,\mathbb{E}[\gamma] \sum_{m=1}^{M}\sum_{n=0}^{M} \frac{\mathbb{E}\left[\lambda^{m+n-1}\right]}{\Gamma(\alpha m+1)\Gamma(\alpha n+1)} t^{\alpha(m+n)} \\
& - \mathbb{E}[\beta_0]\,\mathbb{E}[\gamma] \left(\sum_{m=1}^{M} \frac{\mathbb{E}[\lambda^{m-1}]}{\Gamma(\alpha m+1)} t^{\alpha m}\right)\left(\sum_{n=0}^{M} \frac{\mathbb{E}[\lambda^n]}{\Gamma(\alpha n+1)} t^{\alpha n}\right) \\
& + \mathbb{E}\left[\gamma^2\right] \sum_{m=1}^{M}\sum_{n=1}^{M} \frac{\mathbb{E}\left[\lambda^{m+n-2}\right]}{\Gamma(\alpha m+1)\Gamma(\alpha n+1)} t^{\alpha(m+n)} \\
& - (\mathbb{E}[\gamma])^2 \left(\sum_{m=1}^{M} \frac{\mathbb{E}[\lambda^{m-1}]}{\Gamma(\alpha m+1)} t^{\alpha m}\right)\left(\sum_{n=1}^{M} \frac{\mathbb{E}[\lambda^{n-1}]}{\Gamma(\alpha n+1)t^n} t^{\alpha n}\right).
\end{aligned} \tag{10}$$

The next goal is to provide more specific probability information of the solution stochastic process. This will be done by determining approximations of its first probability density function. To this end we will first apply the Random Variable Transformation technique [3, Section 2.4.2] to the finite approximation $Y_M(t)$ by defining the following auxiliary map $\mathbf{r}:\mathbb{R}^3 \longrightarrow \mathbb{R}^3$

$$\begin{aligned} z_1 &= r_1(\beta_0,\gamma,\lambda) = \beta_0 S_1^M(t,\alpha;\lambda) + \gamma S_2^M(t,\alpha;\lambda), \\ z_2 &= r_2(\beta_0,\gamma,\lambda) = \gamma, \\ z_3 &= r_3(\beta_0,\gamma,\lambda) = \lambda, \end{aligned}$$

where

$$S_1^M(t,\alpha;\lambda) = \sum_{m=0}^{M} \frac{\lambda^m}{\Gamma(\alpha m+1)} t^{\alpha m}, \quad S_2^M(t,\alpha;\lambda) = \sum_{m=0}^{M} \frac{\lambda^m}{\Gamma(\alpha(m+1)+1)} t^{\alpha(m+1)}. \tag{11}$$

After some computations, we obtain

$$f_1^M(y,t) = \int_{\mathcal{D}(\gamma)} \int_{\mathcal{D}(\lambda)} f_{\beta_0}\left(\frac{y-\gamma S_2^M(t,\alpha;\lambda)}{S_1^M(t,\alpha;\lambda)}\right) f_\gamma(\gamma) f_\lambda(\lambda) \left|\frac{1}{S_1^M(t,\alpha;\lambda)}\right| \mathrm{d}\lambda \mathrm{d}\gamma, \tag{12}$$

where $\mathcal{D}(\gamma)$ and $\mathcal{D}(\lambda)$ denote the domains of random variables γ and λ, and f_{β_0}, f_γ and f_λ denote the probability density functions of β_0, γ and λ, respectively.

Under assumptions H1 and H2, the convergence of $f_1^M(y,t)$ to the 1-PDF of $Y(t)$, $f_1(y,t)$ can be proved, [12].

The following example illustrates all the theoretical results previously established.

Example 1. *Let us consider IVP* (5)*, where the order of the fractional derivative is* $\alpha = 0.3$*. The initial condition* β_0 *is assumed to be an Exponential distribution with mean* 1 *(*$\beta_0 \sim Exp(1)$*) and* γ *is a Gaussian distribution with mean 0 and standard deviation* 0.05*, (*$\gamma \sim N(0, 0.05^2)$*). Let us consider* λ *as a Beta distribution of parameters 37 and 50 (*$\lambda \sim Be(37, 50)$*), so hypothesis H2 fulfills for* $p = 0$ *(since it is a bounded random variable). As a consequence,* $Y_M(t)$ *converges in mean square sense to the solution stochastic process* $Y(t)$ *for all* t*.*

Figure 1 illustrates the approximations for the mean and for the standard deviation given by expressions (9) *and* (10)*, in the interval* $t \in [0, 12]$ *for different order of truncations* $M = \{5, 7, 10, 12, 15\}$*. We can see their convergence as* M *increases. In Fig. 2 we show the approximation of the 1-PDF, given by* (12)*, at the time instants* $\hat{t} = \{1, 5, 10\}$ *considering different order of truncations* $M = \{2, 4, 6, 8, 10\}$*. These approximations have been constructed via the Random Variable Transformation technique.*

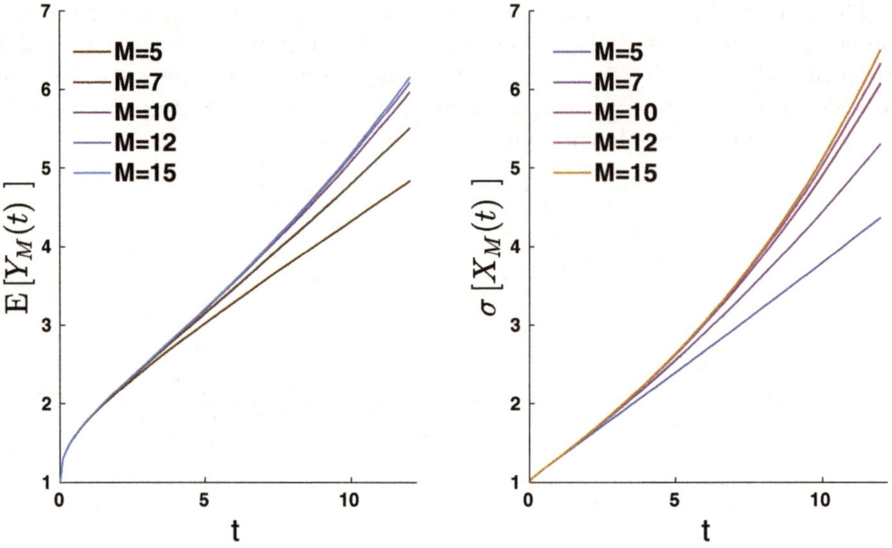

Fig. 1. Mean and standard deviation of the truncated solution stochastic process given (9) and (10), respectively, using different orders of truncation $M = \{5, 7, 10, 12, 15\}$ in the interval $t \in [0, 12]$. Example 1.

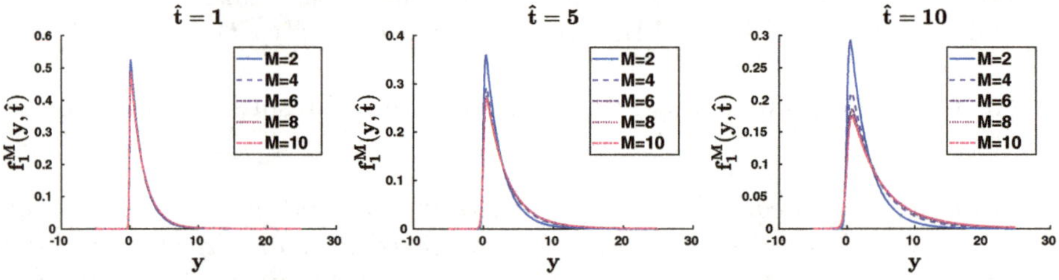

Fig. 2. Plots of the 1-PDF, $f_1^M(y, t)$, given by (12), for different orders of truncations M at the time instants $\hat{t} = 1$ (left), $\hat{t} = 5$ (center) and $\hat{t} = 10$ (right). Example 1.

3.2 Solving a Non-autonomous Random Fractional Initial Value Problem

In this subsection, we go further by constructing the solution stochastic process of a non-autonomous random fractional IVP where the order of the fractional derivative, α, is arbitrary. We study the following random fractional IVP

$$\begin{cases} \left({}^{C}D_{0+}^{\alpha} Y\right)(t) - B\, t^{\beta} Y(t) = 0, & t > 0, \quad n-1 < \alpha \leq n, \;\; \beta > 0, \\ \qquad Y^{(j)}(0) = A_j, & j = 0, 1, \ldots, n-1. \end{cases} \tag{13}$$

Observe that the notation of the IVP (13) refers to different IVPs. Indeed, depending on the order of the fractional derivative, different initial conditions must be considered. If $0 < \alpha \leq 1$, there is only one initial condition, say $Y(0) = A_0$, but if $1 < \alpha \leq 2$, then there are two initial conditions, $Y(0) = A_0$ and $Y'(0) = A_1$. Generalizing, if $n-1 < \alpha \leq n$, where $n = -[-\alpha]$, the IVP is formulated via n initial conditions, $Y(0) = A_0, Y'(0) = A_1, \ldots, Y^{n-1}(0) = A_{n-1}$. It is important to remark that IVP (13) with $\beta = 1$ and $\alpha = 2$ is the well known random fractional Airy equation [13].

For the analysis of the fractional IVP (13) we will assume the following hypotheses

$\hat{H}1$: For every j, $j = 0, 1, \ldots, n-1$, A_j and B are independent random variables.
$\hat{H}2$: There exist constants η, $\mathcal{H}$, and $p \geq 0$ and an integer m_0 such that

$$\|B^m\|_2 \leq \eta \mathcal{H}^{m-1}((m-1)!)^p, \quad \forall m \geq m_0 \text{ integer}. \tag{14}$$

Hypotheses $\hat{H}1$ and $\hat{H}2$ are the analogous ones to assumptions $H1$ and $H2$ defined in Subsect. 3.1, but adapted to IVP (13). To obtain the solution of IVP (13) we will apply the following generalized version of Fröbenius method

$$Y(t) = \sum_{j=0}^{n-1} \left(\sum_{m=0}^{\infty} B^m A_j G_{m,j} t^{\gamma m + j} \right), \quad \gamma = \alpha + \beta, \tag{15}$$

where

$$G_{m,j} = \prod_{k=1}^{m} \frac{\Gamma(k\gamma + j + 1 - \alpha)}{\Gamma(k\gamma + j + 1)}, \quad m = 0, 1, 2, \ldots, \tag{16}$$

for $t \in \mathcal{T} = [0, b] \subset \mathbb{R}$. In [14] the mean square convergence of this series is carefully studied. The analysis concludes that if $p < \alpha$, convergence fulfils for $t > 0$, while if $p = \alpha$, the convergence can be only guaranteed in the interval $t \in \left[0, \frac{(\alpha+\beta)^{\frac{\alpha}{\alpha+\beta}}}{H^{\frac{1}{\alpha+\beta}}}\right[$. To construct approximations for the mean and for the variance, it is convenient to truncate the series solution

$$Y_M(t) = \sum_{j=0}^{n-1} \left(\sum_{m=0}^{M} B^m A_j G_{m,j} t^{\gamma m + j} \right). \tag{17}$$

Then using the properties of the expectation and the covariance operators, one gets the following values to approximate the mean

$$\mathbb{E}[Y_M(t)] = \sum_{j=0}^{n-1} \left(\sum_{m=0}^{M} \mathbb{E}[B^m] \mathbb{E}[A_j] G_{m,j} t^{\gamma m + j} \right). \tag{18}$$

and the variance

$$\begin{aligned}\mathbb{V}\left[Y_M(t)\right] &= \mathbb{C}\mathrm{ov}\left[Y_M(t), Y_M(t)\right] \\ &= \sum_{j=0}^{n-1}\sum_{k=0}^{n-1}\left[\sum_{r=0}^{M}\sum_{m=0}^{M} \mathbb{E}\left[A_j A_k\right]\mathbb{E}\left[B^{m+r}\right] G_{r,j} G_{m,k} t^{\gamma(r+m)+j+k}\right] \\ &- \sum_{j=0}^{n-1}\sum_{k=0}^{n-1}\left[\sum_{r=0}^{M}\sum_{m=0}^{M} \mathbb{E}\left[A_j\right]\mathbb{E}\left[A_k\right]\mathbb{E}\left[B^{m}\right]\mathbb{E}\left[B^{r}\right] G_{r,j} G_{m,k} t^{\gamma(r+m)+j+k}\right].\end{aligned} \tag{19}$$

Just to stress in different potential apporaches to approximate the 1-PDF, apart from the Random Variable Transformation technique, hereinafter we present an alternative method that takes advantage of the previous computation of the mean and the variance and which is based on the Maximum Entropy Principle, [10]. Under this approach, fixed t_0, the PDF of the solution stochastic process at t_0 is seeking of the form

$$f(y, t_0) = e^{-1-\lambda_0-\lambda_1 y-\lambda_2 y^2}, \tag{20}$$

where λ_0, λ_1 and $\lambda_2 \in \mathbb{R}$ to be determined. To this end, we impose that $f(y, t_0)$ is a PDF, so its integral is the unit and that the available information about the approximate mean and variance (or equivalently about the second order moment)

$$\begin{aligned}\int_{\mathcal{D}_{Y_M(t_0)}} e^{-1-\lambda_0-\lambda_1 y-\lambda_2 y^2}\, dy &= 1, \\ \int_{\mathcal{D}_{Y_M(t_0)}} y e^{-1-\lambda_0-\lambda_1 y-\lambda_2 y^2}\, dy &= \mathbb{E}[Y_M(t_0)], \\ \int_{\mathcal{D}_{Y_M(t_0)}} y^2 e^{-1-\lambda_0-\lambda_1 y-\lambda_2 y^2}\, dy &= \mathbb{E}[Y_M(t_0)]^2 + \mathbb{V}[Y_M(t_0)].\end{aligned} \tag{21}$$

Next, parameters λ_i, $i = 0, 1, 2$ are calculated solving the above nonlinear system using numerical methods. This approach is carried out for any time t_0 where the PDF must be approximated. On this manner, an approximation of the 1-PDF, $f_1(y, t)$, is constructed.

The following example illustrates all foregoing ideas.

Example 2. *Let us assume that the order of the fractional derivative in the IVP* (13) *is* $\alpha = 1.7$, *so, according to* (13), $n = 2$ *and we require two initial conditions* A_0 *and* A_1. *We will assume that their two first moments are given by*

$$\mathbb{E}[A_0] = 1 \qquad \mathbb{E}[A_0^2] = 1 \qquad \mathbb{E}[A_1] = 2 \qquad \mathbb{E}[A_1^2] = 4.$$

Let us consider $\beta = 0.8$ *and the random variable* B *has a Beta distribution with parameters* 20 *and* 30*,* $(B \sim Be(20, 30))$*.*

Now we prove that the approximations for the mean and the variance of the solution converge for all $t > 0$ *to the corresponding exact values. To this end, we will apply property* (1)*, so we will show that the solution stochastic process* $Y(t)$*, defined by* (15)–(16) *is mean square convergent for all* $t > 0$*. This can be straightforwardly seen majorizing in norm the solution. First observe that* $\|Y(t)\|_2 \leq \|Y_1(t)\|_2 + \|Y_2(t)\|_2$ *and this two norms can be majorized by the following series of positive numbers*

$$\|Y_j(t)\|_2 \leq \sum_{m \geq 0} \delta_{m,j}(t), \; \delta_{m,j}(t) := \eta \mathcal{H}^{m-1}((m-1)!)^p \|A_j\|_2 \, G_{m,j} t^{\gamma m + j}, \; j = 0, 1, \tag{22}$$

where we have used hypotheses $\hat{H}1$ *and* $\hat{H}2$*. For bounded random variables, it is easy to check that condition* (14) *is satisfied for* $\eta = 1$*,* $\mathcal{H} = 1$ *and* $p = 0$ *[11, Table 1]. So, this applies to input parameter* B*, which is assumed to have a Beta distribution. Moreover, using the Stirling's formula for approximation the gamma function involved in the definition of coefficients* $G_{m,j}$ *(see* (16)*), it can be checked that*

$$\begin{aligned}
\lim_{m \to \infty} \frac{\delta_{m+1,j}(t)}{\delta_{m,j}(t)} &= t^{\alpha+\beta} \mathcal{H} \lim_{m \to \infty} \frac{m^p}{((m+1)(\alpha+\beta)+j)^{\alpha}} \\
&= t^{2.5} \lim_{m \to \infty} \frac{1}{(2.5(m+1)+j)^{1.7}} = 0 < 1, \quad \forall t > 0, \, j = 0, 1.
\end{aligned}$$

This shows that the solution is mean square convergent for every interval $[0, b]$*,* $b > 0$*, and thus its mean and variance. In Fig. 3, we show the approximations for the mean and for the variance in the interval* $t \in [0, 9]$ *for different order of truncations* $M = \{14, 15, 16, 17, 18\}$ *and* $M = \{17, 18, 19, 20, 21\}$*, respectively. These approximations have been computed using expressions* (18) *and* (19)*, respectively. Notice that we have needed higher order of truncation for approximating the variance than for the mean as expected. From these graphical representations, we see that convergence is reaching as* M *increases. From this graphical representation we can also observe that due to the oscillatory nature of the approximations they change rapidly from consecutive truncation orders,* M *and* $M + 1$*. The example is completed with Fig. 4, where several approximations of the 1-PDF,* $f_1(y, t)$*, using the Maximum Entropy Principle has been applied in the interval* $t \in [0, 9]$*, is shown. We can observe that this plot is in agreement with the results shown in Fig. 4.*

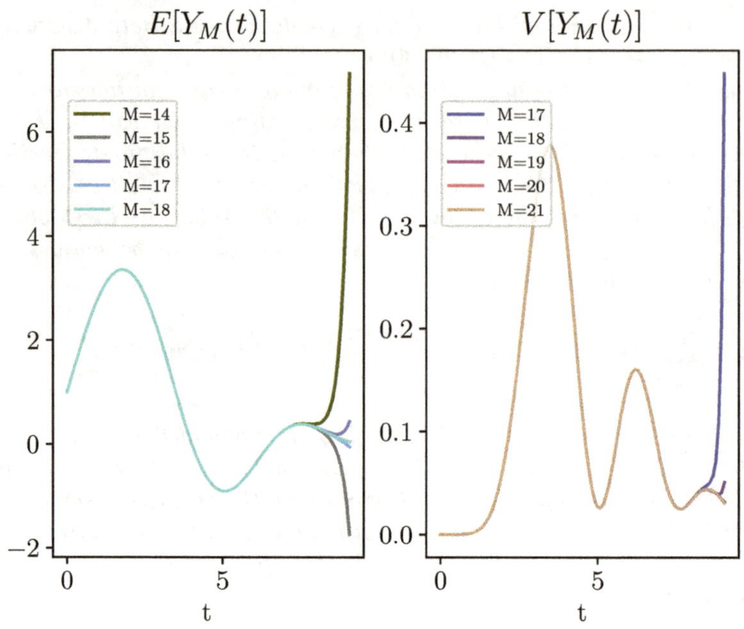

Fig. 3. Mean and variance of the truncated solution stochastic process given in (17) using different orders of truncation $M = \{14, 15, 16, 17, 18\}$ and $M = \{17, 18, 19, 20\}$, respectively on $t \in [0, 9]$. Example 2.

4 Numerical Methods to Approximate the Solution Stochastic Process

It is well known that is not always possible to determine the solution stochastic process of a random fractional IVP via analytical methods. In this section we extend a deterministic fractional method [4] to the random framework. First of all, let us consider the following generalized IVP, where the order of the derivative α lies between 0 and 1

$$\begin{cases} \left({}^C D^{\alpha}_{a^+} Y\right)(t) = f(Y(t), t), & t \in [0, b], \quad 0 < \alpha \leq 1, \\ Y(0) \qquad = Y_0, \end{cases} \tag{23}$$

where $Y_0 \in \mathrm{L}^2(\Omega)$ and $f : S \subseteq \mathrm{L}^2(\Omega) \times [0, b] \to \mathrm{L}^2(\Omega)$ is assumed to satisfy the following conditions:

$\tilde{\mathrm{H}}1$: f is mean square Lipschitz, that is, there exists $\kappa > 0$ such that

$$\|f(X,t) - f(Y,t)\|_2 \leq \kappa \|X - Y\|_2, \quad X, Y \in \mathrm{L}^2(\Omega),$$

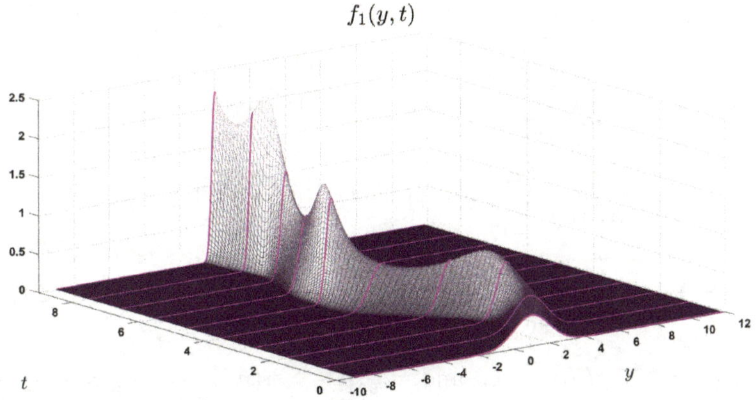

Fig. 4. Approximation of the 1-PDF obtained using Maximum Entropy Principle in the time interval $t = [0, 9]$. Example 2.

$\tilde{\mathrm{H}}2$ f satisfies the mean square modulus of continuity property, i.e.,

$$\lim_{h\to 0} W(S,h) = 0, \quad W(S,h) = \sup_{Y\in S\subseteq \mathrm{L}^2(\Omega)} \sup_{|t-t'|\le h} \|f(Y,t) - f(Y,t')\|_2 ,$$

where S is bounded.

Applying properties of the mean square calculus [3], it can rigorously be proved that $Y(t)$ is a solution of the IVP (23) if and only if $Y(t)$ satisfies the following integral equation, named Volterra fractional integral equation,

$$Y(t) = Y_0 + \frac{1}{\Gamma(\alpha)} \int_0^t (t-s)^{\alpha-1} f(Y(s),s)\,\mathrm{d}s, \quad t\in[0,b], \quad 0<\alpha\le 1. \quad (24)$$

Let us define a mesh of M nodes in the interval $[0,b]$ with length h, i.e. $t_i = ih$ $0\le i\le n$. Evaluating (24) at t_n, $1\le n\le M$, one gets

$$\begin{aligned} Y(t_n) &= Y_0 + \frac{1}{\Gamma(\alpha)} \int_a^{t_n} (t_n-s)^{\alpha-1} f(Y(s),s)\,\mathrm{d}s, \\ &= Y_0 + \frac{1}{\Gamma(\alpha)} \sum_{j=0}^{n-1} \int_{t_j}^{t_{j+1}} (t_n-s)^{\alpha-1} f(Y(s),s)\,\mathrm{d}s, \quad t_n\in[0,b],\ 0<\alpha<1. \end{aligned} \quad (25)$$

Now, taking the following approximation $\int_{t_j}^{t_{j+1}} (t_n - s)^{\alpha-1} f(Y(s),s)\,\mathrm{d}s \approx \int_{t_j}^{t_{j+1}} (t_n-s)^{\alpha-1} f(Y(t_j),t_j)\,\mathrm{d}s$, we obtain that the second order random variable

Y_n approximates the solution at the time instant t_n, i.e. $Y_n \approx Y(t_n)$. Thus,

$$\begin{aligned} Y_n &= Y_0 + \frac{1}{\Gamma(\alpha)} \sum_{j=0}^{n-1} \int_{t_j}^{t_{j+1}} (t_n - s)^{\alpha-1} f(Y_j, t_j)\, \mathrm{d}s \\ &= Y_0 + \frac{h^\alpha}{\alpha\Gamma(\alpha)} \sum_{j=0}^{n-1} [(n-j)^\alpha - (n-(j+1))^\alpha]\, f(Y_j, t_j) \\ &= Y_0 + C(\alpha) \sum_{j=0}^{n-1} a_{n,j}(\alpha) f(Y_j, t_j) \qquad t_n \in [a, b], \quad 0 < \alpha < 1, \end{aligned} \tag{26}$$

where $C(\alpha) = h^\alpha/\Gamma(\alpha+1)$ and $a_{nj}(\alpha) = (n-j)^\alpha - (n-(j+1))^\alpha$. The reason why we approximate only $f(Y(s), s)$ and not the whole integral $\int_{t_j}^{t_{j+1}} (t_n - s)^{\alpha-1} f(Y(s), s)\, \mathrm{d}s$ using the rectangle rule is because if we consider a fine mesh, as $\alpha - 1 < 0$, we realize that in the last term in the sum, corresponding to $j = n-1$, leads to small quotients, $(t_n - t_{n-1})^{\alpha-1}$, and this is not affordable when working with floating point. The mean square convergence of the method can be proved under hypotheses $\tilde{\text{H}}1$ and $\tilde{\text{H}}2$. The approximations for the mean and for the variance can be obtained by (26) as

$$\mathbb{E}[Y_n] = \mathbb{E}[Y_0] + C(\alpha) \sum_{j=0}^{n-1} a_{nj}(\alpha)\, \mathbb{E}[f(Y_j, t_j)], \tag{27}$$

and

$$\begin{aligned} \mathbb{V}[Y_n] &= \mathbb{V}[Y_0] + 2 \sum_{j=0}^{n-1} C(\alpha) a_{nj}(\alpha) \mathbb{C}\mathrm{ov}[Y_0, f(Y_j, t_j)] \\ &+ \sum_{j=0}^{n-1} \sum_{i=0}^{n-1} (C(\alpha))^2 a_{nj}(\alpha) a_{ni}(\alpha) \mathbb{C}\mathrm{ov}[f(Y_j, t_j), f(Y_i, t_i)], \end{aligned} \tag{28}$$

respectively. The next example illustrates numerically the previous results.

Example 3. *Let us consider IVP* (5) *where* β_0 *and* γ *are independent random variables and* $\lambda \in \mathbb{R}$.

The numerical random fractional Euler scheme (26) *for this particular problem is given by*

$$Y_n = Y_0 + \frac{h^\alpha}{\Gamma(\alpha+1)} \sum_{j=0}^{n-1} [(n-j)^\alpha - (n-(j+1))^\alpha] (\lambda Y_j + \gamma), \tag{29}$$

where $Y_0 = \beta_0$. *The numerical scheme for the mean and for the variance of the IVP* (5) *are given by*

$$\mathbb{E}[Y_n] = \mathbb{E}[Y_0] + C(\alpha) \sum_{j=0}^{n-1} a_{nj}(\alpha) (\lambda \mathbb{E}[Y_j] + \mathbb{E}[\gamma]), \tag{30}$$

and

$$\mathbb{V}[Y_n] = \mathbb{V}[Y_0] + 2C(\alpha)\sum_{j=0}^{n-1} a_{nj}(\alpha)\left(\lambda \mathrm{Cov}[Y_0, Y_j] + \mathrm{Cov}[Y_0, \gamma]\right)$$
$$+ (C(\alpha))^2 \sum_{j=0}^{n-1}\sum_{i=0}^{n-1} a_{nj}(\alpha)a_{ni}(\alpha)\left(\lambda^2 \mathrm{Cov}[Y_j, Y_i] + \lambda \mathrm{Cov}[Y_j, \gamma] + \lambda \mathrm{Cov}[Y_i, \gamma] + \mathbb{V}\mathrm{ar}[\gamma]\right), \tag{31}$$

respectively.

Let us consider $\gamma \sim Ga(\frac{1}{2}, \frac{1}{2})$*,* $\beta_0 \sim Exp(4)$*,* $\alpha = 0.5$ *and* $\lambda = 0.6$*. In Tables 1 and 2, we show the approximations for the mean and for the variance, respectively, for different* $t = \{0.2, 0.4, 0.6, 0.8\}$ *and for different nodes of discretization* $M = \{50, 100, 200, 400, 800, 1600, 3200\}$*. In Tables 3 and 4 we show the values of the absolute error, by comparing these values with the ones obtained by means of* (9) *and* (10) *(here we assume these latter values are exact since we have taken an order of truncation until that stabilization is reached). The relative error could be also suitable to compute the error, however, as we can see in Table 1 and Table 2, the approximations are close to zero, so absolute value is a more reliable measure of the error. Observe that taking as truncation order* $M = 50$*, the error of the approximations is of order* 10^{-3}*. This order keeps for higher values of* M*, so the convergence with* M *is slow.*

Table 1. Values for the mean given by expression (30) in the context of Example 3 at the time instants $t = [0.2, 0.4, 0.6, 0.8]$ using different orders of truncation, $M = \{50, 100, 200, 400, 800, 1600\}$.

	$t = 0.2$	$t = 0.4$	$t = 0.6$	$t = 0.8$
$M = 50$	4.25176e−01	5.31352e−01	6.31543e−01	7.32009e−01
$M = 100$	4.26994e−01	5.33904e−01	6.34919e−01	7.36299e−01
$M = 200$	4.31356e−01	5.38890e−01	6.40474e−01	7.42413e−01
$M = 400$	4.25954e−01	5.32494e−01	6.33120e−01	7.34087e−01
$M = 800$	4.31098e−01	5.38685e−01	6.40345e−01	7.42379e−01
$M = 1600$	4.33053e−01	5.40942e−01	6.42879e−01	7.45187e−01

In Figure 5 we show the approximate 1-PDF in the time interval $t = [0, 0.8]$. It has been computed using the Maximum Entropy Principle from the numerical approximations of the mean and the variance collected in Tables 3 and 4, respectively. In Table 5, we can see the values for the parameters λ_0, λ_1 and λ_2 at the time instants $t = [0.2, 0.4, 0.6, 0.8]$ defining the 1-PDF according to (20).

Table 2. Values for the variance given by expression (31) in the context of Example 3 at the time instants $t = \{0.2, 0.4, 0.6, 0.8\}$ using different orders of truncation $M = \{50, 100, 200, 400, 800, 1600\}$.

	$t = 0.2$	$t = 0.4$	$t = 0.6$	$t = 0.8$
$M = 50$	1.27981e−01	1.81679e−01	2.41869e−01	3.11975e−01
$M = 100$	1.30526e−01	1.85908e−01	2.48194e−01	3.20894e−01
$M = 200$	1.27420e−01	1.81513e−01	2.42336e−01	3.13320e−01
$M = 400$	1.27344e−01	1.81654e−01	2.42816e−01	3.14276e−01
$M = 800$	1.26911e−01	1.80492e−01	2.40674e−01	3.10868e−01
$M = 1600$	1.31820e−01	1.87131e−01	2.49234e−01	3.21676e−01

Table 3. Absolute errors for the approximations of the mean tabulated in Table 1. Example 3.

	$t = 0.2$	$t = 0.4$	$t = 0.6$	$t = 0.8$
$M = 50$	3.91803e−03	5.03467e−03	6.23382e−03	7.53972e−03
$M = 100$	2.10033e−03	2.48242e−03	2.85850e−03	3.24967e−03
$M = 200$	2.26251e−03	2.50318e−03	2.69642e−03	2.86445e−03
$M = 400$	3.13952e−03	3.89250e−03	4.65679e−03	5.46176e−03
$M = 800$	2.00382e−03	2.29862e−03	2.56766e−03	2.83050e−03
$M = 1600$	3.95927e−03	4.55503e−03	5.10149e−03	5.63832e−03

Table 4. Absolute errors for the approximations of the variance tabulated in Table 2. Example 3.

	$t = 0.2$	$t = 0.4$	$t = 0.6$	$t = 0.8$
$M = 50$	1.65404e−03	2.87430e−03	4.45046e−03	6.45923e−03
$M = 100$	8.90695e−04	1.35425e−03	1.87428e−03	2.45962e−03
$M = 200$	2.21535e−03	3.04015e−03	3.98360e−03	5.11419e−03
$M = 400$	2.29075e−03	2.89878e−03	3.50368e−03	4.15854e−03
$M = 800$	2.72447e−03	4.06084e−03	5.64501e−03	7.56628e−03
$M = 1600$	2.18535e−03	2.57824e−03	2.91451e−03	3.24174e−03

Table 5. Values of the parameters λ_0, λ_1 and λ_2 of the approximate 1-PDF using Maximum Entropy Principle for the IVP (5) in the context of Example 3 for different time instants $t = \{0.2, 0.4, 0.6, 0.8\}$.

	λ_0	λ_1	λ_2
t=0.2	−3.92424e−01	−3.31002e+00	3.85698e+00
t=0.4	−1.46491e−01	−2.90641e+00	2.70925e+00
t=0.6	4.40506e−02	−2.58923e+00	2.02988e+00
t=0.8	2.05553e−01	−2.32246e+00	1.57018e+00

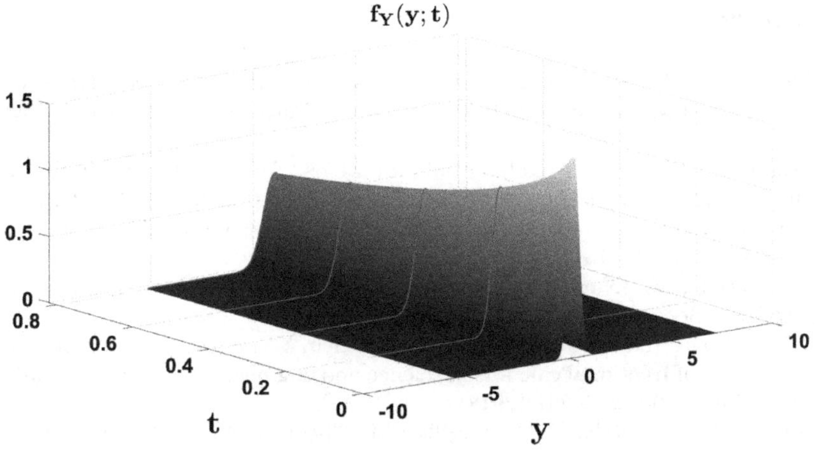

Fig. 5. Approximate 1-PDF for the IVP (5) in the context of Example 3 using Maximum Entropy Principle.

5 Conclusions

In this contribution we have presented several analytic and numerical techniques to deal with fractional differential equations with uncertainties. The analysis has been conducted taking advantage of random mean square calculus since it possesses key properties that guarantees the approximations for the mean and the variance of the solution stochastic process constructed via analytic or numerical methods will converge to the exact one. Furthermore, we have show how to go further by constructing reliable approximations of the first probability density function of the solution using two different techniques, namely, the Random Variable Transformation methods and the Maximum Entropy Principle. The contribution is intended to provide a fair overview of useful techniques to study this types of fractional differential equations whose interest in the scientific community is increasing since a wide range of probability distributions can be considered for their input parameters. This fact make random fractional differential equations particularly interesting for modelling purposes. Finally, we want to point out that there are still many deterministic results belonging to the field or fractional differential equations that can be extended to the random setting using the ideas exhibited throughout this contribution.

Acknowledgements. This work has been supported by the Spanish Ministerio de Economía, Industria y Competitividad (MINECO), the Agencia Estatal de Investigación (AEI) and Fondo Europeo de Desarrollo Regional (FEDER UE) grant MTM2017-89664-P. Computations have been carried thanks to the collaboration of Raúl San Julián Garcés and Elena López Navarro granted by European Union through the Operational Program of the European Regional Development Fund (ERDF)/European Social Fund (ESF) of the Valencian Community 2014–2020, grants GJIDI/2018/A/009 and GJIDI/2018/A/010, respectively.

References

1. Kolmanovskii, V., Shaikhet, L.: Control of Systems with Aftereffect, Translations of Mathematical Monographs, vol. 157. American Mathematical Society, Providence (2006)
2. Øksendal, B.: Stochastic Differential Equations: An Introduction with Applications. Springer, Heidelberg (2003). https://doi.org/10.1007/978-3-642-14394-6
3. Soong, T.T.: Random Differential Equations in Science and Engineering. Academic Press, New York (1973)
4. Li, C., Zeng, F.: Numerical Methods for Fractional Calculus. Chapman and Hall/CRC, Boca Raton (2015)
5. Sun, H., Zhang, Y., Baleanu, D., Chen, W., Chen, Y.: A new collection of real world applications of fractional calculus in science and engineering. Commun. Nonlin. Sci. Numer. Simul. **64**, 213–231 (2018)
6. Kilbas, A.A., Srivastava, H.M., Trujillo, J.J.: Theory and Applications of Fractional Differential Equations. Elsevier, The Netherlands (2006)
7. Burgos, C., Cortés, J.C., Villafuerte, L., Villanueva, R.J.: Extending the deterministic Riemann–Liouville and Caputo operators to the random framework: A mean square approach with applications to solve random fractional differential equations. Chaos, Solitons & Fractals (2017)
8. Kilbas, A.A.A., Srivastava, H.M., Trujillo, J.J.: Theory and Applications of Fractional Differential Equations, vol. 204. Elsevier, Amsterdam (2006)
9. Acedo, L., Burgos, C., Cortés, J.C., Villanueva, R.J.: Probabilistic prediction of outbreaks of meningococcus W-135 infections over the next few years in Spain. Phys. A: Stat. Mech. Appl. **486**, 106–117 (2017)
10. Michalowicz, J.V., Nichols, J.M., Bucholtz, F.: Handbook of Differential Entropy. Chapman and Hall/CRC, New York (2013)
11. Burgos, C., Cortés, J.C., Villafuerte, L., Villanueva, R.: Solving random mean square fractional linear differential equations by generalized power series: analysis and computing. J. Comput. Appl. Math. **339**, 94–110 (2018). Modern Fractional Dynamic Systems and Applications, MFDSA 2017
12. Burgos, C., Calatayud, J., Cortés, J.C., Navarro-Quiles, A.: A full probabilistic solution of the random linear fractional differential equation via the random variable transformation technique. Math. Meth. Appl. Sci. **41**(18), 9037–9047 (2018)
13. Cortés, J.C., Navarro-Quiles, A., Romero, J.V., Roselló, M.D.: Solving second-order linear differential equations with random analytic coefficients about ordinary points: a full probabilistic solution by the first probability density function. Appl. Math. Comput. **331**, 33–45 (2018)
14. Burgos, C., Cortés, J.C., Debbouche, A., Villafuerte, L., Villanueva, R.J.: Random fractional generalized Airy differential equations: a probabilistic analysis using mean square calculus. Appl. Math. Comput. **352**, 15–29 (2019)

Global Sensitivity Analysis of Offshore Wind Turbine Jacket

Chao Ren(✉), Younes Aoues, Didier Lemosse, and Eduardo Souza De Cursi

Laboratory of Mechanics of Normandy, INSA Rouen Normandie, Rouen, France
chao.ren@insa-rouen.fr

Abstract. Nowadays, offshore wind turbine (OWT) energy is considered as one of the most promising among renewable energies. Many researches have been done to study offshore wind turbine foundations like jacket and monopile. The sensitivity analysis of the foundation can also be found in several researches. But there are always some limitations in these related studies. For example, the sea current is not considered and the directions of wind and wave are assumed to be aligned. The behavior of one specific element or part of jacket is studied. In this paper, the global sensitivity of the maximum stress and displacement with respect to material, local geometry and environmental uncertain parameters are investigated for the offshore wind turbine jacket used in Code Comparison Collaboration Continuation (OC4) project. Compared to other previous studies, the wind parameters are replaced by the load actions simulated in aerodynamic software FAST developed by National Renewable Energy Laboratory (NREL). The sea current is considered and the directions of the wind, sea wave and current are assumed to be independent. Also, the maximum stresses of static analysis in the different parts of jacket are investigated. The Morris screening and Fourier amplitude sensitivity test methods are applied to perform global sensitivity analysis. The results show that maximum stresses of different parts of jacket are affected by different parameters. The zones of the jacket above the mean sea level (MSL) are more sensitive to the geometry parameters and redefined wind loads. However, the zones below MSL are affected a lot by the wave parameters. In addition, the results of the two global sensitivity analyses are mostly in good agreement.

Keywords: Offshore wind turbine jacket · Uncertainty · Sensitivity analysis · Morris screening method · Fourier amplitude sensitivity test

1 Introduction

Over the past decades, offshore wind turbine energy has been considered as one of the most promising among renewable energies. To increase the power production, the wind tower and foundation become bigger and requires more investment cost, where the offshore wind turbine foundation cost represents a high part of construction and installation costs [13]. Several research works have been done

J. E. S. De Cursi (Ed.): Uncertainties 2020, LNME, pp. 35–48, 2021.
https://doi.org/10.1007/978-3-030-53669-5_3

to study offshore wind turbine foundations like jacket and monopile. Several parameters are involved in the simulation models of the wind turbine jacket: geometry, material, and geotechnical parameters, soil conditions, loading due to environmental actions (wind, wave, sea current, etc.). Uncertainty propagation and reliability analysis of the wind turbine jacket is an important research field. However, an important number of parameters in the multi-physics simulations models (aerodynamics, fluid-structure interaction, etc.) are involved and makes the probabilistic approaches more complex.

The sensitivity analysis (SA) is widely used to quantify the effects of parameters and reduce the stochastic dimensions. Over the few decades, many global sensitivity analysis (GSA) techniques have been developed. There are some GSA methods widely used including variance-based methods [16], Morris screening [12] and linear regression of Monte Carlo simulations [8]. In addition, the SA of the offshore wind turbines can also be found in many papers. The influences of geotechnical parameters were investigated in [14]. The effect of soil spatial variability was studied in [3]. The impact of the geometrical inputs can be found in [6]. Toft et al. [20] have analysed the influence of uncertain wind parameters on fatigue loads.

For the offshore wind turbine foundations, Glisic et al. [5] performed a sensitivity analysis of offshore monopile fatigue loads with respect to wave and soil parameters. Hubler et al. [10] analyzed the sensitivity of a monopile and a jacket foundation accounting for soil, wind, wave and structural parameter variations by a multi-step approach. Velarde et al. [22] studied the sensitivity of the fatigue load in OWT support structure by linear regression simulation and Morris screening method. While most of the above-mentioned studies did not consider the sea current and the directions of wind, sea wave and current were assumed to be aligned. Also, the focus was always on one specific point or element of the foundation. Thus, the results provided limited information.

In this paper, the GSA of the maximum stress and displacement with respect to material, local geometry and environmental uncertain parameters are investigated for OWT jackets used in Code Comparison Collaboration Continuation (OC4) project [15]. Compared to other previous studies, the wind parameters are replaced by the wind load actions simulated in aerodynamic software FAST developed by NREL [11]. The sea current is considered and the direction of wind, sea wave and current are assumed to be independent. In addition, the maximum stresses of static analysis in different parts of the jacket are considered. The Morris screening method [12] and the Fourier amplitude sensitivity test method [2] are applied to perform the global sensitivity analysis.

2 Methodology

In this part, the general procedure for structural modeling and sensitivity analysis is presented. The two sensitivity methods used, Morris Screening and Fourier amplitude sensitivity test methods are also briefly discussed.

2.1 Wind Turbine Model and Support Structure

The wind turbine model used in this paper is the fixed-bottom offshore wind turbine. Our focus is on the jacket foundation used in OC4 project [15]. To save simulation time, the load actions applied on the jacket are uncoupled and simulated separately like [7]. The aerodynamic simulation is performed in aerodynamic software FAST. The maximum loads on the transition piece are taken as the bound references for the redefined aerodynamic forces and moments. The redefined aerodynamic loads are applied on the top of jacket models developed in the finite element software ANSYS, combined with different geometrical, material parameters of the jacket and ocean environmental parameters. The maximum stresses of static analysis in different zones of the jacket are considered. All the process is showed in Fig. 1. The random variables are firstly sampled for each sensitivity analysis method by using different sampling method. The sensitivity analysis will be conducted after enough simulations. The different zones of jacket are presented in Fig. 2. There are 5 zones (Yups, K1s, K2s, K3s and Muds) and two specific parts (Nu, Es). The sensitivity analysis of the maximum stress of these 5 zones and Es part are studied. The sensitivity analysis of the maximum displacement of Nu part is also investigated.

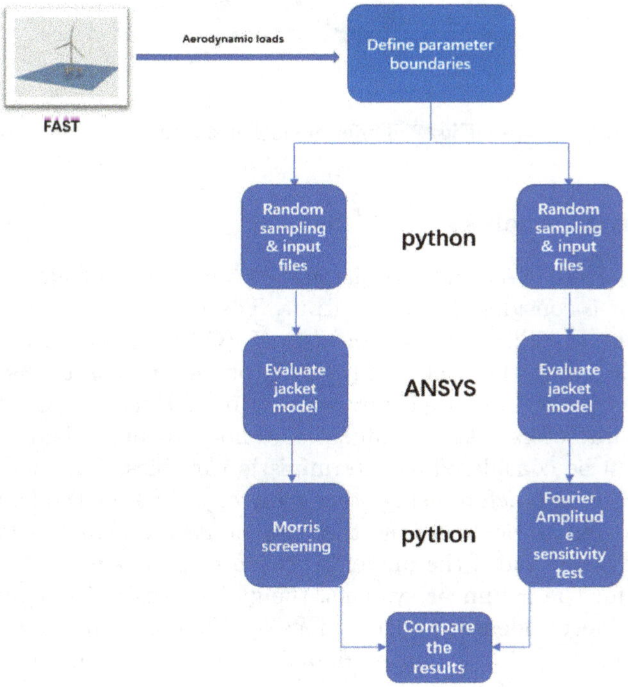

Fig. 1. General workflow

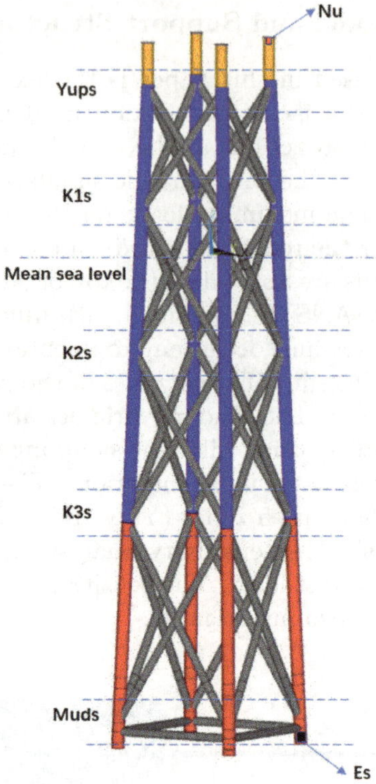

Fig. 2. Different jacket zones

2.2 Sensitivity Analysis

Sensitivity analysis (SA) aims to quantify the influence of all uncertain input parameters for a considered model output. The most engineering-used method is based on the so-called "one-factor-a-time" (OAT) design, where each input is varied while fixing the others. The OAT procedure has the advantage that the model has to be evaluated only a few times. Hence, it is widely used to identify the parameters that are influential or no-influential. Those no-influential parameters can be considered as deterministic variables. The model complexity is therefore simplified before using other more costly SA methods. However, this approach leads to the conclusions that are limited within the local sampling space and are only valid if the model is proven to be linear [17].

However, for the nonlinear systems, the global sensitivity analysis (GSA) is needed. It can better identify contributions of different inputs to the uncertainty of model output in full distribution domain of the inputs and comprehensively consider the average effect of the inputs on the output. Many GSA methods have been developed over the past decades. In this paper, the two GSA methods are applied namely the (1) Morris Screening method, (2) Fourier amplitude sensitivity test method.

Morris screening method known as Elementary Effects method [12], it is also based on one-factor-a-time, but it overcomes the limitation related to local variation. It allows to classify the inputs in three groups: inputs having negligible effects, input having overall effects without interaction and inputs having large non-linear and/or interaction effects. The method consists in discretizing the input space for each variable, then performing a given number of OAT designs. Such design of experiment is randomly chosen in the input space, and the variation direction is also random. The elementary effects for each input can be estimated by repeating these steps. Sensitivity indices are derived from these effects.

In Morris screening method, two sensitivity measures are calculated: μ, which estimates the overall influence of the factor on the output, and σ, which assesses the totality of the factor's higher order effects,i.e. non-linear and interaction effects. Let us denote by r the number of OAT designs (r usually between 4 and 10). Let us discretize the input space in a d-dimension grid with n levels of inputs. The elementary effect of the j-th variable obtained as the i-th repetition is defined as:

$$E_j^{(i)} = \frac{f\left(\mathbf{X}^{(i)} + \Delta e_j\right) - f\left(\mathbf{X}^{(i)}\right)}{\Delta}$$

where Δ is a predetermined multiple of $\frac{1}{(n-1)}$ and e_j a vector of the canonical base. Indices are obtained as follows:

- $\mu_j^* = \frac{1}{r}\sum_{i=1}^{r}\left|E_j^{(i)}\right|$(mean of the absolute value of the elementary effects [1]).
- $\sigma_j = \sqrt{\frac{1}{r}\sum_{i=1}^{r}\left(E_j^{(i)} - \frac{1}{r}\sum_{i=1}^{r}E_j^{(i)}\right)^2}$ (standard deviation of the elementary effects).

Here, μ_j^* is a measure of influence of the j-th input on the output. A high value of μ_j^* indicates that the j-th input parameter gives an important overall influence on the output. σ_j is a measure of nonlinear and/or interaction effects of the j-th input. A high value σ_j means that the j-th input is involved in interaction with other parameters or whose effect is nonlinear. These sensitivity measures, the absolute expected value μ^* and σ^*, are normalized between 0 and 1 using the equations bellow:

$$\mu_{n_j}^* = \frac{\mu_j^*}{\max\left(\mu^*, \sigma\right)}$$

$$\sigma_{n_j} = \frac{\sigma_j}{\max\left(\mu^*, \sigma\right)}$$

Fourier amplitude sensitivity test was firstly introduced by Cukier et al. [2]. It was improved by Saltelli et al.[18], who proposed the extended Fourier amplitude sensitivity test which is more robust and efficient and allows the computation of the total contribution of each input factor to the output variance. The term "total" here means that the factor's main effect, as well as all interaction terms involving that factor, are included. Compared to the GSA method of

Sobol, the obvious advantage of the Fourier amplitude sensitivity test is small sample size. In the Fourier amplitude sensitivity test method, the first-order indice S_{1_i} and the total indice S_{T_i} are calculated as bellow:

$$S_{1_i} = \frac{\sum_{l=1}^{+\infty} |c_{l\omega_i}|^2}{\sum_{\omega=1}^{+\infty} |c_\omega|^2}$$

$$S_{T_i} = \frac{\sum_{\omega=\frac{\omega_i}{2}+1}^{N/2} |\hat{c}_\omega|^2}{\sum_{\omega=1}^{N/2} |\hat{c}_\omega|^2}$$

where c_ω is the Fourier coefficient and more details can be found in [21].

The values of S_{1_i} and S_{T_i} are always between 0 and 1. With high value of S_{1_i}, that means the effect of i-th input on the output is important. With high value of S_{T_i}, that means the main effect of i-th input, including all interaction terms involving i-th input, is significant.

3 Results of Sensitivity Analysis

Before conducting the SA analysis, as the process showed in (Fig. 1 (a)), the aerodynamic loads are firstly simulated in FAST. Only the load case of mean wind velocity ($U_w = 12$ m/s with normal turbulence model) is simulated. The simulation time is of 10 min as proposed in current standards. Additional 60 s at the beginning of each simulation are used to account for transient starts-ups. The maximum loads on the transition piece are taken as the reference for the wind load actions. The bounds of the parameters are defined and the two GSA methods are applied. A static linear analysis of the jacket is performed by using the finite element software ANSYS. The maximum stresses and displacements for different jacket zones are then obtained.

3.1 Parameter Selection and Bounds

The SA of this paper is focused on the environmental conditions like wind, wave and material and local geometrical parameters of the foundation of jacket. For the local geometry parameters of the foundation, the different diameters and thicknesses are in Table 1. All the parameters and bounds are listed in Table 2. They are all assumed to be uniform distribution and independent to each other. Here, E, v, ρ_j and ρ_o are the Young's modulus, Poisson's ratio, density of jacket, density of ocean. Di and Ti are the diameters and thicknesses of the jacket as depicted in Table 1. Hw, Tw, Dw, Vcb, Dcb, Vcm and Dcm are, respectively, the wave height, the wave period, the wave direction, the velocity and the direction of current in bottom of the ocean and the velocity and the direction of current in the mean sea level. Fwi and Mwi are the redefined aerodynamic forces and moments applied on the top of jacket in the finite element models. $MaxFwi$ and $MaxMwi$ are the maximum aerodynamic forces and moments on the transition piece simulated in FAST.

Table 1. Properties of jacket members

Color in (Fig. 1 (b))	Outer diameter(m)	Thickness(m)
Grey	$D1 = 0.8$	$T1 = 0.02$
Red	$D2 = 1.2$	$T2 = 0.05$
Blue	$D2 = 1.2$	$T3 = 0.035$
Orange	$D2 = 1.2$	$T4 = 0.04$

Table 2. Parameter names and bounds

Parameter names	Parameter Bounds
$D1$	$[0.9 \times 0.8,\ 1.1 \times 0.8]\ (m)$
$D2$	$[0.9 \times 1.2,\ 1.1 \times 1.2]\ (m)$
$T1$	$[0.95 \times 0.02,\ 1.05 \times 0.02]\ (m)$
$T2$	$[0.95 \times 0.05,\ 1.05 \times 0.05]\ (m)$
$T3$	$[0.95 \times 0.035,\ 1.05 \times 0.035]\ (m)$
$T4$	$[0.95 \times 0.04,\ 1.05 \times 0.04]\ (m)$
E	$[0.9 \times 2.1e11,\ 1.1 \times 2.1e11]\ (Pa)$
v	$[0.9 \times 0.3,\ 1.1 \times 0.3]$
ρ_j	$[0.9 \times 7850,\ 1.1 \times 7850]\ (kg/m^3)$
ρ_o	$[0.9 \times 1025,\ 1.1 \times 1025]\ (kg/m^3)$
Hw	$[1,\ 8]\ (m)$
Tw	$[5,\ 15]\ (s)$
Dw	$[-90°,\ 90°]$
Vcb	$[0,\ 2]\ (m/s)$
Dcb	$[-90°,\ 90°]$
Vcm	$[0,\ 0.5]\ (m/s)$
Dcm	$[-90°,\ 90°]$
Fwx	$[0.9 \times MaxFwx,\ 1.1 \times MaxFwx]\ (N)$
Fwy	$[0.9 \times MaxFwy,\ 1.1 \times MaxFwy]\ (N)$
Fwz	$[0.9 \times MaxFwz,\ 1.1 \times MaxFwz]\ (N)$
Mwx	$[0.9 \times MaxMwx,\ 1.1 \times MaxMwx]\ (N * m)$
Mwy	$[0.9 \times MaxMwy,\ 1.1 \times MaxMwy]\ (N * m)$
Mwz	$[0.9 \times MaxMwz,\ 1.1 \times MaxMwz]\ (N * m)$

The bounds of the diameters of the jacket, material parameters, density of ocean and the aerodynamic loads simulated in FAST are between 90% and 110% of the original value. The thicknesses of the jacket are between 95% and 105% of the original value to respect the relations between the different thicknesses in the original design. The bounds of parameters of wave and current are defined by the literature [4,19]. In the meantime, the hydrodynamic loads are simulated in the ocean load part of the finite element software ANSYS. All the sensitivity analysis are performed in SALib module of Python [9].

3.2 Morris Screening Results (Elementary Effects)

The total number of evaluations (N_{morris}) of Morris screening is defined by $N_{\text{morris}} = r(k+1)$. r, is the number of trajectories which is normally between 10 and 50 [1] and k is the number of parameters. To get better results, $r = 15, 21, 42, 84, 126$ with 6-level grids are simulated to observe the convergence trend. Hence, the simulations of $N_{\text{morris}} = 360, 504, 1008, 2016, 3024$ are performed and the results of $N_{\text{morris}} = 3024$ are normalized as showed in Fig. 3 and Fig. 4.

As demonstrated in Fig. 3, the maximum stress in different parts of jacket is sensitive to different parameters. Here, the word "sensitive" means that one or both of the two Morris indexes ($\mu^*_{n_j}$ and σ_{n_j}) should be greater than the value of 0.2. In the Es and Muds zones as showed Fig. 3 (a) and (c), it is found that the wave period (Tw) and wave height (Hw) have significant overall effect and nonlinear effect on the maximum stress. Also, the diameter ($D2$) and the velocity of bottom sea current (Vcb) have overall effect on the maximum stress. In the meantime, the wave direction (Dw) has interactions with other factors. In the K3s, K2s and K1s zones as presented in Fig. 3 (d), (e) and (f), it is clear that the wave period (Tw) and wave height (Hw) still have big effect on the maximum stress. The diameter ($D2$) has big overall influence in these 3 zones. The wave direction (Dw) also has interactions with other parameters except in K2s zone. In addition, the thickness ($T3$), the force (Fwx) and the moment (Mwy) start to have a big overall effect. It is also obvious that the differences exist in these 3 zones. The effect of the wave height (Hw) and wave period (Tw) decrease a lot in K1s zone, because the K1s zone is above the mean sea level. For the diameter ($D2$), the thickness ($T3$) and the moment (Mwy), they have more important overall effect in K1s zone than in K2s and K3s zone. In Yups zone as showed in Fig. 3 (b), the diameter ($D2$), the thickness ($T4$), the force (Fwx) and the moment (Mwy) have big overall influence and the impact of wave parameters has reduced nearly to 0. For the displacement as showed in Fig. 4, the wave height (Hw) and wave period (Tw) have important overall effect and nonlinear effect. Also, the wave direction (Dw) has significant interactions with other factors. The Young's modulus (E), the diameter ($D2$), the force (Fwx) and the moment (Mwy) have overall effect on the displacement in Nu part.

3.3 Fourier Amplitude Sensitivity Test Results

Compared to Morris screening method, the Fourier amplitude sensitivity test takes more simulation time. However, compared to other variance-based sensitivity analysis method like Sobol, it is more efficient. To choose the necessary frequencies to make the Fourier amplitude sensitivity test, in the SALib module of Python, that only depends on the number of random variables (D), the number of samples (N) to generate and the interference parameters (M). Once

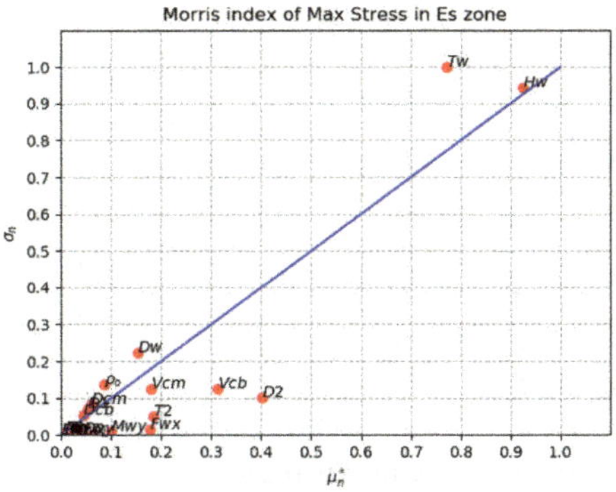

(a) Es part

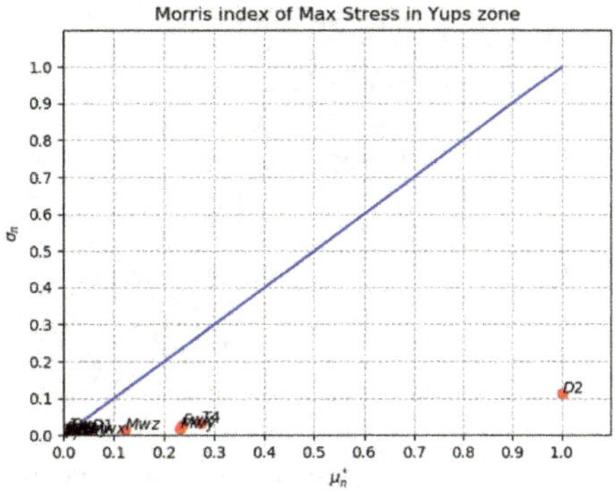

(b) Yups zone

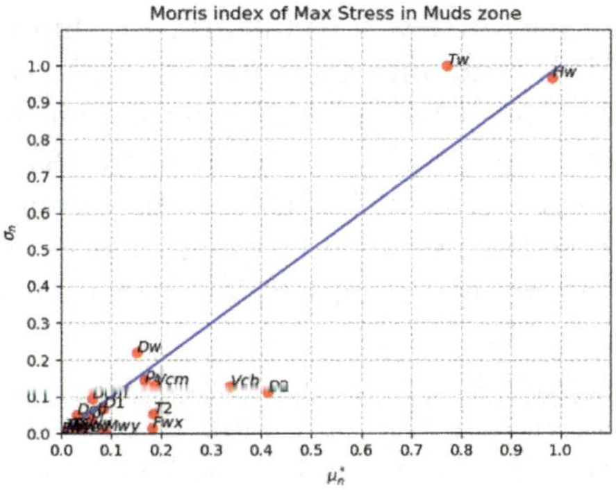

(c) Muds zone

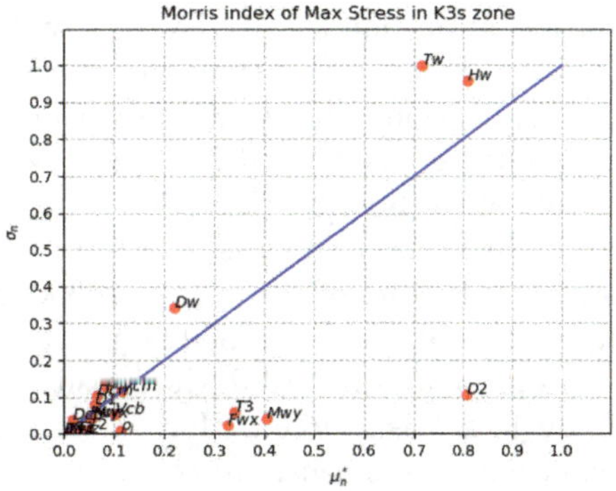

(d) K3s zone

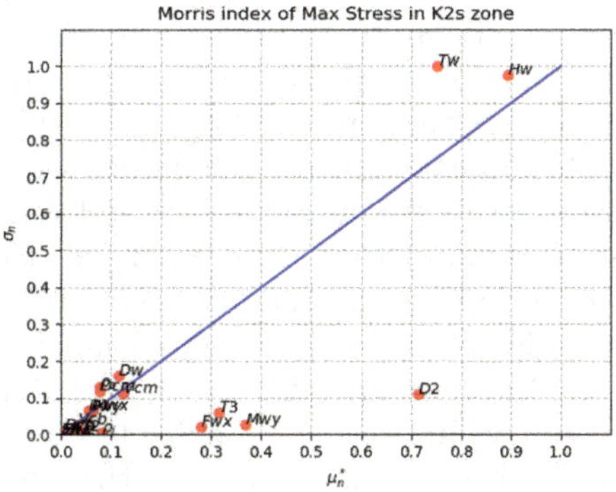

(e) K2s zone

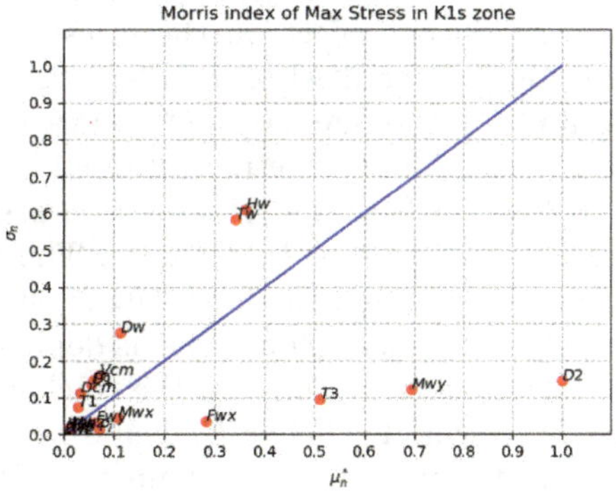

(f) K1s zone

Fig. 3. SA of the maximum stress in different parts of jacket

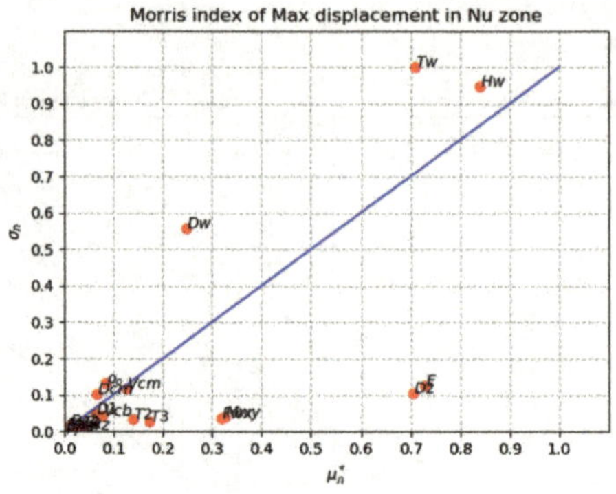

Fig. 4. SA of the maximum displacement in Nu part

these three key parameters (D, N, M) are decided, the frequencies used for Fourier amplitude sensitivity test will be generated with built-in algorithm of SALib module. In addition, the phase shift is randomly generated on $[0, 2\pi)$ at each replication. The total number of evaluations (N_{FAST}) of Fourier amplitude sensitivity test is defined $N_{FAST} = N \times D$. More details can be found in SALib module of Python [9]. Here, $N = 450, 900, 1310, 1750, 2175$ with the interference parameter equal to 4 are simulated to observe the convergence trend. Hence, the total simulations of $N_{FAST} = 10350, 20700, 30130, 40250, 50025$ are performed and results of $N_{FAST} = 50025$ are showed in Fig. 5 and Fig. 6.

As presented in Fig. 5, the sensitivity of maximum stresses in different parts of the jacket is studied. Here, it is supposed that the maximum stresses and displacement are sensitive to the parameter whose total index (S_{T_i}) of Fourier amplitude sensitivity method should be greater than or equal to 0.05. In the Es and Muds zones as presented in Fig. 5 (a) and (c), the wave height (Hw), wave period (Tw), the diameter ($D2$) and the velocity of bottom sea current (Vcb) have significant effect on the stress, which is similar to the results of Morris screening method. Also, in K3s, K2s and K1s zones as presented in Fig. 5 (d), (e) and (f), the diameter ($D2$) always has a big effect on the stress. The wave height (Hw) and wave period (Tw) are still important to the maximum stress in K2s and K3s zones. However, in K1s zone, the wave height (Hw) and wave period (Tw) have little influence. The parameters like the moment (Mwy), the thickness ($T3$) and the force (Fwx) start to have effect on the maximum stress in these 3 zones. In Yups zone as depicted in Fig. 5 (b), the geometry parameters like the diameter ($D2$) and the thickness ($T4$) are important to the stress. The redefined wind loads like Fwx and Mwy also have impact. For the displacement as depicted in Fig. 6, the diameter ($D2$), the Young's modulus (E), the wave height (Hw) and wave period (Tw) have main effect and the wind loads like Fwx and Mwy begin to affect the displacement.

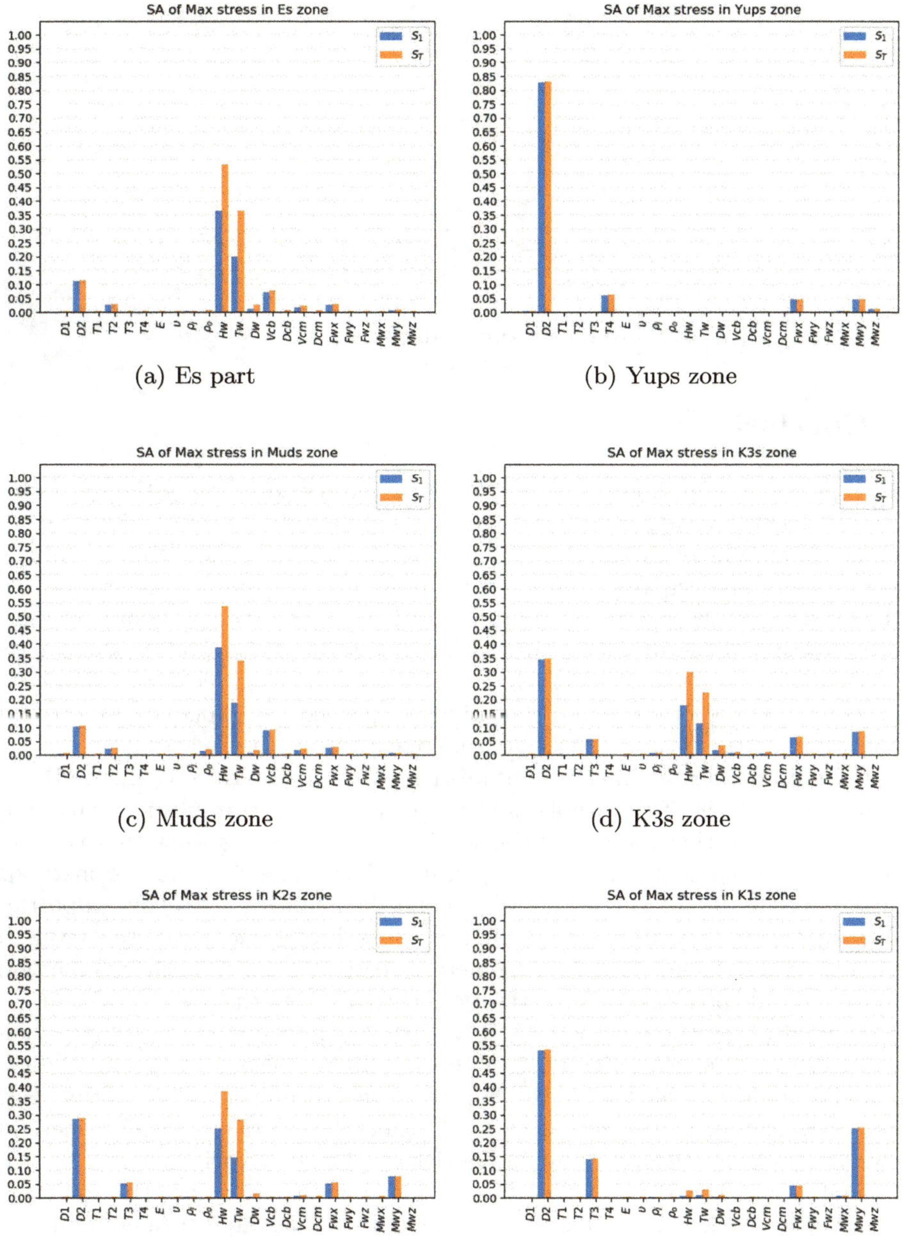

(a) Es part

(b) Yups zone

(c) Muds zone

(d) K3s zone

(e) K2s zone

(f) K1s zone

Fig. 5. SA of the maximum stress in different parts of jacket

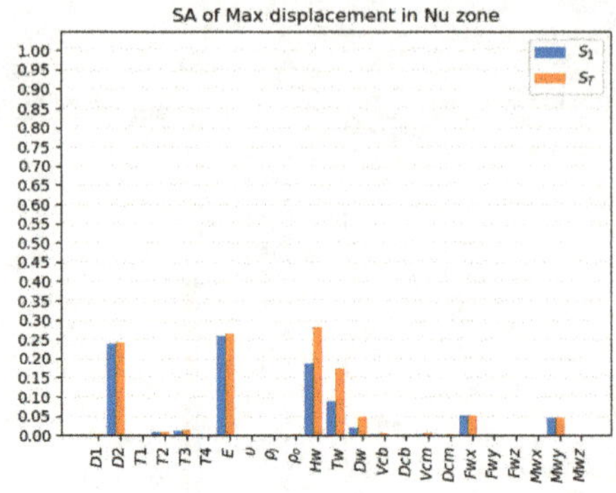

Fig. 6. SA of the maximum displacement in Nu part

4 Conclusion

In this paper, two methods of GSA have been performed to study the sensitivity of environmental, local geometry and material parameters of the maximum stress and the displacement in the jacket. The maximum stresses of different parts of the jacket are investigated. The parameters with main and nonlinear effects in different parts of the jacket are identified. It is clear that the maximum stress is more sensitive to the wave parameters in the bottom of the jacket. But for the zone above the mean sea level, the geometry parameters and redefined wind loads will play a big role. It is noticed that different parameters should be considered, while studying the local behaviors of different parts of the jacket.

In addition, the two methods applied, one (Morris screening method) is a qualitative analysis of sensitivity and the other (Fourier amplitude sensitivity test) is a quantitative analysis of the sensitivity. The results show that they are mostly in good agreement. For example, it is obvious that in the two methods, the wave height (Hw), wave period (Tw) and the diameter ($D2$) have important influence on the maximum stress in the zones below the mean sea level. In K1s zone, the influence of wave parameters decrease a lot in both two method. Also, for the displacement, both the two method show the wave height (Hw), wave period (Tw), the diameter ($D2$) and the Young's modulus (E) have main effects. However, a few differences of the results can also be found in the two methods. For example, the wave direction shows significant interaction with others parameters in the K1s and Muds zones by Morris Screening method. However, it seems not important in Fourier amplitude sensitivity test.

Finally, our works also suffer from some limitations, where the uncertain soil parameters are not considered and the wind actions are considered only for one case of mean value wind velocity. Future works allow to considering these drawbacks. Ongoing works are to consider the reliability analysis and reliability-based optimization.

Acknowledgements. Financial support for this work was partly provided by the CSC program (China Scholarship Council) between the People's Republic of China and the INSA group. This support is gratefully acknowledged.

References

1. Campolongo, F., Cariboni, J., Saltelli, A.: An effective screening design for sensitivity analysis of large models. Environ. Model. Software **22**(10), 1509–1518 (2007)
2. Cukier, R., Fortuin, C., Shuler, K.E., Petschek, A., Schaibly, J.: Study of the sensitivity of coupled reaction systems to uncertainties in rate coefficients i theory. J. Chem. Phys. **59**(8), 3873–3878 (1973)
3. Damgaard, M., Andersen, L.V., Ibsen, L.B., Toft, H.S., Sørensen, J.D.: A probabilistic analysis of the dynamic response of monopile foundations: Soil variability and its consequences. Prob. Eng. Mech. **41**, 46–59 (2015)
4. Digre, K.A., Zwerneman, F., et al.: Insights into using the 22nd edition of API RP 2a recommended practice for planning, designing and constructing fixed offshore platforms-working stress design. In: Offshore Technology Conference. Offshore Technology Conference (2012)
5. Glisic, A., Ferraz, G.T., Schaumann, P.: Sensitivity analysis of monopiles' fatigue stresses to site conditions using monte carlo simulation. In: Proceedings of the Twenty-seventh. International Ocean and Polar Engineering Conference, pp. 305–311. International Society of Offshore and Polar Engineers (ISOPE), Cupertino (2017)
6. Hansen, M., et al.: Probabilistic safety assessment of offshore wind turbines. Annual Report (2010)
7. Haselbach, P., Natarajan, A., Jiwinangun, R.G., Branner, K.: Comparison of coupled and uncoupled load simulations on a jacket support structure. Energy Procedia **35**, 244–252 (2013)
8. Helton, J.C., Davis, F.J.: Latin hypercube sampling and the propagation of uncertainty in analyses of complex systems. Reliab. Eng. Syst. Safety **81**(1), 23–69 (2003)
9. Herman, J.D., Usher, W.: Salib: an open-source python library for sensitivity analysis. J. Open Source Software **2**(9), 97 (2017)
10. Hübler, C., Gebhardt, C.G., Rolfes, R.: Hierarchical four-step global sensitivity analysis of offshore wind turbines based on aeroelastic time domain simulations. Renew. Energy **111**, 878–891 (2017)
11. Jonkman, J.M., Buhl Jr., M.L., et al.: Fast user's guide (2005)
12. Morris, M.D.: Factorial sampling plans for preliminary computational experiments. Technometrics **33**(2), 161–174 (1991)
13. Morthorst, P.E., Auer, H., Garrad, A., Blanco, I.: The economics of wind power. Wind Energy: The Facts, Part Three. www.wind-energy-the-facts.org (2009)
14. Phoon, K.K., Kulhawy, F.H.: Characterization of geotechnical variability. Can. Geotech. J. **36**(4), 612–624 (1999)
15. Popko, W., et al.: Offshore code comparison collaboration continuation (oc4), phase I-results of coupled simulations of an offshore wind turbine with jacket support structure. Technical report, National Renewable Energy Lab. (NREL), Golden, CO (United States) (2012)
16. Saltelli, A., et al.: Global Sensitivity Analysis: The Primer. Wiley, Chichester (2008)

17. Saltelli, A., Ratto, M., Tarantola, S., Campolongo, F., Commission, E., et al.: Sensitivity analysis practices: strategies for model-based inference. Reliab. Eng. Syst. Safety **91**(10–11), 1109–1125 (2006)
18. Saltelli, A., Tarantola, S., Chan, K.S.: A quantitative model-independent method for global sensitivity analysis of model output. Technometrics **41**(1), 39–56 (1999)
19. Shi, W., Park, H., Han, J., Na, S., Kim, C.: A study on the effect of different modeling parameters on the dynamic response of a jacket-type offshore wind turbine in the korean southwest sea. Renew. Energy **58**, 50–59 (2013)
20. Stensgaard Toft, H., Svenningsen, L., Moser, W., Dalsgaard Sørensen, J., Lybech Thøgersen, M.: Wind climate parameters for wind turbine fatigue load assessment. J. Solar Energy Eng. **138**(3), (2016)
21. Tarantola, S., Mara, T.A.: Variance-based sensitivity indices of computer models with dependent inputs: the fourier amplitude sensitivity test. Int. J. Uncertainty Quant. **7**(6), (2017)
22. Velarde, J., Kramhøft, C., Sørensen, J.D.: Global sensitivity analysis of offshore wind turbine foundation fatigue loads. Renew. Energy **140**, 177–189 (2019)

Uncertainty Quantification and Stochastic Modeling for the Determination of a Phase Change Boundary

Juan Manuel Rodriguez Sarita, Renata Troian, Beatriz Costa Bernardes, and Eduardo Souza de Cursi(✉)

LMN, Normandy University, INSA Rouen, 76801 Saint-etienne du Rouvray Cedex, France
souza@insa-rouen.fr

Abstract. We consider the determination of solid/liquid interfaces by the solution of the Stefan problem, involving two heat equations in unknown domains (two-phase problem). We establish a regularized formulation of the Stefan problem, which is used to characterize approximated values of the field of temperatures as means of convenient stochastic processes, using Feynman-Kac representations. The results of the resulting stochastic method are compared to Finite Element Approximations and show to be comparable to T2 finite element approximations. We present an example of variability of the domains occupied by the phases. In future work, methods for the uncertainty quantification of infinite dimensional objects will be applied to characterize the variability of the regions.

Keywords: Stefan problem · Free boundary · Feynman-Kac representation

1 Introduction

Stefan problem is a classical model describing the fusion/solidification, formulated in 1891 [1]. Previously, in 1831, Lamé and Clapeyron considered analogous situations [2]. It belongs to the family of Free Boundary Problems (FBP), id est, problems involving the solution of Partial Differential Equations (PDE) over unknown regions to be determined – one of the unknowns is a boundary separating regions where the PDEs are satisfied. In the case of Stefan's problems, the regions occupied by the solid phase and the liquid phase are unknowns, so that the physical properties are different in two unknown regions, separated by an unknown free boundary, usually called *Stefan boundary*.

Stefan problem was object of intensive research and many works propose methods of solution. Nowadays, numerous researchers are still interested in the Stefan problem, namely by the possibility of carrying out energy storage by phase change – what is an important topic for the development of renewable energies.

In the framework of numerical methods, it is usual to distinguish one-phase Stefan problems and two-phase Stefan's problem. In the first case, one of the phases (solid or liquid) is supposed isothermal, so that there is no equation to solve in this region. The

J. E. S. De Cursi (Ed.): Uncertainties 2020, LNME, pp. 49–68, 2021.
https://doi.org/10.1007/978-3-030-53669-5_4

other phase is supposed to verify the classical heat equation, introduced by Jean-Baptiste-Joseph Fourier [3] in an article published under the name of Simeon-Denis Poisson [4]. This situation was solved by using variational inequalities – see, for instance, [5–9].

In two-phase Stefan problems, both the phases are supposed to verify the heat equation, but with different parameters - changes in the thermophysical properties of the material occur when the temperature varies [10]. In this case, a second approach involving the use of an enthalpic transformation was introduced leading to variational inequalities – see, for instance [11–15]. Regularization [16] and shape optimization [17] approaches were introduced as alternatives to the variational inequality approach.

Despite a very large number of works about the Stefan problem, the application of Uncertainty Quantification (UQ) methods and Stochastic Modeling (SM) techniques remains rare: many works consider the solution of PDE by SM methods (see, for instance, [18–21]), but it is rare to find the solution of Stefan problem by these approaches [22]. In addition, analysis of the variability of the solutions and of the impact of errors in the parameters remain rare. In this work, we present a stochastic method for the solution of a Stefan problem and a comparison with Finite Elements solution by regularization. We study also the variability of the results as function of some parameters.

2 The Stefan Model

We present the model in a bidimensional situation, without loss of generality: the equations extend to three-dimensional situations straightly. In the numerical results, we present some 3D experiments.

Let us consider a region $\Omega \subset \mathbb{R}^2$, occupied by the melting/solidificating material. To simplify the presentation, we consider a rectangular region with dimensions $H_1 \times H_2$ – again, no loss of generality in the equations. The material is partially in liquid phase and partially in solid phase, so that, the region Ω is divided into two regions Ω_S (the solid part) and Ω_L (the liquid part). These regions are separated by a moving region Σ - in general a surface (the Stefan boundary or phase change boundary).

Let us denote by θ the field of temperatures in the material and θ_C be the temperature of phase change: for $\theta < \theta_C$, the material is solid; for $\theta > \theta_C$, it is liquid. Thus,

$$\Omega_S = \{\boldsymbol{x} \in \Omega : \theta < \theta_c\};\ \Omega_L = \{\boldsymbol{x} \in \Omega : \theta > \theta_c\};\ \sum = \{\boldsymbol{x} \in \Omega : \theta = \theta_c\}$$

In a general situation, Σ may be a mushy region, id est, may have a strictly positive surface (or volume, in 3D). The regions are unknown and must be determined: $\Omega_S, \Omega_L, \Sigma$ are unknowns. In addition, these regions vary with the time: the material melts or solidifies, so that the regions change in time. Indeed, let $T > 0$ be the maximal time under consideration. The complete domain of study is $Q = \Omega \times (0, T)$ and

$$Q_S = \{(\boldsymbol{x}, t) \in \mathrm{Q} : \theta < \theta_c\};\ Q_L = \{(\boldsymbol{x}, t) \in \mathrm{Q} : \theta > \theta_c\};\ \mathrm{S} = \{(\boldsymbol{x}, t) \in Q : \theta = \theta_c\}$$

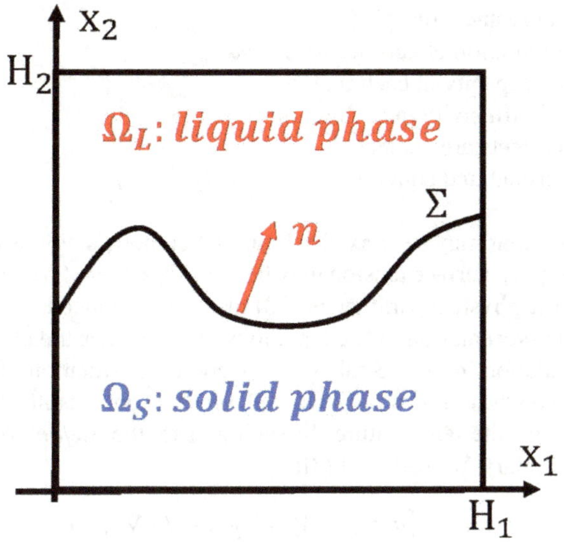

Fig. 1. The domain under consideration

Fourier's law establishes that the heat flux $\boldsymbol{q}$ by conduction is proportional to the opposite of the gradient of the temperature (Fig. 1):

$$\boldsymbol{q} = -k\nabla\theta \tag{1}$$

k is the thermal conductivity of the material. Let c be the specific heat capacity and ρ be the volumic mass of the material. The balance of the energy leads to the classical heat equation:

$$\rho c\frac{\partial\theta}{\partial t} - div(k\nabla\theta) = 0 \tag{2}$$

which is satisfied in the region occupied by any phase. As previously remarked, the thermophysical properties differ between phases solid and liquid phase, so that the ρ, c, k differ in each region – we have a different classical heat equation in each region:

$$\rho_S c_S\frac{\partial\theta}{\partial t} - div(k_S\nabla\theta) = 0 \text{ in } Q_S \tag{3}$$

$$\rho_L c_L\frac{\partial\theta}{\partial t} - div(k_L\nabla\theta) = 0 \text{ in } Q_L \tag{4}$$

The basic Stefan model is based on the following assumptions:

- The material is pure and isotropic.
- Convection and radiation effects are neglected
- Constant thermal capacity in each region
- Constant thermal diffusivity in each region
- Surface tension effects are neglected.
- Densities are constant and equal in each region ($\rho_S = \rho_L = \rho$).

Each of these restrictions may be relaxed to build richer models. For instance, convection may be introduced [15]; surface tension may be considered [24, 25]; anisotropy [26, 27] and variability of the physical parameters [28] may be introduced.

To simplify the presentation of the equations, we consider the classical framework [29], but the formulation in Sect. 3 takes into account the situation where the physical parameters depend on the temperature – we shall present some results where dependance of the properties with the temperature. In such a case, the *Stefan condition*, which is satisfied on the boundary Σ, reads as [28]:

$$\lambda v = -\left[\boldsymbol{q}.\boldsymbol{n}\right]_S = k_S \, \nabla\theta|_S.\boldsymbol{n} - k_L \, \nabla\theta|_L.\boldsymbol{n} \tag{5}$$

Here, v is the normal velocity of a point in Σ. In addition to these conditions, we must consider initial and boundary conditions: the basic Stefan's model with Dirichlet conditions reads as

$$\rho c_S \frac{\partial\theta}{\partial t} - div(k_S \nabla\theta) = 0, \; in \; \mathrm{Q}_S;$$

$$\rho c_L \frac{\partial\theta}{\partial t} - div(k_L \nabla\theta) = 0, \; in \; \mathrm{Q}_L;$$

$$\lambda v = k_S \, \nabla\theta|_S.\boldsymbol{n} - k_L \, \nabla\theta|_L.\boldsymbol{n} on \; S$$

$$\theta(\boldsymbol{x}, 0) = \theta_0(x) on \; \Omega$$

$$\theta(\boldsymbol{x}, t) = \theta_{\partial\Omega}(x, t) on \; \partial\Omega$$

To set the coefficient of the time derivative to the unity, some authors introduce the thermal diffusivity $\alpha = k/(\rho c)$ and rewrite in terms of this variable. In this work, we do not adopt such a point of view.

3 Transformations of the Stefan Model

By considering

$$r(\theta) = \begin{cases} \rho_s c_s, if \theta < \theta_c \\ \rho_L c_L, if \theta > \theta_c \end{cases}, \quad k(\theta) = \begin{cases} k_S, if \theta < \theta_c \\ k_L, if \theta > \theta_c \end{cases} \tag{6}$$

We have

$$r(\theta)\frac{\partial \theta}{\partial t} - div(k(\theta)\nabla\theta) = 0 \ in \ Q_S \cup Q_L \tag{7}$$

We may find in the literature two popular transformations of this equation, which are used namely when the physical quantities depend upon the temperature. The first one introduces a function R such that $R'(\theta) = r(\theta)$. Then

$$\frac{\partial}{\partial t} R(\theta) - div(k(\theta)\nabla\theta) = 0 \ in \ Q_S \cup Q_L \tag{8}$$

The second one introduces K such that $K'(\theta) = k(\theta)$. Then

$$\frac{\partial}{\partial t} R(\theta) - div(\nabla K(\theta)) = 0 \ in \ Q_S \cup Q_L \tag{9}$$

Finally, we may consider $u = R(\theta)$, $\beta(u) = K(R^{-1}(u))$. Then

$$\frac{\partial u}{\partial t} - div(\nabla\beta(u)) = 0 \ in \ Q_S \cup Q_L \tag{10}$$

Notice that these formulations take into account the situation where the physical parameters depend on the temperature. In order to apply the stochastic approach, we make a supplementary transformation in the Stefan model, described in the sequel. In this work, we adopt the point of view (9).

3.1 Transformation by Characteristic Function

Let us denote by $\boldsymbol{n}$ the unit normal to Σ, oriented inwards Ω_L. Assume that the Stefan boundary Σ is given by the equation $\eta(\boldsymbol{x}, t) = 0$. Then $\boldsymbol{n}\|\nabla\|\eta = \nabla\eta$ and

$$0 = \frac{\partial \eta}{\partial t} + \frac{\partial \eta}{\partial x_1}\frac{dx_1}{dt} + \frac{\partial \eta}{\partial x_2}\frac{dx_2}{dt} = \frac{\partial \eta}{\partial t} + \frac{d\boldsymbol{x}}{dt}.\boldsymbol{n}\|\nabla\|\eta$$

Thus, the normal velocity v of a point $\boldsymbol{x} \in \Sigma$ of the moving boundary is

$$v = -\frac{1}{\|\nabla\eta\|}\frac{\partial \eta}{\partial t}. \tag{11}$$

Let us introduce

$$\chi(\theta) = \begin{cases} 0, \textit{if}\ (x,t) \in Q_s \\ 1, \textit{if}\ (x,t) \in Q_L \end{cases} \tag{12}$$

We shall establish that

$$\frac{\partial}{\partial t}(R(\theta)) - div(\mathrm{k}(\theta)\nabla\boldsymbol{\theta}) = -\lambda\frac{\partial}{\partial t}(\chi(\theta)),\ in\ \mathrm{Q}; \tag{13}$$

Notice that Eq. (9) is written on $Q_S \cup Q_L$, while this equation is written on the whole domain $Q = Q_S \cup Q_L \cup S$, so that it must take into account the discontinuity through S. The proof is made by treating t as a space variable: let us denote $X_3 = t, \boldsymbol{X} = (x_1, x_2, t)$. For $\boldsymbol{\Psi} = (\psi_1, \psi_2, \psi_3)$ and : $Q \to \mathbb{R}$:

$$div_X(\boldsymbol{\Psi}) = \frac{\partial\psi_1}{\partial X_1} + \frac{\partial\psi_2}{\partial X_2} + \frac{\partial\psi_3}{\partial X_3}; \boldsymbol{grad}_X(\varphi) = \left(\frac{\partial\varphi}{\partial X_1}, \frac{\partial\varphi}{\partial X_2}, \frac{\partial\varphi}{\partial X_3}\right). \tag{14}$$

Let denote by $\mathcal{D}(Q)$ the set of the infinitely many differentiable functions with compact support on Q. Then, for $\varphi \in \mathcal{D}(Q)$ and $\boldsymbol{\Psi} = (0, 0, \varphi)$

$$\left\langle\frac{\partial\chi}{\partial X_3}, \varphi\right\rangle = -\int_Q \chi\frac{\partial\varphi}{\partial X_3}d\boldsymbol{X} = -\int_{Q_L}\frac{\partial\varphi}{\partial X_3}d\boldsymbol{X} = -\int_{Q_L} div_X(\boldsymbol{\Psi})d\boldsymbol{X} = \int_S \varphi N_3 dS, \tag{15}$$

where $\boldsymbol{N} = (N_1, N_2, N_3)$ is the unitary normal to S., oriented inwards Q_L. We have

$$N = \frac{\boldsymbol{grad}_X(\eta)}{\left\|\boldsymbol{grad}_X(\eta)\right\|} = \frac{\|\nabla\eta\|}{\left\|\boldsymbol{grad}_X(\eta)\right\|}(n_1, n_2, -v)$$

Let $\boldsymbol{q} = (q_1, q_2)$ be the heat flux. Analogously, for $\boldsymbol{\Psi} = (q_1, q_2, R)$, we have

$$\langle div_X(\boldsymbol{\Psi}), \varphi\rangle = -\int_Q \boldsymbol{\Psi}.\boldsymbol{grad}_X(\varphi)d\boldsymbol{X}$$

Or,

$$-\int_{Q_L} \boldsymbol{\Psi}.\boldsymbol{grad}(\varphi)d\boldsymbol{X} = \int_{Q_L} \varphi\underbrace{div_X(\boldsymbol{\Psi})}_{=0}d\boldsymbol{X} - \int_{\mathrm{S}} \varphi\boldsymbol{\Psi}_{\mathbf{L}}.\boldsymbol{N}dS$$

$$-\int_{Q_S} \boldsymbol{\Psi}.\boldsymbol{grad}(\varphi)d\boldsymbol{X} = \int_{Q_S} \varphi\underbrace{div_X(\boldsymbol{\Psi})}_{=0}d\boldsymbol{X} + \int_{\mathrm{S}} \varphi\boldsymbol{\Psi}_{\mathbf{S}}.\boldsymbol{N}dS$$

Thus, the continuity of R shows that

$$\langle div_X(\boldsymbol{\Psi}), \varphi\rangle = -\int_S \varphi[\boldsymbol{\Psi}].N dS = -\int_S \varphi[\mathbf{q}.\mathbf{n}]\frac{\|\nabla\eta\|}{\|\boldsymbol{grad}_X(\eta)\|}dS$$

The Stefan condition shows that

$$\langle div_X(\boldsymbol{\Psi}), \varphi\rangle = \int_S \varphi\lambda v\frac{\|\nabla\eta\|}{\|\boldsymbol{grad}_X(\eta)\|}dS = -\int_S \lambda\varphi \mathrm{N}_3 dS.$$

Thus,

$$\langle div_X(\boldsymbol{\Psi}), \varphi\rangle = -\lambda\frac{\partial\chi}{\partial t}, \varphi.$$

Since φ is arbitrary in $\mathcal{D}(Q)$, we have

$$\frac{\partial}{\partial t}(R(\theta)) - div(\mathrm{k}(\theta)\nabla\theta) = div_X(\boldsymbol{\Psi}) = -\lambda\frac{\partial\chi}{\partial t} in\ Q.$$

Consequently, the Stefan's model with Dirichlet conditions reads as

$$\rho(\theta)c(\theta)\frac{\partial\theta}{\partial \mathrm{t}} - div(k(\theta)\nabla\theta) = -\lambda\frac{\partial}{\partial t}\chi(\theta)\ in\ Q. \tag{16}$$

$$\theta(\boldsymbol{x}, 0) = \theta_0(x)\ in\ \Omega \tag{17}$$

$$\theta(\boldsymbol{x}, t) = \theta_{\partial\Omega}(x, t)\ on\ \partial\Omega \tag{18}$$

with

$$\rho(\theta) = \begin{cases}\rho_S, if \theta < \theta_c \\ \rho_L, if \theta > \theta_c\end{cases}, \quad c(\theta) = \begin{cases}c_S, if \theta < \theta_c \\ c_L, if \theta > \theta_c\end{cases} \tag{19}$$

This form of the model is convenient for the numerical solution by regularization, domain optimization or algebraic equation methods.

4 Regularized Model

Since χ is discontinuous, its derivative is not a function, but a distribution. Thus, its numerical manipulation may introduce numerical difficulties. To obtain a regular problem, we may use a regularization: for instance, let $\varepsilon > 0$ and

$$\alpha(s; \varepsilon, s_0, a_0, a_1) = a_0 + \left[3\left(\frac{s - s_0}{\varepsilon}\right)^2 - 2\left(\frac{s - s_0}{\varepsilon}\right)^3\right]a_1$$

We have

$$\alpha(s_0) = a_0, \alpha(s_0 + \varepsilon) = a_1, \alpha'(s_0) = \alpha'(s_0 + \varepsilon) = 0$$

Thus

$$\chi_\varepsilon(\theta) = \alpha(s; \varepsilon, \theta_c, 0, 1)$$

is a differentiable function such that $\chi_\varepsilon(\theta) = \chi(\theta)$ if $\theta \leq \theta_c$ or $\theta \geq \theta_c + \varepsilon$ and $\chi_\varepsilon(\theta) \to \chi(\theta)$ a.e. when $\varepsilon \to 0+$: χ_ε is a regularization of χ. Analogously, let us consider

$$\beta(s; \varepsilon, s_0, a_0, a_1) = \frac{1}{2}(a_1 + a_0) + \frac{1}{4}\left[3\left(\frac{s - s_0}{\varepsilon}\right) - \left(\frac{s - s_0}{\varepsilon}\right)^3\right](a_1 - a_0)$$

Then

$$\beta(s_0 - \varepsilon) = a_0, \beta(s_0 + \varepsilon) = a_1, \beta'(s_0 - \varepsilon) = \beta'(s_0 + \varepsilon) = 0$$

and

$$k_\varepsilon(\theta) = \beta(s; \varepsilon, \theta_c, k_S, k_L)$$

is a differentiable function such that $k_\varepsilon(\theta) = k(\theta)$ if $|\theta - \theta_c| \geq \varepsilon$ and $k_\varepsilon(\theta) \to k(\theta)$ almost everywhere (a.e.) when $\varepsilon \to 0+$: k_ε is a regularization of k. We consider the approximated problem

$$\rho(\theta_\varepsilon)c(\theta_\varepsilon)\frac{\partial \theta_\varepsilon}{\partial t} - div(k_\varepsilon(\theta_\varepsilon)\nabla\theta_\varepsilon) = -\lambda\frac{\partial}{\partial t}\chi_\varepsilon(\theta_\varepsilon) \ in \ Q. \tag{20}$$

$$\theta_\varepsilon(\boldsymbol{x}, 0) = \theta_0(x) \ in \ \Omega \tag{21}$$

$$\theta_\varepsilon(\boldsymbol{x}, t) = \theta_{\partial\Omega}(x, t) on \ \partial\Omega \tag{22}$$

A result of convergence $\theta_\varepsilon \to \theta$ may be found in [16]. Regularizations of ρ and c may be introduced by the same way, if necessary.

5 Stochastic Approach for the Determination of the Stefan Boundary

Let us consider the regularized problem (20)–(22). In the sequel, we alleviate the notation by dropping the index ε. The field of temperatures may be characterized by using a Feynman-Kac formula involving stochastic diffusions [37]. Indeed, let us consider the generic parabolic equation

$$\ell(\boldsymbol{x}, t)\frac{\partial u}{\partial t} - div(A(\boldsymbol{x}, t)\nabla u) = 0, u = u_{\partial\Omega} \ on \ \partial\Omega, u(x, 0) = u_0(x) \tag{23}$$

Let $u_{\partial Q}$ verify

$$u_{\partial Q}(x,t) = u_0(x),\ if\ t = 0;\ u_{\partial Q}(x,t) = u_{\partial\Omega}(x,t),\ if\ t > 0. \tag{24}$$

Then

$$u(\boldsymbol{x},t) = E(Y_\tau), \tag{25}$$

where

$$Y_\tau = u_{\partial Q}(\boldsymbol{X}_\tau, \tau),\ \tau = \inf\{t \geq 0 : (\boldsymbol{X}_t, t) \notin Q\}. \tag{26}$$

Here, $(\boldsymbol{X}_s, T_s)$ is a couple of stochastic processes such that $\boldsymbol{X}_0 = \boldsymbol{x}$, $T_0 = t$ and

$$d\boldsymbol{X}_t = (2A(\boldsymbol{X}_t,t))^{\frac{1}{2}} d\boldsymbol{W}_t + \nabla A(\boldsymbol{X}_t,t)dt,\ dT_t = \ -\ell(\boldsymbol{X}_t,t)dt, \tag{27}$$

where $\boldsymbol{W}$ is the bidimensionnal Wiener process. Indeed, Ito's formula show that

$$dY_t = \frac{\partial u}{\partial t} dT_t + \sum_i \frac{\partial u}{\partial x_i} dX_{it} + \frac{1}{2}\sum_{i,j} \frac{\partial^2 u}{\partial x_i \partial x_j} dX_{it} dX_{jt}.$$

Since

$$-\ell(\boldsymbol{X}_t,t)\frac{\partial u}{\partial t} + \sum_i \frac{\partial u}{\partial x_i}\frac{\partial A}{\partial x_i} + A(\boldsymbol{X}_t,t)\sum_{i,j}\frac{\partial^2 u}{\partial x_i \partial x_j} = 0,$$

we have

$$dY_t = (2A(\boldsymbol{X}_t,t))^{\frac{1}{2}} \sum_i \frac{\partial u}{\partial x_i} dW_{it}.$$

In addition, the Wiener process is non anticipative, so that $E(dW_{it}) = 0$ and we have

$$u(\boldsymbol{x},t) = Y_0 = E(Y_t), \forall t.$$

Considering $t = \tau$ furnishes (23). In practice, we generate a sample $Y^1, \ldots, Y^{ns}$ of ns variates from Y_τ and we approximate $E(Y_t)$ by the empirical mean on the sample.

5.1 Using the Stefan Condition to Track the Free Boundary

A first method for the determination of the free boundary consists in evaluating the normal velocity v on Σ and, then, using it to determine the new position of the Stefan boundary. Assume that the position of Σ is known at $t \geq 0$ and we are interested in the position at $t + \Delta t$. From the Stefan condition:

$$v = \frac{1}{\lambda}(k_S \nabla\theta|_S.\boldsymbol{n} - k_L \nabla\theta|_L.\boldsymbol{n}).$$

Let $\boldsymbol{x} \in \Sigma$, $\boldsymbol{n}$ be the normal at $\boldsymbol{x}$, $\delta > 0$. Thus, we may evaluate

$$\nabla\theta|_S.\boldsymbol{n} \approx -\frac{\theta(\boldsymbol{x} - \delta\boldsymbol{n}) - \theta_c}{\delta},\ \nabla\theta|_L.\boldsymbol{n} \approx \frac{\theta(\boldsymbol{x} + \delta\boldsymbol{n}) - \theta_c}{\delta}.$$

The values of $\theta(\boldsymbol{x} \pm \delta\boldsymbol{n})$ may be determined by (25)–(27), with $\ell(\boldsymbol{x},t) = \rho(\theta(\boldsymbol{x},t))c(\theta(\boldsymbol{x},t)) + \lambda\chi'(\theta(\boldsymbol{x},t))$, $A(\boldsymbol{x},t) = k(\theta(\boldsymbol{x},t))$. Once v is determined, it is used to furnish the new position of the free boundary.

5.2 Solving the Equation Describing the Stefan Boundary

Assume again that the position of Σ is known at $t \geq 0$ and we are interested in the position at $t + \Delta t$. The free boundary satisfies algebraic equations having the form

$$E(\theta(\boldsymbol{x}, t + \Delta t) - \theta_c) = 0.$$

For instance, we may use $E(s) = s$ or $E(s) = sign(s)$. We may solve this equation for $\boldsymbol{x}$: let us consider a discretization of $(0, H_1)$, formed by the points $0 = x_1^1 < x_1^2 < \ldots < x_1^{n_1} < x_1^{n_1+1} = H_1$. For each x_1^i, we determine x_2^i such that

$$f\left(x_2^i\right) = E\left(\theta\left(x_1^i, x_2^i, t + \Delta t\right) - \theta_c\right) = 0.$$

Then, the Stefan boundary at time $t + \Delta t$ is approximated by connecting the points determined: $\boldsymbol{x}^i = \left(x_1^i, x_2^i\right)$, $1 \leq i \leq n_1 + 1$. Given a trial value $x_2^{i,k}$, $\theta\left(x_1^i, x_2^{i,k}, t\right)$ may be evaluated by (25)–(27). This value is used to generate a new value $x_2^{i,k+1}$ by an adapted numerical procedure, until convergence. For instance, we may use the Robbins-Monro procedure

$$x_2^{i,k+1} = x_2^{i,k} - \omega_k f\left(x_2^{i,k}\right), \omega_k > 0, \sum \omega_n = +\infty, \sum \omega_n^2 < \infty.$$

Robbins-Monro procedure is adapted to situations where errors in the evaluation of f must be considered. The values of $\theta\left(x_1^i, x_2^{i,k}, t\right)$ may be generated by using the stochastic approach (25)–(27) with $\ell(\boldsymbol{x}, t) = \rho(\theta(\boldsymbol{x}, t))c(\theta(\boldsymbol{x}, t)) + \lambda\chi'(\theta(\boldsymbol{x}, t))$, $A(\boldsymbol{x}, t) = k(\theta(\boldsymbol{x}, t))$.

5.3 Finding the Stefan Boundary by Domain Optimization

Let $\phi : Q \to \mathbb{R}$ be a function and

$$\rho(\phi) = \begin{cases} \rho_S, \text{ if } \phi < 0 \\ \rho_L, \text{ if } \phi < 0 \end{cases}, c(\phi) = \begin{cases} c_S, \text{ if } \phi < 0 \\ c_L, \text{ if } \phi < 0 \end{cases}, k(\phi) = \begin{cases} k_S, \text{ if } \phi < 0 \\ k_L, \text{ if } \phi < 0 \end{cases}.$$

We may define a map $\phi \to \theta(\phi)$, with

$$\rho(\phi)c(\phi)\frac{\partial\theta}{\partial t} - div(k(\phi)\nabla\theta) = -\lambda\frac{\partial}{\partial t}\chi(\phi) \text{ in } Q. \tag{28}$$

$$\theta(\boldsymbol{x}, 0) = \theta_0(x) \text{ in } \Omega \tag{29}$$

$$\theta(\boldsymbol{x}, t) = \theta_{\partial\Omega}(x, t) \text{ on } \partial\Omega \tag{30}$$

Then, we look for ϕ such that

$$sign(\phi) = sign(\theta(\phi) - \theta_c) \text{ on } Q \tag{31}$$

An alternative is looking for ϕ such that

$$\phi = \theta(\phi) - \theta_c \text{ on } Q \tag{32}$$

Equation (31) or (32) may be solved by adapted methods – one of them is the minimization of the different between the left and right sides of the equality, what corresponds to a domain optimization problem. Again, we may use a discretization in time: $\phi(t)$ being known, we look for $\phi(t + \Delta t)$ and the values of may be generated by the stochastic approach (25)–(27) with adapted choices of $\ell(\boldsymbol{x}, t)$ and $A(\boldsymbol{x}, t)$: for instance, an iterative approach may be used, where $\phi^{(n+1)}(t + \Delta t)$ is determined from $\phi^{(n)}(t + \Delta t)$, by using $\ell(\boldsymbol{x}, t) = \rho\big(\theta\big(\phi^{int}\big)\big)c\big(\theta\big(\phi^{int}\big)\big) + \lambda\chi'\big(\theta\big(\phi^{int}\big)\big), A(\boldsymbol{x}, t) = k\big(\theta\big(\phi^{int}\big)\big)$, where ϕ^{int} furnishes interpolated values, such as, for instance,

$$\phi^{int}(t + s\Delta t) = s\phi^{(n)}(t + \Delta t) + (1 - s)\phi(t), 0 \leq s \leq 1$$

5.4 Using a Discretization in Time for the PDE

The preceding approaches use a discretization in time to determine the Stefan boundary, but do not consider discretization of the PDE: the stochastic method solves Eq. (23) without any discretization in time. Nevertheless, the equation

$$\xi(\boldsymbol{x})u - div(\zeta(\boldsymbol{x})u) = f \text{ on } \Omega \text{ , } u = u_{\partial\Omega} \text{ on } \partial\Omega \tag{33}$$

may be solved by stochastic methods. We have

$$u(\boldsymbol{x}) = E(Y_\tau), \tag{34}$$

where

$$Y_\tau = u_{\partial\Omega}(\boldsymbol{X_\tau}) \exp\left(-\int_0^\tau \xi(a)da\right) + \int_0^\tau \left[f(\boldsymbol{X}_s) \exp\left(-\int_0^s \xi(a)da\right)\right] ds, \tag{35}$$

$$\tau = \inf\{t \geq 0 : \boldsymbol{X}_t \notin \Omega\}.$$

Here, $\boldsymbol{X}_s$ is a c stochastic process such that $\boldsymbol{X}_0 = \boldsymbol{x}$ and

$$d\boldsymbol{X}_t = (2\zeta(\boldsymbol{X}_t))^{\frac{1}{2}} d\boldsymbol{W}_t + \nabla\zeta(\boldsymbol{X}_t)dt \tag{36}$$

where $\boldsymbol{W}$ is the bidimensionnal Wiener process. Let the time interval $(0, T)$ be discretized in moments $0 = t_1 < t_2 < \ldots < t_n < t_{n+1} = T$ – for instance, with a step Δt: $t_i = (i - 1)\Delta t$. We may discretize Eq. (23) to obtain equations having the form (33). For instance, we may consider the discretization

$$\ell(\boldsymbol{x}, t_i)\frac{u(\boldsymbol{x}, t_{i+1}) - u(\boldsymbol{x}, t_i)}{\Delta t} - div(A(\boldsymbol{x}, t_i)\nabla u(\boldsymbol{x}, t_{i+1})) = 0,$$

which is equivalent to Eq. (33) with

$$\xi(\boldsymbol{x}) = \ell(\boldsymbol{x}, t_i), \zeta(\boldsymbol{x}) = \Delta t A(\boldsymbol{x}, t_i), f = \ell(\boldsymbol{x}, t_i)u(\boldsymbol{x}, t_i), u_{\partial\Omega}(\boldsymbol{x}) = u_{\partial Q}(\boldsymbol{x}, t_{i+1})$$

Thus, we may determine $\theta(\boldsymbol{x}, t_{i+1})$ by generating a sample from Y_τ given by (35) and estimating $E(Y_\tau)$ by the empirical mean on the sample.

6 Finite Element Approach

Finite Element approximations (FEA) of Eqs. (20)–(22) are generated by considering

$$\theta(x,t) \approx \sum_{i=1}^{nel} \theta_i(t)\varphi_i(x),$$

where *nel* is the number of elements and $\{\varphi_i : 1 \le i \le nel\}$ are the associated shape functions. Denoting

$$\Theta(t) = (\theta_1(t), \ldots, \theta_{nel}(t))^t,$$

The general form of the FEA of Eq. (7) is

$$M(\Theta)\frac{d\Theta}{dt} + K(\Theta)\Theta = F(\Theta), \tag{37}$$

while the general form of the FEA of Eq. (9) is

$$\frac{d}{dt}(M(\Theta)\Theta) + K(\Theta)\Theta = F(\Theta), \tag{38}$$

In the sequel, we consider both the forms. These systems of differential equations must be discretized in time: we consider moments $0 = t_1 < t_2 < \ldots < t_n < t_{n+1} = T$ – for instance, with a step Δt: $t_i = (i-1)\Delta t$. Let us denote $\Theta^i = \Theta(t_i)$, $M^i = M\left(\Theta^i\right)$, $K^i = K\left(\Theta^i\right)$, $F^i = F\left(\Theta^i\right)$. We set

$$\Theta^{n+\frac{1}{2}} = \frac{\Theta^{n+1} + \Theta^n}{2},$$

$$M^{n+1/2} = M\left(\Theta^{n+1/2}\right), K^{n+1/2} = K\left(\Theta^{n+1/2}\right), F^{n+1/2} = F\left(\Theta^{n+1/2}\right)$$

Equation (37) or (38) may be discretized according to various schemes. Table 1 below gives some of the discretizations we used in our experiments. Notice that the equation is, in principle, nonlinear. The first 3 lines concern temporal discretizations of (37), while the last 3 lines concern temporal discretizations of (38).

In Table 1, M, K, F depend on the way which we approximate the left-hand side of Eq. (20). For instance, we may consider

$$\frac{d}{dt}\chi(\theta) = \chi\prime(\theta)\frac{d\theta}{dt} \tag{39}$$

In this case, in the temporal discretizations corresponding to (37) (first 3 lines of Table 1), $M(\Theta)$ is the mass matrix associated to the density $\rho(\theta)c(\theta) + \lambda\chi\prime(\theta)$ and $F(\Theta) = 0$. We may also consider

$$\frac{d}{dt}\chi \approx \frac{\chi\left(\theta^{i+1}\right) - \chi\left(\theta^i\right)}{\Delta t} \tag{40}$$

In this case, in the temporal discretizations corresponding to (37) (first 3 lines of Table 1), $M(\Theta)$ is the mass matrix associated to the density $\rho(\theta)c(\theta)$ and $F(\Theta)$ is the force matrix associated to $\frac{\chi(\theta)-\chi\left(\theta^i\right)}{\Delta t}$..

We observe that, when $\rho(\theta)c(\theta)$ is constant in each phase, the same values of $M(\Theta)$ may be used for discretizations corresponding to (38) (last 3 lines of Table 1).

Table 1. Some possible discretizations of (37)

Number	Discretization
1	$\left(\frac{M^{n+1}+M^n}{2}\right)\left(\frac{\Theta^{n+1}-\Theta^n}{\Delta t}\right)+\left(\frac{K^{n+1}+K^n}{2}\right)\Theta^{n+1/2}=\frac{\mathrm{F}^{n+1}+\mathrm{F}^n}{2}$
2	$M^{n+1/2}\left(\frac{\Theta^{n+1}-\Theta^n}{\Delta t}\right)+K^{n+1/2}\Theta^{n+1/2}=\mathrm{F}^{n+1/2}$
3	$M^n\frac{\Theta^{n+1}-\Theta^n}{\Delta t}+K^n\Theta^{n+1}=\mathrm{F}^n$
4	$\frac{M^{n+1}\Theta^{n+1}-M^n\Theta^n}{\Delta t}+\left(\frac{K^{n+1}+K^n}{2}\right)\Theta^{n+1/2}=\frac{\mathrm{F}^{n+1}+\mathrm{F}^n}{2}$
5	$\frac{M^{n+1}\Theta^{n+1}-M^n\Theta^n}{\Delta t}+K^{n+1/2}\Theta^{n+1/2}=\mathrm{F}^{n+1/2}$
6	$\frac{M^{n+1}\Theta^{n+1}-M^n\Theta^n}{\Delta t}+K^n\Theta^{n+1}=\mathrm{F}^n$

7 A Numerical Example

We consider a numerical example [16] with the following coefficients, initial and boundary conditions:

$$\lambda = 1, \rho_S c_S = \rho_L c_L = 1, k_S = 11, k_L = 10, \theta_c = 0$$

$$f(\boldsymbol{x}, t) = \begin{cases} e^{-t} - 44 & if\ x_1^2 + x_2^2 - e^{-t} > 0 \\ e^{-t} - 40 & if\ x_1^2 + x_2^2 - e^{-t} < 0 \end{cases}$$

$$\theta_{\partial Q}(\boldsymbol{x}, t) = x_1^2 + x_2^2 - e^{-t}$$

The exact solution is $\theta_{\text{th}} = \theta_{\partial\Omega}$. In this situation, $\rho(\theta)c(\theta)$ is constant in each phase, so that we may use the matrices $M(\Theta)$ and $F(\Theta)$ as defined in the last section. By limitation of the room, we do not present all the experiments realized and focus on some significant results.

The Stefan problem was solved by the Stochastic Methods (SM) presented in the preceding. In Fig. 2, we exhibit the results furnished by the approach 5.3, with a step $\Delta t = 1$ – a single time step goes from the initial situation to the situation presented in the Fig. 2. The rectangular domain has dimensions $H1 = H2 = 1$. The temperatures were determined by using samples of 100 variates at each evaluation point and (27) was discretized by Euler's method with a step 1E-5.

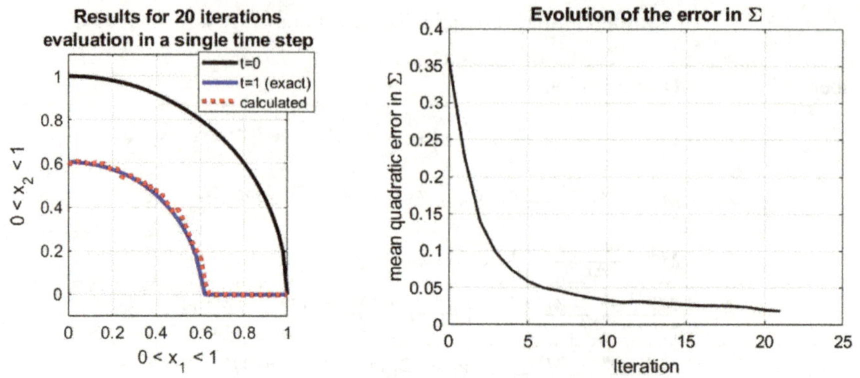

Fig. 2. An example of result furnished by stochastic methods. We use approach 5.3. At left, the Stefan boundary for $t = 1$. At right, the evolution of the error in Σ.

Table 2. Comparison between stochastic modeling (SM) and FEA

Method	Mesh/Element	$err(1)$ with (39)	$err(1)$ with (40)
SM	No mesh	4.6E−3	4.6E−3
FEA 1	T2	5.4E−3	5.4E−3
FEA 2	T2	5.9E−3	3.0E−3
FEA 3	T2	7.3E−3	7.3E−3
FEA 4	T2	5.7E−3	5.7E−3
FEA 5	T2	5.9E−3	5.9E−3
FEA 6	T2	7.3E−3	7.3E−3
FEA 1	Q4	3.2E−4	4.2E−4
FEA 2	Q4	3.0E−4	9.2E−4
FEA 3	Q4	2.8E−4	2.8E−4
FEA 4	Q4	1.4E−4	3.2E−4
FEA 5	Q4	1.6E−4	3.0E−4
FEA 6	Q4	2.8E−4	2.8E−4

For the same situation, we determined the FEA using a time step $\Delta t = 0.05$ and a spatial grid generated by 20 interval on each axis (21 points, equally spaced with distance 0.05). All the discretizations furnish good results, whether (39) or (40) is used. Figure 3 shows the results furnished by FEA with elements Q4 (quadrangles with 9 nodes). Red region is solid and blue region is liquid. The time discretization follows scheme 6. We considered also T2 elements (triangles with 6 nodes).

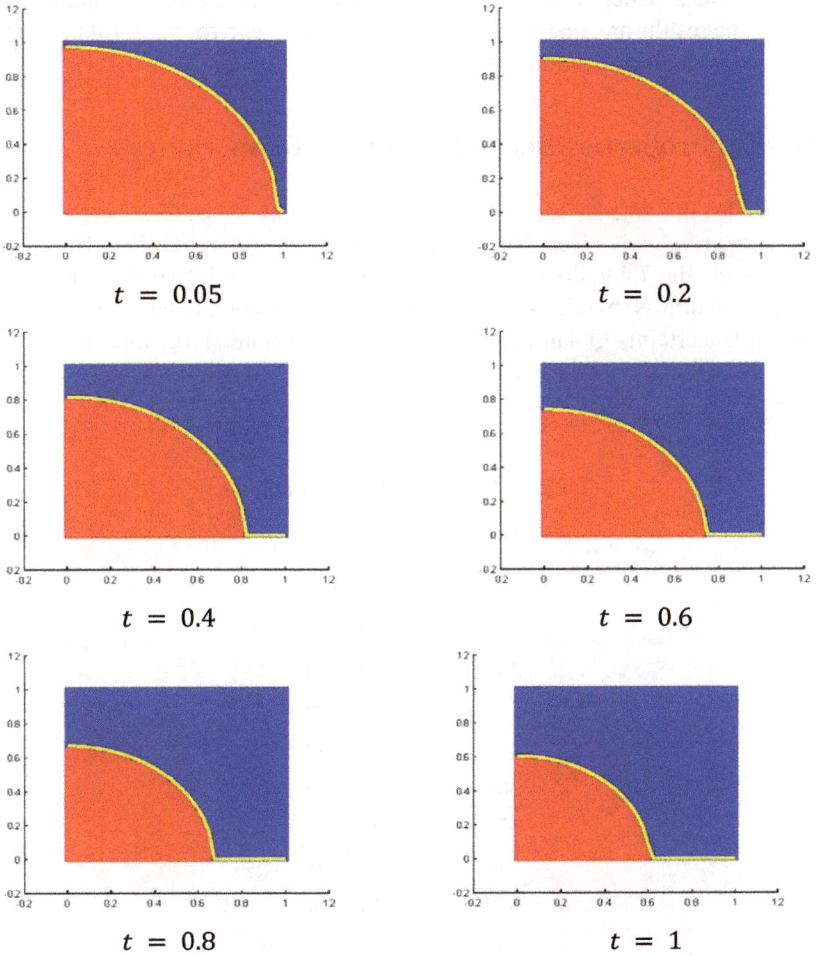

Fig. 3. Results furnished by the FEA Q4, temporal discretization 6. The yellow curve is the exact position of the Stefan boundary. Red region is solid, blue region is liquid. The problem is deterministic.

In Table 2, we show the results obtained for different discretizations and types of elements. $err(1)$ is the mean squared error in the temperatures at the final time.

SM furnishes results that are similar to the results of T2. FEA approaches using Q4 furnish better results. Nevertheless, we underline that SM does not require a mesh and furnishes the result in a single time step. In addition, SM is naturally adapted to distributed computation and parallel calculations, since each point is evaluated independently. FEA with Q4 furnishes better results, but its parallelization is difficult. Among the FEA considered, the results are analogous, with a slight advantage for the formulations using (39) and the last 3 discretization schemes.

8 Physical Properties Depending on the Temperature

Let us consider the data in Table 3: experimental values of ρ, C, k corresponding to different temperatures is given. The critical temperature is $\theta_c = 247 \pm 10$ K and $\lambda = 270 \pm 20$. Since the Table does not contain enough data, we interpolate the data using an Artificial Neural Network (ANN), trained with data completed by the collocation approach of Uncertainty Quantification (see [31], for instance). Examples of results for a 40×40 mesh and a boundary temperature $\theta_{\partial\Omega} = 250$ are shown in Fig. 4. The initial temperature is $\theta_{\partial\Omega} = 200$

Table 3. Experimental data (SI units)

θ	$\rho(\theta)$	$c(\theta)$	$k(\theta)$
230	1430	1.25	112
240	1400	1.27	107
250	1370	1.29	103
260	1340	1.31	98
280	1270	1.36	89
300	1200	1.43	80
320	1115	1.54	72
340	1015	1.75	63
360	870	2.44	54

In this experiment, the variability caused by changes in λ was small and may be neglected. The effects of variations of θ_c are significant and must be considered.

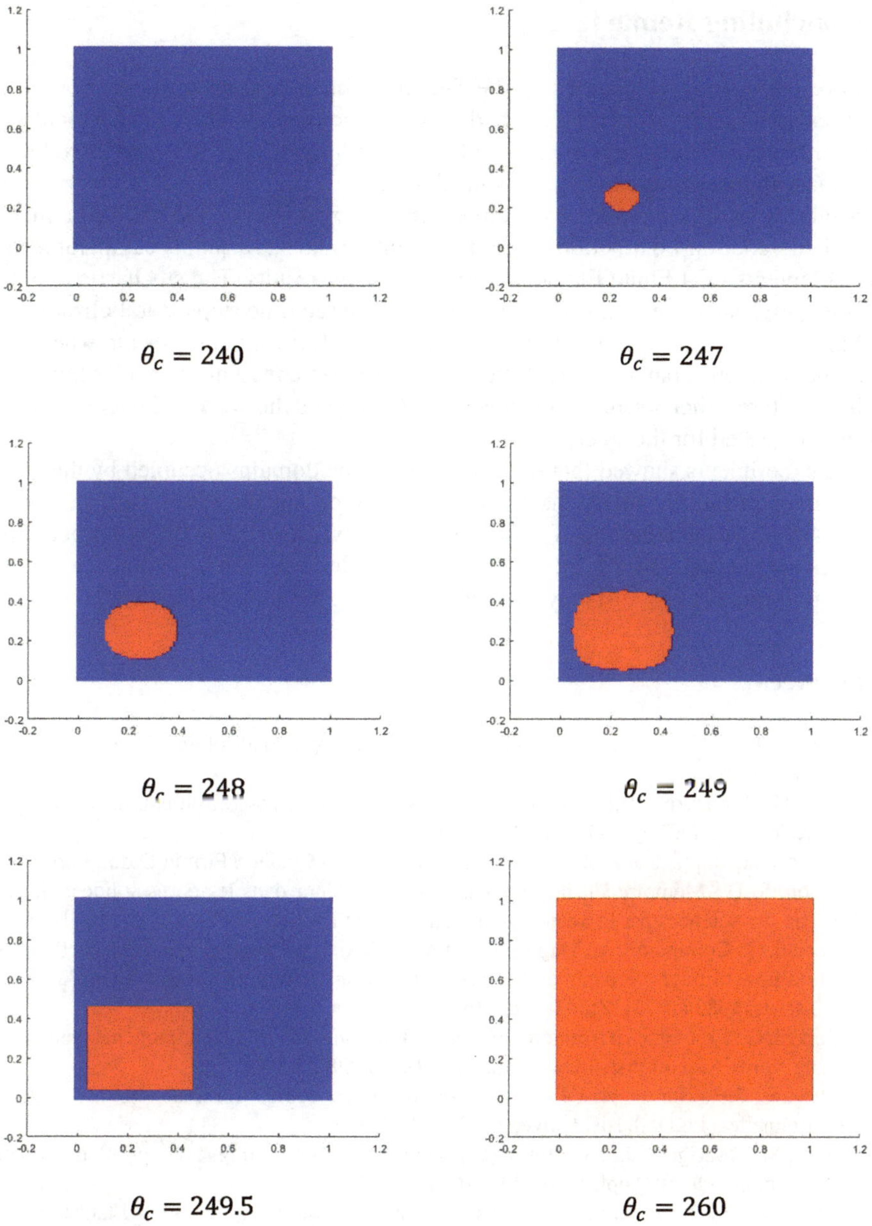

Fig. 4. Variability of the liquid (blue) and solid (red) domains at $t = 1$ for $\lambda = 250$ and different values of θ_c. The boundary temperature is $\theta_{\partial\Omega} = 250$, while $\theta_0 = 200$. The regions appear as regular due to the properties of the partial differential equation considered.

9 Concluding Remarks

We have presented formulations of the Stefan problem leading to, on the one hand, numerical solution by stochastic methods and, on the other hand, Finite Element solutions. The original Stefan problem involves a non regular function and introduces a Dirac mass in the equations. We introduced a regularization approach which preserves the regularity of all the terms of the model. In our experiments, the stochastic method fournished a good approximation of the exact solution and had a quality camparable to T2 Finite Elements - Q4 Finite Elements furnished better results. The stochastic approach has some advantages in terms of flexibility, use of large time steps, parallelization and meshless characteristics. Finite Elements are able to furnish better results when using high order elements, but are more difficult to parallelize and request smaller time steps.

In the future, other discretization methods [32, 33] and the Walk on Spheres approach [34] will be tested for the stochastic method.

The experiments showed that the variability in the domains occupied by the phases may be significant. However, the domains are regions in the space (surfaces in 2D, volumes in 3D), so that the UQ of these regions request the manipulation of probabilities in spaces having domains (id est, subsets of $\mathbb{R}^n$) as elements. An approach analogous to the one presented in [35, 36] may be used and will be matter of future work.

References

1. Stefan, J.: Über die Theorie der Eisbildung im Polarmeere. Annals of Physics and Chemistrie, 269–281 (1891)
2. Lamé, G., Clapeyron, B.P.: Mémoire sur la solidification par refroidissement d'un globe solide. Ann. Chem. Phys. **47**, 250–256 (1831)
3. Fourier, Jean-Baptiste-Joseph: Théorie Analytique de la Chaleur. Firmin Didot, Paris (1822)
4. Poisson, S.-D.: Mémoire sur la propagation de la chaleur dans les corps solides. Nouveau Bulletin des sciences par la Société philomathique de Paris, t. I, n°6, p. 112–116 (1808)
5. Baiocchi, C., Comincioli, V., Magenes, E., Pozzi, G.A.: Free boundary problems in the theory of fluid flow through porous media. Existence and uniqueness theorems. Annali Mat. Pura App. **4**(97), 1–82 (1973). Zbl0343.76036
6. Baiocchi, C.: Problèmes à frontière libre en hydraulique: milieu non homogène. Annali della Scuola Norm. Sup. di Pisa **28**, 429–453 (1977). Zbl0386.35044
7. Rodrigues, J.F.: Sur la cristallisation d'un métal en coulée continue par des méthodes variationnelles. Ph.D. thesis, Universite Paris 6 (1980)
8. Chipot, M., Rodrigues, J.F.: On the steady-state continuous casting stefan problem with non-linear cooling. Quart. Appl. Math. **40**, 476–491 (1983)
9. Saguez, C.: Contrôle optimal de systèmes à frontière libre, Thèse d'État, Université Technologie de Compiègne (1980)
10. Alexiades, V., Solomon, A.: Mathematical Modeling of Melting and Freezing Processes. CRC Press (Taylor and Francis), Boca Raton (1992)
11. Ciavaldini, J.F.: Resolution numerique d'un problème de Stefan à deux phases. - Ph.D. thesis, Rennes, France (1972)
12. El Bagdouri, M.: Commande optimale d'un système thermique non-lineaire. Thèse de Doctorat d'Etates-Sciences, Ecole Nationale Superieure de Mecanique, Universite de Nantes (1987)

13. Tarzia, D.A.: Etude de l'inequation variationnelle proposée par Duvaut pour le problème de Stefan à deux phases. I. Boll. Unione Mat. Ital. **6**(1-B), 865–883 (1982)
14. Tarzia, D. A.: Étude de l'inéquation variationnelle proposee par Duvaut pour le problème de Stefan à deux phases. II. Bollettino della Unione Matemàtica Italiana. Serie VI. B.. (1983)
15. Péneau, S., Humeau, J.P., Jarny, Y.: Front motion and convective heat flux determination in a phase change process. Inverse Probl. Eng. **4**(1), 53–91 (1996). https://doi.org/10.1080/174159796088027633
16. Souza de Cursi, J.E., Humeau, J.P.: Regularization and numerical resolution of a bidimensional Stefan problem. J. Math. Syst. Estimat. Control **3**(4), 473–497 (1992)
17. Haggouch, I., Souza de Cursi, J.E., Aboulaich, R.: Affordable domain optimization for Stefan's model of phase change systems. In: Advanced Concepts and Techniques in Thermal Modelling, Mons, Belgium, pp. 183–190 (1998)
18. Souza de Cursi, J.E.: Numerical methods for linear boundary value problems based on Feyman-Kac representations. Math. Comput. Simul. **36**(1), 1–16 (1994)
19. Morillon, J.P.: Numerical solutions of linear mixed boundary value problems using stochastic representations. Int. J. Numer. Meth. Eng. **40**(3), 387–405 (1997)
20. Milstein, G.N.: The probability approach to numerical solution of nonlinear parabolic equations. Numer. Meth. Partial Differ. Eqn. **18**(4), 490–522 (2002)
21. Milstein, G.N., Tretyakov, M.V.: Numerical solution of the Dirichlet problem for nonlinear parabolic equations by a probabilistic approach. IMA J. Numer. Anal. **21**(4), 887–917 (2001)
22. Souza de Cursi, J.E.: A Feynman-Kac method for the determination of the Stefan's free boundary. In: Inverse Problems in Engineering – vol. I, e-papers, Rio de Janeiro, Brazil (2002). ISBN 85-87922-42-4
23. Hambly, B., Kalsi, J.: Stefan problems for reflected SPDEs driven by space-time white noise. Stochastic Processes Appl. (2019). https://doi.org/10.1016/j.spa.2019.04.003
24. Visintin, A.: Stefan problem with surface tension. In: Rodrigues, J.F. (eds.) Mathematical Models for Phase Change Problems. International Series of Numerical Mathematics, vol. 88, Birkhäuser Basel (1989)
25. Plotnikov, P.I., Starovoitov, V.N.: Stefan Problem with surface tension as a limit of the phase field model. In: Antontsev, S.N., Khludnev, A.M., Hoffmann, K.H. (eds.) Free Boundary Problems in Continuum Mechanics. International Series of Numerical Mathematics/ Internationale Schriftenreihe zur Numerischen Mathematik/ Série Internationale d'Analyse Numérique, vol 106. Birkhäuser, Basel (1992)
26. Fremond, M.: Non-Smooth Thermomechanics. Springer, Heidelberg (2001)
27. Roubicek, T.: The Stefan problem in heterogeneous media. Annales de l'I. H. P., section C, tome 6, no. 6, p. 481–501 (1989)
28. Myers, T.G., Hennessy, M.G., Calvo-Schwarzwälder, M.: The Stefan problem with variable thermophysical properties and phase change temperature. Int. J. Heat Mass Transfer **149**, 118975 (2020)
29. Gupta, S.C.: The Classical Stefan Problem: Basic Concepts, Modelling and Analysis with Quasi-Analytical Solutions and Methods, vol. 45. Elsevier, Amdterdam (2017)
30. Frank, P., Dewitt, D.P.: Fundamentals of Heat and Mass Transfer. Wiley, Hoboken (2001)
31. Souza de Cursi, E., Sampaio, R.: Uncertainty Quantification and Stochastic Modeling with Matlab. Elsevier Science Publishers B. V., NLD (2015)
32. Iacus, S.M.: Simulation and Inference for Stochastic Differential Equations: With R Examples. Springer, New York (2009)
33. Kloeden, Peter E., Platen, Eckhard: Numerical Solution of Stochastic Differential Equations, vol. 23. Springer, Heildelberg (2013)
34. Muller, M.E.: Some continuous monte carlo methods for the Dirichlet problem. Ann. Math. Statist. 27(3), 569–589 (1956). https://doi.org/10.1214/aoms/1177728169. https://projecteuclid.org/euclid.aoms/1177728169

35. Bassi, M., Souza de Cursi, E., Pagnacco, E., Rachid, E.: Statistics of the pareto front in multi-objective optimization under uncertainties. Latin Am. J. Solids Struct. 15 (2018). https://doi.org/10.1590/1679-78255018
36. Bassi, M., Pagnacco, E., Souza de Cursi E.S.: uncertainty quantification and statistics of curves and surfaces. In: Llanes Santiago, O., Cruz Corona, C., Silva Neto, A., Verdegay, J. (eds.) Computational Intelligence in Emerging Technologies for Engineering Applications. Studies in Computational Intelligence, vol 872. Springer, Cham (2020)
37. Dautray, R., et al.: Méthodes Probabilistes pour les Equations de la Physique, Eyrolles, Paris (1989)

Multiscale Method: A Powerful Tool to Reduce the Computational Cost of Big Data Problems Involving Stick-Slip Oscillations

Mariana Gomes(✉), Roberta Lima(✉), and Rubens Sampaio(✉)

Laboratório de Vibrações, Pontifícia Universidade Católica do Rio de Janeiro, Rua Marques de São Vicente 255, Gávea, Rio de Janeiro, RJ, Brazil
marigomes.santos@hotmail.com, {robertalima,rsampaio}@puc-rio.br

Abstract. Nonlinear initial values problems are often used to model the dynamics of many different physical phenomena, for example, systems with dry friction. Usually, these nonlinear IVP do not present a known analytical solution. Then, in order to study these problems, a possible approach is to use approximation methods. The literature dealing with different types of approximation techniques is extensive. Usually, the methods are classified as numerical or analytical. Both can be accurate and provide approximations with any desired precision. However, their efficiencies in terms of computational cost can be very different when they are applied in problems involving big data, for example, stochastic simulations. With analytical methods it is possible to obtain an analytical expression as an approximation to the solution to the IVP, which may be very useful. For example, these analytical expressions can applied to speed up Monte Carlo simulations. The Monte Carlo method is an important tool, which permits to construct statistical models for random object transformations. To build an accurate statistical model (often histograms and sample statistics), several realizations of the transformation output are usually required, a big data problem. If each realization is obtained by a numerical integration, the computation of the Monte Carlo simulations can become a task with high computational and temporal costs. This paper shows that an option to reduce those costs is to use analytical approximations instead of numerical approximations. By doing this, instead of to perform a numerical integration for each realization, which is time consuming task, a simple substitution of values in the analytical expressions can be done. This article aims to compare the computational costs of the construction of statistical models by Monte Carlo simulations, using numerical and analytical approximations. The objective is to show the gain in terms of CPU time when analytical approximations are used instead of numerical ones. To exemplify the gain, an interesting big data problem involving stick-slip phenomenon is developed. The system analyzed has a mass moving over a belt involving random dry friction.

J. E. S. De Cursi (Ed.): Uncertainties 2020, LNME, pp. 69–79, 2021.
https://doi.org/10.1007/978-3-030-53669-5_5

Keywords: Analytical approximation · Multiscale method · Monte Carlo method · Computational cost · Stick slip phenomenon · Big data

1 Introduction

Nonlinear initial values problems are often used to model the dynamics of many different physical phenomena, for example, systems with dry friction. Usually, these nonlinear IVP do not present a known analytical solution. Then, in order to study these problems, a possible approach is to use approximation methods. Numerical and analytical methods can be accurate and can provide approximations with any desired precision. Due to ease of implementation and accuracy, numerical methods have been widely applied in this type of problems. Many computational packages have been developed and included in famous softwares, as for example in MATLAB. One of the most used package in MATLAB is the ode45 function which uses 4th and 5th order Runge-Kutta method. However, analytical approximations have an advantage in relation to numerical ones: they allow a deeper understanding into how the solution depends on the problem parameters. There are many analytical techniques discussed in the literature that help us in this assignment [4,6,13,14]. One of this methods is the Multiscale method, a powerful technique to compute analytical approximations. The main idea of this method is to transform a nonlinear IVP with ordinary differential equation into a series of linear-IVP family of partial differential equations. To make this transformation, we substitute the original scale t by new scales defined as functions of a perturbation parameter.

A common type of tool used in the study of the response of dynamical systems is parametric analysis. This type of analysis, when carried out by means of numerical approximations, has a high computational and temporal costs, since for each value of the parameter it is necessary to compute a numerical integration [10]. The cost to do such analysis could be even higher if we take a stochastic approach as we can see in [7–9]. In the paper [9], a parametric study of a stochastic nonlinear IVP was performed. The influence of combinations of two parameters was analyzed by numerical approximations. To make the combinations, it was considered that one of the parameters assumed 40 different values and the other 8, totalizing 320 combinations. As the article considered uncertainties and aimed to build a statistical model, for each combination of the parameters it was necessary to do a Monte Carlo simulation. The Monte Carlo method is an important tool, which permits us to construct statistical models of random objects transformations. To obtain an accurate statistical model (often histograms and sample statistics), usually several samples of the transformation output are required. In the paper [9], for each Monte Carlo simulation the IVP was integrated 2,000 times, totalizing 640,000 numerical integrations. To perform all these calculations sequentially, 2,5 years would be required. Alternatively, the parallelization strategy was used. Integrations have been distributed into 16 computers, reducing the simulation time for 55 days. Note that even with parallelization, the computational cost remained high and took almost 2 months. An alternative to decrease more the computation time would be to replace the numerical integrations by analytical approximations. Doing this, instead of computing numerical integrations, we only need to do substitutions into an analytical expression for each realization in the Monte Carlo method. A much less costly task. The objective of this paper is to show how analytical approximations, particularly the ones obtained by multiscale method, are a powerful tool to reduce the computational cost of big data problems, as for example stochastic problems. To show the efficiency of the tool we compare the CPU time to compute analytical and numerical approximations to a IVP involving a random stick-slip phenomenon.

2 System Dynamics and Definition of the Stick and Slip Phases

The system considered in this paper has a mass, a spring and a damper. The mass moves over a belt with constant velocity v, as shown in Fig. 1. The mass is modeled as a particle, and there is dry friction between the mass and the belt.

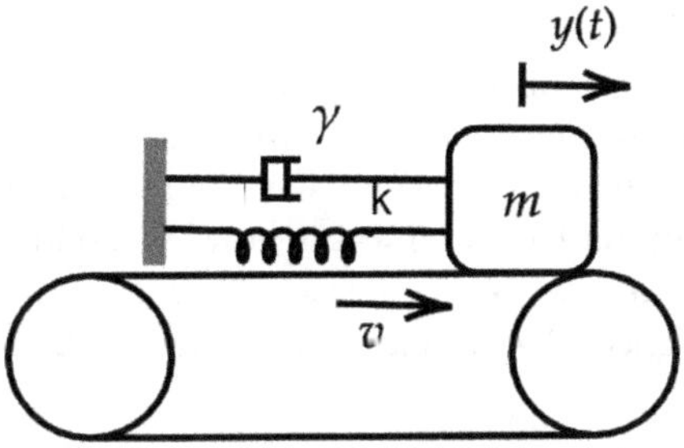

Fig. 1. Mass-spring-damper system with dry friction.

With the Lagrangian method we can construct the IVP that governs the system dynamics. This IVP has one ordinary differential equation given by

$$m\,\ddot{y}(t) + \epsilon\,\gamma\,\dot{y}(t) + k\,y(t) = f_{at}(t)\,, \tag{1}$$

with initial conditions $y(0) = r$ and $\dot{y}(0) = b$. It is considered that y is the particle position, m is particle mass, γ is the damping coefficient, k is the spring constant, ϵ is a perturbation parameter, f_{at} is the friction force between the mass and the belt and V is a multiple of the relative velocity between them. The friction force is modeled for $V \neq 0$ as

$$f_{at}(V) = \frac{1}{3}\,a\,V\,(V^2 - 3) + f_d\,sign(V) \tag{2}$$

where $V = (\epsilon\,v - \epsilon\,\dot{y})$, a is constant and f_d is dynamic friction coefficient. Figure 2 presents the dry friction model. Observe that when $V = 0$, the dry friction force can assume values between $[-f_e, f_e]$, where f_e is coefficient of static friction. For $V \neq 0$, f_{at} is described by Eq. (2).

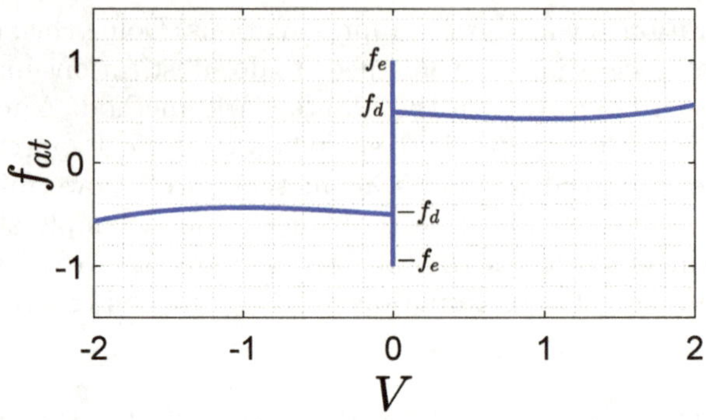

Fig. 2. Friction force model with $a = 0.1$ and $f_d = 0.5$.

The system response with dry friction is composed by 2 phases, stick and slip. These phases alternate with abrupt transition, and compose the stick-slip phenomenon. An important characteristic of these phases is that they have non-null duration, that is, they are not instantaneous. During the stick phase, for a positive belt velocity, the position of the mass, y, grows and consequently f_{at} also increases. This growth lasts until f_{at} reaches the value f_e. At that moment, the stick phase ends and a slip phase starts. Figure 3 shows schematically a sequence of the sticks and slips phases in a system response the sequence starts with a stick phase and it helps us to understand how stick-slip phases alternate. The figure illustrates a possible sequence of sticks and slips during the interval $[0, T]$ considering the occurrence of st sticks and sl slips. The following variables can be defined:

- h_i is the duration of the i-th stick where i can assume values between 1 and st;
- p_i is the duration of the i-th slip where i can assume values between 1 and sl;
- l_i means the instant of beginning of the i-th slip phase, where i can assume values between 1 and sl;
- d_i means the instant of beginning of the i-th slip phase, where i can assume values between 1 and st.

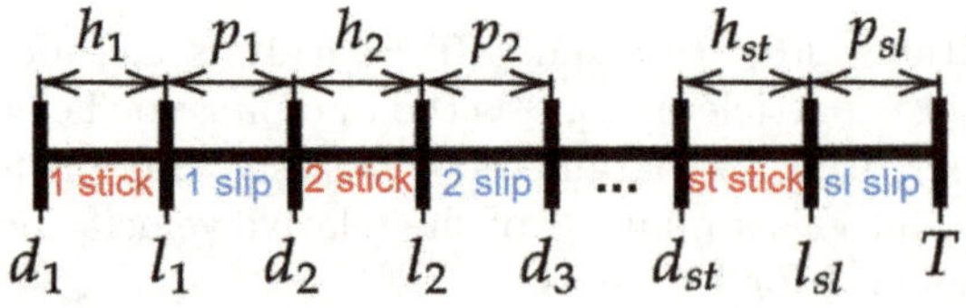

Fig. 3. Sequence of stick and slip phases in the system response.

The stick phase exhibits the following properties: relative velocity zero, $V = 0$, and the friction force with value between $[-f_e, f_e]$, in other words,

$$-f_e \leqslant f_{at} \leqslant f_e \,. \tag{3}$$

During this phase, to determine f_{at} value at each instant, it is necessary to make a force balance using Eq. (1). As mentioned, during at this phase the mass velocity is constant and equal to the belt velocity, $\dot{y}(t) = v$, therefor the mass acceleration is equal to zero, $\ddot{y}(t) = 0$. Applying these conditions into Eq. (1), we have

$$\epsilon \gamma v + k y = f_{at} \,. \tag{4}$$

Applying it into Eq. (3), we have

$$-f_e \leqslant \epsilon \gamma v + k y \leqslant f_e \,. \tag{5}$$

Thus, when the mass velocity satisfies $\dot{y} = v$ and when

$$\frac{-f_e - \epsilon \gamma v}{k} \leqslant y \leqslant \frac{f_e - \epsilon \gamma v,}{k} \,, \tag{6}$$

during an interval of non-null duration, the system is considered to be in stick. Otherwise, it is considered to be in slip. During stick phase the differential equation of the IVP that describes the system dynamics becomes an algebraic equation, see Eq. (4). Thus, during a stick phase the mass has an uniform motion with analytical solution given by

$$y(t) = \Delta_i + v\,t \,, \qquad i = 1, \ldots, st \tag{7}$$

where Δ_i represents the mass displacement at the beginning of each stick phase.

During the slip phase the relative velocity between the mass and the belt is different to zero and the friction forces is defined by Eq. (2), so Eq. (1) becomes

$$m\,\ddot{y}(t) + \epsilon\,\gamma\,\dot{y}(t) + k\,y(t) = \frac{1}{3}\,a\,V\,(V^2 - 3) + f_d\,sign(V) \,. \tag{8}$$

As we can observe in the equation above, the slip phase is governed by a nonlinear IVP. So it does not has a known analytical solution. In this way, for each slip phase, we will compute an analytical approximation using the multiscale method.

3 Analytical Approximation

The multiscale method is a perturbation method which allow us to compute hierarchical analytical approximations to a nonlinear IVP. Through algebraic manipulations with the perturbation parameter ϵ, this method transforms the nonlinear IVP, with one or more ordinary differential equations (ODE), in a linear-IVP family with N partial differential equations (PDE) [1,2]. The method assumes as solution to the nonlinear IVP an uniform expansion in relation to the parameter ϵ, as

$$y(t) = y_0(t) + \epsilon\,y_1(t) + \epsilon^2\,y_2(t) + \ldots + \epsilon^N\,y_N(t) \,. \tag{9}$$

The method proposes to replace the original scale t by new scales that are function of the perturbation parameter, defined as

$$T_0 = t \tag{10}$$
$$T_1 = \epsilon t \tag{11}$$
$$T_2 = \epsilon^2 t \tag{12}$$
$$\vdots$$
$$T_N = \epsilon^N t\,. \tag{13}$$

With the new scales, we apply it into Eq. (9), which becomes

$$\begin{aligned}\bar{y}(T_0, T_1, \ldots, T_N) = \bar{y}_0(T_0, T_1, \ldots, T_N) + \epsilon\, \bar{y}_1(T_0, T_1, \ldots, T_N) \\ + \epsilon^2\, \bar{y}_2(T_0, T_1, \ldots, T_N) + \ldots + \epsilon^N\, \bar{y}_N(T_0, T_1, \ldots, T_N)\,. \end{aligned} \tag{14}$$

Since during the slip phase the IVP that governs the system dynamics, Eq. (8), does not have a known solution, we use the multiscale method to compute an approximation to it.

The first step to apply the method is to define the desired approximation order. After we truncate the uniform expansion according with the order defined. In this paper, we decide to compute a first order approximation, so two new scales are defined as $T_0 = t$, $T_1 = \epsilon t$ and the Eq. (14) is truncated in the second term, as

$$\bar{y}\,(T_0, T_1) \approx \bar{y}_0(T_0, T_1) + \epsilon\, \bar{y}_1(T_0, T_1)\,. \tag{15}$$

Since Eq. (15) is function of the new scales, it is required to derive the $\bar{y}$ solution in relation to the new scales. To do this, we considered the new scales as independent terms, and using the chain rule, we have

$$\frac{dy}{dt} = \frac{\partial T_0}{\partial t}\frac{\partial \bar{y}}{\partial T_0} + \frac{\partial T_1}{\partial t}\frac{\partial \bar{y}}{\partial T_1} = \frac{\partial \bar{y}}{\partial T_0} + \epsilon\,\frac{\partial \bar{y}}{\partial T_1}\,, \tag{16a}$$

$$\frac{d^2 y}{dt^2} = \frac{\partial^2 \bar{y}}{\partial T_0^2} + 2\,\epsilon\,\frac{\partial^2 \bar{y}}{\partial T_0\, T_1} + \epsilon^2\,\frac{\partial^2 \bar{y}}{\partial T_1^2}\,. \tag{16b}$$

Now, we apply the Eqs. (16a), (16b) and Eq. (15) into Eq. (8). After this, we shall expand and collect the coefficients according to the degree of ϵ, so we obtain

$$m\,\frac{\partial^2 \bar{y}_0}{\partial T_0^2} + k\,\bar{y}_0 + \epsilon\left[m\,\frac{\partial^2 \bar{y}_1}{\partial T_0^2} + 2\,\frac{\partial^2 \bar{y}_0}{\partial T_0\, T_1} + \gamma\,\frac{\partial \bar{y}_0}{\partial T_0} + k\,\bar{y}_1\right] = -a\,\epsilon\left[v - \frac{\partial \bar{y}_0}{\partial T_0}\right] + f_d\,. \tag{17}$$

For simplicity, we omitted the term $sign(V)$, and the notation to the partial differential will be represented by $\partial_t \bar{y} = \partial_{T_0} \bar{y} + \epsilon\, \partial_{T_1} \bar{y}$. As we computed a first order approximation, only the coefficients of ϵ^0 and ϵ^1 will be considered. Doing this, the IVP family with linear partial differential equations becomes

$$m\,\partial^2_{T_0}\,\bar{y}_0 + k\,\bar{y}_0 = f_d\,, \tag{18a}$$

$$m\,\partial^2_{T_0}\,\bar{y}_1 + k\,\bar{y}_1 = -2\,\partial_{T_0}\,\partial_{T_1}\,\bar{y}_0 - \gamma\,\partial_{T_0}\,\bar{y}_0 + a\,\partial_{T_0}\,\bar{y}_0 - a\,v\,. \tag{18b}$$

Since now, the solution of $\bar{y}$ is function of the new scales, the differential equations become partial, and the integration constants become functions. The next step is to solve hierarchically the Eqs. (18a) and (18b). So, the solution to the first one is

$$\bar{y}_0(T_0,T_1) = c_1(T_1)\, e^{\sqrt{k\,m}\,T_0\,I/m} + c_2(T_1)\, e^{-\sqrt{k\,m}\,T_0\,I/m} + \frac{f_d}{k}\,. \tag{19}$$

To define the function $c_1(T_1)$ and $c_2(T_1)$, it is required to solve the next linear-IVP. To do this, we apply the Eq. (19) into Eq. (18b),

$$\begin{aligned} m\,\partial^2_{T_0}\,\bar{y}_1 + k\,\bar{y}_1 = & \left[-2\,\frac{\sqrt{k\,m}}{m}\,I\,\partial_{T_1} c_1 - \gamma\,\frac{\sqrt{k\,m}}{m}\,I\,c_1 + a\,\frac{\sqrt{k\,m}}{m}\,I\,c_1\right] e^{\sqrt{k\,m}\,T_0\,I/m} \\ & \left[2\,\frac{\sqrt{k\,m}}{m}\,I\,\partial_{T_1} c_2 + \gamma\,\frac{\sqrt{k\,m}}{m}\,I\,c_2 - a\,\frac{\sqrt{k\,m}}{m}\,I\,c_2\right] e^{-\sqrt{k\,m}\,T_0\,I/m}\,. \end{aligned} \tag{20}$$

To solve $c_1(T_1)$ e $c_2(T_1)$, we eliminate the terms that causes resonance. So the coefficients of $e^{\sqrt{k\,m}\,T_0\,I/m}$ and $e^{-\sqrt{k\,m}\,T_0\,I/m}$ should be equated to zero,

$$-2\,\frac{\sqrt{k\,m}}{m}\,I\,\partial_{T_1} c_1 - \gamma\,\frac{\sqrt{k\,m}}{m}\,I\,c_1 + a\,\frac{\sqrt{k\,m}}{m}\,I\,c_1 = 0\,, \tag{21a}$$

$$2\,\frac{\sqrt{k\,m}}{m}\,I\,\partial_{T_1} c_2 + \gamma\,\frac{\sqrt{k\,m}}{m}\,I\,c_2 - a\,\frac{\sqrt{k\,m}}{m}\,I\,c_2 = 0\,. \tag{21b}$$

The Eqs. (21a) and (21b) are linear ODE and have the following analytical solutions

$$c_1(T_1) = \alpha_1\, e^{T_1(a-\gamma)/2}\,, \tag{22}$$

$$c_2(T_1) = \alpha_2\, e^{T_1(a-\gamma)/2}\,, \tag{23}$$

with α_1 and α_2 being integrations constant which will be determined according to the initial conditions. With the function $c_1(T_1)$ and $c_2(T_1)$ defined, the first order approximation for each slip phase is

$$\bar{y}(T_0,T_1) \approx \alpha_1\, e^{T_1(a-\gamma)/2}\, e^{\sqrt{k\,m}\,T_0\,I/m} + \alpha_2\, e^{T_1(a-\gamma)/2}\, e^{-\sqrt{k\,m}\,T_0\,I/m} + \frac{f_d}{k}\,. \tag{24}$$

We can put the Eq. (24) again in the domain of time t, and applying the Euler law, change the exponential form to the sin and cos form, we have

$$y(t) \approx e^{z\,t}\,[d\,\cos(\sqrt{k}\,t) + e\,\sin(\sqrt{k}\,t)] + \frac{f_d}{k}\,, \tag{25}$$

where $z = \epsilon\,(a-\gamma)/2$, $d = s - f_d/k$ e $e = [q - z\,d]/\sqrt{k}$, with s is the initial position and q the initial velocity for each slip phase.

4 Stochastic Approach - Monte Carlo Method

The Monte Carlo method is an important tool to deal with stochastic problems. It is used to construct statistical models of stochastic objects and is grounded in two theorems, the law of large numbers and the central limit theorem [12].

We chose to model the initial position as a random variable, R, with Beta distribution. Its probability density function (PDF) is

$$p_R(x) = \frac{x^{\alpha-1}\,(x-1)^{\beta-1}}{B(\alpha,\beta)}\,, \tag{26}$$

where $B(\alpha,\beta) = \frac{\Gamma(\alpha)\,\Gamma(\beta)}{\Gamma(\alpha+\beta)}$, Γ is the Gamma function and α and β are two parameters. We fixed $\beta = 2$ and choose 22 different values for α between $[1.5, 8]$. The support of each pg these Beta distributions is $[0, 200]$. It is important to remark that the shape of the PDF changes with α, as shown in Fig. 4. So, for values of α closer to 1.5 we have more probability to have values of R under 100, while for values of α closer to 8 there is a higher probability of having values of R above 100.

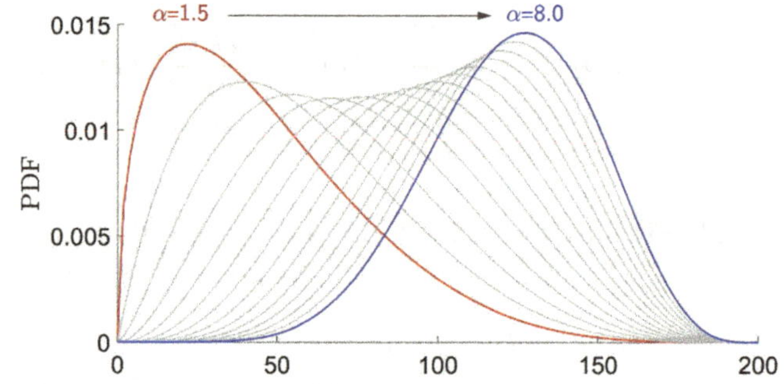

Fig. 4. PDF of R for different values of the parameter α.

5 Computational Race

As explained in the introduction, the objetive of this paper is to compare the temporal cost in terms of CPU time to compute Monte Carlo simulations, considering analytical and numerical approximations, i.e, the objetive is to make a race between the methods. The analysis is carried out by varying the parameter α. For the numerical approximations, we used the Runge-Kutta method of 4th and 5th order to integrate Eq. (8), and for the analytical approximation we used Eq. (25). We considered 22 different Beta distributions with α between $[1.5, 8]$, and 10^4 realizations for each Monte Carlo of each distribution. That is, for each Monte Carlo simulation, we computed 10^4 approximations to the system response with the analytical method and 10^4 approximations with the numerical method. In total we computed $22 \, . \, 10^4$ numerical integrations. The parameters values used into the simulations are in Table 1. Given that this stick-slip oscillator is not a trivial problem due to the discontinuity in the friction force, we need to use a small time step in the numerical approximations. We used 10^4 realizations in each Monte Carlo to highlight the efficiency of the analytical approximations in a big data problem.

Table 1. Parameters values used in the simulation

Parameter values	
m	1 [kg]
γ	1 [N.s/m]
k	0.1[N/m]
v	−1 [m/s]
T	100 [s]
b	−1 [m/s]
ϵ	0.0001
f_e	2.0
f_d	1.5
a	0.1

To perform this race between the numerical and analytical approximations, we used a computer with 2 Intel Xeon CPU E5-2590 v4 @2.60 ghz (14 cores each) and 32 GB memory. To reduce the real (elapsed) time, we parallelized the numerical integrations into the 28 cores of the computer. Figure 5 shows the CPU time in seconds taken for the analytical and numerical approximations for

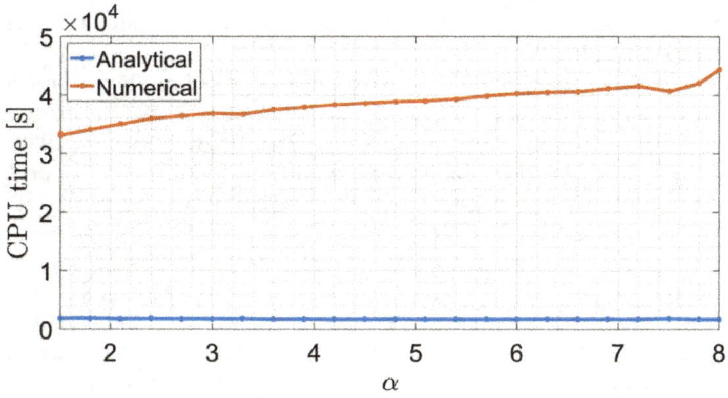

Fig. 5. CPU time of the analytical and numerical approximation for each of the 22 samples of Beta distribution.

the 22 Beta distributions considered. The numerical approximation took almost 10 days of CPU time, and the analytical approximation required only 11 min of CPU time. There is a CPU time difference of two orders of magnitude between the analytical and numerical simulations, therefore the improvement is highly appreciable, being the analytical approach much faster.

6 Conclusions

Monte Carlo simulations in nonlinear IVP usually lead to a big data problem, as an elevated number of realizations is often necessary to construct an accurate statistical model. Sometimes, due to computational and time limitations, to compute an accurate statistical model becomes an infeasible task. In the paper [5], we did a similar comparison using the Duffing equation, a simpler problem, and the numerical approximation took much longer than the analytical one. In complex problems, an option to perform such analysis would be to reduce the number of realizations and parallelize. However, even using these strategies the CPU time can be significant. Given this, this paper presents a new strategy to reduce the CPU time of big data problems involving nonlinear IVP, the use of analytical approximation. In this work, the use of analytical approximations is explored and compared in terms of CPU time against numerical approximations. We show with an example that while the CPU time to compute the numerical approximations takes days, the same problem with an analytical approach can be solved in hours. It could also be observed that the CPU time to compute a numerical approximation changes with respect to the value of the parameter considered, while the computational cost of the analytical approximation does not present a significant change.

References

1. Awad, A.I. Advances and Applications of Multiple Scale Methods in Complex Dynamical Systems. Ph.D. thesis, 2017, University of Washington, Aeronautics and Astronautics (2017)
2. Bender Carl, M., Orszag, S.A.: Advanced mathematical methods for scientists and engineers I: asymptotic methods and perturbation theory. Springer, New York (1999). 175 Fifth Avenue, New York, NY, 10010, USA, 1 edition, 1999
3. Gomes, M.: Estratégias de aproximações analíticas hierárquicas de problemas não lineares: métodos de perturbação. Master's Thesis, Pontifícia Universidade Católica do Rio de Janeiro (2019)
4. Gomes, M., Lima, R., Sampaio, R.: Método de múltiplas escalas aplicado em um problema de valor inicial com atrito seco. CNMAC 2019 XXXVIII Congresso Nacional de Matemática Aplicada e Computacional, Uberlândia (2019)
5. Gomes, M., Lima, R., Sampaio, R.: A Race in the Monte Carlo Method: numerical and analytical methods. ENIEF 2019. XXIV Congreso sobre Métodos Numéricos y sus Aplicaciones. Mecánica Computacional Vol XXXVII, págs. 649–655. Santa Fé, Argentina (2019)

6. Gomes, M., Lima, R., Sampaio, R.: Multiscale method applied in a stick-slip oscillator. In: COBEM 2019 Proceedings of the 25th International Congress of Mechanical Engineering, Uberlândia (2019)
7. Lima, R., Sampaio, R.: Stick-mode duration of a dry-friction oscillator with an uncertain model. J. Sound Vib. **353**, 259–271 (2015)
8. Lima, R., Sampaio, R.: Construction of a statistical model for the dynamics of a base-driven stick-slip oscillator. Mech. Syst. Signal Process. **91**, 157–166 (2017a)
9. Lima, R., Sampaio, R.: Parametric analysis of the statistical model of the stick-slip process. J. Sound Vib. **397**, 141–151 (2017b)
10. Pasquetti, E.: Métodos Aproximados de Soluçâo de Sistemas Dinâmicos Não-Lineares. Ph.D. thesis, PUC-Rio, Engenharia Civil (2008)
11. Radhika, T.S.L., Iyengar, T.R.T.R.: Approximate Analytical Methods for Solving Ordinary Differential Equations. Taylor and Francis Group/CRC, Boca Raton (2015). 6000 Broken Sound Parkway NW, Suite 300 Boca Raton, FL 33487-2742, 2015
12. Sampaio, R., Lima, R.: Modelagem estocástica e geraçâo de amostras de variáveis e vetores aleatórios. Notas de Matemática Aplicada, SBMAC, vol. 70 (2012)
13. Sanchez, N.E.: The method of multiple scales: asymptotic solutions and normal forms for nonlinear oscillatory problems. J. Symbol. Comput. **21**(2), 245–252 (1996)
14. Shonkwiler, R.W., Mendivil, F.: Explorations in Monte Carlo Methods. Springer, New York (2009)

Coupled Lateral-Torsional Drill-String with Uncertainties

Lucas P. Volpi[1,2], Daniel M. Lobo[1,2], and Thiago G. Ritto[1,2](✉)

[1] Department of Mechanical Engineering, Federal University of Rio de Janeiro, Rio de Janeiro, Brazil
lp.volpi@mecanica.coppe.ufrj.br, tritto@mecanica.ufrj.br
[2] Laboratório de Acoustica e Vibrações, Federal University of Rio de Janeiro, Av. Horacio Macedo, 2030, 21941-914 Rio de Janeiro, Brazil

Abstract. This paper treats the nonlinear problem of a coupled lateral-torsional drill string. A simplified 3 DOF system is used to model the system. The nonlinearity has two main sources: (1) the nonlinear bit-rock interaction, and (2) the impact between the column and the borehole. We classified different dynamics regimes (stick-slip, forward whirl, backward whirl, etc) and construct a map varying the rotational speed imposed at the top and the weight-on-bit. The Maximum Entropy Principle is employed to construct a probabilistic model for the borehole wall friction, and the stochastic response is approximated using the Monte Carlo Method. A probabilistic map is built with the stochastic model used, where the probability of the system to reach a given dynamic regime is shown.

Keywords: Drill-string · Uncertainties · Lateral vibrations · Torsional vibrations

1 Introduction

The drill-string is a key component in the oil and gas industry. It is an extremely slender rotating structure, that can present kilometers in length. Due to that, it is prone to several undesirable vibrations. Those often lead to efficiency loss and, in extreme cases, may jeopardize the operation.

The drill-string is usually divided in two main parts. The former, are the drill-pipes, which are slender structures that occupy most of the drill-string's extension. The latter, is the Bottom Hole Assembly (BHA), which is a stiffer structure, where most of the equipment is located. The BHA presents stabilizers that minimize transversal vibrations and enables higher drilling loads (weight on bit).

Those vibrations are usually divided according to it's respective direction. In the axial direction, in severe cases, there is the bit-bounce, associated with the intermittent contact between the bit and the rock formation. In the torsional direction, the critical vibration is the stick-slip. It is assumed to be a consequence

J. E. S. De Cursi (Ed.): Uncertainties 2020, LNME, pp. 80–88, 2021.
https://doi.org/10.1007/978-3-030-53669-5_6

of the bit-rock interface, which leads to the sticking of the bit, which implies in the torsion of the drill-string, followed by the release of the bit, which causes it to reach high speeds. Finally, lateral vibrations are a consequence of imperfections in the drill-string, which causes it to oscillate transversely. If this translation is in the same direction of the rotation of the drill-string, it is named forward whirl. Otherwise, it is known as backward whirl. A condition for occurrence of the backward whirl is the contact with the borehole wall, which leads to a rolling motion. In this paper, lateral-torsional vibrations are analyzed through a coupled model.

The modeling of each vibration is often a key aspect, thus, throughout the years, several models were proposed for bit-rock interaction, lateral contact, fluid-structure interaction and so on. Contact dynamics, for instance, are commonly modeled according to the Hertizian contact law [3,17] or from a simplified linear elastic relation [1,2,6,9,10]. The bit-rock interface, on the other hand, presents models based in dry-friction [8], in a combination of dry-friction with an exponential decaying law [12] or even a smooth interaction, such as provided in [13,19].

Due to the complex nature of the process, uncertainties are frequently taken in account in order to provide a robust analysis. In [14,15], uncertainties in the applied rotating speed and in bit-rock interaction. Also, [11] considered uncertainties in the transition between rock formation.

The objectives of this work are: (i) to model uncertainties in the friction coefficient between the BHA and the borehole wall and (ii) analyze the overall dynamic through maps containing the probability of occurrence of different phenomena.

This work starts with a deterministic base model in Sect. 2, a stochastic model in Sect. 3, results in Sect. 4 and concluding remarks in Sect. 5.

2 Deterministic Model

In this section, the deterministic base model is briefly explained. It is divided in two main parts, one for lateral vibrations and another for torsional vibrations. In summary, torsional vibrations are modeled as a torsional pendulum, while lateral vibrations are modeled as a rotor. In Fig. 1, a sketch of the overall model is presented, where the torsion spring is attached to a rotor-like structure (a BHA section).

As a consequence of the forces in the axial direction, the drill-pipes are under traction and the BHA is mostly under compression. Hence, lateral vibrations are often analyzed in the BHA.

The lumped parameter model considers stabilizers as rigid bearings, restricting the model to a region between them. Also, the rotating speed in the BHA is considered rigid to torsion – the BHA rotates at the same as the bit speed.

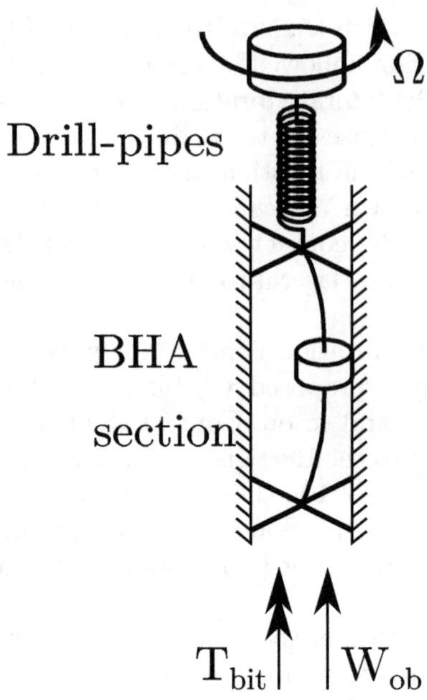

Fig. 1. Sketch of the overall model, where torsional dynamics is modeled as a linear torsional spring and lateral dynamics are modeled as a rotor.

In this region, the vibration is modeled as a consequence of unbalance forces acting in the Jeffcot rotor as in [4]:

$$(m+m_f)(\ddot{r}-\dot{\theta}^2 r)+c_h|\nu|\dot{r}+k(T_{bit})r \tag{1}$$
$$=(m+m_f)[e\dot{\phi}^2\cos(\phi-\theta)+e\ddot{\phi}\sin(\phi-\theta)]-F_n$$
$$(m+m_f)(\ddot{\theta}r+2\dot{r}\dot{\theta})+c_h|\nu|r\dot{\theta}=(m+m_f)[e\dot{\phi}^2\sin(\phi-\theta)-e\ddot{\phi}\cos(\phi-\theta)]-F_{fat},$$

where r and θ are the radial displacement and the angle of whirl of the BHA geometric center,ϕ the torsion angle, e is the eccentricity and $|\nu|$ is the modulus of the velocity of the BHA geometric center. m and m_f are the mass and added fluid mass, c_h is the hydraulic damping coefficient, modeled as the drag between the BHA section and the surrounding fluid and $k(\dot{\phi})$ the lateral stiffness:

$$k(\dot{\phi})=\frac{EI_a\pi^4}{2L_c^3}-\frac{T_{bit}(\dot{\phi})\pi^3}{2L_c^2}-\frac{W_{ob}\pi^2}{2L_c}, \tag{2}$$

where E the Young's Modulus, I_a the section's inertia, $T_{bit}(\dot{\phi})$ and W_{ob} the torque to the bit-rock interaction and weight on bit and L_c is the BHA's length. It can be rewritten as:

$$k(\dot{\phi})=k_0-\frac{T_{bit}(\dot{\phi})\pi^3}{2L_c^2}-\frac{W_{ob}\pi^2}{2L_c}, \tag{3}$$

where k_0 is a constant stiffness.

Forces F_n and F_{fat} are the contact forces, modeled by a linear elastic force and a smooth Coulomb friction model:

$$F_n = H(r - r_c)k_s(r - r_c) \quad \text{and} \quad F_{at} = \mu \tanh{(v_{rel}/V_{ref})}F_n, \tag{4}$$

where $H(r - r_c)$ is the step function, r_c is the clearance between the BHA and the borehole wall. v_{rel} is the relative speed between the BHA and the wall, k_s is the wall's stiffness and μ is the Coulomb friction coefficient. V_{ref} is a model constant.

Torsional lumped parameter considered a torsional pendulum [7]:

$$I_m\ddot{\phi} + c_t\phi + k_t\phi = c_t\Omega + k_t\Omega t + T_{bit} + T_{lat}, \tag{5}$$

where I_m is the moment of inertia of the drill-pipes, Ω is the rotation speed. c_t and k_t the damping and stiffness coefficients. T_{bit} is the bit-rock interaction torque:

$$T_{bit} = W_{ob}b_0\left(\tanh{(b_1\dot{\phi})} + \frac{b_2\dot{\phi}}{1 + b_3\dot{\phi}^2}\right) \tag{6}$$

and T_{lat} the torques due to lateral coupling:

$$T_{lat} = -F_n e\sin{(\phi - \theta)} - F_{fat}\left(\frac{D_{co}}{2} - e\cos{(\phi - \theta)}\right)$$

$$- c_h|\nu|(\dot{r}e\sin{(\phi - \theta)} - r\dot{\theta}e\cos{(\phi - \theta)}), \tag{7}$$

where b_0, b_1, b_2 and b_3 are model constants.

3 Stochastic Model

In this section, the proposed stochastic model is presented. During drilling operations, the borehole wall properties depend on the type of rock formation such as geology and surface regularity. Hence, a robust approach would be considering an unknown friction coefficient.

In [5], experiments regarding steel friction coefficient with several rocks and other surfaces presented coefficients from $\mu = 0.13$ to $\mu = 0.5$. Hence, to consider en-globe a frictionless to an extreme situation, the support chosen is $\boldsymbol{\mu} \in [\,0\;,\;0.6\,]$, where $\boldsymbol{\mu}$ is the random friction coefficient. It is reasonable to consider a mean value of this variable equal to the deterministic: $\bar{\boldsymbol{\mu}} = \mu = 0.30$. Finally, in order to guarantee a robust analysis the coefficient of variation chosen was $c_v = 0.50$.

Given that the friction coefficient is known to lie between two values, $\mu \in [0, 0.6]$, its mean is given by the average of the support limits, $\bar{\mu} = 0.3$, and its variance is known and less than the variance of an uniform distribution with the same characteristics, $\sigma_\mu^2 = 0.15^2 = 0,0225 < \frac{1}{3}[(0.6 - 0)/2]^2 = 0.03$, the maximization of the Shannon Entropy leads to a truncated normal distribution [20]:

$$\boldsymbol{\mu} \approx N(\bar{\boldsymbol{\mu}}, \sigma^2, [\,0\;,\;0.6\,]) \tag{8}$$

4 Numerical Results

In this section, both deterministic and stochastic results will be presented and discussed. In Table 1, the parameters used are presented – based on the drill-string analyzed in [16] and with each constant calculated in accordance with the shown in [4]. In order to solve the system a 4th–5th-order Runge-Kutta-Felbergh method is used.

Table 1. Summary of drill-string properties.

Parameter	Symbol	Values	Units
Lateral stiffness	k_0	698140	N/m
Torsional stiffness	k_t	324.86	N/rad
Mass	m	633.78	kg
Added fluid mass	m_f	275.75	kg
Hydraulic damping	c_h	465.39	Ns/m
Moment of inertia	I_m	518.05	kg m^2
Contact stiffness	k_s	10×10^8	N/m
Wall friction coeff	μ	0.30	-
Friction constant	V_{ref}	1×10^{-4}	m/s
T_{bit} model constant	b_0	0.024	m
T_{bit} model constant	b_1	1.910	s
T_{bit} model constant	b_2	8.500	s
T_{bit} model constant	b_3	5.470	s^2
Clearance	r_c	0.022	m
BHA outer diameter	D_o	0.17	m
BHA section length	L_c	8.55	m

4.1 Deterministic Results

In this section, the mean behavior is briefly explored through the deterministic model. The system is modeled in different rotational speeds and weight on bits. In each configuration, the whirl frequency is analyzed through a fast Fourier transform in the complex polar coordinates [18] where positive frequencies indicate forward whirl whereas negative frequencies indicate backward whirl. The mean radial displacement is then compared with maximum radial displacement and clearance between the BHA and the borehole wall to evaluate whether the BHA is in contact or not and the type of contact. Finally, the amplitude in torsional speed is calculated and normalized to evaluate the presence of intense torsional vibrations.

In Fig. 2, each identified regimen is defined in accordance with the legend provided in Table 2. From the twelve possible regimen, five are observed.

Table 2. Combination of possible regimens. Where: (i) 'ss' stands for severe torsional vibrations; (ii) 'fw' indicates forward whirl and 'bw' backward whirl; (iii) 'nc' define contact-less vibration, 'im' an impact vibration and 'ro' a rolling regimen.

Torsional vibration	ss	ss	ss	ss	ss	ss	-	-	-	-	-	-
Whirl	fw	fw	fw	bw	bw	bw	fw	fw	fw	bw	bw	bw
Contact	nc	im	ro	nc	im	ro	nc	im	ro	nc	im	ro

The first presents intense torsional vibration ('ss'), the second a safe zone with forward whirl. In sequence, a backward whirl region with rolling motion ('bw, ro'). The forth region presents a combination of phenomena, with intense torsional vibrations, backward whirl and impact ('ss, bw, im'). Finally, there is a transition region with severe torsional vibrations and impact ('ss, im').

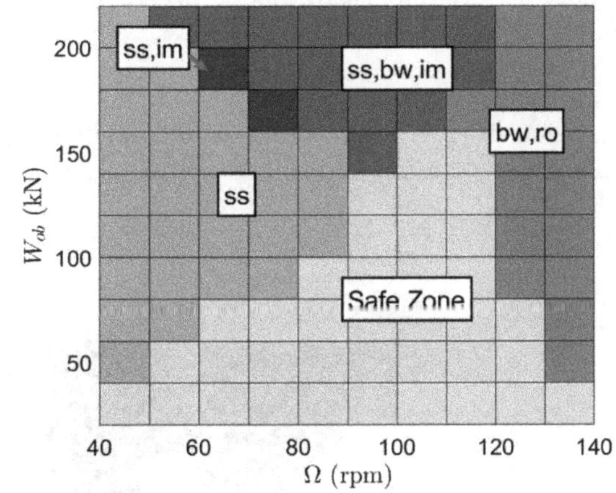

Fig. 2. Regimen map with deterministic parameters where, in accordance with Table 2, 'ss' stands for severe torsional vibration, 'bw' for backward whirl, 'im' for impact, 'ro' for rolling. In the central region, is defined a safe zone, where there is no intense torsional vibration, no contact and forward whirl.

4.2 Stochastic Results

In this section, the results provided by the stochastic model is explored. Each point of the map represents the probability of occurrence of impact calculated through Monte Carlo Simulations. In Fig. 3, the quadratic mean convergence is shown each point of the map.

Figure 4 depicts the map of probability of backward whirl. It can be seen the it is almost unaffected by the stochastic model. In Fig. 5, however, the probability of impact is presented and, in addition, the deterministic region of impact

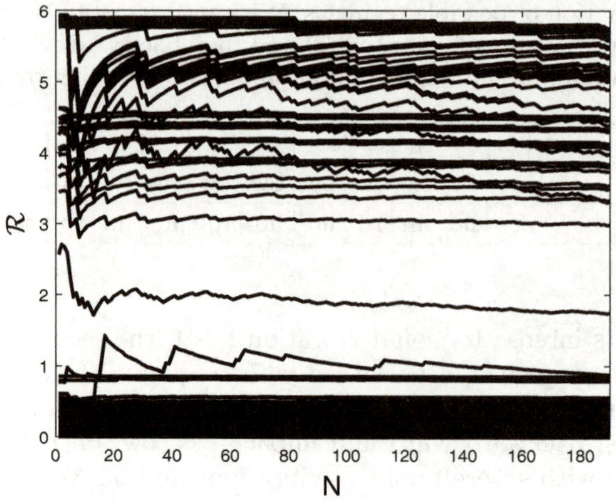

Fig. 3. Convergence curves for the stochastic models, where $\mathcal{R}$ represents the quadratic mean convergence in the Nth simulation.

is highlighted. It is clear that a region that would formerly present a rolling dynamic presents a high probability of impact – specially near the transition of regimen.

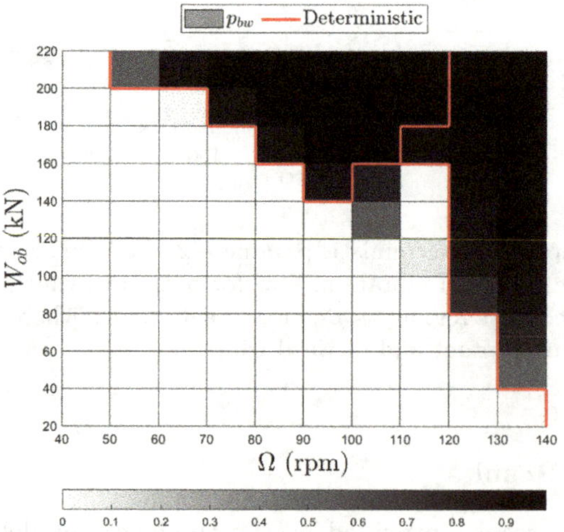

Fig. 4. Map containing the probability of occurrence of backward whirl where the threshold that contains the deterministic backward whirl region is highlighted.

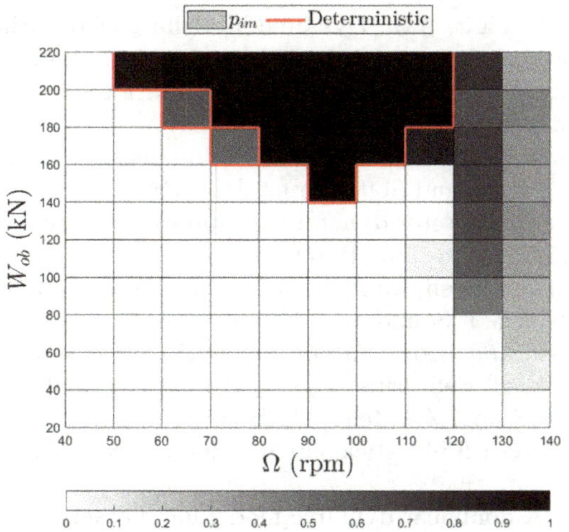

Fig. 5. Map containing the probability of occurrence of impact where the threshold that contains the deterministic impact region is highlighted.

5 Concluding Remarks

Initially, a lumped parameter deterministic model is presented for coupled lateral-torsional vibrations. In sequence, a stochastic model is proposed for the Coulomb friction constant. With the deterministic model, a map of different regimen is created, where only a few of the possible regimens can be seen. In it, a safe zone is well defined whereas different critical phenomena can be seen. The stochastic model, on the other hand, provided a significantly higher impact region, where formerly a rolling motion was expected. Hence a region where the – already critical – vibration of backward whirl was formerly defined presents high incidence of a possibility of a backward whirl vibration with impact.

Acknowledgements. This study was financed in part by the Coordenacao Aperfeicoamento de Pessoal de Nível Superior (CAPES) - Finance code 001 - Grant PROEX 803/2018, and the Brazilian agencies: Conselho Nacional de Desenvolvimento Científico e Tecnologico (CNPQ) - Grants 303302/2015-1, 400933/2016-0, Fundação Carlos Chagas Filho de Amparo a Pesquisa do Estado do Rio de Janeiro (FAPERJ) - Grant E-26/201.572/2014. R. A. Borges is thankful for the CNPq (grant #439126=2018 - 5).

References

1. Al-Hiddabi, S.A., Samanta, B., Seibi, A.: Non-linear control of torsional and bending vibrations of oilwell drillstrings. J. Sound Vib. **265**(2), 401–415 (2003)
2. Ambrus, A., Skadsem, H.J., Mihai, R.G.: Similarity analysis for downscaling a full size drill string to a laboratory scale test drilling rig. In: ASME 2018 37th International Conference on Ocean, Offshore and Arctic Engineering, pp. V008T11A005–V008T11A005. American Society of Mechanical Engineers (2018)

3. Christoforou, A.P., Yigit, A.S.: Dynamic Modelling of Rotating Drillstrings With Borehole Interactions. J. Sound Vib. **206**(2), 243–260 (1997)
4. Christoforou, A.S., Yigit, A.P.: Fully coupled vibrations of actively controlled drillstrings. J. Sound Vib. **267**(5), 1029–1045 (2003)
5. Gaffney, E.S.: Measurements of dynamic friction between rock and steel. Technical report, Systems Science and Software La Jolla CA (1976)
6. Jansen, J.D.: Non-linear rotor dynamics as applied to oilwell drillstring vibrations. J. Sound Vib. **147**(1), 115–135 (1991)
7. Jansen, J.D., van den Steen, L.: Active damping of self-excited torsional vibrations in oil well drillstrings. J. Sound Vib. **179**(4), 647–668 (1995)
8. Johan Dirk Jansen and Leon van den Steen: Active damping of self-excited torsional vibrations in oil well drillstrings. J. Sound Vib. **179**(4), 647–668 (1995)
9. Liu, X., Vlajic, N., Long, X., Meng, G., Balachandran, B.: Nonlinear motions of a flexible rotor with a drill bit: stick-slip and delay effects. Nonlinear Dyn. **72**(1–2), 61–77 (2013)
10. Liu, Y., Gao, D.: A nonlinear dynamic model for characterizing downhole motions of drill-string in a deviated well. J. Nat. Gas Sci. Eng. **38**, 466–474 (2017)
11. Lobo, D.M., Ritto, T.G., Castello, D.A.: Stochastic analysis of torsional drill-string vibrations considering the passage from a soft to a harder rock layer. J. Brazilian Soc. Mech. Sci. Engi. **39**(6), 2341–2349 (2017)
12. Navarro-López, E.M., Cortés, D.: Avoiding harmful oscillations in a drillstring through dynamical analysis. J. Sound Vib. **307**(1–2), 152–171 (2007)
13. Nogueira, B.F.: Robust rotational stability analysis of a vertical drill-string. In: ABCM (ed.) 23rd International Congress of Mechanical Engineering, number 1985, Rio de Janeiro. ABCM (2012)
14. Ritto, T.G., Soize, C., Sampaio, R.: Non-linear dynamics of a drill-string with uncertain model of the bit-rock interaction. Int. J. Non-Linear Mech. **44**(8), 865–876 (2009)
15. Ritto, T.G., Soize, C., Sampaio, R.: Probabilistic model identification of the bit-rock-interaction-model uncertainties in nonlinear dynamics of a drill-string. Mech. Res. Commun. **37**(6), 584–589 (2010)
16. Ritto, T.G., Aguiar, R.R., Hbaieb, S.: Validation of a drill string dynamical model and torsional stability. Meccanica **52**(11–12), 2959–2967 (2017)
17. Spanos, P.D., Chevallier, A.M., Politis, N.P.: Nonlinear stochastic drill-string. J. Vib. Acoust. **124**, 512–518 (2002)
18. Tiwari, R.: Rotor Systems: Analysis and Identification. CRC Press, Boca Raton (2017)
19. Tucker, R.W., Wang, C.: The excitation and control of torsional slip-stick in the presence of axial-vibrations, pp. 1–5 (1997)
20. Udwadia, F.E.: Response of uncertain dynamic systems. I. Appl. Math. Comput. **22**(2–3), 115–150 (1987)

A Stochastic Surrogate Modelling of a NonLinear Time-Delay Mechanical System

Emanuel Cruvinel[1], Marcos Rabelo[1], Marcos L. Henrique[2], and Romes Antonio Borges[1(✉)]

[1] Mathematics and Technology Institute - School of Industrial Mathematics, Federal University of Catalão, Catalão, GO 75704-020, Brazil
romes@ufg.br

[2] Interdisciplinary Nucleus of Exact Sciences and Technological Innovation, Federal University of Pernambuco, Campus Agreste, Caruaru, Recife, PE 55002-971, Brazil

Abstract. Nonlinear time-delay dynamic is present in a wide range of engineering problems. This is due to the modernization of structures related to the need of using lighter, more resistant and flexible materials. In mechanical systems, nonlinearities may have physical or geometric characteristics. Most of these systems may possess complex equations that demands a significant computer processing time in order to solve them. In addition, these systems may be subject to uncertainties, such as material properties, random forces, dimensional tolerances and others. The complexity and the time required to solve the equations will be increased with the addition of uncertainties to the inputs of the dynamic system model. In this case, a surrogate model based on Karhunen-Loève decomposition or polynomial chaos of dynamic system is a viable choice to reduce the complexity and the computational time of the problem, as well as obtaining the statistical responses of the model. Surrogate modeling (also known as metamodeling) is employed to replace the original model of high complexity by a simpler model whose computation cost is reduced. In the field of uncertainty quantification, the statistical moments of a complex model can be easily obtained once a surrogate model is created. Methods like KLD (Karhunen-Loève Expansion), which relies on the covariance function of the system and decompose the model into a set of eigenvalues and eigenvectors which represents the surrogate model, or PCE (polynomial chaos expansion), that uses a set of multivariate orthogonal polynomials to build the surrogate model are applied to represent the system output. The purpose of this paper is to build a surrogate model of a nonlinear mechanical system with time delay using PCE and KL. A comparison between the original model response will be made against the surrogate model.

Keywords: Karhunen-Loève Decomposition · Polynomial chaos expansion · Radial basis functions · Nonlinear time-delay dynamic systems · Surrogate modeling

J. E. S. De Cursi (Ed.): Uncertainties 2020, LNME, pp. 89–102, 2021.
https://doi.org/10.1007/978-3-030-53669-5_7

1 Introduction

Nonlinear time delay (NTD) arise in many systems and industrial processes, such as biological systems, acoustics, machining, metal forming, thermal systems, and others.

Although many of these systems are modeled by nonlinear differential equations with deterministic parameters, it is known that these systems are subjected to uncertainties, such as noise in sensor measurement, random spatial variability of material properties, geometric dimensioning and random forces [1].

Buckwar et al. [2] show the effect of random variation in material parameters in a stochastic delay differential equation for machine tool vibrations. The introduction of this random parameter is important to describe the regenerative chatter phenomenon of machining process.

Barrio et al. [3] studied the oscillatory regulation of the protein *Hes1*. The authors concluded that the utilization of a stochastic delay modelling rather than a continuous deterministic model was able to capture the dynamics of oscillation of this protein.

Therefore, solutions obtained from NTD models with deterministic parameters may not capture the real dynamics of the system, forcing to introduce uncertainties in the model.

Propagating uncertainties in these problems may increase the computational cost, due to the complexity of the model and the number of simulations that one must run in order to assess its statistics. One way to circumvent this issue is by using an approximation of the problem, namely rewrite the original problem as a metamodel or surrogate model [4], which has the advantage of being fast to compute and good accuracy [5] compared to the original model. Types of surrogate modelling can be found in the literature, such as polynomial chaos [6], radial basis functions [7] and kriging [8].

The main purpose herein is to present that stochastic models can be taken as an option to approximate the response of the deterministic model while also reducing the computational cost.

This paper is organized as follow: in Sect. 2, is presented a brief overview of surrogate modelling, introducing two methods: Polynomial Chaos Expansion (PCE) and Karhunen-Loève Decomposition combined with Radial Basis Function (KLD-RBF). Then, in Sect. 3, is shown a short description of a nonlinear time delay mechanical system, and the computational implementation procedure of the surrogate modelling methods described in Sect. 2. Finally, numerical results are presented, comparing the accuracy obtained by the surrogate model.

2 Surrogate Modelling

Surrogate modelling is an approach used when there is a computational model that often requires a high computational cost to solve it. A vast number of applications can be considered, like solving a finite element model of a vehicle frame, weather forecasting [9], design of an airfoil [10] and so on. In the domain of stochastic modelling, a given model has to be solved dozen of times in order to evaluate the mean and standard deviation, usually applying the Monte Carlo Method [11]. Consequently, it might be impractical for the analyst to run these simulations if the model becomes too-time consuming, hence the necessity of implementing a less time-consuming model that emulates the original one. In the next section is presented two approaches that can be used in order to generate surrogate models.

2.1 Polynomial Chaos Expansion

Polynomial chaos expansion (PCE) can be viewed as a method to substitute a complex model by a simplified one based on a set of orthogonal polynomials [12]. Two main approaches can be used for solving stochastic systems with PCE, respectively, intrusive and non-intrusive methods.

The intrusive methods have as weaknesses, the difficulty in implementation and the high computational cost. Already in the non-intrusive methods, the PCE construction process is straightforward and can be exemplified by: run the deterministic code for each sample, collect the results and use a fitting algorithm to calculate the PCE coefficients. For the applications in this article, the non-intrusive PCE will be used.

Let y be the system output, x an input random variable following a probability distribution, t the time dimension and f the model, so that $y = f(x, t)$. The main idea of PCE is to approximate $f(x, t)$ according to:

$$f(x,t) = \sum_{\alpha=0}^{N} c_\alpha \psi_\alpha(x) \quad \alpha = \{0, 1, 2, \ldots, N\} \tag{1}$$

where $\psi_\alpha(x)$ are multivariate polynomials orthonormal with respect to the probability distribution of x, c_α are the polynomial coefficients, α is an index to map the components of $\psi_\alpha(x)$ and N is a finite integer. Note that Eq. (1) must be truncated (the number of terms depends on the desired accuracy of the metamodel). To obtain the orthonormal polynomials, the Wiener-Askey scheme is used [13].

In order to calculate the coefficients of the PCE, we sample n times the random variable $\boldsymbol{x} = (x^{(0)}\ x^{(1)} \ldots x^{(n)})^T$ and obtain n evaluations of the system output $\boldsymbol{y} = (y^{(0)}\ y^{(1)} \ldots y^{(n)})^T$. Then, an approximation method is employed to find the coefficients. For this specific problem, the ordinary least squares method will be used as follows:

$$c(\boldsymbol{x}) = \left(A^T A\right)^{-1} A^T \boldsymbol{y} \tag{2}$$

Where:

$$A = \begin{pmatrix} 1 & \psi_0\left(x^{(0)}\right) & \cdots & \psi_N\left(x^{(0)}\right) \\ 1 & \psi_0\left(x^{(1)}\right) & \cdots & \psi_N\left(x^{(1)}\right) \\ \vdots & \vdots & \ddots & \vdots \\ 1 & \psi_0\left(x^{(n)}\right) & \cdots & \psi_N\left(x^{(n)}\right) \end{pmatrix} \tag{3}$$

and:

$$c = (c_0 \ldots c_N)^T \tag{4}$$

Once the coefficients are obtained, our surrogate model is ready. Also, the statistical moments can be calculated directly from this approximation.

2.2 Karhunen-Loève Decomposition

Karhunen-Loève Decomposition is a technique used to construct a low dimensional approximation of a given model. It is based on the modal decomposition of data set (from an experiment or simulation) to approximate the model in question. This technique was originally developed by Karhunen [14] and Loève [15] to represent a stochastic process as a linear combination of orthogonal functions. In the literature it is also common for some authors to use other nomenclatures for this method, such as Proper Orthogonal Decomposition (POD) [16] and Principal Component Analysis (PCA) [17]. According to Liang et al. [18], all these methods are mathematically equivalent. Some applications of KLD can be found in thermal analysis [19], economics [20] and vibration [21].

The following is a brief introduction to the method.

Initially we denote a function *f(x, t)*, where x is a vector of inputs $x = [x_1, .., x_n]$ and t is the discretized time dimension vector $t = [t_1, ..., t_m]$. We then construct the snapshot matrix *S* of size (M × N):

$$S = \begin{bmatrix} f(x_1, t_1) \dots f(x_n, t_1) \\ \vdots \quad \ddots \quad \vdots \\ f(x_1, t_m) \dots f(x_n, t_m) \end{bmatrix} \tag{5}$$

The next step is to calculate the matrix *C* defined below:

$$C = S.S^T \tag{6}$$

After obtaining the matrix C, we compute its eigenvalues $\lambda_i = [\lambda_1, \lambda_2, \ldots, \lambda_n]$, where $\lambda_n > \lambda_{n+1}$ and eigenvectors ν_n associated to λ_n.

Now we compute the amplitude matrix A:

$$A = \phi^T.S \tag{7}$$

In order to reduce the model dimensionality, we pick the $k = \{1, 2, \ldots, N\}$ first columns of the matrix Φ corresponding to the largest eigenvalues and the *k* first rows of the amplitude matrix A. The reduced model is obtained in Eq. (8):

$$S_{approx} = \phi_k.A_k \tag{8}$$

Although the KLD method is able to create an approximation of the original model, it does not allow us to run simulations if different inputs to the approximate model are provided. In fact, this method only expresses the system response in a new basis [22]. However, interpolation techniques can be applied on top of the approximate model to enable us to run simulations with different input parameters. A detailed procedure to apply KLD is given in [22].

2.3 Radial Basis Functions

Radial Basis Function (RBF) is a mathematical tool used to approximate multivariable functions by interpolation of existing data. Applications of RBF can be found in tourism growth forecasting [23], solution of partial differential equations [24] and mineral prospectivity prediction [25].

The approximation of a function $f(x)$ using the RBF approach is given below:

$$f(\boldsymbol{x}) = \sum_{i=1}^{N} c_i g_i(\boldsymbol{x}) \tag{9}$$

Where the function $f(x)$ is approximated by summing N radial basis functions $g_i(x)$ multiplied by coefficients c_i. The variable $\boldsymbol{x}$ can be a multidimensional vector of size M. Each basis function $g_i(x)$ is associated with a known value x_i, hence $f(x_i)$ is also known. Some choices of basis functions are given below:

$$g_i(x) = ||x - x_i||, \quad linearspline \tag{10}$$

$$g_i(x) = ||x - x_i||^3, \quad cubicspline \tag{11}$$

$$g_i(x) = \exp\left(-c^2.||x - x_i||\right), \quad gaussian \tag{12}$$

Therefore, for a set of given points $\boldsymbol{x_i} = \{x_1, \ldots, x_N\}$, knowing the values of $f(\boldsymbol{x_i})$ is sufficient for calculating the coefficients c_i.

Combined with RBF interpolation, the KLD-RBF is capable of creating a surrogate model. In the next section we demonstrate how to implement the KLD-RBF method.

2.4 KLD – RBF Method

The KLD RBF method proposed by Buljak [22] is resumed below.

- Obtain the snapshot matrix according to Eq. (5)

  Calculate the matrix C according to Eq. (6)

- Calculate the eigenvalues λ_i and eigenvectors Φ of C. Sort the eigenvectors based on the eigenvalues magnitude.
- Keep the first k eigenvector columns, where k is the number of eigenmodes. The selection of the number of eigenmodes is used to reduce the dimension of our model.
- Obtain the reduced amplitude matrix $A_k = \Phi_k.S$
- Choose a RBF interpolation formula, for instance, the Euclidian distance: $g_i(x) = \|x - x_i\|$
- Compute matrix G:

$$G = \begin{bmatrix} g_1(x_1) & \ldots & g_1(x_m) \\ \vdots & \ddots & \vdots \\ g_m(x_m) & \ldots & g_m(x_m) \end{bmatrix} \tag{13}$$

- Calculate the matrix of coefficients of interpolation B, defined in Eq. (14):

$$B = G^{-1}.A_k \tag{14}$$

- Given a vector of known parameters x, calculate g(x):

$$g(x) = \begin{bmatrix} g_1(x) \\ \vdots \\ g_n(x) \end{bmatrix} \tag{15}$$

- Finally, obtain the system approximated response:

$$u(x) = \phi_k.B.g(x) \tag{16}$$

In Eq. (16), we developed a general formula to approximate the response of the system given a vector of parameters x. If a new vector x is given, we must calculate g(x) according to Eq. (15) prior to utilizing Eq. (16).

3 Mathematical Model of a Nonlinear Time Delay Mechanical System

We now present in this section the mathematical model of a nonlinear time delay mechanical system whose analytical solution (deterministic) of the system was described in [26].

The model is represented by a two degree of freedom mechanical system with linear and nonlinear dampers and springs, and it is composed of two subsystems: the first subsystem is attached to the ground, supported by nonlinear springs and dampers, whereas the second subsystem is also composed of linear and nonlinear springs and dampers, and submitted to a time delay [26].

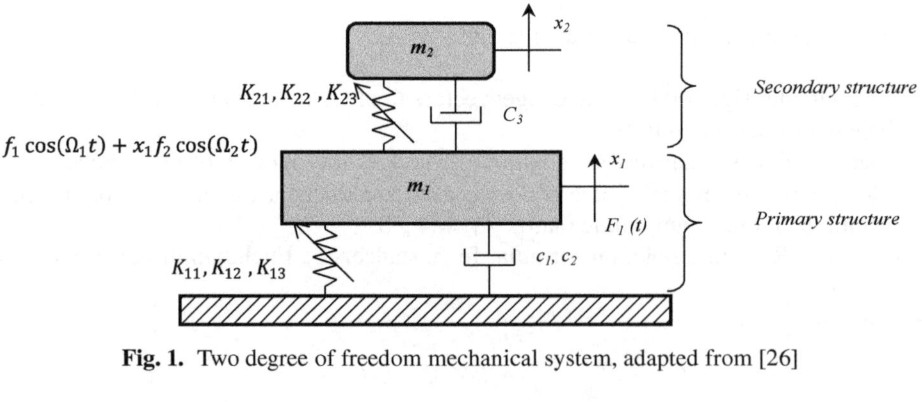

Fig. 1. Two degree of freedom mechanical system, adapted from [26]

The equations of motion for the system in Fig. 1 are given below:

$$\begin{aligned} &\ddot{x}_1 + \omega_1^2 x_1 + \alpha_{12} x_1^2 + \alpha_{13} x_1^3 + \zeta_1 \dot{x}_1 + \zeta_2 \dot{x}_1 x_1^2 - \alpha_{21} x_2 + \zeta_3 (\dot{x}_1(t-\tau) \\ &- \dot{x}_2(t-\tau)) + \alpha_{22}(x_1 - x_2)^2 + \alpha_{23}(x_1 - x_2)^3 = F_1 \cos(\Omega_1 t) + x_1 F_2 \cos(\Omega_2 t) \end{aligned} \tag{17}$$

$$\ddot{x}_2 + \omega_2^2(x_2 - x_1) - \beta_{22}(x_2 - x_1)^2 + \beta_{23}(x_2 - x_1)^3 + \zeta_4(\dot{x}_2(t-\tau) - \dot{x}_1(t-\tau)) = 0 \tag{18}$$

where:

$$\omega_1^2 = \frac{k_{11}+k_{21}}{m_1}, \alpha_{12} = \frac{k_{12}}{m1}, \alpha_{13} = \frac{k_{13}}{m_1}, \zeta_1 = \frac{c_1}{m_1}, \zeta_2 = \frac{c_2}{m_1}, \zeta_3 = \frac{c_3}{m_1}, \zeta_4 = \frac{c_3}{m_2}, \alpha_{21} = \frac{k_{21}}{m1}, \alpha_{22} = \frac{k_{22}}{m1}, \alpha_{23} = \frac{k_{23}}{m1}, F_1 = \frac{f_1}{m_1}, F_2 = \frac{f_2}{m1}, \omega_2^2 = \frac{k_{21}}{m2}, \beta_{22} = \frac{k_{22}}{m2}, \beta_{23} = \frac{k_{23}}{m2} \tag{19}$$

with the following variables: m_1 and m_2 are the masses of the two subsystems, x_1 and x_2 are the displacements, ω_1 and ω_2 are natural frequencies, k_{11} and k_{22} are the stiffness linear parameters, k_{i2} and k_{i3} are the quadratic and cubic stiffness nonlinear parameters, f_1 and f_2 are the external excitation force amplitudes, c_1 and c_2 are the linear and nonlinear dampers, c_3 the controller linear damper and τ the time delay, and their values can be found in [26].

In this paper, three different surrogate models from each method (PCE and KLD-RBF) are created by choosing some variables from the model above: the time delay (τ), the ratio of stiffness in the secondary system ($\kappa_2 = k_{23}/k_{21}$) and the ratio between damping and mass of the primary system ($\zeta_2 = c_2/m_1$). The first model will comprise only the time delay, whereas the second model will use κ_2 and ζ_2, and finally, for the third model, the three variables will be used to build the surrogate model.

4 Numerical Results and Discussion

This section presents the comparative results using Polynomial Chaos Expansion (PCE) and Karhunen-Loève Decomposition combined with Radial Basis Function (KLD-RBF). The mean square error (MSE) of the PCE and KLD-RBF surrogate models are calculated for evaluation of their accuracy.

4.1 Study of Polynomial Degree

For the PCE method, the uniform probability distribution was chosen for each variable for sampling and construction of the polynomials. The probability density function of the uniform distribution is defined in Eq. (20):

$$f(x) = \begin{cases} \frac{1}{b-a}, & for\, a \le x \le b \\ 0, & for\, x < a\, or\, x > b \end{cases} \tag{20}$$

Where a and b are the lower and upper boundary of the uniform distribution. In this model, the variables have the boundaries defined in Table 1:

Table 1. Lower and upper boundaries of the variables used in the PCE surrogate model following the uniform distribution

Variable	a	b
τ	0.1	0.3
κ_2	0.01	0.1
ζ_2	0.01	0.1

Initially, a study related to the order of the polynomial is presented. In this sense, we try to find which order of the PCE presents better results when compared to the deterministic approach. Using a uniform probability distribution, the number of samples for each variable was fixed at 500. Figure 2 shows the results of the approximations calculated for several orders of the polynomial (order ranging from 1 to 8):

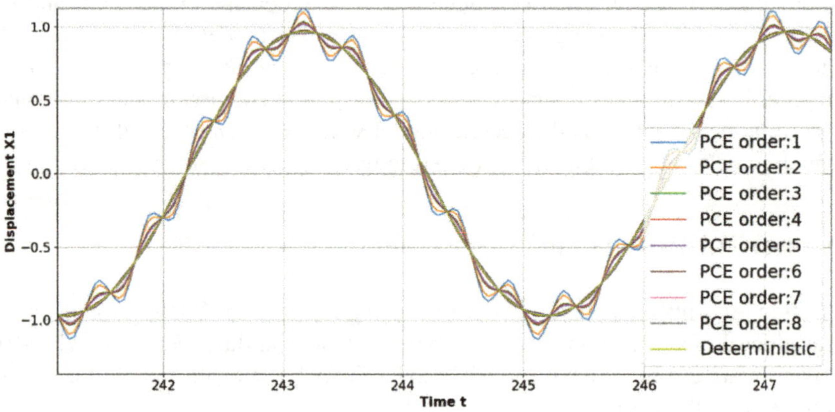

Fig. 2. Deterministic values and stochastic approximation of the response

In Fig. 2 it can be seen that the approximation improves as the degree of the polynomial increases. In this case, it can be said that the degree eight polynomial has a better response (better approximation with the exact solution) when compared to the others. For this reason, in the results that follow, the degree eight polynomial will be used.

4.2 Results of the Stochastic Approach

For the KLD-RBF approach, the vector of parameters to build our snapshot matrix is defined as $x = [a{:}b{:}c]$, where x is the variable (or parameter) of the model, a is the minimum value of the vector, b is the maximum value of the vector and c is the number of equally spaced points inside the vector. Hence, the variables defined are $\tau = [0.1{:}0.5{:}10]$, $\kappa_2 = [0.01{:}0.1{:}10]$ and $\zeta_2 = [0.01{:}0.1{:}10]$.

In order to reduce the model and capture the dynamics of the system, the number of eigenmodes utilized was 15.

In Fig. 3 we show a comparison between the displacement of the mass of the primary system using the original model, the PCE surrogate model and the KLD-RBF model, using $\tau = 0.25$:

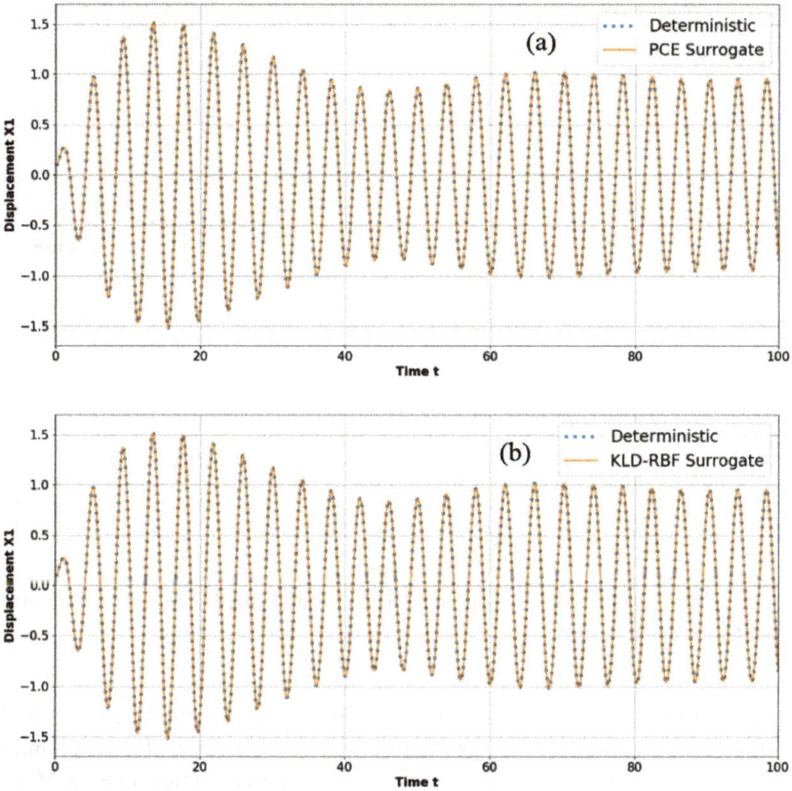

Fig. 3. Comparison between deterministic model response x_1 displacement and surrogate model (a) PCE and (b) KLD-RBF built from time delay, with $\tau = 0.25$

We observe in Fig. 3 (a) and (b) that the displacement of the surrogate model fits the displacement of the original model. The mean square error (MSE) of the PCE model obtained was 74×10^{-7} whereas the MSE of the KLD-RBF model was 66×10^{-8}. Even though both models have a good MSE score, the KLD-RBF model yields a better result compared to PCE.

The next PCE and KLD-RBF surrogate model is built using κ_2 and ζ_2. In Fig. 4 we compare the displacement of the primary system using the deterministic model with the surrogate model, using $\kappa_2 = 0.03$ and $\zeta_2 = 0.08$:

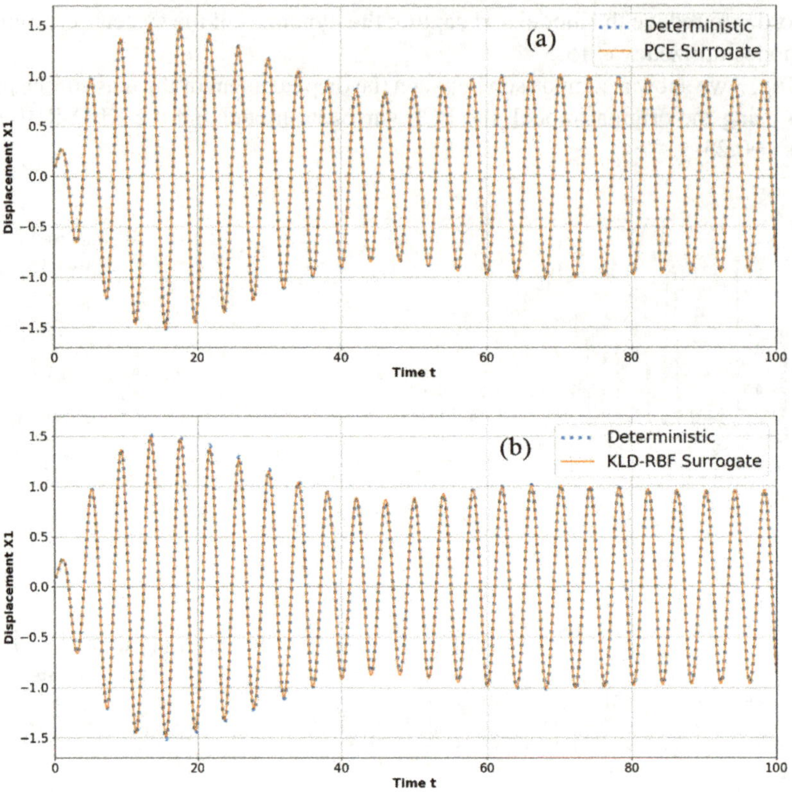

Fig. 4. Comparison between deterministic model response x_1 displacement and surrogate model (a) PCE and (b) KLD-RBF built using κ_2 and ζ_2, with $\kappa_2 = 0.03$ and $\zeta_2 = 0.08$

It is worth noticing in Fig. 4 (a) and (b) that the response of the surrogate model also fits the deterministic model, with a MSE of 5×10^{-11} for the PCE model and a MSE of 3×10^{-4} for the KLD-RBF model. Here, the PCE model yields a better MSE score.

The last surrogate model is built using τ, κ_2 and ζ_2. In Fig. 5 we compare the displacement of the primary system using the deterministic model with the surrogate model, using $\tau = 0.25$, $\kappa_2 = 0.03$ and $\zeta_2 = 0.08$:

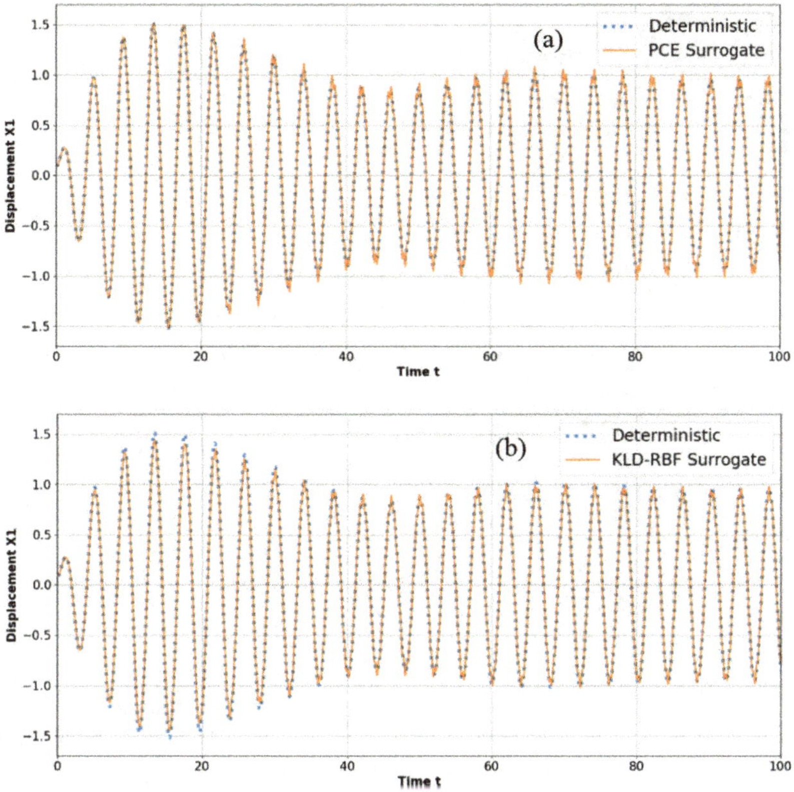

Fig. 5. Comparison between deterministic model response x_1 displacement and surrogate model (a) PCE and (b) KLD-RBF built using τ, κ_2 and ζ_2, with $\tau = 0.25$, $\kappa_2 = 0.03$ and $\zeta_2 = 0.08$

In this case, both models yielded a MSE score of 2×10^{-3}. Looking closely at both models in Fig. 6, we can observe small fluctuations in the response of the system. This suggests that, in order to increase the accuracy of the models, we must increase the order of the polynomial in the case of PCE [27], and increase the number of points in the vector of parameters in the case of the KLD-RBF [22].

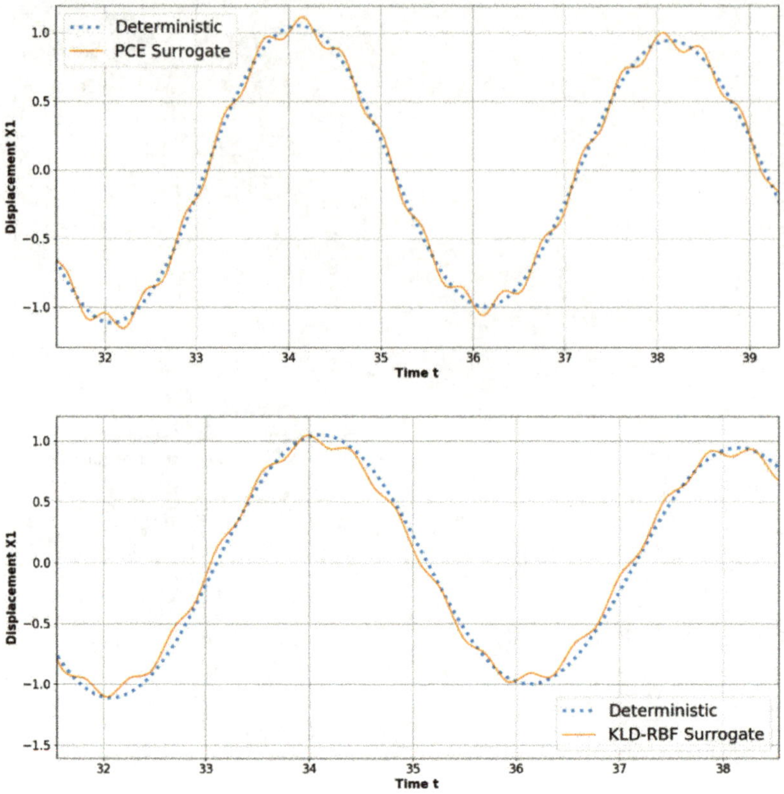

Fig. 6. Small fluctuations in displacement for both models compared to Deterministic Model

4.3 Computation Time of the Models

One of the advantages is the reduced computation time of the surrogate model. Table 2 below shows the computation time for each model:

Table 2. Computation time of original and surrogate models

Model	Input parameters	%
Deterministic		100%
PCE	τ	9.43%
PCE	κ_2, ζ_2	10.07%
PCE	τ, κ_2, ζ_2	19.50%
KLD-RBF	τ	30.82%
KLD-RBF	κ_2, ζ_2	35.85%
KLD-RBF	τ, κ_2, ζ_2	32.08%

The PCE models computation time are faster than KLD-RBF models. Using only τ as the input parameter, the computation time was reduced to 9.43% compared to the deterministic model. The KLD-RBF are slightly slower than PCE due to the computation of Eqs. (15) and (16).

5 Conclusions

In this paper a stochastic surrogate model of a nonlinear time delay mechanical system was created using Polynomial Chaos Expansion and the Karhunen-Loève Decomposition combined with Radial Basis Functions. It was shown that both methods can create surrogate models of nonlinear systems with good accuracy.

Considering the study carried out, it can be concluded that the two stochastic approaches are adequate to solve this system, both with respect to the approximation and the reduction of computational cost.

On the other hand, it is possible to conclude that the PCE models have a lower computational cost than the KLD-RBF models.

References

1. Pukl, R., Jansta, M., Červenka, J., Vořechovský, M., Novák, D., Rusina, R.: Spatial variability of material properties in nonlinear computer simulation. Comput. Model. Concr. Struct. - Proc. EURO-C 2006, pp. 891–896 (2006)
2. Buckwar, E., Kuske, R., L'esperance, B., Soo, T.: Noise-sensitivity in machine tool vibrations. Int. J. Bifurc. Chaos. **16**, 2407–2416 (2006). https://doi.org/10.1142/S021812740601615X
3. Barrio, M., Burrage, K., Leier, A., Tian, T.: Oscillatory regulation of Hes1: discrete stochastic delay modelling and simulation. PLoS Comput. Biol. **2**, e117 (2006). https://doi.org/10.1371/journal.pcbi.0020117
4. Lataniotis, C., Marelli, S., Sudret, B.: Extending classical surrogate modelling to ultrahigh dimensional problems through supervised dimensionality reduction: a data-driven approach, pp. 1–38 (2018)
5. Forrester, A.I.J., Sóbester, A., Keane, A.J.: Engineering Design via Surrogate Modelling. Wiley, Chichester (2008)
6. Ghanem, R., Spanos, P.: Stochastic Finite Elements – A Spectral Approach. Springer, New York (1991)
7. Chen, W., Fu, Z.-J., Chen, C.S.: Recent Advances in Radial Basis Function Collocation Methods. Springer, Heidelberg (2014)
8. Stein, M.L.: Interpolation of Spatial Data. Springer, New York (1999)
9. Maity, S., Bonthu, S., Warrior, H.V., Sasmal, K., Warrior, H.: Role of Parallel Computing in Numerical Weather Forecasting Models. 2012 (2014)
10. Lee, D.S., Gonzalez, L.F., Periaux, J., Srinivas, K.: Robust design optimisation using multi-objective evolutionary algorithms. Comput. Fluids **37**, 565–583 (2008). https://doi.org/10.1016/j.compfluid.2007.07.011
11. Graham, C., Talay, D.: Stochastic Simulation and Monte Carlo Methods. Springer, Heidelberg (2013)
12. Ghanem, R.G., Spanos, P.D.: Stochastic Finite Elements: A Spectral Approach. Springer, New York (1991)

13. Xiu, D., Karniadakis, G.E.: The wiener-askey polynomial chaos for stochastic differential equations. SIAM J. Sci. Comput. **24**, 619–644 (2002)
14. Karhunen, K.: Uber linear Methoden fur Wahrscheiniogkeitsrechnung. Ann. Acad. Sci. Fennicae Series Al Math. Phys. **37**, 3–79 (1946)
15. Loève, M.: Probability Theory. Springer, Princeton (1977)
16. Pearson, K.: LIII. On lines and planes of closest fit to systems of points in space. London, Edinburgh, Dublin Philos. Mag. J. Sci. **2**, 559–572 (1901). https://doi.org/10.1080/14786440109462720
17. Abdi, H., Williams, L.J.: Principal component analysis. Wiley Interdiscip. Rev. Comput. Stat. **2**, 433–459 (2010). https://doi.org/10.1002/wics.101
18. Liang, Y.C., Lee, H.P., Lim, S.P., Lin, W.Z., Lee, K.H., Wu, C.G.: Proper orthogonal decomposition and its applications - Part I: theory. J. Sound Vib. **252**, 527–544 (2002). https://doi.org/10.1006/jsvi.2001.4041
19. Del Barrio, E.P., Dauvergne, J.L., Pradere, C.: Thermal characterization of materials using Karhunen-Loève decomposition techniques - Part I. Orthotropic materials. Inverse Probl. Sci. Eng. **20**, 1115–1143 (2012). https://doi.org/10.1080/17415977.2012.658388
20. Yao, J., Laurent, S., Bénaben, B.: Managing volatility risk: an application of karhunen-loève decomposition and filtered historical simulation, pp. 1–22 (2017)
21. Bellizzi, S., Sampaio, R.: Smooth Karhunen-Loève decomposition to analyze randomly vibrating systems. J. Sound Vib. **325**, 491–498 (2009). https://doi.org/10.1016/j.jsv.2009.03.044
22. Buljak, V.: Inverse Analyses with Model Reduction. Springer, Heidelberg (2012)
23. Hu, S.F., Zhu, H. Bin, Zhao, L.: Radial basis function and its application in tourism management. Mod. Phys. Lett. B. **32**, 1–5 (2018). https://doi.org/10.1142/S0217984918400547
24. Bhatia, G.S., Arora, G.: Radial basis function methods for solving partial differential equations-A review. Indian J. Sci. Technol. **9** (2016). https://doi.org/10.17485/ijst/2016/v9i45/105079
25. Ghezelbash, R., Maghsoudi, A., Carranza, E.J.M.: Performance evaluation of RBF- and SVM-based machine learning algorithms for predictive mineral prospectivity modeling: integration of S-A multifractal model and mineralization controls. Earth Sci. Inform. **12**, 277–293 (2019). https://doi.org/10.1007/s12145-018-00377-6
26. Rabelo, M., Silva, L., Borges, R., Henrique, M.: Computational and numerical analysis of a nonlinear mechanical system with bounded delay. Int. J. Non-Linear Mech. **91**, 36–57 (2017)
27. Lu, F., Morzfeld, M., Tu, X., Chorin, A.J.: Limitations of polynomial chaos expansions in the Bayesian solution of inverse problems. J. Comput. Phys. **282**, 138–147 (2015). https://doi.org/10.1016/j.jcp.2014.11.010

A Stochastic Approach for a Cosserat Rod Drill-String Model with Stick-Slip Motion

Hector Eduardo Goicoechea[1,2(✉)], Roberta Lima[2], Rubens Sampaio[2], Marta B. Rosales[1], and F. S. Buezas[1]

[1] Laboratório de Vibrações, Pontifícia Universidade Católica do Rio de Janeiro, Rua Marques de São Vicente 255, Gávea, Rio de Janeiro, RJ, Brazil
h.e.goicoechea@gmail.com

[2] Instituto de Física del Sur (IFISUR), Universidad Nacional del Sur, CONICET, Av. L. N. Alem 1253, B8000CPB Bahía Blanca, Argentina

Abstract. Drill-strings employed in the oil extraction process present complex dynamics. Due to their slenderness (with length surpassing 1000 m), geometrical aspects of oil-wells, and contact forces, drill-strings exhibit a strong non-linear response. Many unwanted phenomena, like stick-slip oscillations, may occur under certain conditions. Moreover, there are several intrinsic uncertainties, e.g. in modelling the soil, which may lead to unwanted operation regimes and affect performance, i.e. degraded rate-of-penetration (ROP), and stick-slip. In this work, a stochastic approach to study drill-string dynamics is employed. The physical model is based on the Cosserat rod theory. Non-linearities of the drill-string problem may occur due to geometrical factors, such as finite displacements and rotations in deviated wells, as well as to contact and friction at the borehole wall and bit. In this particular paper, uncertainties are considered within the friction parameters. The random response is analysed through the trajectories of the drill-string axis and other quantities, such as the evolution of the angular velocities.

Keywords: Drill-string dynamics · Horizontal drill-string · Cosserat rods · Stick-skip oscillations

1 Introduction

A drill-string is the structure employed in the drilling process of rocks in the search of oil. It is a slender structure whose main elements are thin tubes (drill-pipes), thick tubes (drill-collars), a bit fixed at one of the ends, and other complementary devices such as downhole motors, rotary steerable systems, measurement while drilling tools, etc.

There are many papers available in the literature that treat drill-string dynamics, based upon different theoretical frameworks. For this task, some works

J. E. S. De Cursi (Ed.): Uncertainties 2020, LNME, pp. 103–110, 2021.
https://doi.org/10.1007/978-3-030-53669-5_8

consider lumped models as in [1–3], while others employ continuous beam theories [4–6] or rod theories [7,8]. The latter can also be based on different theories employed in solid mechanics, such as the Bernoulli-Euler beam and Cosserat rod theories.

Horizontal drilling has been analysed in [3,5]. Particularly, in [5], where a bar model (longitudinal wave equation) has been employed to analyse the dynamics of a drill-string during the penetration process. This means that the formulation only accounts for longitudinal displacements, and that no lateral and torsional motion is allowed. The model also includes the bit-rock interaction force between the bit and the borehole wall.

In the present paper, the problem treated in [5] is studied by means of a Cosserat rod theory based model. This approach is an improvement over the original proposal as it allows studying the lateral dynamics as well as torsional behaviour, if rotation is included. A comparison between the approach in [5] and the current model is made with a deterministic problem. Finally, the stochastic problem considering a random field for the friction coefficient is studied.

2 The Deterministic Model

The dynamics of a horizontal drill-string considering contact at the bit as well as friction and contact within the drill-string length are studied in [5], where the authors employed a bar model (longitudinal wave equation). It was assumed that contact occurs along the whole length of the drill-string under static conditions, that is, the contact (normal) force equals the self-weight of the structure. Also, the frictional force was considered to be proportional to the self-weight.

In this paper, the problem dealt in [5] is revisited. A Cosserat rod model is employed to simulate the dynamics of a horizontal drill-string, considering the forces shown in Fig. 1. The model in this paper can describe longitudinal, lateral, and torsional dynamics, as opposed to the original model where only longitudinal effects were considered.

In this model, three distributed forces are considered: a contact force $\mathbf{f}_c$, a friction force $\mathbf{f}_{fric}$ and the self-weight $\mathbf{f}_g$. The self-weight ($\mathbf{f}_g$) is taken into account in the dynamics of the drill-string, and in the calculation of the frictional force ($\mathbf{f}_{fric}$), which is proportional to the contact force ($\mathbf{f}_c$) that now depends on the actual dynamics of the structure. The geometric and material properties used in the simulations are shown in Table 1.

The expression for the distributed self-weight is given by

$$\mathbf{f}_g = -\rho A g \hat{\mathbf{k}}, \quad g = 9.81 \text{ m/s}^2 \tag{1}$$

The contact force $\mathbf{f}_s$ is modelled based on the penalty method, as explained in [8]. It is supposed to be proportional to a penetration constant k_s, and it can be stated as

$$\mathbf{f}_s = -k_s \, f_{sp} \, \hat{\mathbf{r}}_r \tag{2}$$

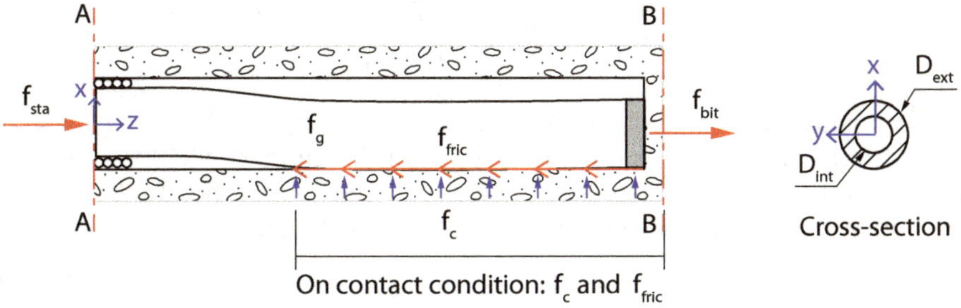

Fig. 1. Sketch of the drill-string model employed. In the graphic, $\mathbf{f}_{sta}$ stands for the force transmitted by the pipes before the simulated segment of drill-string, $\mathbf{f}_c$ is the contact force, the friction force is given by $\mathbf{f}_{fric}$, and the self-weight is $\mathbf{f}_g$. D_ext and D_in are the internal and external diameters of the cross-section.

Table 1. Material and geometric properties used in the modelling of the horizontal drill-string.

Property	Value	Description
L	60 m	Length of the drill-string
D_{ext}	0.15 m	Cross-section external diametre
D_{int}	0.10 m	Cross-section internal diametre
ρ	7850 kg/m^3	Material density
E	210 GPa	Young modulus
m_{bit}	20 kg	Mass at the bit
k_s	$1 \cdot 10^6$ N/m^2	Soil constant
c_1	1.4 10^3 N	Coefficient of the bit-rock interaction model
c_2	400	Coefficient of the bit-rock interaction model
b	10	Bit-rock cross-correlation decay length
ω_f	$100 \cdot 2\pi/60$ rad/s	Frequency of harmonic excitation
t	$\in [0, 10]$ s	Simulation time
f_{sta}	5500 N	Transmitted force
f_0	500 N	Mean magnitude of the oscillatory force
μ	–	Mean value of the friction coefficient
σ	–	Standard deviation for the friction coefficient

In the previous expressoin, $\mathbf{r}(x, t)$ is the current position vector of the centreline of the drill-string; $\mathbf{r}_p(s)$ is position vector of the centreline curve of the borehole; $\mathbf{r}_r = \mathbf{r} - \mathbf{r}_p$ is a relative position vector that describes the position of the centreline of the drill-string with respect to the borehole centreline; $\hat{\mathbf{r}}_r$ is a unitary vector in the same direction; and f_{sp} is a penetration function defined by

$$f_{sp} = \begin{cases} |\mathbf{r}_r| - c_{gap}, & \text{if } |\mathbf{r}_r| - c_{gap} \geq 0. \\ 0, & \text{otherwise.} \end{cases} \tag{3}$$

with c_{gap} being the radius of the borehole.

The friction force is given by the expression hereunder, where $\hat{\mathbf{v}}_c$ denotes a unitary vector in the direction of the velocity at the contact point. It is assumed that contact occurs only at one point of the cross-section.

$$\mathbf{f}_{fr} = -k_{fr}|f_s|\hat{\mathbf{v}}_c \tag{4}$$

The boundary conditions for this problem are: at the points lying in section A-A of Fig. 1, the rotation of the section in any direction is restricted, and no movement around the x-axis and y-axis is allowed, while in the z-axis direction a load $\mathbf{f}_{sta} = f_{sta}\hat{\mathbf{k}}$ is applied.

On the other hand, at the points within section B-B, a free-end condition with an applied load $\mathbf{f}_{bit}$ is considered. The force applied to the bit actually describes the following three effects: a contact force with the wall (f_{wall}), a harmonic force (f_{har}) due to the mud motor, and the inertial effect of the mass of the bit (f_m), which is modelled as a lumped mass.

$$\mathbf{f}_{bit} = \mathbf{f}_{har} - \mathbf{f}_{wall} - \mathbf{f}_m \tag{5}$$

The contact force at the bit f_{wall} is expressed as in [5]. The constants c_1 and c_2 are two constants related to the bit-rock interaction.

$$\mathbf{f}_{wall} = \begin{cases} (c_1 e^{-c_2 \dot{u}(L,t)} - c_1)\hat{\mathbf{k}}, & \text{if } \dot{u}(L,t) > 0 \\ \mathbf{0}, & \text{otherwise} \end{cases} \tag{6}$$

A harmonic force applied at the bit is also imposed on the system, as the driving source of the horizontal drilling is the mud motor, which rotates about a given nominal rotational speed (in steady operation).

$$\mathbf{f}_{har}(x,t) = F_0 \sin(\omega_f t)\hat{\mathbf{k}} \tag{7}$$

Finally, the initial conditions are such that the structure is at rest at the initial time $t = 0$.

3 The Stochastic Model

The nature of the soil as a material is a source of uncertainties in any mechanical problem, given that properties such as the friction coefficient may vary greatly from one point to another. For this reason, following [5], the soil is simulated as a random force, by taking the friction coefficient k_{fr} as a random field.

It is be assumed that k_{fr} follows a truncated gaussian field with support $[0, 0.6]$, and exponential autocorrelation given by

$$R(x_1, x_2) = \sigma^2 e^{\left(-\frac{|x_2 - x_1|}{b}\right)} \tag{8}$$

where b is the correlation length that measures the rate of decay of the autocorrelation function.

A Karhunen-Loàve expansion is employed to approximate the friction field k_{fr}, by taking the first N terms of the series as follows:

$$k_{fr} \approx \mu(x) + \sum_{k=1}^{N} \sqrt{\lambda_k} Z_k(\zeta)\phi_k(x) \tag{9}$$

In the previous expression, μ is the mean value of the friction coefficient, λ_k and ϕ_k are the eigenvalues and eigenvectors of the autocorrelation function R. Z_k are independent standard Gaussian random variables.

4 Results

In this section, the results for the simulations are presented. Two different cases are considered. Firstly, a deterministic problem with the current approach is compared against the model in [5]. Secondly, a numerical approximation to the solution of the stochastic problem is shown.

4.1 The Deterministic Case

The Cosserat model in this paper is firstly solved considering the parameters in Table 1. For the time being, the friction field is not considered stochastic.

For this problem, the friction coefficient k_{fr} takes the value $k_{fr} = 0.10$. Two different cases are considered. In case A the contact force is a direct consequence of the dynamics of the model. It is obtained by considering the penalty approach given as stated in (2). Case B considers the contact force to be constant and equal to the self-weight along the entire length of the rod. The latter coincides with the hypothesis provided in [5]. The solution is shown in Fig. 2 and compared against that of [5].

A stick-slip motion in a mechanical system implies that two very distinct phases can be appreciated in the dynamics. Namely, the stick and the slip phases. On the one hand, stick is characterized for presenting zero relative velocities between the two surfaces in contact for a given period (not just an instant). On the other hand, the slip phase is defined for those instants that do not comply with the previous conditions.

In a qualitative comparison between the solutions, it is observed that Case A and B differ greately. Due to the use of a numerical model, the system exhibits vibrations in which the velocities go to almost zero, but not exactly zero, as observed in Case B from Fig. 2. For this reason, the motion is said to be pseudo stick-slip. In case A contact acts only on a section of the structure, as shown in Fig. 3. Therefore, the friction force varies in such way that case A does not exhibit any stick-slip like behaviour, while case B exhibits pseudo stck-slip motion. This implies that, under certain conditions, it is important to consider the lateral dynamics of the structure. Moreover, case A shows a penetration δ_z that is 100 times higher in order than that of case B.

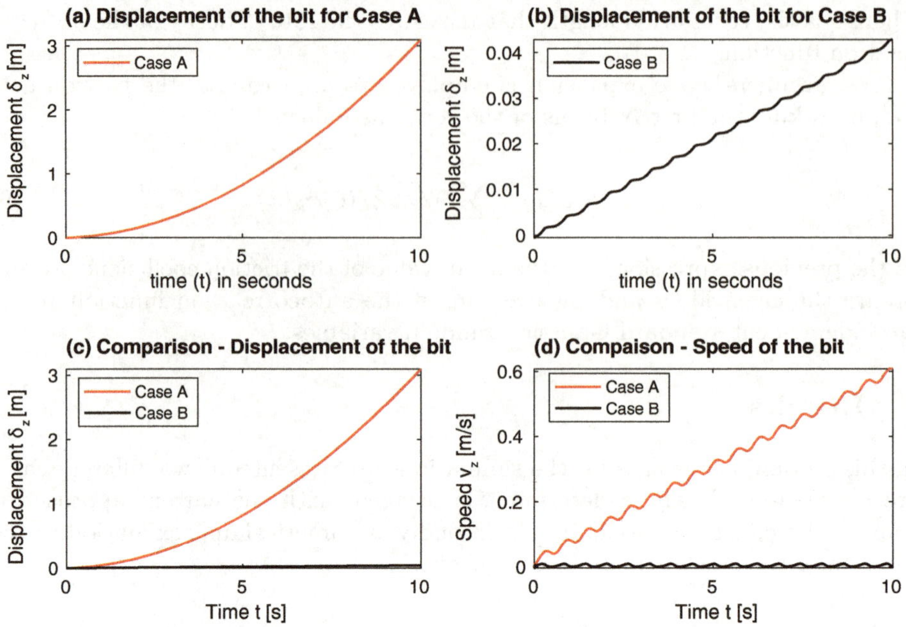

Fig. 2. Results for the deterministic problem. Case A corresponds to the hypothesis that contact occurs due to the deflection and contact of the beam. Case B considers contact on the whole length of the beam, with the hypothesis that the distributed normal force equals the self weight of the structure. (a), (b) and (c) show the displacement of the bit in the direction of the z-axis δ_z, while (d) shows the velocity v_z at the bit.

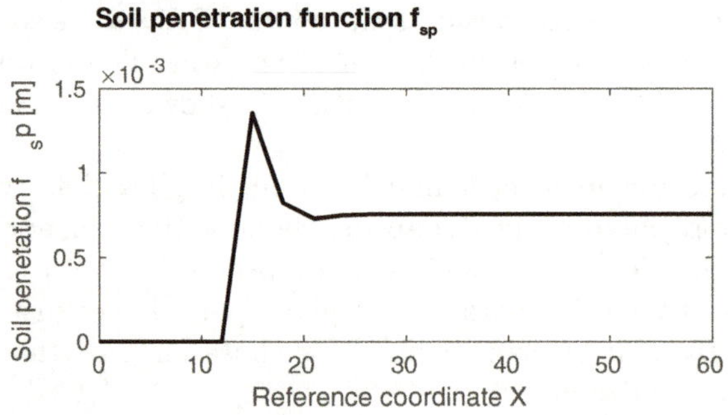

Fig. 3. Soil penetration function f_{sp} [m].

4.2 Stochastic Results

In this section, a stochastic problem considering friction as a random field (see Sect. 3) is presented. The parameters employed are those provided in Table 1 are employed in the simulations, with $\mu = 0.12$, $\sigma = 0.1 \cdot \mu$, and $N = 100$ terms in the Karhunen-Loàve expansion.

The power ratio $p_r = p_{out}/p_{in}$ is used as a measurement of the efficiency of the system. It is defined as the ration between the input power (p_{in}) and the output power (p_{out}), given by the following expressions:

$$p_{in} = \frac{1}{t_1 - t_0} \int_{t_0}^{t_1} (f_{sta}\dot{r}_z(0,t)) + f_{har}\dot{r}_z(L,t)\, dt \tag{10}$$

$$p_{out} = \frac{1}{t_1 - t_0} \int_{t_0}^{t_1} f_{bit}\dot{r}_z(L,t)\, dt \tag{11}$$

The results from the Monte-Carlo simulations are shown in what follows (Fig. 4).

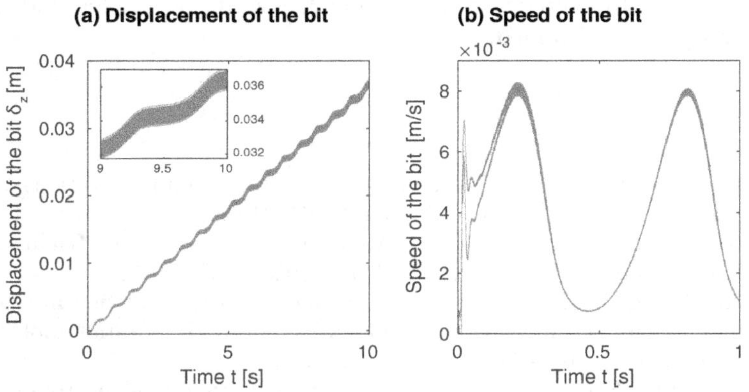

Fig. 4. Some realisations from the Monte-Carlo simulations. (a) Displacement of the bit. (b) Speed of the bit.

The PDF of the power ratio is shown in Fig. 5, as well as a convergence graph for the standard deviation of the response.

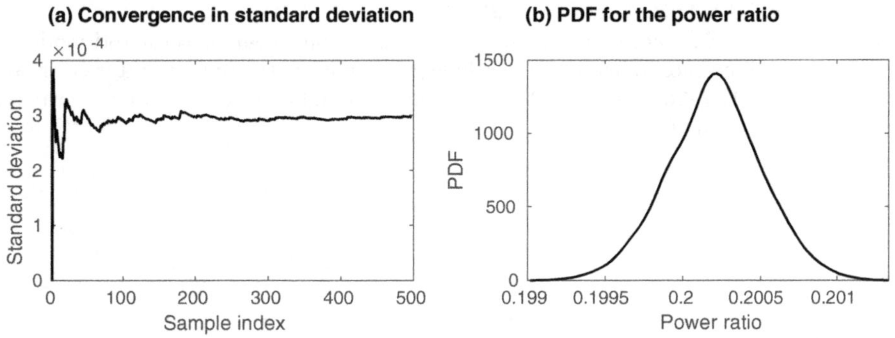

Fig. 5. Results for the stochastic problem. (a) Convergence graph for the standard deviation of the power ratio. (b) PDF for the power ratio.

5 Conclusions

In this paper a model capable of capturing the longitudinal, lateral and torsional dynamics of a drill-string has been employed. The Cosserat rod model has been compared with a bar model to show the importance of considering lateral displacements in the calculation of the contact and friction forces involved. Firstly, it is observed that the mechanical system changes its behaviour depending on the value of the friction coefficient. In fact, under some conditions pseudo stick-slip motion is shown, which greatly affects the rate-of-penetration. Moreover, when the contact and friction forces are obtained as a result of the dynamics of the system, the solution can be different than that obtained by considering contact to be equal to the self-weight (at an equilibium state). Secondly, the stochastic problem is solved by employing the current Cosserat rod model, and the PDF of the power ratio is obtained.

References

1. Yigit, A.S., Christoforou, A.P.: Coupled torsional and bending vibrations of actively controlled drillstrings. J. Sound Vib. **234**(1), 67–83 (2000)
2. Ritto, T.G., Aguiar, R.R., Hbaieb, S.: Validation of a drill string dynamical model and torsional stability. Meccanica **52**(11–12), 2959–2967 (2017)
3. Wang, R., Wang, X., Ni, H., Zhang, L., Wang, P.: Drag-reduction and resonance problems of a jointed drillstring in the presence of an axial excitation tool. J. Energy Resour. Technol. Trans. ASME **141**(3), 1–8 (2019)
4. Aarsnes, U.J.F., Shor, R.J.: Torsional vibrations with bit off bottom: modeling, characterization and field data validation. J. Petrol. Sci. Eng. **163**, 712–721 (2018)
5. Ritto, T.G., Escalante, M.R., Sampaio, R., Rosales, M.B.: Drill-string horizontal dynamics with uncertainty on the frictional force. J. Sound Vib. **332**(1), 145–153 (2013)
6. Tikhonov, V., Valiullin, K., Nurgaleev, A., Ring, L., Gandikota, R., Chaguine, P., et al.: Dynamic Model for Stiff-String Torque and Drag (see associated supplementary discussion). SPE Drill. Compl. **29**(03), 279–294 (2014)
7. Belyaev, A.K., Eliseev, V.V.: Flexible rod model for the rotation of a drill string in an arbitrary borehole. Acta Mech. **229**(2), 841–848 (2018)
8. Goicoechea, H.E., Buezas, F.S., Rosales, M.B.: A non-linear cosserat rod model for drill-string dynamics in arbitrary borehole geometries with contact and friction. Int. J. Mech. Sci. **157-158**, 98–110 (2019). https://linkinghub.elsevier.com/retrieve/pii/S002074031834030X

Stochastic Aspects in Dynamics of Curved Electromechanic Metastructures

Lucas E. Di Giorgio[1] and Marcelo T. Piovan[1,2(✉)]

[1] Centro de Investigaciones en Mecánica Teórica y Aplicada, Universidad Tecnológica Nacional, Facultad Regional Bahía Blanca, 11 de abril 461, 8000 Bahía Blanca, Argentina
ldigiorgio,mpiovan@frbb.utn.edu.ar
[2] Conicet, Buenos Aires, Argentina
http://frbb.utn.edu.ar

Abstract. In this work we perform uncertainty quantification in the dynamic response of curved electromechanic metastructures made up of piezoelectrics in Bimorph configuration. The study is performed through a new reduced 1D finite element model derived from the theory of linear elasticity and general piezoelectricity, obtained through the Hamilton's Principle and Gauss's Law, who serves as a mean deterministic reference. Parametric uncertainty is taken into account through different constructive and constitutive parameters. Probabilistic model is constructed with the basis of the deterministic model and both are calculated within finite element approaches. Monte Carlo Method is employed to calculate random realizations. A number of scenarios are evaluated in order to identify the parameter sensitivity. Also the reduced model is contrasted with different models for validation.

Keywords: Metastructure · Curved beam · Piezoelectric · Band gap · Stochastic model

1 Introduction

Locally resonant metastructures have been studied for many years in order to generate stop bands for mechanical wave propagation. An interesting way of having the same effect without using considerable mases is replicating the resonators with piezoelectric damping [5]. Piezoelectric metastructures are made up of an array of tuned resonator cells in order to obtain the desired dynamic. Typical approach to the study of these structures is to consider them of infinite length, thus allowing the use of Bloch's theorem to obtain their band structure [1]. However, finite dimensions require the study of dynamics through models that incorporate actual dimensions and border conditions [2,12].

In this work a reduced 1D finite element model of a constant radius curved metastructure is presented, consisting of an arrangement of piezoelectric cells

J. E. S. De Cursi (Ed.): Uncertainties 2020, LNME, pp. 111–120, 2021.
https://doi.org/10.1007/978-3-030-53669-5_9

in bimorph configuration, feedback through a load inductance. Shunt connexion of the internal piezoelectric capacity with the inductor generates an LC resonant circuit, leading to a local electromechanical resonator in each cell [11]. The reduced model will be able to reproduce the dynamics of the curve metastructure in all its dimensions in order to quantify the propagation of uncertainty produced by some constructive parameters towards the dynamics through a Probabilistic Parametric approach PPA [7,9,10]. The stochastic model is created, incorporating uncertainty into the parameters of interest, turning them into random variables whose probability density functions PDF are deduced from the Principle of Maximum Entropy [4]. Finally, the Monte Carlo method is used to perform the statistical analysis, where the results are presented in different graphs.

2 Deteministic Model

2.1 Description

Figure 1 represents the curved metastructure consisting of piezoelectric layers and a bearing substrate. Piezoelectric electrodes are segmented and insulated creating an array of independent bimorph series blocks with same dimensions, each one fed back with its corresponding electrical inductance Lc. The number of blocks is considered variable. The cross section is regular and the reference system origin c is located in the centroid. Theoretical model is based on the following hypotheses:

- The cross section is rigid.
- Small displacements and linear elasticity is considered.
- The shear stress produced by bending and torsion is considered.
- A warping function referred to the centroid is defined.
- The electric field E is in the direction of y.

The field of displacements follows from the hypotheses:

$$\begin{Bmatrix} u_x \\ u_y \\ u_z \end{Bmatrix} = \begin{Bmatrix} u_{xc} - \omega\phi_w \\ u_{yc} \\ u_{zc} \end{Bmatrix} + \begin{bmatrix} 0 & -\phi_z & \phi_y \\ \phi_z & 0 & -\phi_x \\ -\phi_y & \phi_x & 0 \end{bmatrix} \begin{Bmatrix} 0 \\ y \\ z \end{Bmatrix} \tag{1}$$

where

$$\phi_x = \theta_x,\ \phi_y = \theta_y,\ \phi_z = \theta_z - \frac{u_{xc}}{R},\ \phi_w = \theta_w + \frac{\theta_y}{R} \tag{2}$$

being $u_{xc}, u_{yc}, u_{zc}, \theta_y, \theta_z, \theta_w$ the kinematic variables and the warping function defined as follows:

$$\omega = \bar{\omega}\mathcal{F} \text{ with } \mathcal{F} = \frac{R}{R+y} \text{ and } \bar{\omega} = -xy \tag{3}$$

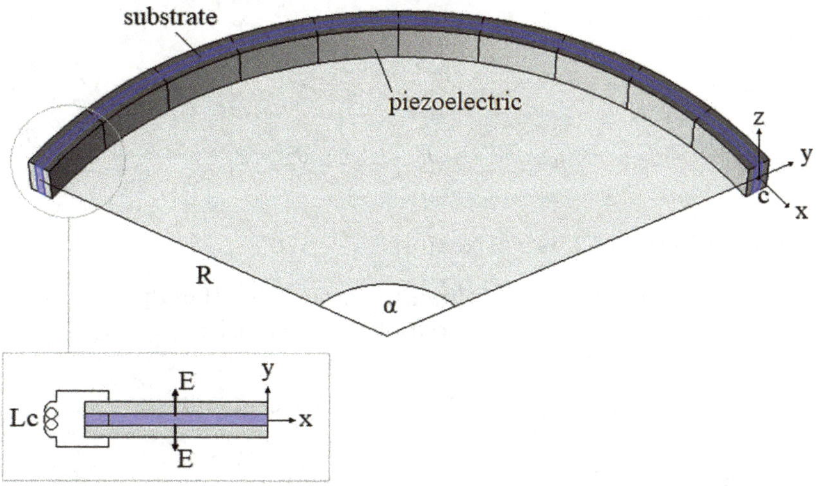

Fig. 1. Curve metastructure.

2.2 Constitutive Equations and Deformation Stress Fields

Considering the crystallographic axes match the cartesian axes, and taking into account that $S_{yy} = S_{zz} = S_{yz} = 0$, the reduced constitutive equations for the piezoelectric are expressed according to Eq. 4 [3].

$$\begin{Bmatrix} T_{xx} \\ T_{xy} \\ T_{xz} \\ D_{yy} \end{Bmatrix} = \begin{bmatrix} c_{11}^E & 0 & 0 & -e_{31} \\ 0 & c_{44}^E & 0 & 0 \\ 0 & 0 & c_{55}^E & 0 \\ e_{31} & 0 & 0 & \epsilon_{33}^S \end{bmatrix} \begin{Bmatrix} S_{xx} \\ S_{xy} \\ S_{xz} \\ E_{yy} \end{Bmatrix} \tag{4}$$

where T_i is the component of the voltage vector $[N/m^2]$, D_i the electric displacement $[Cb/m^2]$, E_i the electric field vector component $[V/m]$, c^E the elasticity constant under constant electric field (N/m^2), e the coupling constant $[Cb/m^2]$, ϵ^S the electrical permittivity under constant stress $[Cb^2/Nm^2]$ and S_i the deformation.

Deformations of the curved structure as a function of the displacement field of the Eq. (1) are expressed as follows [8]:

$$\begin{Bmatrix} S_{xx} \\ S_{xy} \\ S_{xz} \end{Bmatrix} = \frac{1}{1+\frac{y}{R}} \begin{bmatrix} 1 & z & -y & -\omega & 0 & 0 & 0 & 0 \\ 0 & 0 & 0 & 0 & 1 & 0 & \frac{\partial \bar{\omega}}{\partial y} & -(z+\frac{\partial \bar{\omega}}{\partial y}) \\ 0 & 0 & 0 & 0 & 0 & 1 & \frac{\partial \bar{\omega}}{\partial z} & (y-\frac{\partial \bar{\omega}}{\partial z}) \end{bmatrix} \begin{Bmatrix} \varepsilon_{D1} \\ \varepsilon_{D2} \\ \varepsilon_{D3} \\ \varepsilon_{D4} \\ \varepsilon_{D5} \\ \varepsilon_{D6} \\ \varepsilon_{D7} \\ \varepsilon_{D8} \end{Bmatrix} \tag{5}$$

where:

$$\varepsilon_{D1} = u'_{xc} + \frac{u_{yc}}{R} \quad \varepsilon_{D2} = \theta'_y - \frac{\theta_x}{R} \quad \varepsilon_{D3} = \theta'_z - \frac{u'_{xc}}{R} \quad \varepsilon_{D4} = \theta'_w + \frac{\theta'_y}{R}$$

$$\varepsilon_{D5} = u'_{yc} - \theta_z \quad \varepsilon_{D6} = u'_{zc} + \theta_y \quad \varepsilon_{D7} = \theta'_x - \theta_w \quad \varepsilon_{D8} = \theta'_x + \frac{\theta_y}{R}$$

2.3 Finite Elements Discretization

The finite elements formulation is obtained through the Hamilton's Principle for electromechanical systems, whose variational indicator is:

$$V.I. = \int_{t_1}^{t_2} [\delta(T^* + W_e^*) + \delta W_{nc}]dt = 0 \tag{6}$$

where the terms of the Lagrangian $(T^* + W_e^*)$ are:

$$\begin{aligned} T^* &= \frac{1}{2}\int_{\Omega} \rho\{\dot{u}\}^T\{\dot{u}\}d\Omega \\ W_e^* &= \frac{1}{2}\int_{\Omega} (\{E\}^T[\epsilon]\{E\} + 2\{S\}^T[e]\{E\} - \{S\}^T[c]\{S\})d\Omega \end{aligned} \tag{7}$$

and the work of non conservative forces (external applied forces):

$$W_{nc} = \int_{\Omega} \{\delta u\}^T f_i \; d\Omega + \int_S \{\delta u\}^T t_i \; dS = 0 \tag{8}$$

Replacing Eq. 7 and Eq. 8 in Eq. 6 the weak formulation is obtained:

$$\begin{aligned} \int_{\Omega} (\{\delta E\}^T[\epsilon]\{E\} + \{\delta E\}^T[e]\{S\} + \{\delta S\}^T[e]\{E\} - \{\delta S\}^T[c]\{S\} \\ - \rho\{\delta u\}^T\{\ddot{u}\} - \{\delta u\}^T f_i)d\Omega - \int_S \{\delta u\}^T t_i = 0 \end{aligned} \tag{9}$$

where f_i y t_i are the external volumetric and surface forces.

The finite element formulation is obtained through the discretization of Eq. 9 considering the electric field $E_3 = V/h_p$, where h_p is the height of the piezoelectric. Discretization was carried out using 3-node isoparametric elements and quadratic functions [6]. The vector of displacement variables can be expressed:

$$\bar{\mathbf{U}}_e = \left\{\bar{\mathbf{U}}_e^{(1)}, \bar{\mathbf{U}}_e^{(2)}, \bar{\mathbf{U}}_e^{(3)}\right\}, \quad \bar{\mathbf{U}}_e^{(j)} = \{u_{xc_j}, u_{yc_j}, \theta_{z_j}, u_{zc_j}, \theta_{y_j}, \phi_{x_j}, \theta_{x_j}\}, \quad j = 1, 2, 3 \tag{10}$$

The finite element method derives in the following matrix equations:

$$\begin{aligned} \mathbf{K}\bar{\mathbf{W}} + \mathbf{C}_{RD}\dot{\bar{\mathbf{W}}} + \mathbf{M}\ddot{\bar{\mathbf{W}}} - \mathbf{T}V_0 &= \bar{\mathbf{F}} \\ \mathbf{T}^T\dot{\bar{\mathbf{W}}} + C_p\dot{V}_0 + \frac{1}{L_c}\int_0^t V_0 \; dt &= 0 \end{aligned} \tag{11}$$

where $\mathbf{M}$ and $\mathbf{K}$ are the matrices of mass and elasticity, $\mathbf{C}_{RD} = \eta_1\mathbf{M} + \eta_2\mathbf{K}$ is the damping matrix of Raleigh [7] and $\mathbf{T}$ the coupling matrix, C_p is the piezoelectric capacitance of the cell, and L_c is the electric inductance, also connected in every cell.

Steady state equation is obtained clearing V_0 in Eq. 11.b and replacing in Eq. 11.a:

$$\begin{aligned} \widehat{\mathbf{W}}(\omega) &= [\mathbf{K} - \mathbf{M}\omega^2 + j\omega\mathbf{C}_{RD} + (j\omega\mathbf{T}\mathbf{T}^T)(\frac{1}{j\omega L_c} + j\omega C_P)^{-1}]^{-1}\widehat{\mathbf{F}}(\omega) \\ &= [\mathbf{K} - \mathbf{M}\omega^2 + j\omega\mathbf{C}_{RD} + \frac{\omega^2\omega_t^2 L_c\mathbf{T}\mathbf{T}^T}{\omega^2 - \omega_t^2}]^{-1}\widehat{\mathbf{F}}(\omega) \end{aligned} \tag{12}$$

being $\widehat{\mathbf{W}}$ and $\widehat{\mathbf{F}}$ the Fourier transforms of the displacement and force vectors respectively and $\omega_t = 1/\sqrt{L_c C_p}$ is the resonators tuning frequency.

3 Stochastic Model

The stochastic model is constructed from the deterministic model finite elements formulation, considering as random variables the parameters that can vary due to uncertainty in the construction of the structure. These random variables are represented by probability density functions PDF obtained by the principle of Maximum Entropy [4]. Although the model has a large number of uncertain parameters, the radius of curvature R, the coupling constant e_{31}, and the electrical permittivity ϵ_{33} are of interest for the present work.

The random variables V_i are considered bounded, whose limit values are known. It is assumed that their average value coincides with the deterministic value of each parameter in order to check convergence. In addition, since there is no correlation or dependence between them, independent random variables are assumed. From the above, the PDFs of the variables can be expressed:

$$p_{V_i}(v_i) = \mathcal{S}_{[L_{V_i}, U_{V_i}]}(v_i)\frac{1}{2\sqrt{3}\underline{V}_i\delta_{V_i}}, i = 1, 2, 3 \tag{13}$$

where $\mathcal{S}_{[L_{V_i}, U_{V_i}]}(v_i)$ is the support, L_{V_i} and U_{V_i} lower and upper bounds, $\underline{V}_i$ the expected value and δV_i the coefficient of variation.

Through the Monte Carlo method, considering the FEM model of Eq. 12 with the PDFs defined in Eq. 13, the stochastic process can be defined as follows:

$$\widehat{\mathbb{W}}(\omega) = [\mathbb{K} - \mathbb{M}\omega^2 + j\omega\mathbb{C}_{RD} + (j\omega\mathbb{T}\mathbb{T}^T)(\frac{1}{j\omega L_c} + j\omega C_P)^{-1}]^{-1}\widehat{\mathbf{F}}(\omega) \tag{14}$$

The convergence of the method is analyzed in a quadratic mean through the expression:

$$conv\,(N_S) = \sqrt{\frac{1}{N_{MS}}\sum_{j=1}^{N_{MS}}\int_W \left\|\widehat{\mathbb{W}}_j(\omega) - \widehat{\mathbf{W}}(\omega)\right\|^2 d\omega} \tag{15}$$

where N_S is the number of Monte Carlo iterations and W is the frequency band of the analysis.

4 Computational Studies

4.1 Preliminary Validations

In this section, the reduced FEM 1D model is compared with the analytical model proposed by Sugino in [11]. Both the metastructures are straight (i.e $R \to \infty$) and built with PZT-5A piezoelectrics ($C_{11} = 61\,\text{GPa}$, $C_{44} = C_{55} = 21\,\text{GPa}$, $\epsilon_{33} = 13.3e^{-9}\,\text{F/m}$, $e_{31} = -12.3\,\text{C/m}^2$, $\rho = 7750\,\text{Kg/m}^3$) and an aluminum substrate ($E = 70\,\text{GPa}$, $G = 26.32\,\text{GPa}$, $\nu = 0.33$, $\rho = 2700\,\text{Kg/m}^3$). The dimensions of the structure are: length $L = 100\,\text{mm}$, width $b = 10\,\text{mm}$, piezo height $hp = 0.3\,\text{mm}$, substrate height $hs = 0.1\,\text{mm}$. The tuning frequency of the resonators is $\omega_t = 100\omega_1$.

Figure 2 shows the computational results for the propagation or transmissibility, defined as the ratio of displacements at some output location x_L to some input location x_0, of a structure with clamped-free boundary condition (clamped at $x = 0$ and free at $x = L$), exited with base motion at $x = 0$ for different number of resonators S. Clearly the similarity between the responses with the band gap zone can be observed, however it should be noted that Sugino's analytical

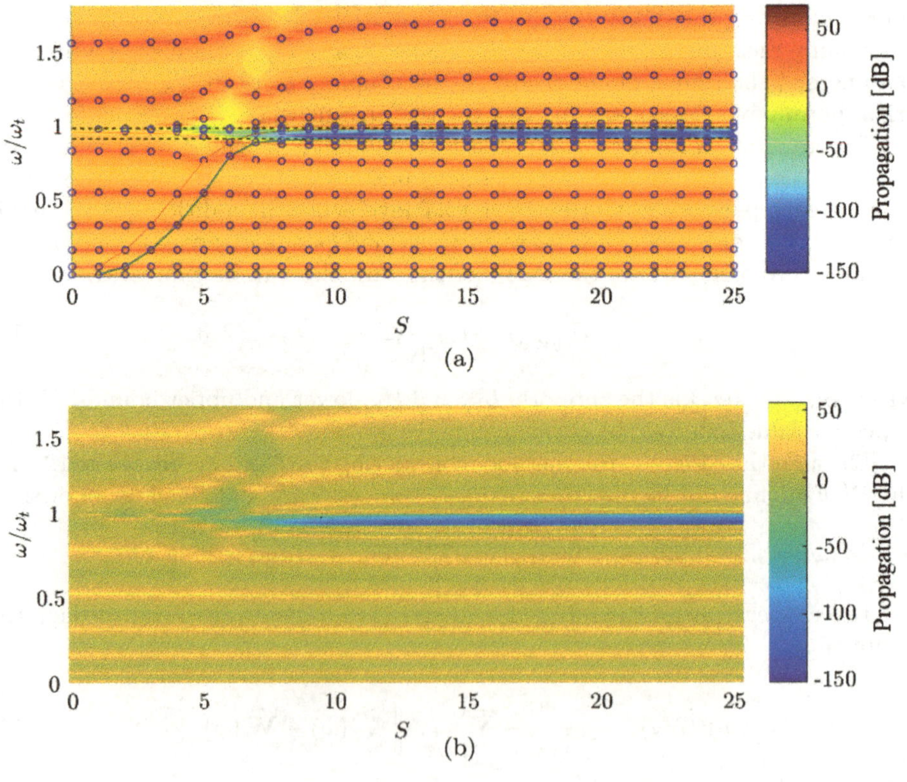

Fig. 2. Propagation and frequencies versus S. (a) Sugino's model, (b) 1D reduced FEM model.

model only contemplates movements in a single plane for exclusively straight structures, while the reduced FEM 1D extends for curves structures and can represent displacements in all 3 directions.

4.2 Uncertainty in the Dynamic Response

In this section the study of uncertainty propagation through the dynamics of the structure is carried out. The chosen parameters are the radius of curvature R, the coupling constant e_{31}, and the electrical permittivity ϵ_{33}. The study is performed for a clamped-free structure, evaluating the propagation frequency response for a base motion BM of $\|U(0)\| = 1\,\text{mm}$ and the frequency response by applying a force of $\|F(L)\| = 1N$ at the free end FRF. The propagation of parametric uncertainty is carried out as follows: first considering the variation of one parameter at a time, keeping the remaining fixed at the expected value (deterministic value), then varying them from 2 in 2, and finally varying the 3 parameters in simultaneous. The structure for this analysis is curved, built with 25 resonators, whose dimensions are: length $L = 100\,\text{mm}$, width $b = 5\,\text{mm}$, piezo height $hp = 1\,\text{mm}$ and substrate height $hs = 0.5\,\text{mm}$. The expected values of the variables are: $\bar{R} = 100\,\text{mm}$, $\bar{e}_{31} = -12.3\,\text{C/m}^2$, $\bar{\epsilon}_{33} = 13.3e^{-9}\,\text{F/m}$. Most of the realizations have a stable convergence after 500 iterations, as shown in Fig. 3.

An example of the confidence interval of the propagation frequency response for the random variables ϵ_{33} and R is presented in Fig. 4. It can be seen that ϵ_{33} greatly influences the band gap zone while R does not practically affect the response. This is because ϵ_{33} controls the electrical capacity of the piezoelectric, responsible of tuning the resonator together with de inductance L_c. The curvature does not seem to affect the response since the structure was excited in the same plane. Examples of FRF type simulation can be seen in Fig. 5, where the response for e_{31} is sensitive in frequencies of the ban gap and R has influence in higher modes than the tuning frequency of the resonators.

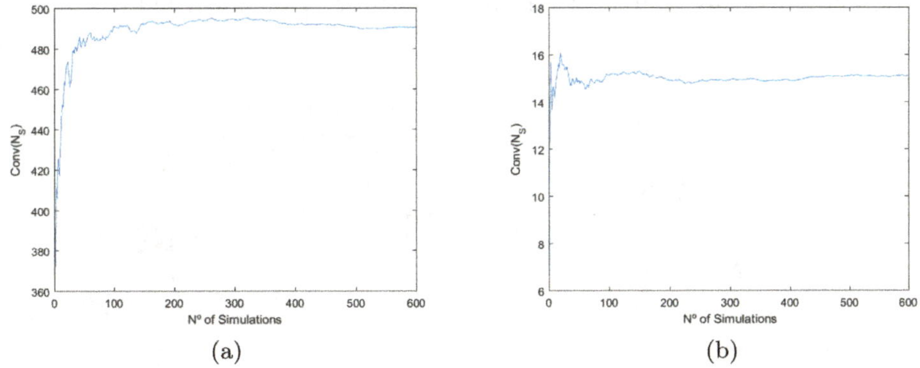

Fig. 3. Convergence. (a) All random variables for $CoV = 0.05$, BM, (b) all random variables for FRF, $CoV = 0.10$.

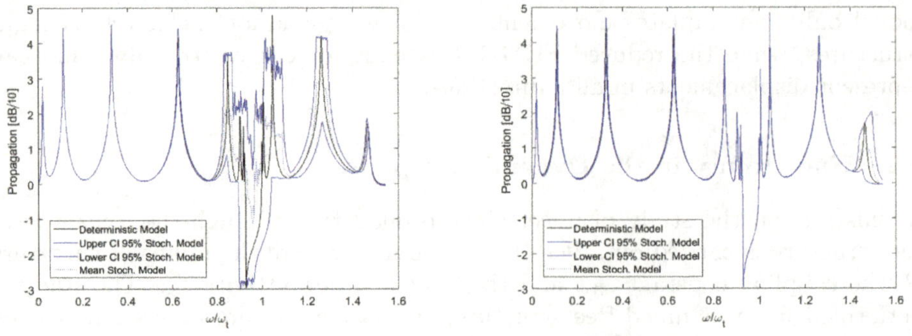

(a) (b)

Fig. 4. Confidence intervals for $CoV = 0.05$, BM. (a) Random ϵ_{33}, (b) random R.

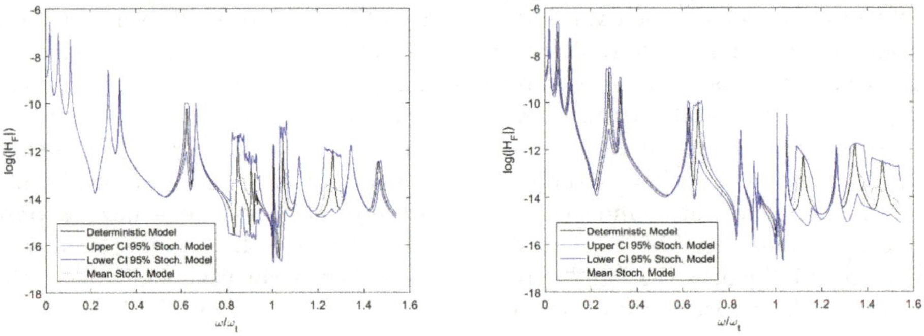

(a) (b)

Fig. 5. Confidence intervals for $CoV = 0.10$, FRF. (a) Random e_{31}, (b) random R.

Figure 6 shows that the frequency response of the propagation is highly sensitive but practically bounded to the band gap for ϵ_{33}, and lower but distributed

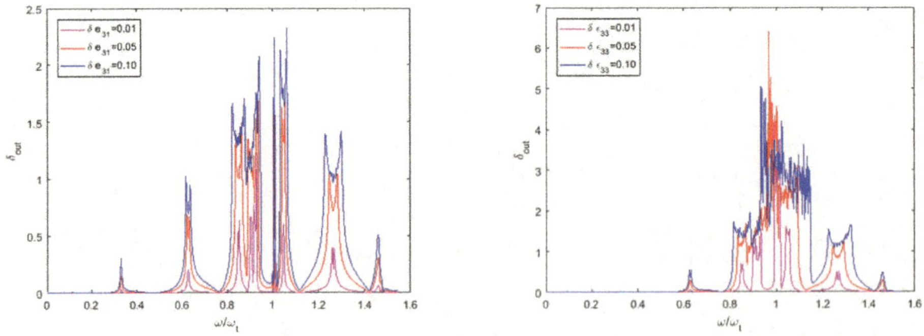

(a) (b)

Fig. 6. Output δ_i for different inputs, BM. (a) Random e_{31}, (b) random ϵ_{33}.

over adjacent the modes to the band gap, for coupling constant e_{31}. Finally, some comparative histograms for BM and FRF are shown in Fig. 7.

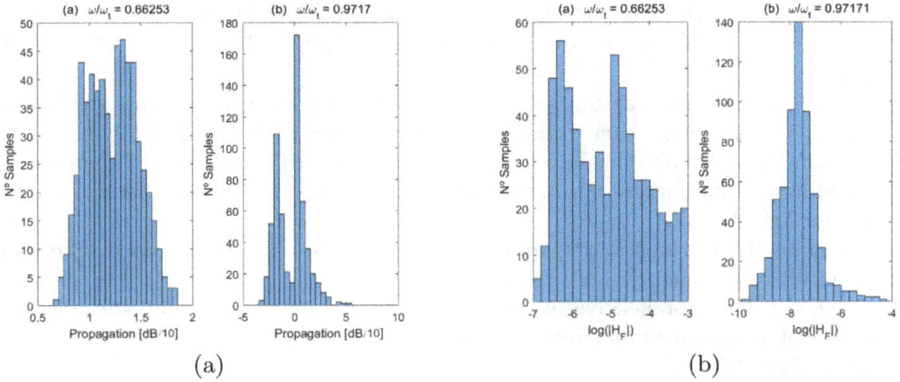

Fig. 7. Histograms at given frequencies. (a) All random variables for $CoV = 0.10$, BM, (b) All random variables for $CoV = 0.10$, FRF.

5 Conclusions

In this work, a reduced 1D FEM model of an electromagnetic metastructure based on Sugino's work [11] was introduced, with the possibility of being extended to curved structures and with the ability to reproduce the dynamics in 3 dimensions, based on Timoshenko's theory. This model allows, due to its low computational cost (approximately 20 s on a desktop computer) to perform analysis with recursive methods, such as the case of uncertainty analysis using the Monte Carlo method.

A computational example of parametric uncertainty was performed for constitutive parameters of the piezoelectric e_{31} and ϵ_{33}, and for the geometric parameter R, showing a high sensitivity of the response exclusively in the band gap area due variations of ϵ_{33}, and a lower but extended sensitivity, beyond the band gap zone, due to e_{31} variations. The curvature radius R belong the resonators plane, so variations of it are practically insensitive for dynamics in BM mode, however it affects the modes outside this plane in FRF, mostly in higher frequencies than ω_t.

References

1. Brillouin, L.: Wave Propagation in Periodic Structures: Electric Filters and Crystal Lattices. Courier Corporation, North Chelmsford (2003)
2. Erturk, A., Inman, D.J.: Piezoelectric Energy Harvesting. Wiley, Hoboken (2011)
3. IEEE: An American National Standard: IEEE Standard on Piezoelectricity Standard. IEEE (1988). https://books.google.com.ar/books?id=YCs5nQEACAAJ

4. Jaynes, E.: Information theory and statistical mechanics I and II. Phys. Rev. **106**, 1620–1630 (1957)
5. Lesieutre, G.A.: Vibration damping and control using shunted piezoelectric materials. Shock Vib. Dig. **30**(3), 187–195 (1998)
6. Piovan, M., Cortinez, V.: Mechanics of thin-walled curved beams made of composite materials, allowing for shear deformability. Thin-Walled Struct. **45**, 759–789 (2007)
7. Piovan, M., Ramirez, J., Sampaio, R.: Dynamics of thin-walled composite beams: analysis of parametric uncertainties. Compos. Struct. **105**, 14–28 (2013)
8. Piovan, M.T., Domini, S., Ramirez, J.M.: In-plane and out-of-plane dynamics and buckling of functionally graded circular curved beams. Compos. Struct. **94**(11), 3194–3206 (2012)
9. Sampaio, R., Cataldo, E.: Comparing two strategies to model uncertainties in structural dynamics. Shock Vib. **17**, 171–186 (2011)
10. Soize, C.: A comprehensive overview of a non-parametric probabilistic approach of model uncetainties for predictive models in structural dynamics. J. Sound Vib. **289**, 623–652 (2005)
11. Sugino, C., Leadenham, S., Ruzzene, M., Erturk, A.: An investigation of electroelastic bandgap formation in locally resonant piezoelectric metastructures. Smart Mater. Struct. **26**(5), 055029 (2017)
12. Surabhi, A.: Finite element beam model for piezoelectric energy harvesting using higher order shear deformation theory. Ph.D. thesis, Graduate School of Clemson University (2014)

Local Interval Fields for Spatial Inhomogeneous Uncertainty Modelling

Robin Callens(✉), Matthias Faes, and David Moens

Department of Mechanical Engineering, KU Leuven,
Campus De Nayer, Jan De Nayerlaan 5, St.-Katelijne-Waver, Leuven, Belgium
robin.callens@kuleuven.be

Abstract. In an engineering context, design optimization is usually performed virtually using numerical models to approximate the underlying partial differential equations. However, valid criticism exists concerning such an approach, as more often than not, only partial or uninformative data are available to estimate the corresponding model parameters. As a result hereof, the results that are obtained by such numerical approximation can diverge significantly from the real structural behaviour of the design. Under such scarce data, especially interval analysis has been proven to provide robust bounds on the structure's performance, often at a small-to-moderate cost. Furthermore, to model spatial dependence throughout the model domain, interval fields were recently introduced by the authors as an interval counterpart to the established random fields framework. However, currently available interval field methods cannot model local inhomogeneous uncertainty. This paper presents a local interval field approach to model the local inhomogeneous uncertainty under scarce data. The method is based on the use of explicit interval fields [1] and the recently used inverse distance weighting function [2]. This paper presents the approach for one dimension of spatial uncertainty. Nonetheless, the approach can be extended to an n-dimensional context. In the first part of the paper, a detailed theoretical background of interval fields is covered, and then the local interval fields approach is introduced. Furthermore, an academic case study is performed to compare the local interval field approach with inverse distance weighting.

Keywords: Interval analysis · Non-probabilistic analysis · Local interval fields

1 Introduction

Interval analysis is becoming popular when there is only limited or incomplete data of the true parameter data. In comparison to probabilistic techniques, which require distributions of the uncertain parameter data, intervals bounds the uncertainty on the actual parameter value by an upper and a lower bound. Given these bounds, interval computation methods quantify the best and worst-case behaviour of the structure. By definition, intervals are independent, and hence,

J. E. S. De Cursi (Ed.): Uncertainties 2020, LNME, pp. 121–135, 2021.
https://doi.org/10.1007/978-3-030-53669-5_10

the joint description of several interval-valued parameters is given by a hyper-rectangle [3]. For interval Finite Element Analysis, the independent intervals are defined on locations in the FE-model (e.g. element centres, element nodes, Gauss points) to describe spatial uncertainty [4]. The independent intervals neglect all correlation present in nature, this results in over-conservative results and unrealistic interval fields. In order to obtain less conservative results Moens et al. [1] proposed a method to represent spatial uncertainty in Finite Element analyses: interval fields. This by limiting the number of intervals and describing the dependency of each interval to other location of the FE-model (e.g. Element centres, Element nodes, gaussian points) with basis functions [1]. The big advantage of this method is that the dimension of the hyper-rectangle is reduced from the number of all locations in the FE-model to a limited number of interval scalars. This interval fields where recently used for several cases: modelling of dynamic phenomena [5–7] and additive manufactured plastic parts [8]. Another approach to model interval fields is via an affine arithmetic representation of the interval uncertainty in combination with the Karhunen-Loeve expansion by Sofi et al. [9]. Also, other authors [10–14] have proposed other formulations in the context of spatial interval fields. Following the explicit interval fields [1], Faes et al. [2] introduced the use of inverse distance weighting function. Where the interval scalars are defined on control points, locations in de model domain. The spatial dependency of the control points to all locations a control point.

When considering local inhomogeneous uncertainty all those interval field techniques miss the possibility to model realistic local uncertainty. Local inhomogeneous uncertainty is presented in parts which are for instance deep-drawn or casted. When deep-drawing, there is e.g. local uncertainty in the thickness of the part introduced by local micro cracks and voids. For the cased part, local uncertainty is present inside thicker sections of the part as micro cracks and voids are more likely to be present in those areas due to cooling effects.

In this paper, the description of explicit interval fields [1] is used to model local inhomogeneous spatial uncertainty. As in [2], the interval scalars are defined on locations in the model domain (control points) and the spatial dependency is described with basis functions. The modelling of local inhomogeneous uncertainty is achieved by defining the basis function such that the dependency is limited to a zone around there control point. To build these basis functions the technique of inverse distance weighting from Shepard [15] is modified and combined with another weighting function, which is often used in the domain of Meshless Methods [16].

The paper is structured as follows: Sect. 2 presents interval field Finite Element Analysis. Then, in Sect. 3 this technique is extended to model intervals fields with basis functions and a limited set of interval scalars: explicit interval fields. Next, in Sect. 4 the method of inverse distance weighting (IDW) is described. Section 5 proposes a new local interval field method to model local uncertainty. An academic case study is used to illustrate the difference between the basis functions of IDW [15] and the local interval fields in Sect. 6. At the end of this paper, the conclusions are summarised in Sect. 7.

2 Interval Finite Element Analysis

Today Finite Element Analysis is a popular technique used to approximate the solution of a partial differential equation (PDE). Finite Element Analysis uses a numerical model $\mathcal{M}(\boldsymbol{x})$, parameterized by a vector $\boldsymbol{x}(\boldsymbol{r}) \in \mathcal{X} \subset \mathbb{R}^{d_i}$ with $\mathcal{X}$ the admissible set of input parameters and $d_i \in \mathbb{N}$ the number of input parameters. For instance $\boldsymbol{x}(\boldsymbol{r})$ contains material parameters as a functions of the spatial coordinate $\boldsymbol{r} \in \Omega \subset \mathbb{R}^{d_\Omega}$ with $d_\Omega \in \mathbb{N}, d_\Omega \leq 4$ the number of dimensions (max. 3 cartesian dimensions and 1 time dimension).

Using Finite Element Analysis the model domain Ω is discretised in N_e Finite Elements $\Omega_e \subset \Omega$. As a result, the numerical model is built from N_e Finite Elements yielding in d_d degrees of freedom (DOF) and the PDE is approximated by the solution of a set of algebraic equations.

The model $\mathcal{M}(\boldsymbol{r})$ provides a vector of model responses $\boldsymbol{y}(\boldsymbol{r}) \in \mathcal{Y} \subset \mathbb{R}^{d_o}$ with $\mathcal{Y}$ the admissible set of output parameters and $d_o \in \mathbb{N}$ the number of output parameters. This is defined as:

$$M(\boldsymbol{x}) : y_i(\boldsymbol{r}) = m_i(\boldsymbol{x}(\boldsymbol{r})) \tag{1}$$

with $m_i : \mathbb{R}^{d_i} \mapsto \mathbb{R}$ functions of the Finite Element Analysis and $i = 1, 2, \ldots, d_o$. This is only valid when the output $\boldsymbol{y}$ is generated on element or nodal level $\boldsymbol{r}$ (e.g. displacement, strains), not when the output is defined on a global level (e.g. eigenfrequencies).

Spatial uncertainty is introduced in Finite Element Analysis with random field [17] or eider interval fields [18]. Interval fields have the advantage they require only limited data of the uncertain parameter data, where random fields require that the uncertainty of the parameter data is fully known.

An interval scalar $x^I \in \mathbb{IR}$, $\mathbb{IR}$ is the domain of closed real-valued intervals, is defined as:

$$x^I = [x_{min}\ x_{max}] = [\underline{x}\ \overline{x}] = \{x \in \mathbb{R} | \underline{x} \leq x \leq \overline{x}\}, \tag{2}$$

with x_{min} and x_{max} bounds of the uncertainty. The midpoint of the interval is defined as:

$$x_\mu = \frac{\underline{x} + \overline{x}}{2}. \tag{3}$$

An interval vector $\boldsymbol{x}^I \in \mathbb{IR}^{d_i}$ contains d_i interval scalars which are from definition independent from each other. In general, the interval Finite Element Analysis [18] has an uncertain interval vector $\boldsymbol{x}^I \in \mathbb{IR}^{d_i}$ as input:

$$\boldsymbol{x}^I = \begin{Bmatrix} x_1^I \\ x_2^I \\ \vdots \\ x_{d_i}^I \end{Bmatrix} = \left\{\boldsymbol{x} \in \mathbb{R}^{d_i} | x_i \in x_i^I\right\}. \tag{4}$$

The solution set $\boldsymbol{y}^S \in \mathbb{R}^{d_o}$ of the interval Finite Element Analysis is then formulated as:

$$\boldsymbol{y}^S = \left\{\boldsymbol{y} | \left(\boldsymbol{x} \in \boldsymbol{x}^I\right) (\boldsymbol{y} = \boldsymbol{m}(\boldsymbol{x}))\right\}, \tag{5}$$

with $\boldsymbol{m}(\boldsymbol{x})$ the deterministic function of the Finite Element Analysis and $y \in \mathbb{R}^{d_o}$. The solution set $\boldsymbol{y}^S$ is approximated by the interval vector $\boldsymbol{y}^I \in \mathbb{IR}^{d_o}$:

$$\boldsymbol{y}^I = \begin{Bmatrix} y_1^I \\ y_2^I \\ \vdots \\ y_{d_o}^I \end{Bmatrix}. \tag{6}$$

The components $y_i^I = \left[\underline{y}_i\ \overline{y}_i\right]$ of $\boldsymbol{y}^I$ are found by optimisation:

$$\underline{y}_i = \min_{\boldsymbol{x} \in \boldsymbol{x}^I} m_i(\boldsymbol{x}), \tag{7}$$

$$\overline{y}_i = \max_{\boldsymbol{x} \in \boldsymbol{x}^I} m_i(\boldsymbol{x}), \tag{8}$$

with this optimisation the interval of each component is found independently, such that the solution set $\boldsymbol{y}^S$ is approximated with a hyper-rectangle.

3 Explicit Interval Fields for Spatial Uncertainty

On the input side of an interval Finite Element Analysis, the description of the spatial variation of an input parameter is possible. As on the input side, interval fields are used to model the uncertainty of (e.g. material properties, force distribution). When considering spatial uncertainty in interval Finite Element Analysis with on all location (e.g. element centres, element nodes, Gauss points) an independent interval. The resulting field will be too conservative, as in nature the spatial dependency of an input parameter is present. Moens et al. [1] introduced the explicit interval fields as an interval field which considers spatial dependency on Ω.

With explicit interval fields, there are independent intervals scalars $\boldsymbol{\alpha}^I \in \mathbb{IR}^{n_b}$ with $n_1, n_2, \ldots, n_b \in \mathbb{N}$, such that the input space is reduced from $\mathbb{IR}^{d_i}$ to $\mathbb{IR}^{n_b}, d_i \geq n_b$. From these interval scalars the dependency inside Ω is modelled with basis functions $\boldsymbol{\psi}_i(\boldsymbol{r}) : \Omega \mapsto \mathbb{R}^{k_{FE}}$ with $i = 1, 2, \ldots, n_b$ and k_{FE} the number of locations in Ω. An explicit interval field $\boldsymbol{x}^I(\boldsymbol{r}) : \Omega \times \mathbb{IR}^{n_b} \mapsto \mathbb{IR}^{k_{FE}}$ is as such build as a series expansion of n_b basis functions, multiplied with interval scalars:

$$\boldsymbol{x}^I(\boldsymbol{r}_k) = \sum_{i=1}^{n_b} \boldsymbol{\psi}_i(\boldsymbol{r}_k) \cdot \alpha_i^I \quad k = 1, 2, \ldots, k_{FE}, \tag{9}$$

with $\boldsymbol{\alpha}^I \in \mathbb{IR}^{n_b}$ the interval scalars with $n_1, n_2, \ldots, n_b \in \mathbb{N}$ and k_{FE} the number of elements in $\boldsymbol{\psi}_i$ which is the same as the number of location in the model Ω.

To solve the interval field the number of required deterministic model evaluations is important, especially with industrial-sized FE models (up to a millions degree of freedom). To achieve this, the number of interval scalars d_i must be limited as the number of deterministic model evaluations scale with 2^{d_i} when considering linear monotonic interval analysis. In this perspective, the explicit interval field formulation has the advantage that the input space is reduced from $\mathbb{IR}^{d_i}$ to $\mathbb{IR}^{n_b}, d_i \geq n_b$.

4 Interval Fields with Inverse Distance Weighting

Section 3 introduced the description of explicit interval fields, where the interval field is built with interval scalars and basis functions. As the basis function determines the spatial dependency of the field, it is of great importance to select a basis function that represents the physical nature of the field. Recently inverse distance weighting (IDW) is used by Faes et al. [19] to generate basis functions for explicit interval fields. With IDW the interval scalars are defined on specific locations $\boldsymbol{r}_k$ in the domain Ω and from those locations, the IDW basis functions describe the dependency. The inverse distance weighting basis function $\psi_i(\boldsymbol{r})$ is build from weight functions ${}^{IDW}\boldsymbol{w}_i(\boldsymbol{r})$:

$$\psi_i(\boldsymbol{r}_k) = \frac{{}^{IDW}\boldsymbol{w}_i(\boldsymbol{r}_k)}{\sum_{j=1}^{n_b} {}^{IDW}w_j(\boldsymbol{r}_k)} \quad k = 1, 2, \ldots, k_{FE}, \tag{10}$$

with $\boldsymbol{r}_k \in \Omega$ and k_{FE} the number of elements in ${}^{IDW}\boldsymbol{w}_i$, which is the same as the number of location in the FE-model Ω.

The weight functions ${}^{IDW}\boldsymbol{w}_i(\boldsymbol{r})$ are calculated as:

$${}^{IDW}\boldsymbol{w}_i(\boldsymbol{r}) = \frac{1}{[\boldsymbol{d}(\boldsymbol{r}_k, \boldsymbol{r}_{n_i})]^p}, \tag{11}$$

where $\boldsymbol{d}(\boldsymbol{r}_k, \boldsymbol{r}_{n_i})$ is the euclidean distance between the locations $\boldsymbol{r}_k \in \Omega$ and locations of the control points $\boldsymbol{r}_{n_i} \in \Omega$. The power p defines the gradient and the continuity of the weight function. As a result of the distance calculations, the computational cost of generating the weight functions scales with $k_{FE} \times n_b$, the number of locations in the FE model k_{FE} and the control points n_b.

From Eq. 10 and 11 the inverse distance basis functions has a dependency on all locations of the model (global dependency). Except that from Eq. 11 the basis functions are zero on the location of other control points. The global dependency is there from the weighting functions as they have the following properties:

1. $\dfrac{\partial \, {}^{IDW}\boldsymbol{w}_i}{\partial \boldsymbol{d}(\boldsymbol{r}_k, \boldsymbol{r}_{n_i})} \leq 0$ when $\boldsymbol{d}(\boldsymbol{r}_k, \boldsymbol{r}_{n_i}) \to \infty : \; {}^{IDW}\boldsymbol{w}_i \to 0,$
2. $\begin{cases} {}^{IDW}\boldsymbol{w}_i(\boldsymbol{r}) \neq 0 & \text{if} \quad \boldsymbol{d}(\boldsymbol{r}_k, \boldsymbol{r}_{n_i}) \neq \infty, \\ {}^{IDW}\boldsymbol{w}_i(\boldsymbol{r}) = 0 & \text{if} \quad \boldsymbol{d}(\boldsymbol{r}_k, \boldsymbol{r}_{n_i}) = \infty, \end{cases}$

those properties are only valid when the size of the Finite Element $h_e \to 0$.

The first property comes from the continuity of the weight function: "the weight function and its partial derivative are only decreasing and when the distance is infinity the weight function is zero", this is summarised in property two. As a result, this weighting technique with global dependency is not capable to describe local dependency. An idea could be to truncate the basis functions on a distance from the control points. The disadvantage of truncating is that it introduces discontinuities in the field. As a result, the field is no longer continuous. Another idea is to use other weighting functions that are zero on a finite distance of a control point.

5 Interval Field with Local Dependency

An explicit interval field which has local dependency if it is built from weighting functions $\boldsymbol{w}_i$ that satisfy the following properties:

1. $\dfrac{\partial \boldsymbol{w}_i}{\partial \boldsymbol{d}(\boldsymbol{r}_k, \boldsymbol{r}_{n_i})} \leq 0$ when $\boldsymbol{d}(\boldsymbol{r}_k, \boldsymbol{r}_{n_i}) \rightarrow \boldsymbol{R}_i : \boldsymbol{w}_i(\boldsymbol{r}) \rightarrow 0$,
2. $\begin{cases} \boldsymbol{w}_i(\boldsymbol{r}) \neq 0 & \text{if} \quad \boldsymbol{d}(\boldsymbol{r}_k, \boldsymbol{r}_{n_i}) < \boldsymbol{R}_i, \\ \boldsymbol{w}_i(\boldsymbol{r}) = 0 & \text{if} \quad \boldsymbol{d}(\boldsymbol{r}_k, \boldsymbol{r}_{n_i}) \geq \boldsymbol{R}_i, \end{cases}$

with $\boldsymbol{R}_i$ the width of the support zone $\Omega_i \subset \Omega$ with $i = 1, \ldots, d_\Omega$ and $d_\Omega \leq 4$ around one control point n_i. The first property ensures that the weighting function is monotonic decreasing to zero on the edge of the support zone Ω_i. Property two and three introduces the local description of the weight function, the weight function is zero from a distance $\boldsymbol{R}_i$ around the control point n_i.

The local character of the weight function introduces the following computational advantage in comparison to IDW. The support zone $\Omega_i \in \Omega$ where $\boldsymbol{d}(\boldsymbol{r}_k, \boldsymbol{r}_{n_i}) < \boldsymbol{R}_i \rightarrow \boldsymbol{w}_i(\boldsymbol{r}) \neq 0$ is compact $\mathbb{K}$ such that it is closed and bounded. This means that the weight function must only be calculated inside Ω_i and not on the full domain Ω, yielding a drastic gain in computational efficiency as k_{FE} is reduced.

Further in this paper, the support zone Ω_i of one control points n_i is defined as $\mathcal{K}$. Weight functions are then $\boldsymbol{w}^{\mathcal{K}}(\boldsymbol{r}) : \mathcal{K} \mapsto \mathbb{R}^{k_{FE}}$ and basis functions are $\boldsymbol{\psi}^{\mathcal{K}}(\boldsymbol{r}) : \mathcal{K} \mapsto \mathbb{R}^{k_{FE}}$. Subsection 5.1 describes the construction of basis functions for 1D ($d_\Omega = 1$).

5.1 Local Basis Functions for 1D

To build local basis functions $\boldsymbol{\psi}^{\mathcal{K}}(\boldsymbol{r}) : \mathcal{K} \mapsto \mathbb{R}^{k_{FE}}$ from weight functions Eq. 10 is changed. In Eq. 10, the weight functions are normalised to the weight functions of other control points. As there is only one control point and weight function inside the domain $\mathcal{K}$, the weight function $\boldsymbol{w}^{\mathcal{K}}(\boldsymbol{r})$ is normalised to itself and the basis function $\boldsymbol{\psi}^{\mathcal{K}}(\boldsymbol{r})$ will be a constant function equal to one inside $\mathcal{K}$. This will result in a discontinuous basis function: one inside $\mathcal{K}$ and zero outside $\mathcal{K}$. A continuous basis function is achieved by adding two virtual nodes $n^{\mathcal{K}}$ on the edges of each spatial dimension of the domain $\mathcal{K}$, for a one-dimensional problem the nodes inside $\mathcal{K}$ are located on:

1. the edge of $\mathcal{K}$: $\boldsymbol{r}^{\mathcal{K}}_{n=1} = \boldsymbol{r}_{n_i} - R$,
2. the midpoint of $\mathcal{K}$: $\boldsymbol{r}^{\mathcal{K}}_{n=2} = \boldsymbol{r}_{n_i}$,
3. the edge of $\mathcal{K}$: $\boldsymbol{r}^{\mathcal{K}}_{n=3} = \boldsymbol{r}_{n_i} + R$,

with the first and last virtual node on the edges of $\mathcal{K}$ and $\boldsymbol{r}_{n_i} \in \mathcal{K}$ the location of the control point n_i. On al those virtual nodes weighting functions are defined $\boldsymbol{w}^{\mathcal{K}}_i(\boldsymbol{r})$ with $i = 1, 2, 3$ for 1D. This is visualised in Fig. 1, which illustrates the building of one basis function inside $\mathcal{K}$ from the weight functions (quartic splines

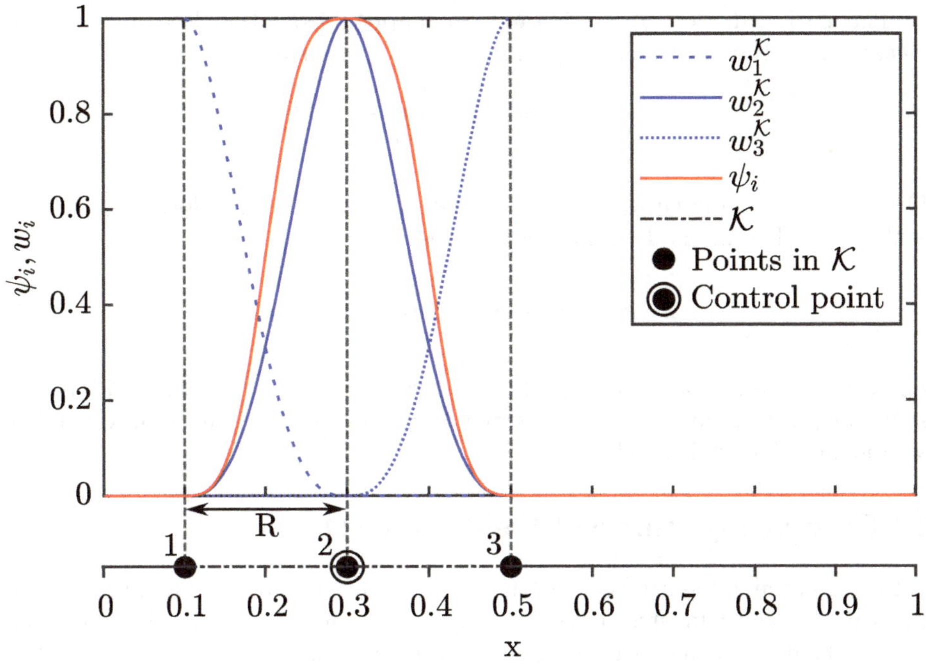

Fig. 1. For a 1D domain, the location of the virtual control points (small black dots) and control point (big black dot). The corresponding weight functions (quartic spline) and basis function are visualised in blue and red. R describes the width of the domain $\mathcal{K}$ around the control point.

see Sect. 6.2). The result is a basis function which is continuous in $\mathcal{K}$ and has a local spatial dependency.

Using those virtual nodes and control point in $\mathcal{K}$ Eq. 10 is changed to:

$$\psi^{\mathcal{K}}(\boldsymbol{r}^{\mathcal{K}}) = \boldsymbol{a}^{\mathcal{K}} \frac{\boldsymbol{w}_i^{\mathcal{K}}(\boldsymbol{r}_k^{\mathcal{K}})}{\sum_{j=1}^{3} w_j^{\mathcal{K}}(\boldsymbol{r}_k^{\mathcal{K}})} \quad k = 1, 2, \ldots, k_{FE}, \tag{12}$$

with $\boldsymbol{r}_k^{\mathcal{K}} \in \mathcal{K} \subset \Omega$, k_{FE} the number of elements in $\boldsymbol{w}_i^{\mathcal{K}}$ (=the number of locations in $\mathcal{K} \subset \Omega$) and $\boldsymbol{a}^{\mathcal{K}}$ defined as:

$$\boldsymbol{a}^{\mathcal{K}} = \begin{cases} 1 \text{ if } \boldsymbol{r}_n^{\mathcal{K}} = \boldsymbol{r}_{n_i} \\ 0 \text{ if } \boldsymbol{r}_n^{\mathcal{K}} \neq \boldsymbol{r}_{n_i}, \end{cases} \tag{13}$$

with $\boldsymbol{r}_n^{\mathcal{K}}$ the locations of the nodes in $\mathcal{K}$ and $\boldsymbol{r}_{n_i}$ the location of the control point n_i in Ω. The definition of $\boldsymbol{a}^{\mathcal{K}}$ is such that the basis function of the control point n_i is retained from all the basis functions of the nodes inside $\mathcal{K}$.

The basis function $\psi^{\mathcal{K}}(\boldsymbol{r}^{\mathcal{K}})$ are then mapped from the domain $\mathcal{K}$ to the domain Ω by the locations of e.g. nodes, element centres, Gauss points:

$$\psi_i(\boldsymbol{r}) = \begin{cases} \psi^{\mathcal{K}}(\boldsymbol{r}^{\mathcal{K}}) & \text{if } \mathcal{K}_e = \Omega_e \\ 0 & \text{if } \mathcal{K}_e \neq \Omega_e. \end{cases} \tag{14}$$

The basis function $\psi_i(\boldsymbol{r}) \in \Omega$ is zero outside $\mathcal{K}$ such that strict local dependency is obtained. The interval field is then represented as:

$$\boldsymbol{x}^{\boldsymbol{I}}(\boldsymbol{r}) = \boldsymbol{x}_{\mu} + \sum_{i=1}^{n_b} \psi_i(\boldsymbol{r}) \cdot (\boldsymbol{\alpha}_i^I - \boldsymbol{x}_{\mu}), \tag{15}$$

where $\boldsymbol{x}_{\mu}$ is the midpoint of the field. The previous methodology is valid as long as the weight function satisfies the three properties. As a result, an interval field with local influence is reached.

6 Case Study: Interval Field on a 1D Bar

In this case, the difference between the local explicit interval field and the explicit interval field with inverse distance weighting is shown. For the local explicit interval fields, the weighting function is quartic splines as it satisfies the two necessary properties for the definition of a local explicit interval field.

6.1 Problem Description

This case study concerns an interval field that is calculated for a 1D Finite Element linear elastic bar, the bar is $L = 1000\,\text{mm}$ long and uniformly discretized with an element size of $h = 2\,\text{mm}$, resulting in 500 finite elements. The boundary condition is fixed displacement for $x = 0 : u = 0\,\text{mm}$ and a force when $x = 100$ mm $F = 1000\,\text{N}$. The bar has a uniform distributed area of $A = 10\,\text{mm}^2$. Figure 2 visualizes the bar with its boundary conditions.

The uncertainty is defined locally on the Youngs Modulus in three locations (control points). Table 1 gives: the coordinates of those control points, the interval values $\boldsymbol{E}_i^I$ and the width R of the domain $\mathcal{K}$ around each control point.

6.2 Local Explicit Interval Fields with Quartic Spline

For the local explicit interval field formulation, the selection of the weighting function is limited by the two properties in Sect. 5. The used weight functions are quartic splines as they satisfy the two properties. As a result, the local explicit interval field formulation can be used to model the local uncertainty. The quartic spline weight function ${}^{Q}\boldsymbol{w}_i^{\mathcal{K}} \in \mathcal{K}$ from [16] is based on the euclidean distance between the location $\boldsymbol{r}_n^{\mathcal{K}}$ of the nodes $n^{\mathcal{K}}$ and the locations $\boldsymbol{r}_k^{\mathcal{K}}$. This distance is then normalised with the support width R:

$$\bar{\boldsymbol{d}}(\boldsymbol{r}_n^{\mathcal{K}}, \boldsymbol{r}_k^{\mathcal{K}}) = \frac{\boldsymbol{d}(\boldsymbol{r}_n^{\mathcal{K}}, \boldsymbol{r}_k^{\mathcal{K}})}{R} \quad k = 1, 2, \ldots, k_{FE}. \tag{16}$$

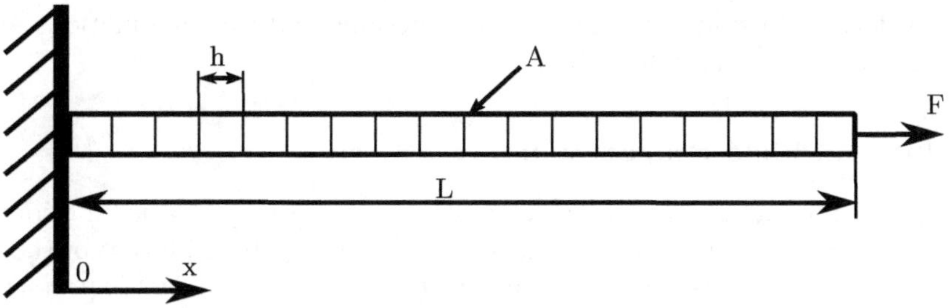

Fig. 2. 1D case: the linear elastic bar of length $L = 1000\,\text{mm}$ is discretized in 500 elements of size $h = 2\,\text{mm}$. Boundary conditions are: fixed displacement for $x = 0$: $u_{x=0} = 0\,\text{mm}$ and an force at $x = 100\,\text{mm}$ of $F = 1000\,\text{N}$. The bar has a uniform distributed area of $A = 10\,\text{mm}^2$

Table 1. Uncertainty in 3 control points with location, intervals $\boldsymbol{E}_i^I$ and width of the local domain $\mathcal{K}$ around each control point i

Control point	Location (mm)	Intervals $\boldsymbol{E}_i^I$ (Gpa)	Width R (mm)
1	200	$[50, 90]$	150
2	500	$[50, 90]$	50
3	800	$[50, 90]$	40

As a result, at a distance R of the nodes $n^{\mathcal{K}}$ the normalised distance equals zero.

Weight functions are then calculated with:

$$
{}^{Q}\boldsymbol{w}_i^{\mathcal{K}}(\bar{\boldsymbol{d}}) = \begin{cases} 1 - 6\bar{\boldsymbol{d}}(\boldsymbol{r}_n^{\mathcal{K}}, \boldsymbol{r}_k^{\mathcal{K}})^2 + 8\bar{\boldsymbol{d}}(\boldsymbol{r}_n^{\mathcal{K}}, \boldsymbol{r}_k^{\mathcal{K}})^3 - 3\bar{\boldsymbol{d}}(\boldsymbol{r}_n^{\mathcal{K}}, \boldsymbol{r}_k^{\mathcal{K}})^4 & \bar{\boldsymbol{d}}(\boldsymbol{r}_n^{\mathcal{K}}, \boldsymbol{r}_k^{\mathcal{K}}) \leq 1 \\ 0 & \bar{\boldsymbol{d}}(\boldsymbol{r}_n^{\mathcal{K}}, \boldsymbol{r}_k^{\mathcal{K}}) > 1. \end{cases} \tag{17}
$$

From the definition of the weight function it is clear that the quartic spline weighting function satisfies the two requirements.

With the values from Table 1 and Eqs. 16 and 17 the local weight functions $\boldsymbol{w}_i^{\mathcal{K}} \in \mathcal{K}$ (quartic splines) for each control point are quantified. From these weight functions the basis functions $\psi^{\mathcal{K}} \in \mathcal{K}$ are then calculated with Eq. 12 and then mapped from the local domain $\mathcal{K}$ to the model domain Ω with Eq. 14. The basis functions and weight functions on the model domain Ω are visualised in Fig. 3.

6.3 Explicit Interval Field with Inverse Distance Weighting

The formulation of an explicit interval field with inverse distance weighting is given in Sect. 4. From Eq. 10 the inverse distance weighting functions ${}^{IDW}\boldsymbol{w}_i \in \Omega$ is calculated and the basis functions $\psi_i \in \Omega$ with Eq. 11. In addition to the values in Table 1 the value of the power factor p of IDW is chosen as $p = 2$. The width of the domain $\mathcal{K}$ is not used with IDW as that the inverse distance weighting function does not use it. The IDW basis functions and weight functions on the model domain Ω are visualised in Fig. 3.

6.4 Interval Field Propagation

The explicit interval field realizations are calculated form the basis functions (see Subsect. 6.2 and 6.3) and Eq. 15 with $\boldsymbol{\alpha}_i^I = \boldsymbol{E}_i^I$ and $\boldsymbol{x}_\mu$ the midpoint of the field. To quantify the interval of the displacement $\boldsymbol{y}^I$ the interval field $\boldsymbol{x}^I$ is propagated. The components of the output interval are the highest and lowest total displacement of the bar. The best-case can either be the low displacement $\underline{y}$ or the high displacement $\overline{y}$, only the analyst knows this. Here the best-case is the low displacement and the worst-case is the high displacement. The bar is linear-elastic and so the problem is monotonic and especially strict negative monotonic, as a high Youngs modulus will result in low displacement. As a result, the interval of the displacement $\boldsymbol{y}^I$ is quantified by solving the FE-model with two realisations of the interval field. Each realisation has different interval field scalars:

- Interval field scalars for the best-case $\underline{y}$:
  1. $\boldsymbol{E}_1^I : \overline{E}_1 = 90$ Gpa,
  2. $\boldsymbol{E}_2^I : \overline{E}_2 = 90$ Gpa,
  3. $\boldsymbol{E}_3^I : \overline{E}_3 = 90$ Gpa.
- Interval field scalars for the worst-case $\overline{y}$:
  1. $\boldsymbol{E}_1^I : \underline{E}_1 = 50$ Gpa,
  2. $\boldsymbol{E}_2^I : \underline{E}_2 = 50$ Gpa,
  3. $\boldsymbol{E}_3^I : \underline{E}_3 = 50$ Gpa.

As a result of this case the interval on the total displacement is for Local explicit interval field with quartic splines:

$$ {}^{Q}\boldsymbol{y}^I = [\underline{y}, \overline{y}] = [1.35, 1.56]\,, \tag{18} $$

for IDW the interval is:

$$ {}^{IDW}\boldsymbol{y}^I = [\underline{y}, \overline{y}] = [1.11, 2]\,. \tag{19} $$

Figures 4 and 5 visualizes the difference in realisations between the local explicit interval field and inverse distance weighting. Additional to the displacements the strains are calculated as output for both methods, this is visualised in Fig. 6.

6.5 Difference in Computational Cost

The computational cost is defined as the number of distance measures required to compute the three basis functions. For IDW the three weighting functions are defined on the full model domain Ω, such that the required distance measures are:

$$ \frac{3L}{h} = \frac{3 \cdot 1000}{2} = 1500 \tag{20} $$

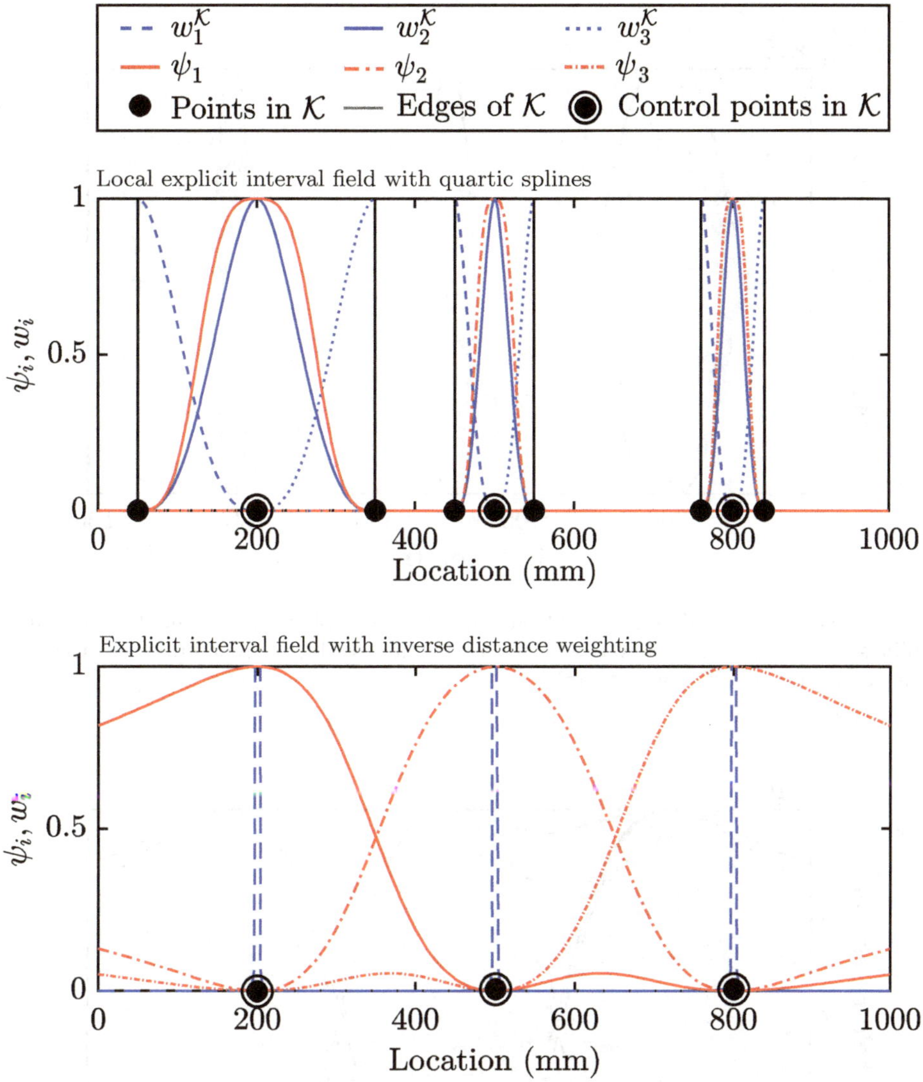

Fig. 3. 1D bar with 3 control points and the corresponding weight and basis function

With Local explicit interval field the three weighting functions are defined on a local domain $\mathcal{K}$ around there control point, the distance measures are:

$$\frac{2R_1}{h} + \frac{2R_2}{h} + \frac{2R_3}{h} = \frac{2 \cdot 150}{2} + \frac{2 \cdot 50}{2} + \frac{2 \cdot 40}{2} = 240 \tag{21}$$

As a result the computational cost is reduced from 1500 to 240 when using the local explicit interval field formulation.

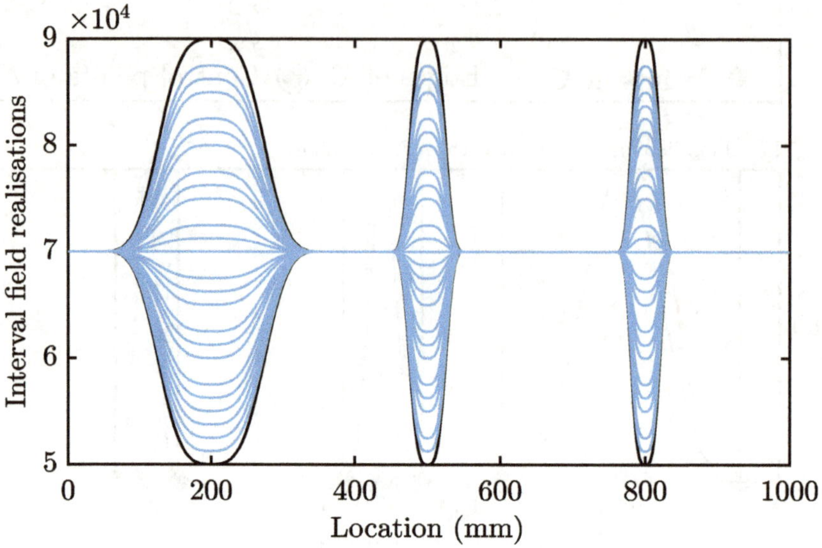

Fig. 4. 1D bar: interval field realisations for the local explicit interval field with quartic splines. The black lines show the vertex realisations of input hypercube and the blue lines realisations of 25 sobol samples.

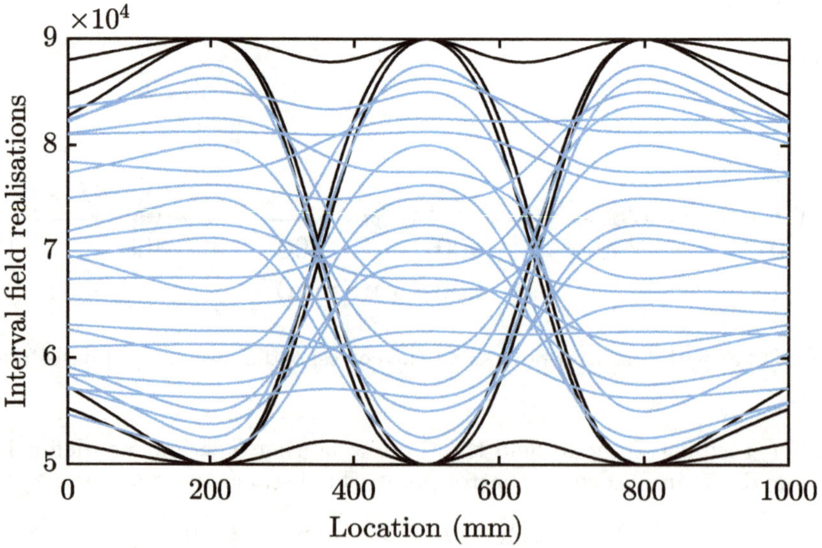

Fig. 5. 1D bar: interval field realisations for the explicit interval field with inverse distance weighting. The black lines show the vertex realisations of input hypercube and the blue lines realisations of 25 sobol samples.

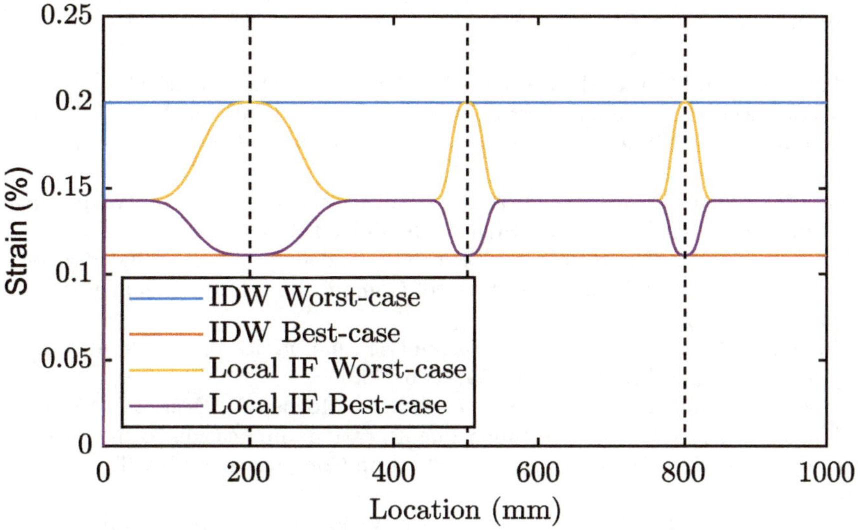

Fig. 6. 1D bar: strain of the worst-case, best-case for IDW as the local explicit interval field with quartic splines, the black dashed line shows the location of the control points

7 Conclusion

This paper presents an approach to generate a local interval field to represent locally spatial uncertain parameters in finite element models for 1D. To get the local character of the interval field this paper proposes two requirements. The use of quartic splines is proposed, as the selection of weighting functions is limited by these requirements. An academic case study is performed to illustrate the difference between the local dependent interval fields and the global dependent interval fields (inverse distance weighting technique). With this test case, it was shown that the local interval fields model the local dependency much better than fields based on inverse distance weighting. Another advantage of the local interval fields is that for this case the computational cost (the number of distance measures) is reduced with a factor of 6.25 in comparison with global interval fields. In this paper, only one spatial dimension and non-overlapping dependency regions in the model domain are considered, the extension to n-dimensions and allowing for overlapping dependency regions requires future research.

Acknowledgements. The authors would like to acknowledge the financial support of the Flemish research foundation in the framework of the research project G0C2218N and the postdoctoral grant 12P359N of Matthias Faes.

References

1. Moens, D., De Munck, M., Desmet, W., Vandepitte, D.: Numerical dynamic analysis of uncertain mechanical structures based on interval fields. In: IUTAM Symposium on the Vibration Analysis of Structures with Uncertainties, pp. 71–83. Springer (2011)
2. Faes, M., Moens, D.: Identification and quantification of spatial interval uncertainty in numerical models. Comput. Struct. **192**, 16–33 (2017)
3. Faes, M., Moens, D.: Recent trends in the modeling and quantification of non-probabilistic uncertainty. In: Archives of Computational Methods in Engineering, pp. 1–39 (2019)
4. Muhanna, R.L., Mullen, R.L.: Uncertainty in mechanics problems-interval-based approach. J. Eng. Mech. **127**(6), 557–566 (2001)
5. Broggi, M., Faes, M., Patelli, E., Govers, Y., Moens, D., Beer, M.: Comparison of bayesian and interval uncertainty quantification: application to the airmod test structure. In: 2017 IEEE Symposium Series on Computational Intelligence (SSCI), pp. 1–8. IEEE (2017)
6. Imholz, M., Vandepitte, D., Moens, D.: Derivation of an input interval field decomposition based on expert knowledge using locally defined basis functions. In: 1st ECCOMAS Thematic Conference on International Conference on Uncertainty Quantification in Computational Sciences and Engineering, pp. 1–19 (2015)
7. Imholz, M., Vandepitte, D., Moens, D.: Analysis of the effect of uncertain clamping stiffness on the dynamical behaviour of structures using interval field methods. Appl. Mech. Mater. **807**, 195–204 (2015). Trans Tech Publ
8. Faes, M., Moens, D.: Identification and quantification of spatial variability in the elastostatic properties of additively manufactured components. In: 19th AIAA Non-Deterministic Approaches Conference, p. 1771 (2017)
9. Sofi, A., Romeo, E.: A novel interval finite element method based on the improved interval analysis. Comput. Methods Appl. Mech. Eng. **311**, 671–697 (2016)
10. Sofi, A., Muscolino, G., Elishakoff, I.: Static response bounds of timoshenko beams with spatially varying interval uncertainties. Acta Mech. **226**(11), 3737–3748 (2015)
11. Wu, D., Gao, W.: Hybrid uncertain static analysis with random and interval fields. Comput. Methods Appl. Mech. Eng. **315**, 222–246 (2017)
12. Wu, D., Gao, W.: Uncertain static plane stress analysis with interval fields. Int. J. Numer. Meth. Eng. **110**(13), 1272–1300 (2017)
13. Xia, B., Wang, L.: Non-probabilistic interval process analysis of time-varying uncertain structures. Eng. Struct. **175**, 101–112 (2018)
14. Li, J., Ni, B., Jiang, C., Fang, T.: Dynamic response bound analysis for elastic beams under uncertain excitations. J. Sound Vib. **422**, 471–489 (2018)
15. Shepard, D.: A two-dimensional interpolation function for irregularly-spaced data. In: Proceedings of the 1968 23rd ACM National Conference, pp. 517–524 (1968)
16. Belytschko, T., Krongauz, Y., Organ, D., Fleming, M., Krysl, P.: Meshless methods: an overview and recent developments. Comput. Methods Appl. Mech. Eng. **139**(1), 3–47 (1996)
17. Vanmarcke, E., Shinozuka, M., Nakagiri, S., Schueller, G., Grigoriu, M.: Random fields and stochastic finite elements. Struct. Saf. **3**(3–4), 143–166 (1986)

18. Moens, D., Hanss, M.: Non-probabilistic finite element analysis for parametric uncertainty treatment in applied mechanics: recent advances. Finite Elem. Anal. Des. **47**(1), 4–16 (2011)
19. Faes, M., Moens, D.: On auto-and cross-interdependence in interval field finite element analysis. Int. J. Numer. Methods Eng. **121**, 2033–2050 (2019)

Processes

Uncertainty Quantification in Subsea Lifting Operations

Luiz Henrique Marra da Silva Ribeiro(✉), Leonardo de Padua Agripa Sales(✉), and Rodrigo Batista Tommasini(✉)

School of Mechanical Engineering, University of Campinas, Campinas, SP 13083-860, Brazil
luiz.marra@outlook.com, leonardopasales@gmail.com, rbtommasini@fem.unicamp.br

Abstract. Subsea lifting operations are dangerous and expensive. One typical problem is the amplification of dynamic forces on the lifting cable at deep water due to resonance of the cable-equipment system. So, it is necessary to guarantee that the cable is always tensioned to prevent slack conditions that lead to snap loads, and at the same time, the cable must be below its structural limit. Several models have been presented to analyze this phenomenon, but they did not consider uncertainties in the determination of the hydrodynamic coefficients (C_a and C_d), which affect considerably the response of the system. Therefore, the objective of this study is to evaluate the influence of the variability of these coefficients via a statistical description of the problem, using Markov Chain Monte Carlo, accept-reject method, maximum likelihood and Monte Carlo simulation in an integrated way. The variability on the structural resistance of the cable is also considered and a reliability study is presented. The stochastic analysis is compared with the deterministic one, and it is concluded that there is a probability of failure and slacking that could be neglected if the deterministic approach is used, which makes the stochastic analysis a more realistic diagnosis of the problem.

Keywords: Bayesian statistics · Reliability analysis · Lifting operation · Crane vessel

1 Introduction

An offshore oil and gas field requires the installation of a considerable amount of equipment on the seafloor in order to control the production. These equipment are installed via subsea lifting operations (Fig. 1) performed by specialized vessels. This activity is expensive due to the high daily rate of these vessels. Therefore, the operations should be carefully analyzed in order to guarantee minimum costs and safety.

J. E. S. De Cursi (Ed.): Uncertainties 2020, LNME, pp. 139–149, 2021.
https://doi.org/10.1007/978-3-030-53669-5_11

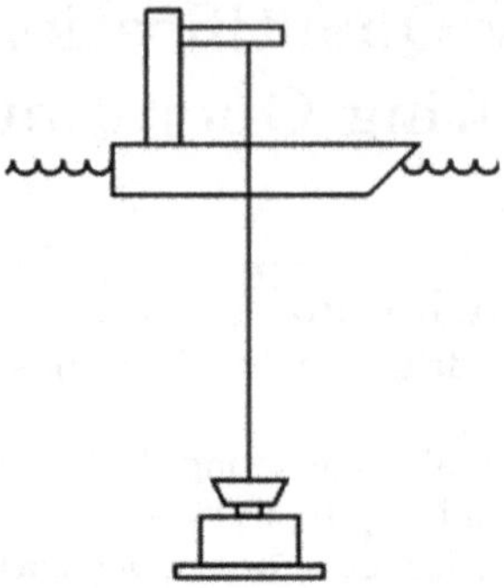

Fig. 1. Typical subsea lifting operation.

A typical problem that is encountered during these operations is the amplification of the dynamic forces on the lifting cable at deep water due to resonance of the cable-equipment system. In this case, it is necessary to assure that the forces on the cable will be less than its structural limit and that it is always keeping a positive tension, in order to prevent slack conditions that lead to snap loads. Several models have been presented to analyze this phenomenon [10,14]. In this study, it has been considered a single degree-of-freedom model that includes the effect of hydrodynamic forces in the system via Morison's equation, as shown below:

$$\left(M + \rho V C_a + \frac{1}{3} m L\right) \ddot{x} + \frac{1}{2} \rho C_d A_p |\dot{x}| \, \dot{x} + \frac{EA}{L} x = \frac{EA}{L} x_0, \tag{1}$$

where M is the mass of the equipment, V is the volume of the equipment, A_p is the vertical projected area of the equipment, C_a is added mass coefficient, C_d is the drag coefficient, EA is the axial stiffness of the cable, m is the linear mass of the cable, L is the suspended length of the cable, ρ is the density of the sea water, x is the displacement of the equipment, and x_0 is the displacement of the hoisting point.

The displacement of the hoisting point is a time signal obtained from the product of the vessel's transfer function (RAO) and the energy spectrum density of the sea waves (S). The energy spectrum of the sea waves is modeled by the JONSWAP formulation, fitted for the Campos Basin, in Brazil, [9]:

$$x_0 = |RAO|^2 S, \tag{2}$$

$$S = \frac{5}{16}(1 - 0.287 ln(\gamma))\gamma^{exp\left[-\frac{1}{2}\left(\frac{(\omega-\omega_p)}{(\sigma\omega_p)}\right)^2\right]} H_s^2 \frac{\omega_p^4}{\omega^5} exp\left[-\frac{5}{4}\left(\frac{\omega}{\omega_p}\right)^{-4}\right], \tag{3}$$

where γ and σ are constant parameters that depends on the location of the operation, ω_p is the peak angular frequency of the waves, and H_s is the significant height of the waves.

The displacement of the equipment is obtained after integration of the motion equation under time-domain and then the dynamic forces on the cable are obtained by:

$$F = \frac{EA}{L}(x - x_0). \tag{4}$$

An important point that affects considerably the response of the system is the hydrodynamic forces. However, the determination of the hydrodynamic coefficients (C_a and C_d) is difficult and leads to high uncertainty. Therefore, the objective of this study is to evaluate the influence of the variability of these coefficients via a statistical description of the problem. The variability on the structural resistance of the cable (SWL) is also considered and a reliability study is presented.

2 Methodology

In offshore recent studies, the ultimate strength has been considered to follow a Weibull distribution [1,2,13]. Such distribution can be estimated, for instance, through the maximum likelihood estimation (MLE) and Bayesian estimation [15]. However, the Bayesian estimation presents a straightforward estimation interpretation and the possibility to use prior information in the analysis [12].

Suppose that a random variable (RV), which follows the probability density function (PDF) $f(x, \alpha, \beta)$, is observed n times. This RV is independent and identically distributed (IID). Then, the likelihood function (LF) is $L(\alpha, \beta, \boldsymbol{x}) = \prod_{i=1}^{n} f(x, \alpha, \beta)$ [8]. If the RV follows a Weibull distribution, the likelihood function is:

$$L(\alpha, \beta, x) = \prod_{i=1}^{n} \frac{\alpha}{\beta}\left(\frac{x}{\beta}\right)^{(\alpha-1)} exp\left(-\left(\frac{x}{\beta}\right)^{\alpha}\right), \tag{5}$$

where the parameters are obtained as follows:

$$max(L(\alpha, \beta, x)), \qquad s.t. : \alpha > 0, \beta > 0. \tag{6}$$

The Bayesian estimation considers the parameter posterior distribution that is proportional to the product of its LF and prior probability distribution [5]:

$$\pi(\alpha, \beta, \boldsymbol{x}) \propto L(\alpha, \beta, \boldsymbol{x}) \times \pi(\boldsymbol{x}). \tag{7}$$

The parameter estimation can be a statistic from the posterior distribution, for instance, the mode. Many combinations of a priori and likelihood functions do not result in a closed analytical solution. However, as stated by [12], simulation through Markov Chain Monte Carlo (MCMC) can be used in Bayesian inference. In order to verify if the simulated distribution represents the "true" one, one has to check the convergence of the MCMC chains. The Geweke $\mid Z \mid$-test [7], and Raftery and Lewis factor [11] can be used to check convergence and independence, respectively.

The reject-accept method can be used to sample from one unknown distribution through one uniform distribution $U(x)$, another distribution $g(x)$, one function $h(x)$ that represents the shape of $f(x)$ and the restriction factor ε [3]:

In i-th iteration:
1- Generate a value from the uniform distribution varying from 0 to 1 ($U_i \sim U[0;1]$);
2- Generate a value from the distribution $g(x)$ (X_i);
3- Accept $X_i \sim f(x)$ if $U_i \leq \varepsilon \frac{h(X_i)}{g(X_i)}$, otherwise, reject X_i.
4- Go to $i = i + 1$;

The Monte Carlo method is traditionally used to generate independent samples from a PDF [8]. In order to understand the response of a model to random variables using the estimated PDF, one can sample from the PDF and run these values in the model, obtaining a response that is also stochastic.

In engineering problems, it is common to assume the admissible force (F_{adm}) being deterministic, but in real circumstances, it has an uncertain value over a range [13]. A latent random variable ($T = F_{max} - F_{adm}$) can be defined and the simulated distribution will better represent the "true" one as the number of simulations n tends to infinity ($n \to \infty$). The proportion of failure will be the number of $T > 0$, and the probability of slacking will be the number of $F_{min} \leq 0$ over n.

The data considered to analyze the model is presented in Table 1.

Table 1. Data considered to analyze the model.

	Parameter	Value
Cable	Linear mass (m)	98 kg/m
	Subsea equivalent mass (m_s)	81.3 kg/m
	Axial stiffness product (EA)	1260 MN
Equipment	Mass (m)	278.5 ton
	Volume (V)	37.5 m^3
	Area (A_p)	142.8 m^2
Environmental condition	Peak angular frequency (ω_p)	0.8976 rad/s
	Significant wave height (H_s)	1.5 m
	Water depth	1,500 m

MLE was used to estimate the F_{adm} distribution [4]. The cable admissible force was considered following a Weibull distribution with mean value as the informed by the supplier 1.3×10^7 N and quantile of 2.5% equal to 3.9×10^6 N. From this distribution, we generated ten observations and these values were used in the MLE estimate.

The data used to estimate the C_a PDF was obtained from [6]. A first analysis of the data via histogram showed a probable bimodal distribution, which is uncommon. Because of this, its PDF was estimated through the accept-reject method because the data presented a bimodal behavior.

The C_d PDF was estimated via a posterior Bayesian estimate. This strategy was adopted because of the prior information used stated that this value should not be higher than 65. Then, after few simulations, it was observed that a Weibull distribution with prior shape and scale parameters following respectively a $U(0;2)$ and $U(0;20)$ presents the C_d posterior distribution with the quantile value of 97,5% approximately equal 65. The Geweke |Z|-test lower than 1.96 indicates absence of non-convergence, and the Raftery and Lewis factor near 1 indicates the absence of non-dependence [12].

The Monte Carlo method was used to generate samples from the estimated distributions to find the answer to the problem F_{max} and F_{min}. A reliability analysis was conducted to verify the probability of failure by counting the proportion of positive values in the latent random variable ($T1 = F_{max} - F_{adm} > 0$) and the probability of slacking through counting the proportion $T2 = F_min - 0 < 0$. The mean and standard deviation were used to verify the Monte Carlo convergence.

Figure 2 summarizes the methodology used in all steps stated above.

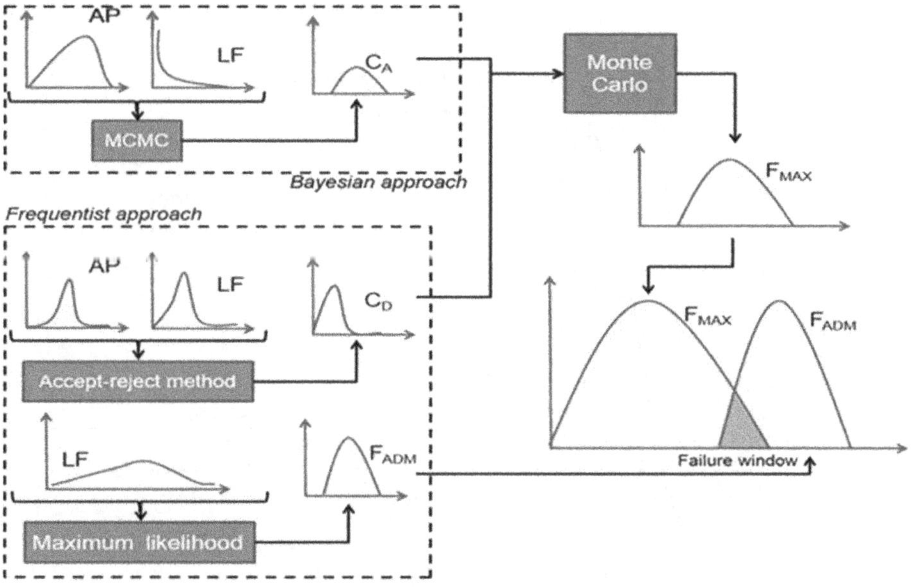

Fig. 2. Overall methodology adopted in the present analysis.

3 Results

Figure 3 represents the Bayesian estimation from the C_d parameter. Firstly, we had only a uniform distribution that was updated with the likelihood function related to the observation.

Even though this uniform distribution does not change significantly the shape of the maximum likelihood, it limits the parametric space because the indicator function and this can affect the results.

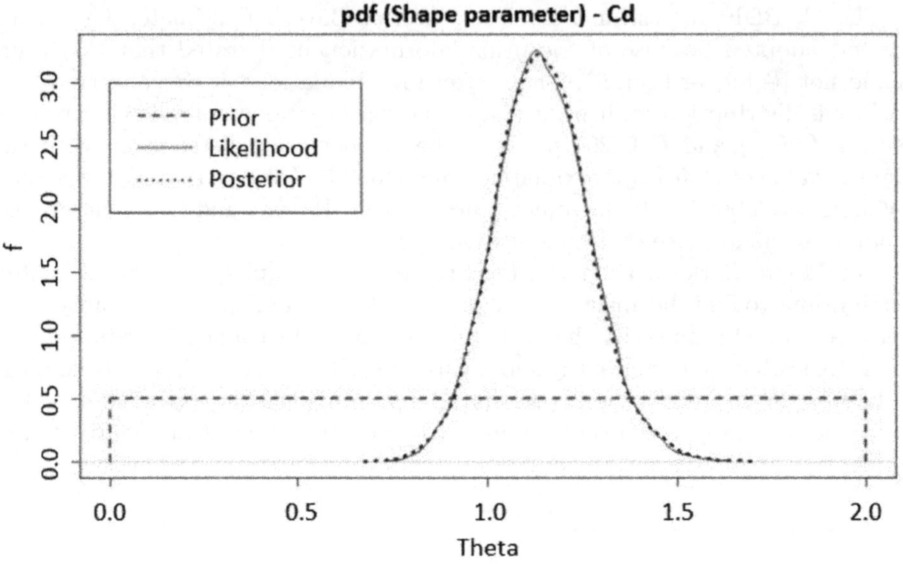

(a) Shape parameter of C_d.

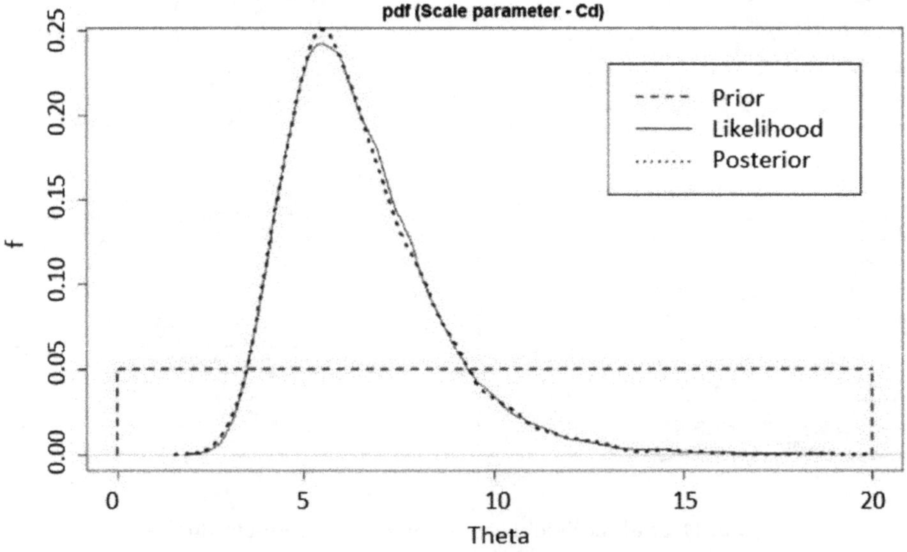

(b) Scale parameter of C_d.

Fig. 3. Prior, likelihood function and posterior distributions of shape and scale parameters of C_d.

Table 2 presents the estimated mode and the values for the convergence and independence tests.

Table 2. Estimated mode, convergence and independence criteria tests values.

Parameter	Estimative	Convergence criteria	
	Posterior mode	ZG	RL factor
Shape	1.1590	0.6223	1.05
Scale	6.6061	0.7728	1.03

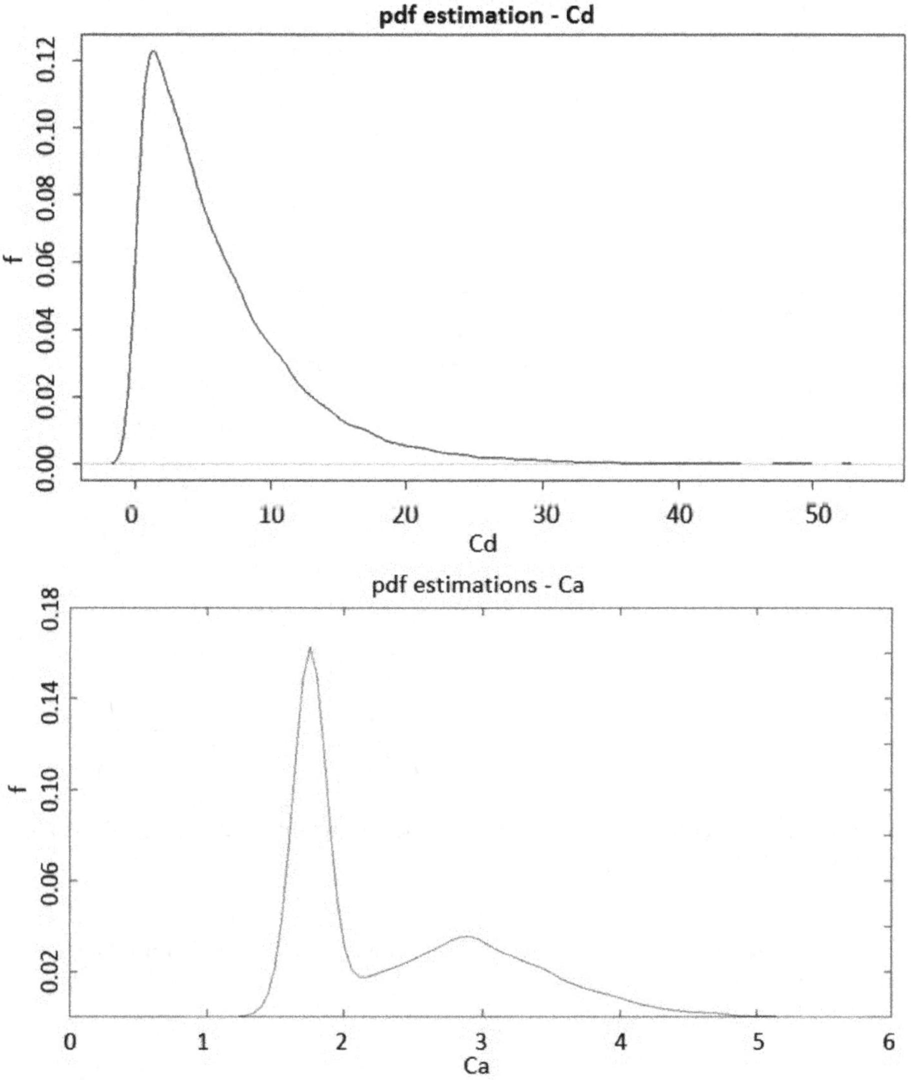

Fig. 4. C_d posterior distribution: mean = 6.12, mode = 0.12 (left), and the representation of the C_a PDF: mean = 2.37, mode = 1.65 (right).

Figure 4 presents the C_d posterior distribution, and the C_a PDF sampled from the accept-reject method.

Figure 5 presents the stochastic answer F_{max}. The red area is the representation of the probability to occur failure in 1200 cycles.

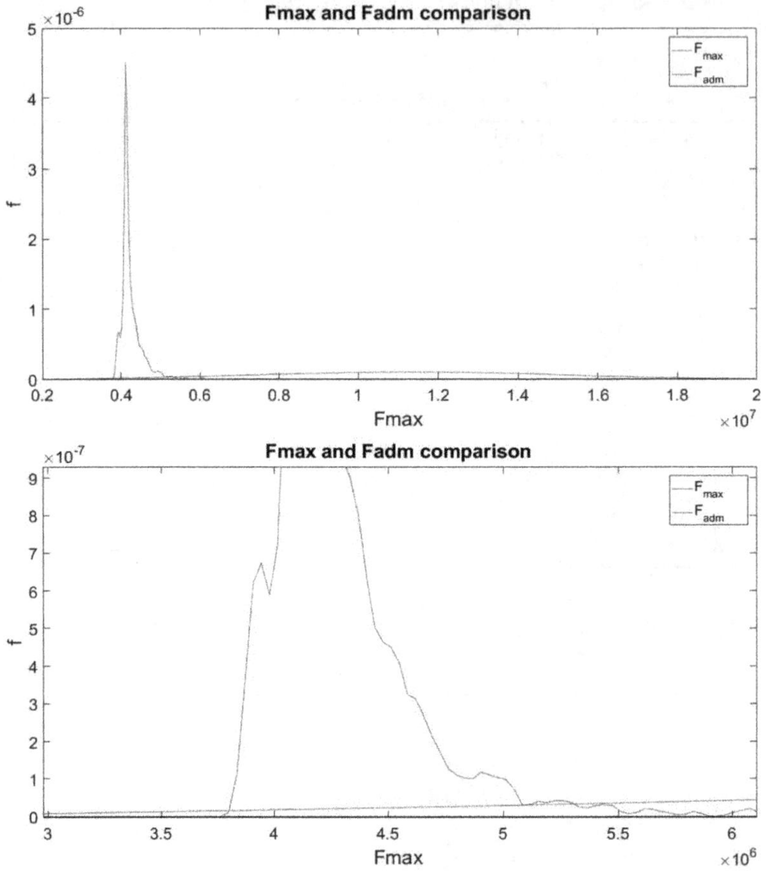

Fig. 5. Graphical representation of the F_{max} being greater than F_{adm} and a zoom in the critical region.

Figure 6 indicates graphically the proportion of F_{min} that is lower than zero, considered in the reliability analysis.

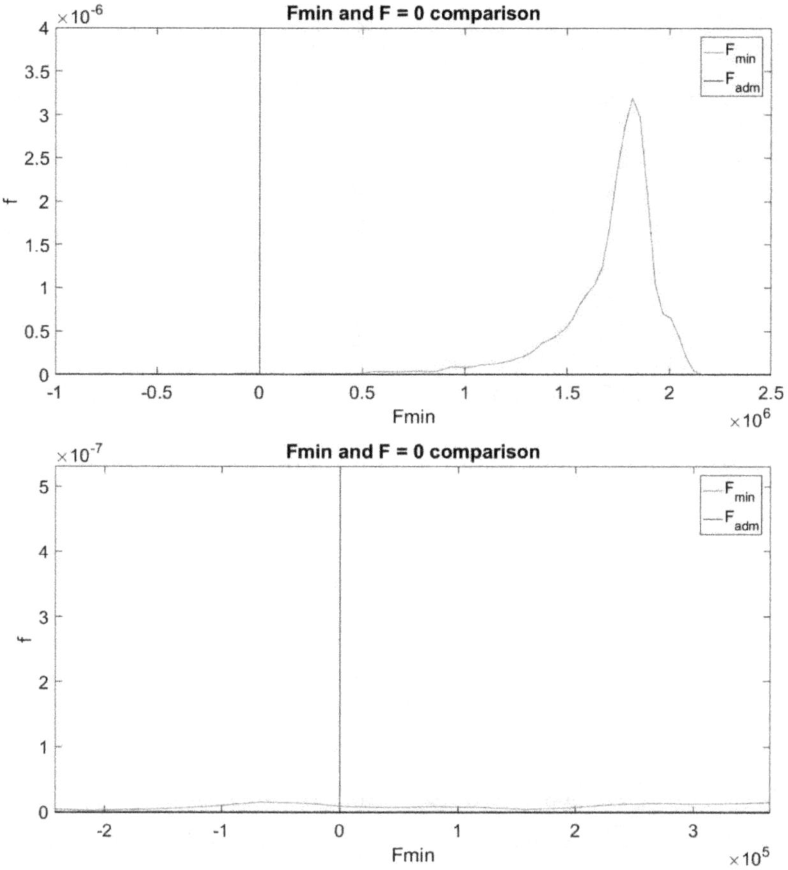

Fig. 6. Graphical representation of the F_{min} being smaller than zero and a zoom in the critical region.

Figure 7 presents the convergence study. In this case, it is plotted the graphs for the mean and standard deviation of F_{min} and F_{max} as a function of the number of iterations of the Monte Carlo method. It is possible to see that a stationary behavior starts to be noticeable after 2000 iterations, indicating that the results obtained have achieved convergence.

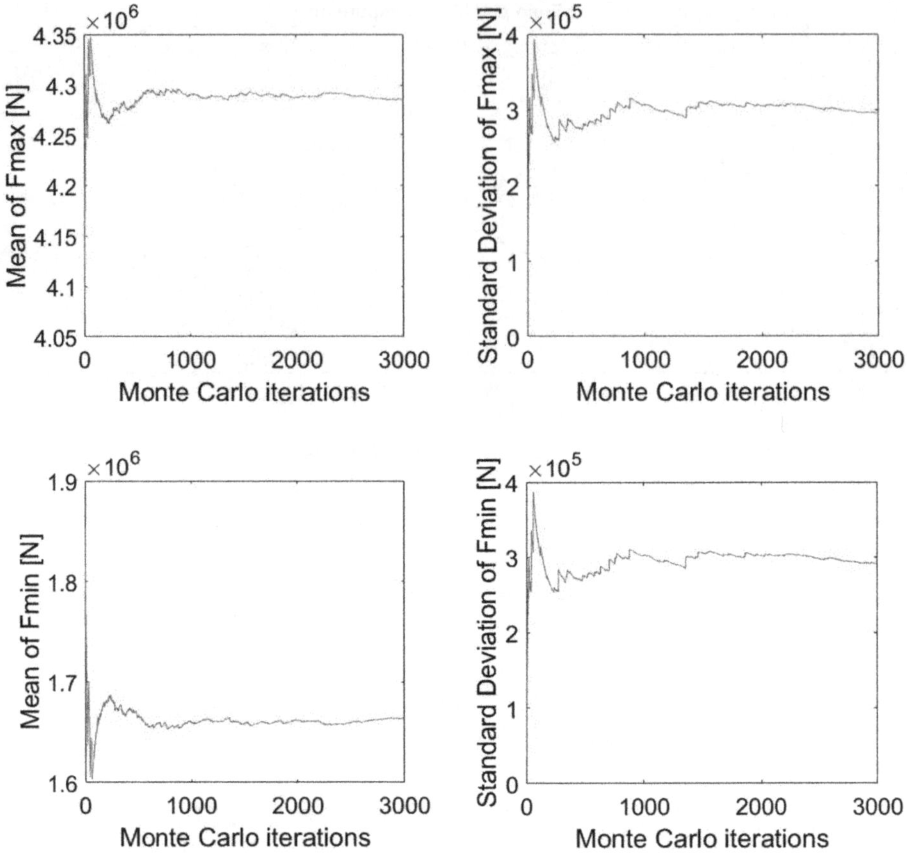

Fig. 7. Convergence study for the Monte Carlo Method.

4 Conclusion

In this study, we evaluated the influence of Ca and Cd via a statistical description of the problem, using Markov Chain Monte Carlo, accept-reject method, maximum likehood and Monte Carlo simulation in an integrated way. The variability on the structural resistance of the cable was also considered and a reliability study was presented.

The proposed method was able to estimate the loads on the cable accordingly, in a stochastic manner. The approach requires a relatively small computational effort and it is able to consider the uncertainties in the C_a and C_d values, along with the structural resistance.

In the case study proposed here, considering metocean conditions, the manifold and the lifting system, 2.28% of the maximum forces in a 180 min window will tear the cable and 0.40% of the minimum forces will be less than zero.

Acknowledgment. The authors gratefully acknowledge: the financial support of the São Paulo Research Foundation (FAPESP) process number 2019/00315-8, and process number 18/15894-0, and the financial support of the National Council for Scientific and Technological Development – CNPq (308551/2017-6) and of the Coordenação de Aperfeiçoamento de Pessoal de Nível Superior (CAPES) for the continuous support for the research.

References

1. Cabrera-Miranda, J.M., Paik, J.K.: On the probabilistic distribution of loads on a marine riser. Ocean Eng. **134**, 105–118 (2017). https://doi.org/10.1016/j.engstruct.2013.05.002
2. Campanile, A., Piscopo, V., Scamardella, A.: Statistical properties of bulk carrier longitudinal strength. Marine Struct. **39**, 438–462 (2014). https://doi.org/10.1016/j.marstruc.2014.10.007
3. Casella, G., Robert, C.P., Wells, M.T.: Generalized accept-reject sampling schemes. In: A Festschrift for Herman Rubin. Institute of Mathematical Statistics, vol. 5, pp. 442-447 (2004). https://doi.org/10.1214/lnms/1196285403
4. CIMAF. Manual Técnico de Cabos de Aço (Brazilian steel cables catalog) (2009). https://www.aecweb.com.br/cls/catalogos/aricabos/CatalogoCIMAF2014Completo.pdf
5. Congdon, P.: Bayesian Statistical Modelling. Wiley, New York (2007)
6. Fernandes, A.C., Mineiro, F.P.S.: Assessment of hydrodynamic properties of bodies with complex shapes. Appl. Ocean Res. **29**, 155–166 (2007). https://doi.org/10.1016/j.apor.2007.04.002
7. Geweke, J.: Evaluating the accuracy of sampling-based approaches to the calculation of posterior moments. In: Bayesian Statistics, vol. 4, pp. 169–193. Oxford University Press (1992). https://pdfs.semanticscholar.org/2e86/50b01dd557ffb15113c795536ea7c6ab1088.pdf
8. Grigoriu, M.: Stochastic Systems: Uncertainty Quantification and Propagation. Springer, New York (2012)
9. Journée, J.M.J., Massie, W.W.: Offshore hydromechanics. TU Delft (2000). http://resolver.tudelft.nl/uuid:31938f94-4e5b-4b69-a249-b88f8c3e59cf
10. Niedzwecki, J.M., Sreekumar, K.T.: Snap loading of marine cable systems. Appl. Ocean Res. **13**(1), 2–11 (1991). https://doi.org/10.1016/S0141-1187(05)80035-5
11. Raftery, A.E., Lewis, S.: How many iterations in the Gibbs sampler? vol. 4, pp. 763–773. Oxford University Press (1991)
12. Ribeiro, L.H.M.S., et al.: Modelling of ISO 9001 certifications for the American countries: a Bayesian approach. Total Qual. Manag. Bus. Excellence **30**, 1–26 (2019). https://doi.org/10.1080/14783363.2019.1696672
13. Teixeira, A.P., Ivanov, L.D., Soares, C.G.: Assessment of characteristic values of the ultimate strength of corroded steel plates with initial imperfections. Eng. Struct. **56**, 517–527 (2013). https://doi.org/10.1016/j.engstruct.2013.05.002
14. Tommasini, R.B., Carvalho, L.O., Pavanello, R.: A dynamic model to evaluate the influence of the laying or retrieval speed on the installation and recovery of subsea equipment. Appl. Ocean Res. **77**, 34–44 (2018). https://doi.org/10.1016/j.apor.2018.05.001
15. Zhang, X., Zhang, L., Hu, J.: Real-time risk assessment of a fracturing manifold system used for shale-gas well hydraulic fracturing activity based on a hybrid Bayesian network. J. Nat. Gas Sci. Eng. **62**, 79–91 (2019). https://doi.org/10.1016/j.jngse.2018.12.001

Product/Process Tolerancing Modelling and Simulation of Flexible Assemblies - Application to a Screwed Assembly with Location Tolerances

Tanguy Moro(✉)

Institut de Recherche Technologique Jules Verne – IRT JV, 44340 Bouguenais, France
tanguy.moro@irt-jules-verne.fr
https://www.irt-jules-verne.fr/en/

Abstract. In industry, the modelling of product/process assemblies is based on the theory of Geometrical Product Specification – GPS – and Tolerancing Analysis. This industrial approach follows several international standards to specify the parts and build stack-ups models of tolerances of an assembly. The main hypothesis of these standards is the rigid workpiece principle. However, for large dimensions thin parts and assemblies as example, the effects of gravity and of the forces and/or displacements imposed by active tools, this rigid bodies assumption is not acceptable and "classic rigid stack-ups" can lead to non-representative results on functional requirements. Thus, this paper proposes an approach to take into account the flexibility of the parts and assemblies in the 3D tolerancing stack-ups. Coupling the tolerancing theory, the structural reliability approaches and FEM simulation, an original approach based on the stochastic polynomial chaos development method, the Sobol's indices and FEM method is developed to build 3D flexible stack-ups and to estimate the main tolerance results. The choice of chaos meta-model is done to be close enough philosophy and form of the linear model of the standard NF E04-008:2013.

Keywords: Tolerance analysis · Flexibles parts and assemblies · Chaos development · Sobol's indices

1 Industrial Standards for Tolerances Analysis

1.1 Tolerances Analysis, Concepts and Standards

In industry, the modelling of product/ process assemblies is based on the theory of Geometrical Product Specification – GPS – and Tolerancing Analysis. This industrial approach follows several international standards to specify the parts and build stack-ups models of tolerances of an assembly. The concept of Geometrical Product Specification provides a global framework for the specification of the parts and the some rules for assembly tolerance analysis:

J. E. S. De Cursi (Ed.): Uncertainties 2020, LNME, pp. 150–158, 2021.
https://doi.org/10.1007/978-3-030-53669-5_12

- ISO GPS fundamental standards for the concepts, principles and definitions of the geometrical specification of the parts: ISO 8015:2011 [1] as example,
- ISO GPS general standards for the geometrical specification of the parts: ISO 14405:Parts 1-3:2016-2017 [2] and ISO 1101:2017 [3] for dimensional, form, orientation, location and run-out tolerances,
- NF E04-008:2013 GPS [4] for tolerance methods based on arithmetics, quadratic statistics and inertial statistics and linearized model.

For each assembly, some product and/or process functional requirements Y are specified, with target and tolerances values. As an example, the total height of an assembly of m parts could be a functional requirement. Then, a geometrical model is built for each functional requirement Yi in relationship with X_j elementary functional characteristics/geometrical specifications. This model is call "stack-up".

$$Y_i = f(X_j) \tag{1}$$

For an easy assembly with simple geometrical parts, the stack-ups could be done "by hand", by summation or subtract X_j elementary functional characteristics. However, for 3D industrial problems, the stack-ups must be done in CAD environment and with dedicated software, as MECAMASTER® used at IRT JV.

1.2 Tolerances Analysis with Independency Principle, Linear Stack-Up Model Assumption and with Rigid Bodies Principle

One of main hypothesis of these standards is the independency principle: "By default, every GPS specification for a feature or relation between features shall be fulfilled independent of other specifications" in compliance with ISO GPS 8015:2011. Extending the independency principle to statistical point view, all random variables, statistical models of X_j elementary functional characteristics, are statically independent. Thus, a linear model of stack-up can be built for each functional requirement in compliance with NF E 04-008:2013:

$$Y_i = c + \sum_{j=1}^{n} \alpha_j.X_j \tag{2}$$

Using previous linear model, tolerances analysis method must be chosen. The standard NF E 04-008:2013 and industrial R&D works [5, 6] propose four main tolerances methods:

- Arithmetic/worst case method [4],
- Quadratic statistics method [4],
- Inertial statistics method [4],
- Process tolerancing method [5, 6].

A second of main hypothesis of these standards is the rigid workpiece principle: "By default, a workpiece shall be considered as having infinite stiffness and all GPS specifications apply in the free state, undeformed by any external forces including the force of gravity", in compliance with ISO GPS 8015:2011. Thus, all the geometrical specifications of parts always keep the same value and are not modified by deformation in the process of assembly.

1.3 Main Tolerancing Results for Quadratic Statistics Method

The main results for a tolerancing quadratic statistics method are summarized in Table 1.

Table 1. Main results of a stack-up for quadratic statistics method and linear model

$\mu(Y_i)/t(Y_i)$	$\sigma(Y_i)=\sqrt{Var(Y_i)}$	$TI(Y_i)=\mathrm{ULS}(Y_i)-\mathrm{LLS}(Y_i)$
$c+\sum_{j=1}^{n}\alpha_j.\mu(X_j)$	$\sqrt{\sum_{j=1}^{n}.\alpha_j^2\sigma(X_j)^2}$	$C_{pk}(Y_i)\sqrt{\sum_{j=1}^{n}\left(\frac{\alpha_j^2.TI(X_j)^2}{C_{pkj}^2}\right)}$
$NCR(Y_i)$ *For X_j normal law & centered hypothesis*	$C_P(Y_i)$ *For X_j normal law & centered hypothesis*	$SRC(X_j)/Y_i$ *Variance contribution of X_j to $Var(Y_i)$*
$P(Y_i \notin TI(Y_i)) = 2\Phi\left(-\frac{\mathrm{TI}(Y_i)}{2\sigma(Y_i)}\right)$	$\frac{\mathrm{TI}(Y_i)}{6\sigma(Y_i)}$	$\frac{\alpha_j^2 Var(X_j)}{Var(Y_i)}$

Where μ is the mean, σ is the standard-deviation, Var is the variance, TI is the tolerance interval, ULS and LLS are the upper and lower specification limits, C_{pk} is the minimum process capability index and NCR is the total fraction nonconforming of a geometrical random variable.

2 Proposal of a Methodology for 3D Tolerances Analysis with Flexible Bodies Assumption

2.1 Tolerances Analysis with Independency Principle, Chaos Stack-up Model Assumption and with Flexible Bodies Principle

However, for large dimensions thin parts and assemblies, under effects of gravity and of the forces and/or displacements imposed by active tools, this rigid bodies assumption is not acceptable and "classic rigid stack-ups" can lead to non-representative results on functional requirements. In industrial assembly workshops, the flexibility of individual parts and of subassemblies is currently used to conform the parts and to allow the assembly.

Thus, this paper proposes an approach to take into account the flexibility of the parts and assemblies in the 3D tolerancing stack-ups. Coupling the tolerancing theory, the structural reliability approaches [7–9] and FEM simulations, an original approach based on the stochastic polynomial chaos development method and FEM method is developed to build 3D flexible stack-ups. The initial idea was proposed in [10] and developed in

[11]. In the past, some academics and industrials have been working on new challenges of 3D flexible tolerancing analysis [10–13].

At first stage, the geometrical model is built with a FEM code. The gravity, the tools loads or displacements, are taken into account in the flexible model. Under all imposed mechanical loads and displacements, the parts and assembly have some deformations, which change the real final geometry of the assembly.

Functional requirement Y is thus developed on a chaos expansion [7, 8] with a maximal degree of polynomial equal to p in relationship with n random geometrical variables X_j, with $\mu(X_j)$ and $\sigma(X_j)$, following normal law assumption and $\xi_j = T(X_j)$. These random variables are independent in compliance with ISO 8015:2011 GPS "Independency principle" and statically independence hypothesis of NF E 04-008:2013:

$$Y = \sum_{k=0}^{J-1} \alpha_k . \Psi_k\left(\{\xi_j\}_{j=1\ldots n}\right) \; with \, J = \frac{(n+p)!}{n!p!} \tag{3}$$

$$\Psi_k\left(\{\xi_j\}_{j=1\ldots n}\right) = \prod_{j=1}^{n} H_{\delta_{kj}}(\xi_j), \sum_{j=1}^{n} \delta_{kj} \leq p \, and \, \delta_{kj} \in IN \tag{4}$$

$$H_m(\xi) = \frac{(-1)^m}{\varphi(\xi)} . \frac{d^m \varphi(\xi)}{d\xi^m} \text{ are Hermite polynomials} \tag{5}$$

The chaos formulation of stack-up model is close to the standard linear stack-up approach. ξ_j, probabilistic transformation of the X_j, geometrical specifications, directly appear in the model, which allows an Expert judgement [11]. The coefficients of the meta-model α_k are estimated by a regression method on a numerical DOE, using the flexible FEM geometrical model at each point of the DOE, realization of the random geometrical variables X_j. The DOE is defined to ensure points with coordinates in normal standard space greater than 3, in coherence with a $C_{pk} = 1$ [5, 6].

2.2 Main Tolerancing Results for Chaos Development Method and Monte Carlo Simulations

The main results of tolerances analysis are then estimated using chaos meta-model and Monte Carlo simulations.

The Sobol's indices [9, 11] are used to estimate the contribution to the variance of the elementary characteristics X_j. The Monte Carlo simulations are done with the chaos meta-model to estimate the total fraction nonconforming of the stack-ups and statistical distribution of Y. The other results of a linear stack-up could be estimated with these mains results (Table 2).

Thus, the choice of chaos meta-model is done to be close enough to philosophy and form of the linear model and to offer same tolerancing results of the standard NF E04-008:2013. In fact, in industry, it is very hard to change methods and the requirements of standards must be applied to validate and certificate products.

Table 2. Main results of a stack-up built on a chaos development

$\mu(Y)/\mathrm{t}(Y)$	$\sigma(Y) = \sqrt{Var(Y)}$	Sobol's Indice $S_j(X_j)/Y$ Variance contribution of X_j to $Var(Y)$
α_0	$\sqrt{\sum_{k=0}^{J-1} .\alpha_k^2 Var\left(\Psi_k\left(\{\xi_j\}_{j=1\ldots n}\right)\right)} = \sqrt{\sum_{k=0}^{J-1} .\alpha_k^2 E\left(\Psi_k\left(\{\xi_j\}_{j=1\ldots n}\right)^2\right)}$	$S_j(X_j) = \frac{\sum_{k \in \Gamma_j} \alpha_k^2 Var\left(\Psi_k\left(\{\xi_j\}_{j=1\ldots n}\right)\right)}{Var(Y)} = \frac{\sum_{k \in \Gamma_j} \alpha_k^2 E\left(\Psi_k\left(\{\xi_j\}_{j=1\ldots n}\right)^2\right)}{\sum_{k=0}^{J-1} .\alpha_k^2 E\left(\Psi_k\left(\{\xi_j\}_{j=1\ldots n}\right)^2\right)}$ *Γ_j correspond to polynomials depending only on ξ_j*
	NCR (Y_i) For X_j normal law & centered hypothesis	
	Monte Carlo Simulations	

3 Application to an Aeronautical Flexible Assembly

The previous flexible 3D tolerancing approach is applied to an aeronautical flexible assembly. Two plates are assembled by a screwing process. Each part is defined with nominal and tolerance values for dimension and position tolerances. Their material is steel E24. The bolts are two Titanium TA6V ASNA 2042 lockbolts (Fig. 1).

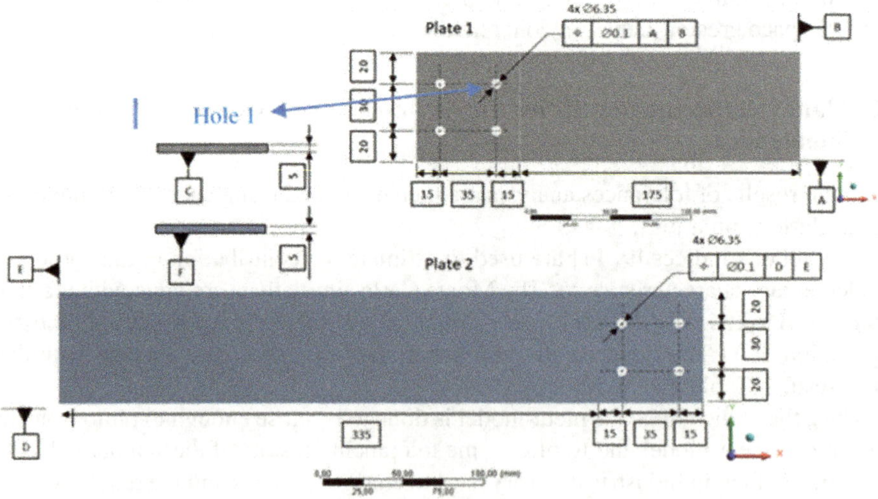

Fig. 1. Parts 1 and 2, nominal and tolerances values.

The FEM model is built with ANSYS® code. 3D hexahedral finite elements are used for plates and 3D tetrahedral finite elements for bolts, with adapted meshes sizes.

Frictional contacts are defined between the interface of the two plates and between holes and bolts. The assembly process of the two structural parts by bolts is defined by:

- To clamp the extremities of the two plates,
- To apply a force $Fz = 200$ N on plate 1 to allow the screwing process,
- To screw with two Ø6.35 bolts.

Some mechanical and geometrical quantities for stochastic tolerances analysis can be computed and linked with functional requirements:

- Maximum Von Mises stresses in part 1, in part 2 and in bolts, "built in stresses" in assembly → Mechanical functional requirements,
- Gap between parts 1 and 2 in process assembly → Geometrical functional requirement (Fig. 2).

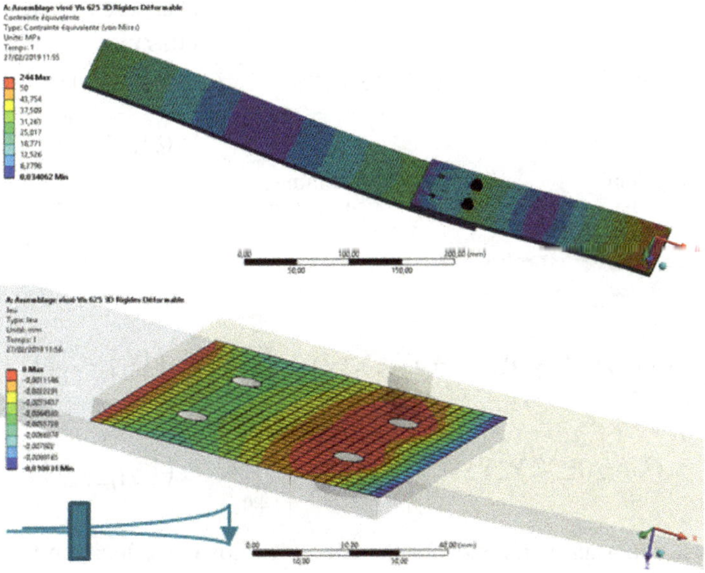

Fig. 2. FEM mechanical and geometrical results on assembly on ANSYS®, functional requirements.

The position tolerance of hole 1's center in assembly, in compliance with ISO 1101:2017 (GPS), is introduced for stochastic problem's solving and stack-up building. Two geometrical random variables P_X and P_Y are defined to model this location tolerance in x and y directions. The stack-up results are computed for Von Mises stresses in plate 2 and for the gap between plates, for $p = 3$ and $n = 2$ (Fig. 3and Table 3):

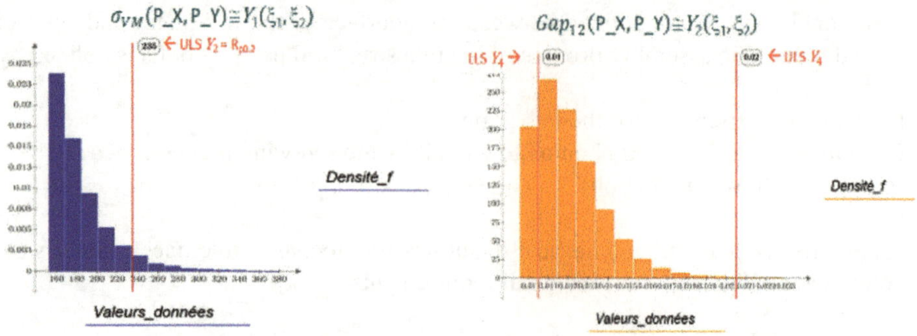

Fig. 3. Chaos statistical distributions for the two functional requirements on MATHCAD®.

Table 3. Main results for two stack-ups built on a chaos development

Y	$\mu(Y)/\text{t}(Y)$	$\sigma(Y) = \sqrt{Var(Y)}$	Sobol's Indice $S_j(X_j)/Y$ Variance contribution of X_j to $Var(Y)$
σ_{VM_P2}	183 MPa $\leq R_{p0,2} = 235$ MPa	29.5 MPa $COV = 16\%$	$S(\xi 1) = 0.155$ $S(\xi 2) = 0.842$ $S(\xi 1, \xi 2) = 0.003$
Gap_{12}	0.012 mm	0.002 mm $COV = 15\%$	$S(\xi 1) = 0.318$ $S(\xi 2) = 0.679$ $S(\xi 1, \xi 2) = 0.003$

$$\sigma_{VM\_P2}(\text{P\_X}, \text{P\_Y}) \cong Y_1(\xi 1, \xi 2) = \sum_{k=0}^{9} \alpha_k . \Psi_k\left(\{\xi_j\}_{j=1\ldots 2}\right) \tag{6}$$

$$Gap_{12}(\text{P\_X}, \text{P\_Y}) \cong Y_2(\xi 1, \xi 2) = \sum_{k=0}^{9} \delta_k . \Psi_k\left(\{\xi_j\}_{j=1\ldots 2}\right) \tag{7}$$

The most important parameter for the two stack-ups is the location tolerance P_Y in y direction. The process load and the off-centering of hole 1 in y direction create shearing in the screw 1 and bending and torsion in plates:

- In x direction by assembly's load Fz,
- + In y direction by position tolerance P_Y deviation.

The statistical distributions, results of Monte Carlo simulations, present non-normal distribution, results which invalidate the linear hypothesis of the stack-up model in the standard E04-008:2013. Then, the nonconformity rate of functional requirements can be estimated. Here, for example, for σ_{VM_P2}, NCR $(Y_1) = 6.42\text{E-}2$.

This industrial example underlines the ability to build a geometrical stack-up with a chaos development to estimate all the product/process quality indicators for a tolerancing analysis. Here, the coefficients of the meta-model are estimated by regression with a numerical DOE. Each point of the DOE is a FEM calculation of the assembly, which allows to take into account the flexibility assumption of the parts and of the assembly.

4 Conclusion

In this paper, a method and numerical tools are proposed to define and to build 3D flexible tolerancing stack-ups for assemblies modeling. Based on tolerancing's principles and objectives, structural reliability methods and FEM simulations, this approach allows to estimate all product/process quality indicators, specific to a tolerance analysis, for functional requirements Y_i of assemblies. The flexible calculations can provide standard geometrical tolerancing's results, with deformations, but also stresses or gaps between parts, coming from deformation and assembly's process. The jigs and tools are modeled and their effects on the stack-up results can be highlighted, as in the example, clamping fixtures and screwing tool.

However, the building of the stack-ups with chaos development imposes a lot of FEM calculations and their minimal number must be greater than J. For a complex industrial stack-up, this is the main issue/limit of this approach. Similarly, meta-models by chaos generally remain limited to taking into account 10 to 15 random variables (816 FEM calculations for $n = 15$ and $p = 3$). This issue is also particularly limiting for a tolerance analysis, which may include several hundred elementary geometrical specifications.

To overcome these limitations, decomposition into sub-assemblies and sub-stack-ups would be a possible solution [5, 6] as well as the implementation of recent advances proposed by Lelièvre & Gayton, AK-HDMR and AK-PCA methods [14, 15]. Another solution to explore would be to favour these approaches for "critical" functional requirements, which require consideration of flexibility to ensure assembly, where the number of random variables would have been reduced by Expert judgment or with a previous stack-up with rigid bodies assumption and a prior sensitivity analysis.

References

1. ISO 8015:2011. Spécification géométrique des produits (GPS) — Principes fondamentaux — Concepts, principes et règles (2011)
2. ISO 14405. Spécification géométrique des produits (GPS) - Tolérancement dimensionnel - Partie 1: Tailles linéaires (2010), Partie 2: Dimensions autres que tailles linéaires (2011), Partie 3: Tailles angulaires (2016)
3. ISO 1101:2017. Spécification géométrique des produits (GPS) - Tolérancement géométrique - Tolérancement de forme, orientation, position et battement
4. NF E04-008:2013 Spécification géométrique des produits (GPS) - Calcul de tolérance, indications et critères d'acceptation - Méthodes arithmétique, statistique quadratique et statistique inertielle
5. Judic, J.-M.: Process tolerancing: a new approach to better integrate the truth of the processes in tolerance analysis and synthesis. In: 14th CIRP Conference on Computer Aided Tolerancing (CAT), Procedia CIRP, pp. 43244–43249 (2016)

6. Judic, J.-M., Gayton, N., Blanc, L.: Objectif smart tolerancing, Des méthodes de tolérancement innovantes et efficientes pour l'usine du futur (2018)
7. Berveiller, M.: Eléments finis stochastiques: approches intrusive et non-intrusive pour des analyses de Fiabilité. Thèse de l'Université BLAISE PASCAL - Clermont II (2005)
8. Blatman, G.: Adaptive sparse polynomial chaos expansions for uncertainty propagation and sensitivity analysis. Thèse de l'Université BLAISE PASCAL - Clermont II (2009)
9. Sudret, B.: Global sensitivity analysis using polynomial chaos expansion. Reliab. Eng. Syst. Saf. **93**(7), 964–979 (2008)
10. Mazur, M.: Tolerance analysis and synthesis of assemblies subject to loading with process integration and design optimization tools. Ph.D. thesis of School of Aerospace, Mechanical and Manufacturing Engineering, RMIT University, Melbourne, Australia (2013)
11. Moro, T.: Analyse et Modélisation Produit/Process d'Assemblages vissés intégrant les Défauts de Tolérances géométriques. 6ème Journée de la conception robuste et fiable, Approches universitaires et industrielles, Toulouse (2019)
12. Homri, L., Goka, E., Levasseur, G., Dantan, J.-Y.: Tolerance analysis — form defects modeling and simulation by modal decomposition and optimization. Comput.-Aided Des. **91**, 46–59 (2017)
13. Falgarone, H., Thiébaut, F., Coloos, J., Mathieu, L.: Variation simulation during assembly of non-rigid components. Realistic assembly simulation with ANATOLEFLEX software. In: 14th CIRP Conference on Computer Aided Tolerancing (CAT), Procedia CIRP, vol. 43, pp. 202–207 (2016)
14. Lelièvre, N.: Développement des méthodes AK pour l'analyse de fiabilité, Focus sur les événements rares et la grande dimension. Thèse de Doctorat de l'Université Blaise Pascal – Clermont II (2018)
15. Gayton, N.: Les méthodes AK pour la classification, Récents développements. 6ème Journée de la conception robuste & fiable, approches universitaires et industrielles, Toulouse (2019)

Methodological Developments for Multi-objective Optimization of Industrial Mechanical Problems Subject to Uncertain Parameters

Artem Bilyk(✉), Emmanuel Pagnacco, and Eduardo J. Souza de Cursi

LMN, EA 3828, INSA Rouen Normandie, Avenue de l'Université, BP 8, Saint-Etienne du Rouvray, France
artem.bilyk@insa-rouen.fr

Abstract. In this paper, we propose a non-intrusive methodology to obtain statistics on multi-objective optimization problems subject to uncertain parameters when using an industrial software design tool. The proposed methodology builds Pareto front samples with low computational cost and proposes a convenient posterior parameterization of the solution set, to enable the statistical analysis and, in perspective, the transformation of small sets of data in large samples, thanks to an Hilbertian approach. The statistics of objects, Hausdorff distance in particular, is applied to Pareto fronts to perform a statistical analysis. This strategy is first demonstrated on a simple test case and then applied to a practical engineering problem.

Keywords: Multi-objective optimization · Uncertainty quantification · Stochastic optimization · Statistics of objects · Hausdorff distance · Abaqus Isight · Python

1 Introduction

During the last decades, optimization has become an increasingly common tool for solving complex engineering problems and it is now conventional to have multi-objective optimization techniques in software design tools such as Abaqus Isight [1]. But multi-objective optimization, and even simple mono-objective optimization, might be time consuming and as problems become more and more complex, much more time is needed to obtain a solution. Nowadays, interest is also on more elaborate optimization methodologies such as stochastic optimization. After a certain point, we cannot rely only on a high number of clusters: we should also think about new methodologies to solve such problems faster, still obtaining a comparable level of accuracy. This insight is even more crucial for industrial problems, where time is precious and the size and complexity of the numerical simulation problems are usually very high. In addition, for multi-objective optimization, the designer is generally interested in statistics of Pareto fronts, such as mean, median or confidence intervals. But uncertainty quantification of the Pareto fronts

J. E. S. De Cursi (Ed.): Uncertainties 2020, LNME, pp. 159–176, 2021.
https://doi.org/10.1007/978-3-030-53669-5_13

introduces new challenges related to probabilities in spaces of infinite dimension, because they are often manifolds belonging to infinite dimensional spaces: for example, a curve in bi-objective optimization. In order to resolve this difficulty when a Monte-Carlo (MC) simulation or other sampling method is used, Bassi et al. [2] introduces a procedure based on the manipulation of solution sets, independent of the basis used to describe the space. In addition, in references [3, 4], it is demonstrated that this approach can be developed further by using stochastic expansions tools to generate large samples and evaluate fine probabilities connected to Pareto fronts. But the works of Bassi et al. are based on an optimization methodology based on the variational method of Zidani-Souza [5] which uses a specific parameterization of Pareto fronts.

Part of the difficulties of the Pareto fronts statistics for industrial problems is addressed in this paper, from a methodology that combines some conventional optimization tools and recent developments in object statistics, when reproducing *a posteriori* the parameterization of Pareto fronts used in Bassi et al. This being the first step to go toward generation of large samples for fine probabilities.

First, some basic definitions related to optimization problems and Pareto fronts statistics are presented in Sect. 2. Next, the proposed theoretical background for parametrization of samples is described in Sect. 3. This methodology is then applied to a theoretical Binh and Korn modified test problem and to a mechanical application solved in Abaqus software, Sect. 4. The conclusions and the potential of the current methodology are finally discussed in Sect. 5.

2 Statistics of Multi-objective Optimization Problems

In this article, we will use the same convention as proposed by Caro [6] in order to describe optimization problems. We will distinguish:

- Performance Functions (PFs);
- Design Variables (DVs) or the decision variables: they are adjustable parameters;
- Design Environmental Parameters (DEPs): they are stochastic variables that are out of control of the decision-maker.

  But we assume for this work that DVs are not subjected to any uncontrollable variations. Thus, only DEPs are the source of randomness in models.

2.1 Deterministic Multi-objective Optimization Problem

A classical multi-objective optimization (MOO) problem reads:

$$\min_{X \in R^{np}} (f_1(\boldsymbol{X}), f_2(\boldsymbol{X}), .., f_{no}(\boldsymbol{X}))$$

$$\text{subject to} \begin{cases} (g_1(\boldsymbol{X}), \ldots, g_{nic}(\boldsymbol{X})) \leq 0 \\ (h_1(\boldsymbol{X}), \ldots, h_{nec}(\boldsymbol{X})) = 0 \\ (\boldsymbol{X})_l \leq X \leq (\boldsymbol{X})_u \end{cases}, \quad (1)$$

where:

- *np*, *no*, *nic* and *nec* are the number of DVs, PFs, inequality and equality constraints respectively;
- $(.)_u$ and $(.)_l$ is an upper and lower acceptable bound for DVs;
- $g_i(\boldsymbol{X})$ and $h_j(\boldsymbol{X})$ are the inequality and equality constraints respectively;
- $\boldsymbol{X} = (x_1, x_2, \ldots, x_{ni}) \in R^{ni}$ is the vector of DVs;
- $f_1(\boldsymbol{X}), f_2(\boldsymbol{X}), .., f_{no}(\boldsymbol{X})$ are the objective functions (PFs).

There exist many different techniques for solving deterministic MOO problems, such as Normal Boundary Intersection (NBI) [7], Non-dominated Sorting Genetic Algorithm-II (NSGA-II) [8] or the variational method of Zidani-Souza [5], as well as many others. However, we do not describe any of these methods in details here. Nowadays, the most commonly used method is the NSGA-II optimization due to its robustness even for highly non-linear design and solution spaces. As a consequence, this optimization technique is available in Abaqus Isight for industrial design. However, one of its main drawbacks is a large number of calculations needed to obtain a high-quality Pareto front. In addition, it is worthwhile mentioning that a continuous description of Pareto fronts is obtained by using the variational method of Zidani-Souza while the NSGA-II gives only discrete points of these Pareto fronts.

2.2 Optimization Problem Under Environmental Uncertainties

In order to take into account uncertainties related to DEPs, the decision-maker must introduce a random variable vector Ξ to the mathematical description and the problem presented in (1) becomes:

$$\min_{X \in R^{ni}} (F_1(\boldsymbol{X}, \Xi), F_2(\boldsymbol{X}, \Xi), .., F_{no}(\boldsymbol{X}, \Xi)),$$

$$\text{subject to} \begin{cases} \text{Prob}[(G_1(\boldsymbol{X}, \Xi), \ldots, G_{nic}(\boldsymbol{X}, \Xi)) \leq 0] \geq \alpha \\ \text{Prob}[(H_1(\boldsymbol{X}, \Xi), \ldots, H_{nec}(\boldsymbol{X}, \Xi)) = 0] \geq \beta, \\ (\boldsymbol{X})_l \leq X \leq (\boldsymbol{X})_u \end{cases} \quad (2)$$

where:

- $F_1(\boldsymbol{X}), F_2(\boldsymbol{X}), .., F_{no}(\boldsymbol{X})$ are the random objective functions;
- $G_i(\boldsymbol{X})$ and $H_j(\boldsymbol{X})$ are the random inequality and equality constraints resp.;
- Prob[.] is the probability operator;
- α and β are the reliability probabilities that are imposed by the decision maker;
- $\Xi = (\xi_1, \ldots, \xi_{ndep})$ is a vector of DEPs with a joint probability distribution for each DEP and *ndep* is a number of DEPs in the optimization problem.

In practice, when a Monte-Carlo (MC) simulation or other sampling method is used, a finite dimension matrix $\Xi = ((\xi_{1,1}, \ldots, \xi_{1,\mathrm{ndep}}), \ldots, (\xi_{\mathrm{ns},1}, \ldots, \xi_{ns,\mathrm{ndep}}))$ is generated, where *ns* is a number of Ξ vectors draw. This yields that for each realization *i* of Ξ matrix, $\Xi_\mathbf{i} = ((\xi_{\mathrm{i},1}), .., (\xi_{\mathrm{i},\mathrm{ndep}}))$, one might solve a deterministic MOO problem and obtain a classic Pareto solution. This leads to a link between the set of DEP values and the obtained solution set composed of design variables and Pareto fronts.

As a consequence, distributions of probabilities and the associated statistics for manifolds - such as curves and surfaces - become of interest. To deal with this problem, we propose to follow the approach proposed by Bassi et al. [2–4], based on Hausdorff distance calculation.

2.3 Statistics of Pareto Fronts

Statistic of objects approach enables to analyze the generated set of Pareto fronts when a Monte-Carlo simulation or other sampling method is used (see [2–4]). This approach is interesting because it enables to find statistics such as the median Pareto front between all generated Pareto fronts, or the confidence intervals of Pareto fronts, where Pareto fronts found for these statistics are real members of the solution set.

As a metric of distance between objects, the Hausdorff distance is selected. This undirected distance between two objects $\mathcal{A}$ and $\mathcal{B}$, known from their N_a and N_b discrete points, is defined as follows:

$$\begin{aligned} \mathcal{HD}(\mathcal{A}, \mathcal{B}) &= \max(d(\mathcal{A}, \mathcal{B}), d(\mathcal{B}, \mathcal{A})), \\ \text{with: } d(\mathcal{A}, \mathcal{B}) &= \min_{a \in \mathcal{A}} d(a, \mathcal{B}) \end{aligned} \tag{3}$$

Then the sum of all distances, between all Pareto fronts S_i^*, is computed:

$$D_i = \sum_{j=1}^{ns} \mathcal{HD}\left(S_i^*, S_j^*\right); \ (i, j) \in [\![1, ns]\!]^2 \tag{4}$$

and the median Pareto front $\tilde{s}$ is defined as:

$$\tilde{s} = S_k^* \text{ such as } \begin{cases} S_k^* \in \left\{S_1^*, S_2^*, \ldots, S_{ns}^*\right\} \\ D_k = \min\{D_i | i \in [\![1, ns]\!]\} \end{cases} \tag{5}$$

while the x-th percentile of Pareto fronts outside the confidence interval is defined as:

$$S_{x-th} = \left\{S_k^* | \left(S_k^* \in S^*, \text{Prob}(\boldsymbol{D} \leq d_k) \geq \frac{x}{100}\right)\right\}, \tag{6}$$

where $\boldsymbol{D} = \{D_1, \ldots, D_{ns}\}$.

In statistic of object, it is advantageous to work with median rather than a mean Pareto front because it is a real existing object, compared to mean. Moreover, there exists an ambiguity for the term “mean” of a set of multiple Pareto fronts, because there is no real convention on how one should calculate such a mean. Detailed discussions and pitfalls on this aspect can be found in references [2–4].

3 Parameterization of the Solution Set

3.1 Parametrisation Methodology

In our strategy, we propose to parameterize PFs and DVs of an optimization solution set with a single parameter ψ, following an Hilbertian approach. This has several advantages: It enables to describe the solution set of interest, as well as its representation, and enables to follow the approach proposed by Bassi et al. [2–4] for the statistics of the MOO solution set. Moreover, it enables the computation of a mean curve in the solutions space and helps to regenerate Pareto front for any combination of DEPs at a low cost to improve the evaluations of the medians and confidence intervals or to have an ideal crowding distance. In addition, this technique might be used as a regularization tool in a MOO solver, with the supplementary advantage to generate Pareto front points with an ideal crowding distance at a marginal cost.

Hence, for an optimization sample composed of PFs $\boldsymbol{F}$ and DVs $\boldsymbol{X}$, we sought coefficients $\boldsymbol{f}$ and $\boldsymbol{x}$ of expansions of the form:

$$\begin{cases} F(\Xi) = \sum_{k\in\mathbb{N}} \boldsymbol{f}_k(\Xi)\varphi_k(\psi) \\ X(\Xi) = \sum_{k\in\mathbb{N}} \boldsymbol{x}_k(\Xi)\varphi_k(\psi) \end{cases} \tag{7}$$

for a basis of total family $\boldsymbol{\varphi}$. In this work, a truncated polynomial basis is used and only the case of continuous curves is addressed.

However, in practice, when using MOO software, we are concerned with a finite number Np of known design points to describe each Pareto front solution that makes up the sample. Hence, for each member of the solution set, and for one realization $\boldsymbol{\Xi_i}$, the representation can be organized in a matrix form as:

$$\begin{pmatrix} F_{1,1} & F_{1,2} & \dots & F_{1,Np} \\ \dots & \dots & \dots & \dots \\ F_{no,1} & F_{no,2} & \dots & F_{no,Np} \\ X_{1,1} & X_{1,2} & \dots & X_{1,Np} \\ \dots & \dots & \dots & \dots \\ X_{np,1} & X_{np,2} & \dots & X_{np,Np} \end{pmatrix} \cong \begin{pmatrix} f_{1,1} & f_{1,2} & \dots & f_{1,degP} \\ \dots & \dots & \dots & \dots \\ f_{no,1} & f_{no,2} & \dots & f_{no,degP} \\ x_{1,1} & x_{1,2} & \dots & x_{1,degP} \\ \dots & \dots & \dots & \dots \\ x_{np,1} & x_{np,2} & \dots & x_{np,degP} \end{pmatrix} \begin{pmatrix} 1 & 1 & \dots & 1 \\ \psi_1 & \psi_2 & \dots & \psi_{Np} \\ .. & .. & .. & .. \\ \psi_1^{degP} & \psi_2^{degP} & \dots & \psi_{Np}^{degP} \end{pmatrix}, \tag{8}$$

what can be written as follows:

$$\boldsymbol{R} \cong \boldsymbol{C.P}, \tag{9}$$

where the $\boldsymbol{R}$ matrix contains the calculated values of all PFs and DPs. This problem of parameterization can be solved by minimizing the residual between the calculated values $\boldsymbol{R}$ and the model $\hat{\boldsymbol{R}} = \boldsymbol{C.P}$, expressed as:

$$\min_{\psi_j, c_{i,j}} \|\boldsymbol{R} - \boldsymbol{C.P}\| = \min_{\psi_j, c_{i,j}} \sum_{i=1;j=1}^{(no+np);Np} [R_{i,j} - \hat{R}_{i,j}]^2$$

$$\text{with boundary conditions: } \begin{cases} \psi = 0, \text{for } \min(R_i) \\ \psi = 1, \text{for } \min(R_j) \end{cases}, i \neq j; i, j \in [\![1, no]\!]$$

$$\text{and subject to:} \begin{cases} \text{tangent of the Pareto front} \leq 0 \\ \min(R_i) \leq \hat{R}_i \leq max(R_i) \quad , \forall i \end{cases} \tag{10}$$

This new optimization problem looks similar to a simple problem of curve fitting, but it is nevertheless quite complex due to the large number of unknowns, being Np parameters and all $c_{i,j}$ coefficients, and due to the simultaneous interaction between these two groups of unknowns of a different nature.

Thus, the analytical expression of the Jacobian is evaluated in order to accelerate the convergence when using a fast and efficient gradient based optimization method. We also suppose that the points describing one Pareto front of the solution set are sorted and thus the end points correspond to the first and last lines of the coefficients matrix $\boldsymbol{C}$, *i.e.* $c_{i,1}$ is known and $c_{i,degP}$ can be expressed in terms of all other values:

$$c_{i,1} = R_{i,1} \quad , i \in [\![1, no + np]\!]$$

$$c_{i,degP} = R_{i,Np} - \sum_{j}^{degP-1} c_{i,j} \ , i \in [\![1, no + np]\!] \tag{11}$$

while the Jacobian matrix (without the square root) is:

$$\frac{\partial\left([\boldsymbol{R} - \hat{\boldsymbol{R}}]^2\right)}{\partial c_{i,j}} = \sum_{k=1}^{Np-2}\left[-2\left(\boldsymbol{R}_{i,k} - \hat{\boldsymbol{R}}_{i,k}\right) * \left(\psi_k^j - \psi_k^{degP}\right)\right],$$
$$i \in [\![1, no + np]\!], \quad j \in [\![2, degP - 1]\!] \tag{12}$$

$$\frac{\partial\left([\boldsymbol{R} - \hat{\boldsymbol{R}}]^2\right)}{\partial \psi_j} = \sum_{i}^{no+np}\{-2\left(\boldsymbol{R} - \hat{\boldsymbol{R}}\right)_{\boldsymbol{i,j}} * (degP * \left[R_{i,Np} - R_{i,1}\right] * \psi_j^{degP-1}$$
$$+ \sum_{k=1}^{degP-1}\left[k * c_{i,k} * \psi_j^{k-1} - degP * c_{i,k} * \psi_j^{degP-1}\right])\},$$
$$j \in [\![1, Np - 2]\!] \tag{13}$$

An example of such parameterization is proposed in Fig. 1. As one can see from the figure, the PFs' and DVs' values might be multiple order of magnitude different. As it is important to achieve the same order of precision for calibration of all PFs as well as DVs, all of them should be normalized. SLSQP gradient based optimisation technique is used for this curve fitting, by using the scipy library in Python. In order to obtain an acceptable quality of the curve fitting, the convergence tolerance value must be small enough. Judging by our experience, it usually varies between 10^{-7} and 10^{-12}.

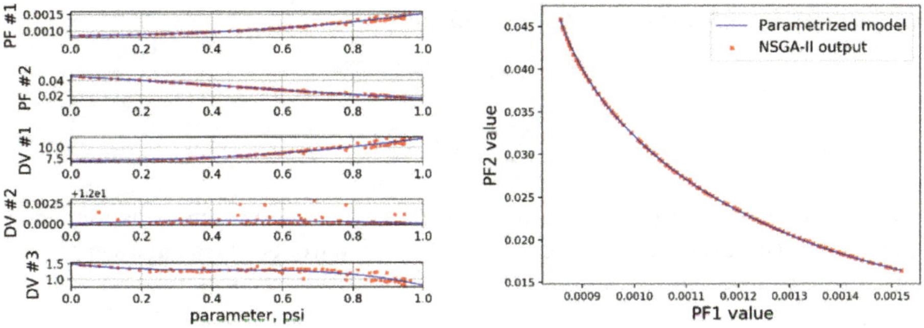

Fig. 1. An example of a Pareto front parameterization of two PFs and three DVs, with a parameterization of each function (left) and a corresponding Pareto front (right)

After the simultaneous identification of the coefficients and parameters of the problem presented in Eq. (10), we improve the precision of curve fitting for each PF and DV individually, solving another small minimization problem described by:

$$\text{For each } i \in [\![1, no + np]\!]:$$

$$\min_{c_{i,j}} \|\boldsymbol{R} - \boldsymbol{C}.\boldsymbol{P}\| = \min_{c_{i,j}} \sqrt{\sum_{j=1}^{N} [R_{i,j} - \hat{R}_{i,j}]^2}$$

$$\text{subject to:} \begin{cases} \text{Tangent of the Pareto front} \leq 0, \textit{ if } i \in [\![1, no]\!] \\ \min(R_i) \leq \hat{R}_i \leq \max(R_i), \forall i \end{cases} \tag{14}$$

The advantage of this second optimization pass is to find a better fit, because in this case a much smaller convergence tolerance is used, while the speed of resolution is much faster than for the initial problem as there are less parameters.

It is worthwhile mentioning that if each Pareto front of the sample are regenerated using the same parameter ψ density (not a case in our study), the points might not be equally distributed in the objectives space.

3.2 Note on Mean Definition for Parametrized Pareto Fronts

Concerning the definition for mean Pareto fronts, special remarks can be drawn in the context of parameterized Pareto fronts. Indeed, it is evident from the adopted expansion that no ambiguity remains for the definition of the mean Pareto since:

$$\mathrm{E}[\boldsymbol{F}] = \sum_{k} \mathrm{E}[\boldsymbol{f}_k]\varphi_k(\psi) \tag{15}$$

where E[∘] is the expectation operator. Moreover, with the adopted parameterization, it becomes possible here to use the simplest definition for the mean Pareto front: supposing that N Pareto fronts exist in a set, each known by their N_p Pareto points:

$$S_i^* = \left\{s_{i,1}^*, \ldots, s_{i,j}^*, \ldots, s_{i,Np}^*\right\}, i \in 1, N \tag{16}$$

then the mean front $\bar{S}$ could be defined as follows:

$$\bar{S} = \{\overline{s_1}, \ldots, \overline{s_j}, \ldots \overline{s_{N_p}}\}$$

$$\overline{s_j} = \frac{1}{N} \sum_{i}^{N} s^*_{i,j}, \; j \in [[1, N_p]] \tag{17}$$

However, no gain is expected here for an evaluation of the mean when a small set of data is considered. This evaluation will be reserved for future works, when large samples will be generated (see perspective section).

4 Applications

4.1 Implementation Details

The strategy adopted here aims to treat stochastic problems by a sample of multiple classical MOO problems, in a non-intrusive approach, based on existing optimization solvers available in industrial software such as NSGA-II in Abaqus Isight and SLSQP in Python (scipy library).

To start the procedure, a sample is generated for the vector Ξ. To obtain a set of random numbers representative of the real variability, we propose to launch a Design of Experiments (DOE) with respect to DEPs, such that the probability density matches the probability density of the DEPs distribution. For example, full factorial or Latin Hypercube (LH) might be used when random variables are assumed to follow a uniform continuous distribution.

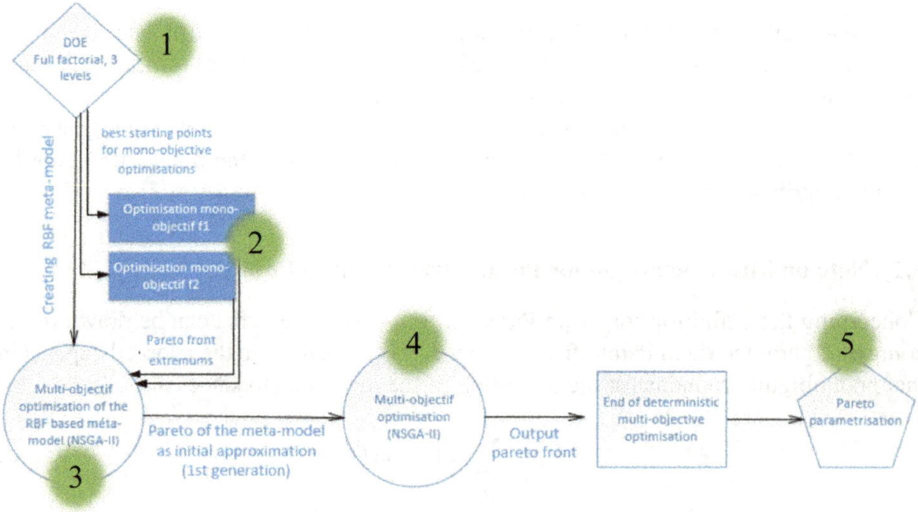

Fig. 2. Methodology workflow adopted for each member of the sample

To obtain the solution set of optimal design variables and Pareto fronts, the workflow schematically described in Fig. 2 is repeated for the desired number of elements in the realization. It is designed to keep the number of evaluations of the original -physical-model as low as possible, without compromise with a good quality of solution. This workflow consists of the five following main steps:

1. Design of Experiments is first performed to calculate the starting guess points for each mono-objective optimization, when only one of the objective is considered at a time (step 2). Each starting guess point is evaluated from a generalization of the representation formula of Pincus, as in Zidani et al. [9];
2. Knowing the starting guess points, mono-objective optimization is performed for each minimization of the objective function of the MOO problem. This enables to determine the true end points of the Pareto front of interest;
3. A meta-model based on Radial Basis Functions is created, using the data generated during the steps 1 and 2. Then, based on this approximation model, NSGA-II MOO is performed to obtain 100 near optimal points that are exported to serve as initial generation for the next step;
4. To find an exact solution, MOO NSGA-II is performed based on the original model and initial generation found in the step 3;
5. Parameterization is performed, as described in Sect. 3, and Pareto front points are regenerated with an ideal crowding distance.

Thus, steps 1 to 5 are repeated for each element of the sample found in the matrix Ξ, in order to obtain the Pareto front sample with respect to the DEP sample. Next, the results are analyzed using the statistics of objects approach.

The built script is designed to be applied to long running industrial problems. That is why the above workflow is implemented in Abaqus Isight, from which we use the RBF approximation, NSGA-II and LSGRG optimization techniques. To code other parts of the methodology, the Isight workflow is extended with some Python code, via Simicode component. Moreover, SLSQP optimization of Scipy library with one function from Pymoo library is used.

4.2 Modified Binh and Korn Test Function

Modified Binh and Korn test problem for multi-objective optimization is used for a validation. The Binh and Korn test problem modified to include uncertainties becomes:

$$\begin{aligned}
&\min_{x \in R^2} \begin{cases} F_1(x) = 4(x_1+\xi_1)^2 + 4(x_2+\xi_2)^2 \\ F_2(x) = (x_1 - 0.2\xi_1 - 5)^2 + (x_2 - 0.2\xi_2 - 5)^2 \end{cases} \\
&\text{with } \Xi = \begin{pmatrix} \xi_1 \\ \xi_2 \end{pmatrix} \sim \begin{pmatrix} U_{[0;1]} \\ U_{[0;1]} \end{pmatrix} \\
&\text{subject to: } \begin{cases} (x_1-5)^2 + x_2^2 \le 25 \\ (x_1-8)^2 + (x_2+3)^2 \ge 7.7 \\ 0 \le x_1 \le 5, 0 \le x_2 \le 3 \end{cases}
\end{aligned} \tag{18}$$

where $U_{[0;1]}$ denotes the continuous uniform distribution in [0; 1].

In Fig. 3, we present each step of the solution:

1. The matrix Ξ is composed of a 5-level full-factorial design of experiments (Fig. 3.a), leading to $ns = 5 \times 5 = 25$ data points;
2. Using the methodology described at the beginning of this section, we obtain the solution set composed of 25 Pareto fronts (Fig. 3.b) where:

   - 5-levels Full Factorial DOE is first performed. Starting point for each mono-objective optimization is calculated using the representation formula of Pincus such that both mono-objective optimization converged in less than 10 iterations;
   - NSGA-II optimization is first performed on RBF approximation model with a population size of 100 for 8000 generations, then 9 generations of NSGA-II are calculated on the original model to obtain a more exact solution;
   - All PFs and DVs in the solution set are parameterized (Fig. 3.c,d). The best result is obtained for a 4-*th* order parameterization with a 10^{-11} precision goal in the stopping criterion of SLSQP optimization. Note that, in this case, $\psi = 0$ not for $\min(F_1)$ as described in Eq. (10), but for $\min(F_2)$. This change is made in order to get a better precision for minimum values of F_2, as it has a wider span over F_1 values.

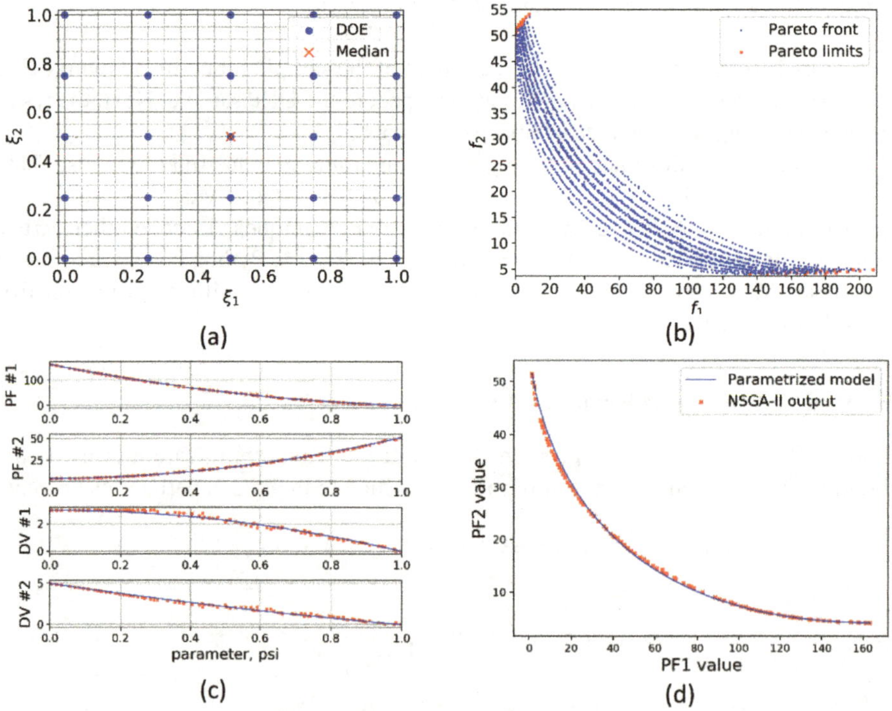

Fig. 3. Full factorial DOE for Ξ (a) as well as 25 corresponding Pareto fronts (b); One typical example of PFs and DVs parameterization (c) and the respective Pareto front (d)

Now, comparing these results (Fig. 4) to the one of reference [2] enables to validate the proposed approach since similar results are obtained. Moreover, the median Pareto front is obtained for $\Xi_i = [0.5; 0.5]$ that corresponds to the true median value. As the uniform distribution is chosen for both DEPs, the median value corresponds also to the mean value in this case.

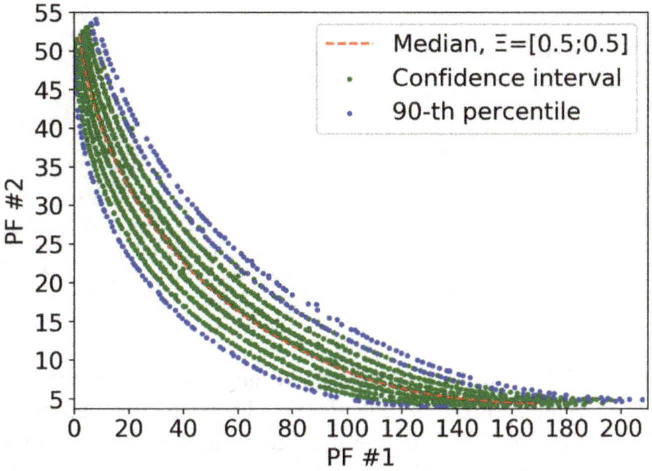

Fig. 4. Statistics for the set of 25 Pareto fronts presented in Fig. 3.b

4.3 Mechanical Application

Mechanical Problem Description. The studied mechanical application consists of a steel mounting bracket of thickness t with a hole, fixed on one side and subjected to a static linear force on its other extremity. Figure 5 shows boundary conditions applied to the 3D bracket and a sketch of the part where the three geometric design variables of $X = (x_1; x_2; x_3)$ are shown.

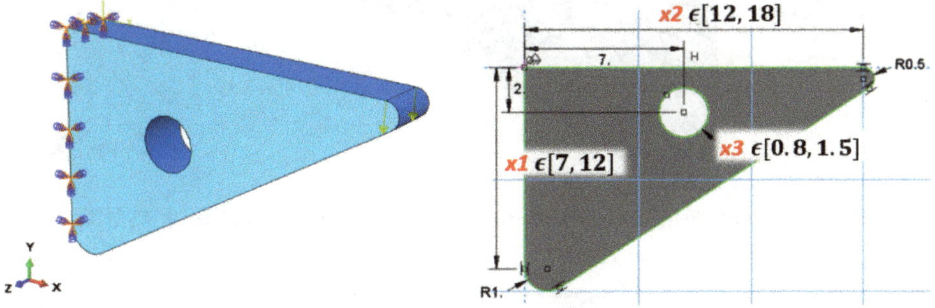

Fig. 5. Boundary conditions applied to the studied 3D bracket (left) and a sketch of the part with the 3 geometric design variables (right)

Isotropic linear elasticity model with parameters E and υ is considered for the material behavior. As small perturbations hypotheses are assumed valid, the linear static solver available in Abaqus/Standard is chosen for the finite element analysis of the mechanical problem. Even though no density ρ is needed to solve this mechanical problem, it is integrated into the model in order to calculate the mass. Nominal values are presented in Table 1.

Table 1. Nominal values of model parameters

	Nominal value
E	210 GPa
ν	0.33
t	3 mm
ρ	$7.85 * 10^{-6}$ kg/mm^3

Mesh quality and the resulting displacement magnitude are presented in Fig. 6. As the force is applied to the extremity of the bracket, the mesh is refined in order to achieve the convergence of the result for all possible combinations of DVs and DEPs. Even though the problem is planar, a 3D linear hexahedral full-integrated element C3D8I with incompatible modes, that does not suffer from shear locking, is used. To reduce the disc space used, only two history results are saved: total mass and a maximum magnitude displacement on bracket extremity, shown in Fig. 6 with the black cross.

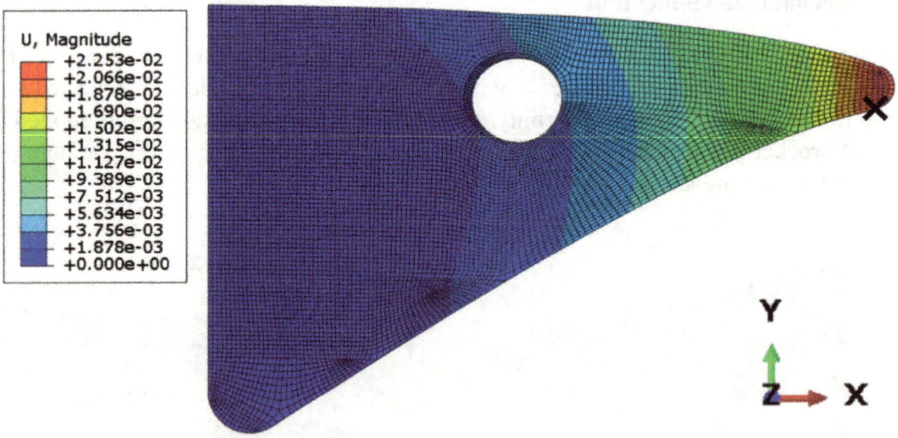

Fig. 6. Magnitude displacement cartography result in [mm] for $\boldsymbol{X} = (9; 15; 1.1)$

Optimization Problem. The goal is to minimize the mass, *i.e.* cost, and maximum displacement magnitude of the mounting bracket. Mathematically, the optimization problem is written as:

$$\min_{X \in R^2} \begin{cases} \text{Mass}(X, \Xi, \rho) \\ \max \text{MAG}(X, \Xi) \end{cases}$$

$$\Xi = \begin{pmatrix} t \\ E \end{pmatrix} \sim \begin{pmatrix} U_{[2;4]}mm \\ U_{[200;220]}GPa \end{pmatrix}, \text{ with } \rho = \frac{C}{E}, C = 1.6485$$

$$\text{subject to:} \begin{pmatrix} 7 \\ 12 \\ 0.8 \end{pmatrix} \leq X \leq \begin{pmatrix} 12 \\ 18 \\ 1.5 \end{pmatrix} \quad (19)$$

The height, the length and the hole radius are DVs (Fig. 5, *right*). Chosen DEPs are the thickness and the Young's modulus. Note that, in this problem, we also suppose that the variation of the rigidity modulus has an impact on the part density.

To start the solution procedure, the matrix Ξ is initialized with 20 design points from the Latin hyper cube procedure (Fig. 7) with a uniform distribution for both DEPs: part thickness t and Young's modulus E. The Pareto fronts of the solution set corresponding to these 20 points are presented in Fig. 7.

Each Pareto front of the solution set is obtained using the proposed methodology:

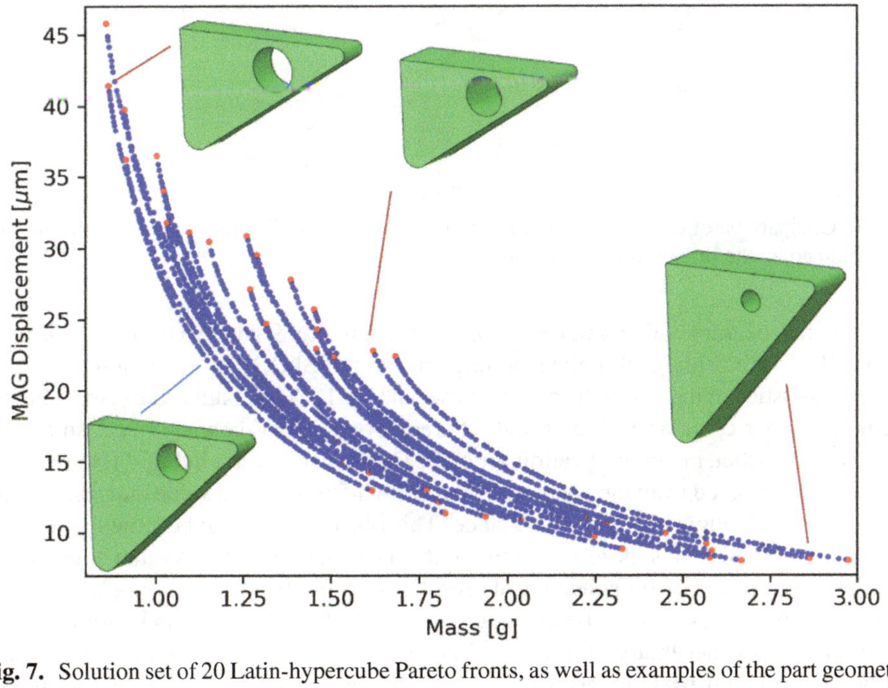

Fig. 7. Solution set of 20 Latin-hypercube Pareto fronts, as well as examples of the part geometry for four designs

- A 3-levels Full Factorial DOE is first performed to evaluate starting point for each mono-objective optimization using the representation formula of Pincus, such that both mono-objective optimizations converged in less than 10 iterations;
- NSGA-II optimization is first performed for 8,000 generations using an RBF approximation model with a population size of 100. Next, using the output population thus generated as the initial, NSGA-II optimization is performed using the mechanical simulation model, such that five generations are enough to obtain a good quality solution.

All these steps take less time to generate a better-quality solution set than the direct application of NSGA-II with more than 30 generations. In Fig. 8, one can compare a Pareto front solution obtained using our methodology and a direct result of the NSGA-II optimization, when the same number of calls to the Abaqus solver is considered. In addition, the Pincus representation formula used to find a good starting point saves a lot of time during single-objective optimizations: a total of less than 20 iterations is enough compared to 160 and more, depending on the starting point.

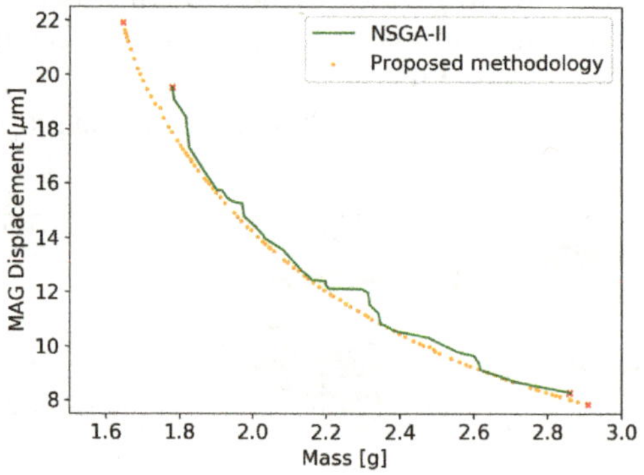

Fig. 8. Comparison of one typical Pareto front using the NSGA-II direct solver and the another one obtained by the proposed methodology

A typical example of parameterization is shown in Fig. 9 where a third order is used with a 10^{-10} precision goal in the stopping criterion of SLSQP optimization.

The statistics of the Pareto fronts are shown in Fig. 10. The result is the same whether we perform our calculations on "rough" Pareto fronts or on parameterized, smoothed data. Note also that, in this application, from the DEPs design space in Fig. 11, the median Pareto front obtained from the generated set is farther from the real value than some other design points if judging by a norm distance. This object is found to be a median instead of any other point, as it is the closest point to the mean and the median value of thickness DEP, that has a greater influence on the resulting Pareto front than Young's modulus. This is the main drawback of the proposed method: to find a true median Pareto front, a large sample Ξ is necessary. But this could be very time consuming or even impossible for large-sized industrial models from a direct approach. So, a methodology to generate a large number of Pareto fronts at a low cost is necessary.

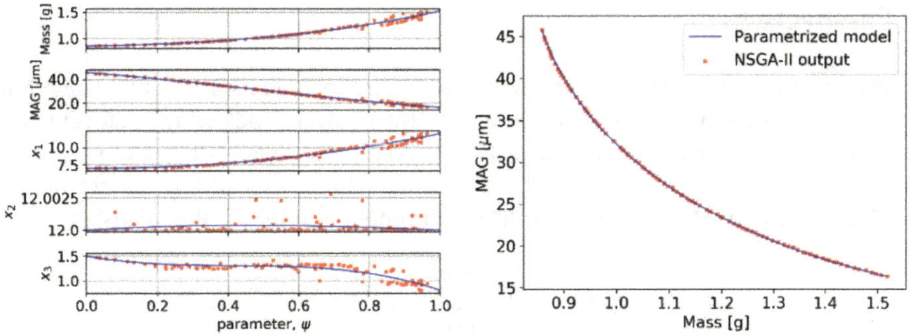

Fig. 9. Example of parameterization at a third order

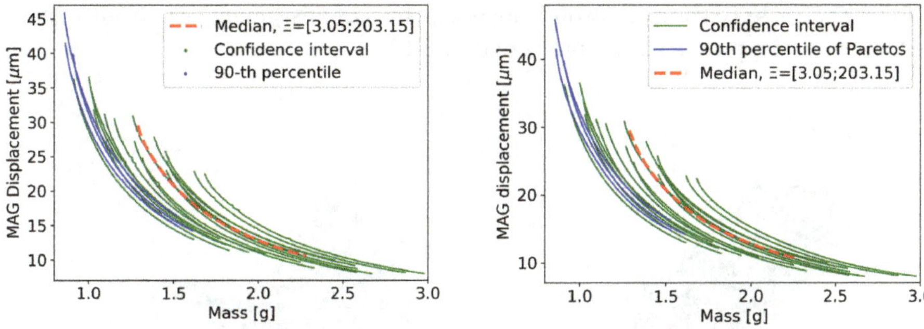

Fig. 10. Statistics of Pareto fronts applied to "rough" Pareto fronts (left) and to parameterized ones (right)

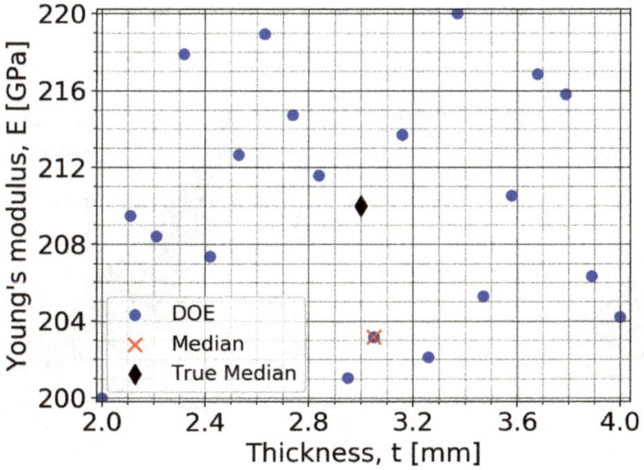

Fig. 11. DOE points in DEP design space, as well the median Pareto front in comparison to the true median (coincides with mean) Pareto front

5 Perspectives

The proposed methodology appears to be well suited for statistics of MOO when long-running mechanical simulations are necessary. But to take full advantage of this methodology, future work to be done is the transformation of small sets of data in large samples, thanks to an Hilbertian approach. This might be achieved by fitting a representation model in order to model coefficient values as a function of DEPs, that reads:

$$f_k(\Xi) = \sum_{j \in \mathbb{N}} c_{k,j} \cdot \varphi_j(\Xi) \tag{20}$$

as it is achieved in references [3, 4]. Using this representation enables to generate a large sample of Pareto fronts and, consequently, smaller percentiles become accessible to further analyses. To fix ideas, promising first results of this strategy for the mechanical problem above are presented in Fig. 12 and Fig. 13.

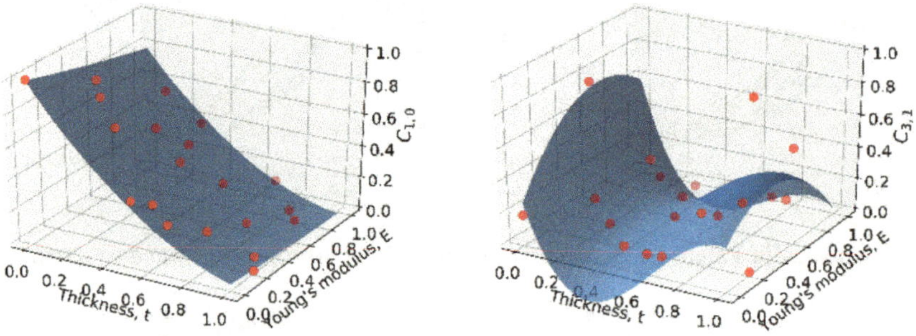

Fig. 12. Example of representation model calibration for $c_{1,0}$(left) and $c_{3,1}$ (right)

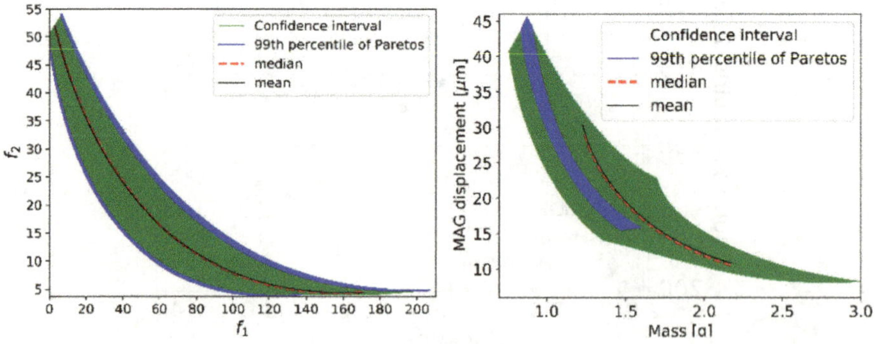

Fig. 13. 10,000 generated Pareto fronts for the modified Binh and Korn problem (left) and the bracket mechanical problem (right)

6 Conclusions

In this work, developments are made toward statistical treatments of MOO problems under uncertainties. As industrial mechanical problems are of interest, the proposed procedure must be designed to be non-intrusive, allowing the use of commercial tools like optimization solvers and mechanical solvers. In this context, our objective is to take advantage of promising recent developments in objects statistics which have proved useful for academic MOO problems when a sampling method is used.

Developments of this work are two-folds: one is the development of a strategy to build at low cost a Pareto front sample with sufficient quality for statistics, and the other is an *a posteriori* parameterization of the built solution set to apply statistics. Major keys that enable to keep the number of evaluations of the original -physical- model as low as possible, without compromise with a good quality of solution, are:

- an efficient search of end points of the Pareto front, thanks to the representation formula of Pincus based on DOE;
- a MOO solver NSGA-II that performs in very few generations, thanks to an almost free starting generation evaluated using a Radial Basis Functions meta-model.

From another side, the parameterization problem that looks *a priori* similar to a simple problem of curve fitting is found quite complex due to the large number of unknowns and interaction between two groups of these unknowns that are of different nature. But it succeeds finally, thanks to a gradient based optimization strategy.

The main benefit of the proposed methodology will be the transformation of small sets of Pareto front data in large samples with low computational cost, as presented in the Perspectives section above. This should allow us to explore the smaller percentiles that are not otherwise accessible, *i.e.* if the set of computed Pareto fronts is too small. Further work needs to be done in order to tune this complete methodology and to apply it with confidence.

References

1. Abaqus Isight User's, Component and Developement Guides (2018)
2. Bassi, M., de Cursi, E.J.S., Pagnacco, E., Ellaia, R.: Statistics of the pareto front in multi-objective optimization under uncertainties. Latin J. Solids Struct. **15**(11), e130 (2018)
3. Bassi, M., Pagnacco, E., de Cursi, E.J.S., Ellaia, R.: Statistics of pareto fronts. In: Le Thi, H.A., Le, H.M., Dinh, T.P. (eds.) Advances in Intelligent Systems and Computing. Optimization of Complex Systems: Theory, Models, Algorithms and Applications, vol. 991. Springer, Switzerland (2019)
4. Bassi, M., Pagnacco, E., de Cursi, E.J.S., Bassi, M.: Uncertainty quantification and statistics of curves and surfaces. In: Llanes Santiago, O., Cruz Corona, C., Silva Neto, A., Verdegay, J. (eds.) Computational Intelligence in Emerging Technologies for Engineering Applications. Studies in Computational Intelligence, vol. 872. Springer, Cham (2020)
5. de Cursi, E.J.S.: Variational Methods for Engineers with Matlab. Wiley, Hoboken (2015)
6. Caro, S.: Comparison of robustness indices and introduction of a tolerance synthesis method for mechanisms. HAL archies-ouvertes.fr (2010)

7. Das, I., Dennis, J.E.: Normal-boundary intersection: a new method for generating Pareto optimal point in nonlinear multicriteria optimization problems. SIAM J. Optim. **8**(3), 631–657 (1998)
8. Deb, K., Agarwal, S., Meyarivan, T.: A fast and elitist multiobjective genetic algorithm: NSGA-II. Evol. Comput. **6**(2), 182–197 (2002)
9. Zidani, H., Sampaio, R., Pagnacco, E., Ellaia, R., de Cursi, E.JS.: Multi-objective optimization by a new hybridized method: applications to random mechanical systems. J. Eng. Optim. **45**, 917–939 (2012)

Uncertainties in Life Cycle Inventories: Monte Carlo and Fuzzy Sets Treatments

Marco Antônio Sabará(✉)

Instituto Federal de Minas Gerais - IFMG, Congonhas, MG 36415-012, Brazil
marco.sabara@ifmg.edu.br

Abstract. The Life Cycle Assessment (LCA) is an impact research methodology that focuses on the life cycle of a product (by extension, services), and is standardized by the ISO 14000 Series. This methodology has been applied in so many areas related to sustainable development, in order to evaluate the environmental, economic and social aspects of the processes of production and distribution of products and service goods. Despite this wide range of applications, the technique still presents weaknesses, especially in the question of the evaluation and expression of the uncertainties present in the various phases of the studies and inherent to the stochastic or subjective variations of the data sources and the generation of models, sometimes reducing the consistency and accuracy of the proposed results. In the present study, we will evaluate a methodology to deal with the best expression of such uncertainties in LCA studies, focusing on the Life Cycle Inventory (LCI) phase. The hypothesis explored is that the application of the Monte Carlo Simulation and Fuzzy Set Theory to the estimation and analysis of stochastic uncertainties in LCA allows a better expression of the level of uncertainty in terms of the Guide to Expression of Uncertainty in Measurements [11], in situations where the original life cycle inventory does not specify the initial uncertainties. The iron ore transport was selected as a process unit by means of an off-road- truck (OHT) with a load capacity of 220 tons and a power of 1700 kW, acting on the route between the mine and the primary crushing of a mining company, in the city of Congonhas (MG). Monte Carlo simulations and Fuzzy Set Theory applications were performed using spreadsheets (MS Excel). The LCA study was conducted in OpenLCA 1.6 (open source) software from data inventories of ELCD database 3.2, also freely accessible. The results obtained were statistically compared using Hypothesis Test and Variance Analysis to identify the effect of the techniques on the results of the Life Cycle Impact Assessment (LCIA) and a Sensitivity Analysis was performed to test the effect of the treatment and function of the distribution of probabilities in the expression of the parameters associated with the items of the original life cycle inventory. Research indicates that inventories with treated data may have their uncertainty expressed to a lesser degree than that expressed in the original inventory, with no change in the final values of the Life Cycle Impact Assessment (LCIA). The treatment of life cycle inventory data through Monte Carlo Simulation and Fuzzy Set Theory resulted in the possibility of expressing the LCI results with a degree of uncertainty lower than that used to express the uncertainty under the standards. Data treatment through Monte Carlo simulation with normal probability distribution showed the lowest values of uncertainty expression with significant difference in relation to the original inventory, at a significance level of 1%.

J. E. S. De Cursi (Ed.): Uncertainties 2020, LNME, pp. 177–197, 2021.
https://doi.org/10.1007/978-3-030-53669-5_14

Keywords: Life Cycle Assessment (LCA) · Analysis of uncertainties · Simulation of Monte Carlo · Theory of fuzzy sets · Sustainability

1 Introduction

Life cycle assessment (LCA) is a scientific approach designed to support decisions by the public authorities and private initiative on issues related to environmental policies, consumption parameters and sustainable production techniques.

According to ISO 14040 [1], a LCA study consists of four phases: definition of objective and scope, inventory analysis, environmental impact assessment and interpretation.

As the technique allows an analysis of potential environmental impacts throughout the life cycle of a product, it has been consolidated as an important tool to assist decision makers, whether agents of the economic community or public policy makers; for example, in choosing the best alternatives in line with the needs of sustainable development [2, 3].

If LCA is used by decision makers in issues involving environmental impacts and the rules of the game are clear in standardization, the quality of the data and results obtained should also be clear. However, the studies already developed point out gaps in databases and software for LCA [4, 5].

Models are simplifications of reality. Therefore, they present levels of inaccuracy and degrees of uncertainty; aspects that require systematic studies and the use of appropriate tools. According to Heijungs e Huijbregts [6] "the idea of statistical analysis is not always part of the standard vocabulary of LCA manufacturers and users".

The uncertainties may be present in the phases of definition of objectives and scope, in inventory analysis and impact assessment. The input data, for example, may be uncertain due to temporal variability or due to the lack of knowledge about the actual values of the emission factors. In addition, data, relationships and choices are part of the various phases of LCA and, in most cases, for each situation there is more than one possible choice.

ISO 14040 [1] defines uncertainty analysis as a "systematic procedure to quantify the uncertainty introduced in the results of a life cycle inventory analysis by the cumulative effects of model inaccuracy, input uncertainty, and data variability."

The construction of methodologies to evaluate the uncertainties present in LCA modeling, then, seeks to ensure the reliability of decisions based on it and serves as a reference to guide future research to reduce uncertainties in all phases of LCA. Uncertainty is important for understanding the reliability and robustness of results in the context of decision making. Indeed, when comparing products or processes, ignoring uncertainties could lead to misguided decisions [3, 7, 9, 10].

As for the input data, uncertainties can be characterized in three types: (i) data for which no value is available; (II) data for which an inappropriate value is available and (III) data for which more than one value is available [6].

In the processing phases, the most commonly used techniques are: (i) parameter variation and scenario analysis; (ii) sampling methods; (iii) analytical methods and (iv) non-traditional methods, such as with the application of fuzzy set theory [6].

Although the standard encourages quantitative analyses of the uncertainties present in LCA studies, it avoids specifying the nature of these studies.

According to the literature of the area, the treatment of uncertainties involves several approaches. Among them, we can mention: scientific approach, constructivist approach, legal approach and statistical approach. The first three seek to reduce the uncertainty associated with the various phases of LCA; the fourth seeks to explain the degree of uncertainty in each phase. Monte Carlo simulation and Fuzzy Set Theories are techniques used in this latter context, allowing the establishment of confidence intervals and robustness indicators [3, 7, 9, 10].

The aim of this research is to analyze the use of Monte Carlo simulation approaches and application of Fuzzy Set Theory in the treatment of uncertainties in Life Cycle Inventory (ICV) modeling.

2 Methodology

The initial stage of development of the research work consisted in the definition of the object of study and the acquisition of its life cycle inventory, which became called base inventory.

In the next step, the uncertainty for each of the items in the Base Inventory was determined, based on the guidelines of the Measurement Uncertainty Guide – GUM ISO (2008) [11].

The third stage consisted in the treatment of the uncertainty of the items of the Base inventory using the Monte Carlo Simulation and the Fuzzy Set Theory. Both techniques were supported by spreadsheet software (MS Excel). For the Monte Carlo Simulation, 10,000 iterations were worked. In relation to the procedures of fuzzification/desfuzzification, the center of gravity method was chosen.

In the fourth stage, the three inventories previously referenced were used in Life Cycle Impact Assessment (LCA) using the OpenLCA software.

Then, the results of impacts obtained were compared through a statistical procedure (Hypothesis Test). Finally, the results were interpreted and discussed.

3 Uncertainties in Life Cycle Assessment

LCA Studies are often comparative, establishing differences between products, processes and other systems. The construction and analysis of the models of these systems is a potential source of uncertainty. To determine if there are statistically significant differences between these options, it is necessary to analyze the uncertainties present in them [1, 12].

And it has three main sources of uncertainty [1, 12, 13] in LCA studies:

- Stochastic uncertainty
- Uncertainty in choices
- Knowledge gaps in the systems studied

The stochastic uncertainties in the ICV can be evaluated by two paths: analytical solution or simulation [12]. The circle of uncertainties is applied to quantify stochastic parameters in the uncertainties of the data.

The Monte Carlo Simulation method is a particularly appropriate method for LCA, as it allows the variation of many factors in parallel and the calculation of general uncertainties of the results at the system level. When performing a simulation of Monte Carlo it is recommended to consider the correlation between the various data values and impact factors considered [12].

The result of the calculation of stochastic uncertainties should not be overestimated, as it may also have a high degree of uncertainty and, mainly, of bias, since it is not capturing systematic uncertainty and gaps in modeling and data [12].

According to Heijung and Huijbregts [6], the uncertainties in the results of an LCA originate from:

- Data that is used in the ICV to represent the elementary flows for all system processes;
- Data that are used in impact analysis to translate elementary flows into environmental impact outcomes;
- The premises/simplifications that are assumed when modeling the system;
- The choices that are made in central decisions such as the form of allocation, the choice of the impact assessment methodology or what future developments are considered in later studies.

Data uncertainty for elementary flows is a statistical uncertainty, that is, of stochastic nature. The same applies to the impact assessment factor within a given evaluation methodology, while the uncertainty introduced by the assumptions and main choices is of a different nature, to the extent that a certain number of discrete results is possible [12].

According to the recommendations of the ILCD Manual (2010), the stochastic uncertainty of process data (such as emissions and input resources) data evaluation (such as characterization factors) means that they must be adequately described by the terms provided by traditional statistics:

- An average measure
- A measure of dispersion around the mean and
- Information about the type of probability distribution of the data.

In contrast to statistical uncertainty, the variations in the choices that make up a LCA study have a discrete nature, that is, several specific options are possible while others are not. There are still potentially a number of methodological choices included [6, 13]:

- ICV modeling principles
- ICV approaches (Basis for standardization and weighting, if included)
- Decisions of cutting parameters and other arrangements at the boundary of the system
- Choose the datasets that make up the system processes
- Choice of impact categories for the application of LCA methods
- Other assumptions.

Within each methodology there are choices to be made from the temporal perspective and in the cultural perspective. Due to their discrete nature, the uncertainties rationed the choices are not described by continuous statistical distributions but are modeled separately, for example, through scenario analysis techniques [1, 12, 13].

Secondly, there are main choices that have the potential to influence the final results of LCA. These significant choices should be identified in different ways in relation to the main contributors: executing the different possible choices as scenarios and comparing their results [6, 12, 13].

A third source of uncertainty is errors attributed to ignorance, that is, gaps in knowledge about the system, leading to omission of data or incorrect assumptions about elementary processes and flows. Ignorance is related to the uncertainty of choice, in the sense that it presents discrete behaviors, but, as it is not performed, it cannot be treated the way choices are treated. This is not ignorance by quantitative assessment of uncertainty, but can be revealed by a qualified peer review [12].

The stochastic uncertainties of inventory and valuation data should be known along with the important uncertainties related to the choice to determine how they propagate in the final results of the LCA. For stochastic uncertainties, their influence on the final results can be evaluated in two different fundamental ways: through an analytical solution or through simulation. Both require knowledge about the types of distribution, mean and variation for process and evaluation data [12].

When inventory results are calculated by disregarding the variation of individual inventory data (i.e., using only average values), the result is the true mean value of the final results, but this approach to meet this challenge is necessary to develop an equation describing the distribution (and therefore also the variation) of the results depending on the distribution of process data for the entire process in the system. The analytical solution becomes very complex, expressing the error even for a simple system, but it can approach a Taylor series by expressing the error in the results as a function of the error in the process data for each process. The analytical solution becomes very complex, expressing the error even for a simple system, but it can approach a Taylor series by expressing the error in the results as a function of the error in the process data for each process [6, 12, 13].

Simulation of the error in the total results of an LCA is typically done using the Monte Carlo Simulation approach. Each subset of inventory data varies independently of the other inventory data around its mean following the distribution that is specified for it (distribution type and variance measure). A calculation of inventory results is performed and stored, and inventory data is again changed randomly within the distributions to arrive at a new set of inventory results. The distribution of the calculated inventory results will approach the distribution of the results when the number of calculations is sufficiently high (often 1000), and thus provides an estimate of the variation around the mean for the final results [12].

The Monte Carlo simulation, part of the standard assumption that all processes and flows are independent and therefore vary independently from each other, both within the system and between systems that are compared in a comparative LCA. This is often not the case as processes can have a technically based mutual dependency or even be the same process that occurs in different system locations (for example, for production or transport processes). In addition to the positive correlation, there is also a negative correlation. Instead of independent variation, these cases may have a high degree of covariance that will tend to reduce or increase the variation of the final results, and should therefore be taken into account when configuring the simulation, which is generally not performed directly [12, 13].

The variation in the final results that is caused by differences related to the choice should be handled by separate calculations for each combination of the relevant choices identified. Where stochastic uncertainties can be addressed and aggregated into a set of outcomes, as described above, the choice-related variation leads to a number of results that can be presented to the decider along with a specification of the underlying choices as possible LCA outcomes, depending on which choices are made. In order to strengthen support for decision-making of LCA results, it is important to reduce the number of choices that are considered to the minimum necessary [6, 12, 13].

The simulation using the Monte Carlo approach is based on information about the distribution of individual elementary flows that are provided by the LCA practitioner. It is often a challenge to provide good information on the statistical distribution of all elementary flows for all processes in the system and this influences the quality of statistical information provided by a Monte Carlo simulation [4, 6, 12].

Sensitivity analysis is a useful tool for identifying where good basic statistical information is most needed. The processes and flows that most contribute to the final results are also those with the strongest potential to contribute to the uncertainty of the final results, and particularly for these key values it is therefore crucial that statistical information is correct [4, 6].

In the absence of tools to support a Monte Carlo simulation, an analysis of the uncertainty of the final results can still be performed along this line, using a sensitivity analysis to identify the main processes, key elementary flows and main options. For each of them, the potential for variation is analyzed and treated basically as discrete options (for stochastic uncertainties, such as the worst realistic case and the best realistic values) in various hypothesis calculations. The result in some cases allows an indicative answer to the question of the definition of the objective. In other cases, the result is inconclusive, which means that a more detailed approach is needed in a new iteration, but this then helps to focus the effort on some of the key data and assumptions [12, 14].

The various processes of treatment of uncertainties, both in the Phase of ICV and of AICV, can help to quantify approximately the range of results and, therefore, confirm and even expand the robustness of the interpretation of the results [12, 14].

4 Monte Carlo Simulation

Monte Carlo simulation is a mathematical simulation technique that makes it possible to evaluate uncertainties in quantitative analyses and decision-making. It is a technique used in many fields of knowledge ranging from simulation of complex physical to economic phenomena. Some examples of application of this method, in different areas, are: finance, project management, energy, industries, engineering, research and development, insurance, oil and gas, transport and environment [10, 15].

The Monte Carlo simulation provides the decision-taker with a range of possible outcomes and the probabilities of occurrences of these results according to the action chosen as the decision. It shows the extreme possibilities

- the results of the boldest and most conservative decisions and all the possible consequences of more moderate decisions [9, 16].

This technique was initially used by scientists who worked on the atomic bomb, and was called Monte Carlo as a reference to the city of Monaco and its casinos. Since its introduction at the time of World War II, Monte Carlo simulation has been used to model a variety of physical and conceptual systems [10, 17].

The Monte Carlo simulation performs simulations by constructing models of possible results, replacing with a range of values – a probability distribution – every factor with inherent uncertainty. The results are then calculated repeatedly for each new set of random values generated by probability functions. Depending on the number of uncertainties and the intervals specified for them, a Monte Carlo simulation can have thousands or tens of thousands of iterations recalculations before it ends. The Monte Carlo simulation then provides distributions of possible results values [16, 18–20].

Therefore, the Monte Carlo Method can be described as a statistical simulation method that uses sequences of random numbers to develop simulations (Fig. 1). In other words, it is seen as a universal numerical method to solve problems through random sampling as an approximation of the solution [14, 16].

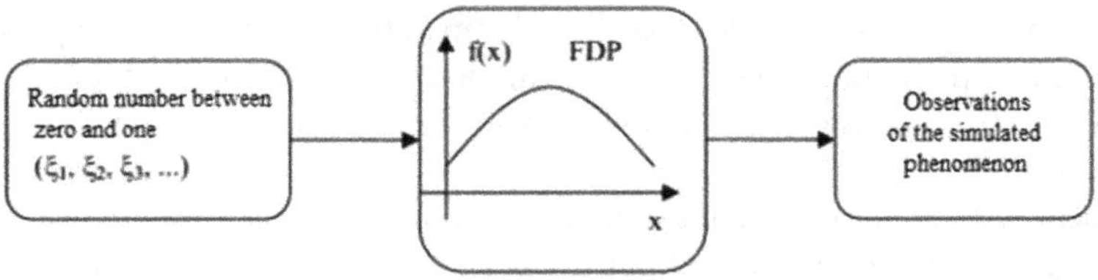

Fig. 1. Overview of the Monte Carlo method

When using probability distributions, variables may present different probabilities of occurrence of different results. Probability distributions represent a very good way to describe uncertainties in risk analysis variables. The most common probability distributions: (i) normal; (ii) lognormal; (iii) uniform; (iv) triangular; (v) PERT and (vi) Discrete [3, 6, 8, 14].

During a Monte Carlo simulation, samples of the values are randomly obtained from input probability distributions (inputs). Each sample set is called an iteration, and the result produced from the sample is recorded. The Monte Carlo simulation does this hundreds or thousands of times, and the product of it is a probability distribution of possible outcomes. In this way, the Monte Carlo simulation provides a much more comprehensive picture of what might happen. It not only tells you what might happen, but also the probability of occurrence [21–23].

5 Fuzzy Set Theory

Fuzzys set theory was introduced by L. Zadeh in 1965 and since then an appropriate approach has been demonstrated to represent epistemic uncertainty in situations of imperfect knowledge of data or process [21].

According to many authors, nebulous logic (Fuzzy logic) is a "tool capable of capturing vague information, usually described in a pseudo natural language and converting it to a numerical format, easily manipulated by computers".

A Fuzzy set defined in a U-speech universe is characterized by a pertinence function A mapped the U elements to the actual range [0, 1], whose notation is represented in Eq. 1 [24]:

$$A : U \rightarrow [0 - 1] \tag{1}$$

Thus ([17]), a pertinence function can associate each element xi - U with an actual number μA (xi) in the range [0, 1], which represents the degree of possibility that element xi will belong to set A. This actual number is given the name of degree of pertinence [24].

A Fuzzy set can then be represented as described by Eq. 2, in which "/" is only a separator [24, 25]:

$$A = \{\mu A\,(x1)/x1,\ \mu A\,(x2)/x2,\quad \mu A\,(x3)/x3,\ \ldots,\quad \mu A\,(xn)/xn\} \tag{2}$$

It is important to emphasize that the degree of relevance cannot always be accurately represented. For this reason, the most plausible is to state that a certain element has a greater or lesser degree of relevance in a comparative relationship [21]. According to the literature, the concept information is modeled or represented much more in the envelopment of the pertinence function than in the accuracy of a specific value of a degree of pertinence.

In practical applications, the most common pertinence functions are gaussian (normal), triangular, trapezoidal, increasing or decreasing [21, 24]. As an example, the triangular pertinence function is described according to Eq. 3:

$$\mu_A(\mathrm{x}) = \begin{cases} 0; & se\, x \le a \\ \frac{x-a}{b-a}; & se\, a < x \le b \\ \frac{x-c}{b-c}; & se\, b < x \le c \\ 0; & se\, x > c \end{cases} \tag{3}$$

Fuzzy Inference interprets the values of the input vector, based on a set of rules, assign a value to the output vector. This is done through a set of "if-then" rules. Fuzzy rules are logical implications that relate fuzzy input sets to output sets. They are usually provided by an expert in the form of linguistic variables [21].

The fuzzifier operator has the function of converting an actual input value into a Fuzzy set. The defuzzification operator performs the reverse process, that is, it estimates the most representative element of the cloudy set.

Among the methods of defuzzification, the most popular is the center of gravity method, represented by Eq. 4, in which element w0 is the estimate that best represents the cloudy set whose pertinence function is w^0:

$$\mathrm{W}^0 = \frac{\sum_i \mu_c(w_i) * W_i}{\sum_i \mu_c(w_i)} \tag{4}$$

The main advantages of Fuzzy inference are the possibility of controlling systems with multiple output variables using a single controller; the ease of using expressions of natural language to draw up linguistic inference rules; possibility of modeling non-linear systems, for which analytical mathematical models are very complex and easy to implement control techniques based on intuitive aspects of inaccurate inputs [21, 24].

On the other hand, the literature points out the following disadvantages in Fuzzy inference: difficulty in evaluating the aspects of optimization, stability and robustness; influence of large amount of parameters in modeling; the accuracy of the system is limited by the experience of the specialist in the configuration and implementation of the model.

6 Selection of the Study Object and Base Inventory

The study object selected was the transport of iron ore by means of an off-road truck with a load capacity of 220 tons and power of 1700 kW, acting on the route between the mine and the primary crushing of a large mining company in the city of Congonhas (MG).

According to Ferreira and Leite [26] although LCA techniques have been used to evaluate environmental impacts associated with the various production processes of the mining industry since the end of the last century, its more intensive use in the evaluation of mineral processing stages is still limited. This is partly due to the difficulty of quantifying the various inputs and outputs involved in the system. In addition, there are significant differences in relation to geological aspects, extraction and processing conditions and various environmental impacts for each type of mineral, requiring specific studies for each case.

Ferreira and Leite [26] describe the product system of iron mining activity as shown in Fig. 2, considered the cradle-to-grave process.

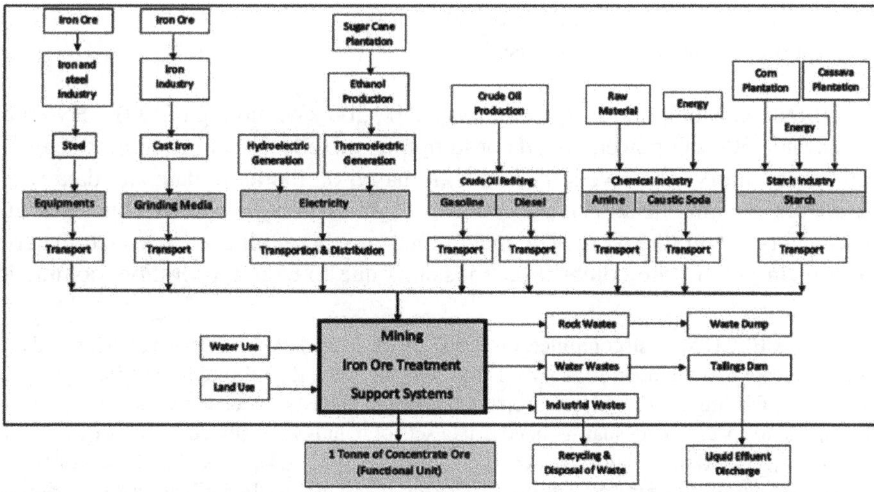

Fig. 2. Iron ore product system (Ferreira and Leite [26])

The approach adopted in the present study was gate-to-gate, considering the initial stage of transport by equipment, as outlined in Fig. 3:

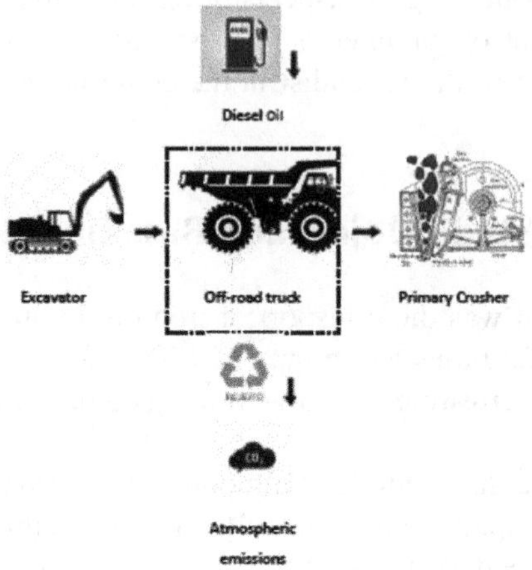

Fig. 3. Gate-to-gate product system (prepared by the author)

The input and output data were obtained from the European Reference Life Cycle Database - ELCD, version 3.2, available free of charge on the https://nexus.openlca.org/database/ELCD website (accessed on October 30, 2018).

The ELCD[1] base was launched in 2006, including Life Cycle Inventory (LCI) data from European Union and other sources of essential materials, energy sources, transportation and waste management. E In addition to this local data, there is also a variety of inventories with global data. Datasets are officially provided and approved by representative associations of the scheduled sectors.

The following inventories were used:

a) Excavator, consumption mix, technology mix, 500 kW, mining (GLO) – Hydraulic Excavator, 500 kW power, mixed consumption, mixed technology, mining segment, global inventory. The inventory is mainly based on literature data and data sheets available machine manufacturers. As for technology, inventoried as and hydraulic excavators for mining work. The main inputs are diesel oil and excavated material. The main outputs are combustion emissions due to engine operation, comprising

[1] A Data base ELCD was discontinued on 29/07/2018, as it data providers have Currently the ability to create and maintain their own Links and share the data through the Life Cycle Data Network (LCDN http://eplca.jrc.ec.europa.eu/LCDN/). The EU launched two new Links, which will respond to specific data sharing needs, data sets of data Icv developed as part of EU-funded research projects and small data providers (i.e. those who need to share less than 10 sets of process data). These entities have the possibility of sharing data without the obligation to develop and maintain of links. The previous bases remain available in https://nexus.openlca.org/database/ELCD for use in OpenLCA initiatives.

regulated emissions (NOx, CO, hydrocarbons and particles), fuel-dependent emissions (CO_2, SO_2, benzene, toluene and xylene) and others such as CH4 and N2O. Emissions due to machinery production, end of life, as well as the fuel supply chain (operating, refinery and transport emissions) are excluded. The data set represents applied technology with a good overall quality.

b) Mining Truck, consumption mix, technology mix, 220 t payload, 1,700 kW (GLO) – Off-Highway Truck (OHT), power 1,700 kW, 220 t load capacity, mixed consumption, mixed technology, mining segment, global inventory. The inventory is mainly based on literature data and data sheets available machine manufacturers. The following combustion emissions (measured data are: benzene, carbon dioxide, carbon monoxide, hydrocarbons, nitrogen oxides, nitrous oxide, PM 2.5 particulate, sulphur dioxide, toluene, xylene. The emissions of hydrocarbons, toluene and xylene result from imperfect combustion and evaporation losses through diffusion through the tank. Diesel-powered vehicle. The dataset represents applied technology with a good overall quality.

Considering as functional unit 1 ton of iron ore transported between the Hydraulic Excavator on the mining front and the primary crusher as the final destination, Table 1 details the entry and exit data of the ICV of the off-road truck with a load capacity of 220 tons and a power of 1700 kW:

Table 1. Life Cycle Inventory Parameters - Mining Truck, consumption mix, technology mix, 220 t payload, 1,700 kW (GLO) - https://nexus.openlca.org/database/ELCD

Inventory item	Value	Unit	Uncertainty
Entries			
Diesel fuel	304,5	Kg/h	Unreported
Outputs			
Benzene	0,45696	g/h	Unreported
Methane	15,6672	g/h	Unreported
Carbon monoxide	1877	g/h	Unreported
Particulate matter (<2,5 μm)	306	g/h	Unreported
Nitrous oxide	45	g/h	Unreported
Non-methane volatile organic compounds	637	g/h	Unreported
Nitrogen oxides	12750	g/h	Unreported
Nitrogen oxides	121,8	g/h	Unreported
Toluene	0,06528	g/h	Unreported
Xylene	5,74464	g/h	Unreported

Source: OpenLCA (Disponível em: https://nexus.openlca.org/database/ELCD. (Acesso em: 01 dez 2018)

7 Results and Discussion

7.1 Estimation of the Uncertainties of the Base Inventory

Table 2 presents the estimated uncertainties for the Mining Truck life cycle inventory, consumption mix, technology mix, 220 t payload, 1,700 kW (GLO) - https://nexus.openlca.org/database/ELCD.

Table 2. Uncertainties estimate for base inventory

Inventory item	Value	Unit	Smallest significant number	Semi-amplitude [a]	Uncertainty $\left[\frac{a}{\sqrt{3}}\right]$
Entries					
Diesel fuel	304,5	Kg/h	1	0,5	0,288675
Outputs					
Benzene	0,45696	g/h	0,0001	0,00005	0,000029
Methane	15,6672	g/h	0,001	0,0005	0,000289
Carbon monoxide	1877	g/h	1	0,5	0,288675
Particulate matter (<2,5 μm)	306	g/h	1	0,5	0,288675
Nitrous oxide	45	g/h	1	0,5	0,288675
Non-methane volatile organic compounds	637	g/h	1	0,5	0,288675
Nitrogen oxides	12750	g/h	1	0,5	0,288675
Nitrogen oxides	121,8	g/h	1	0,5	0,288675
Toluene	0,06528	g/h	0,0001	0,00005	0,000029
Xylene	5,74464	g/h	0,0001	0,00005	0,000029

Source: prepared by the author

7.2 Treatment of the Uncertainties of the Base Inventory Through the Monte Carlo Simulation

Figures of 4 and 5 present, for example, histograms with the results of the Monte Carlo simulation, with 10,000 iterations, for the items Diesel Oil and Benzene from the life cycle inventory.

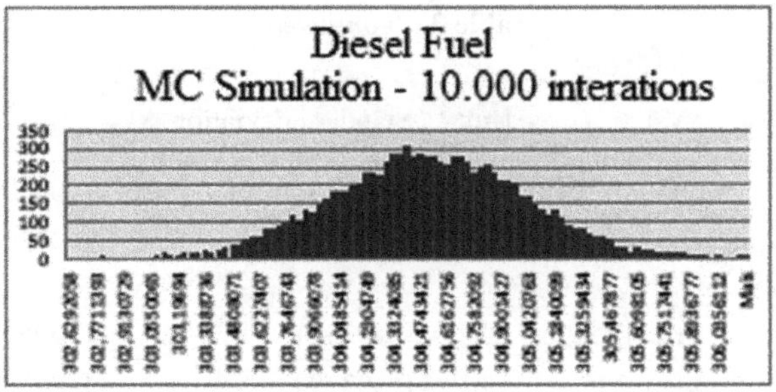

Fig. 4. Histogram with the results of the Monte Carlo simulation for the item Diesel Fuel

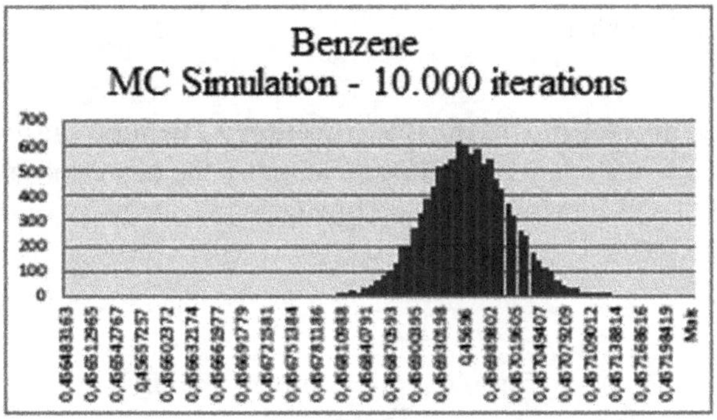

Fig. 5. Histogram with the results of the Monte Carlo simulation for the item Benzene

Simulations for the other inventory items were also calculated. The Table 3 shows the values adjusted by the Monte Carlo simulation for the Mining Truck life cycle inventory, consumption mix, technology mix, 220 t payload, 1,700 kW (GLO) - https://nexus.ope nlca.org/database/ELCD.

Table 3. Base inventory adjusted by Monte Carlo Simulation

Inventory item	Value	Unit	Standard deviation [s]	Uncertainty $\left[\frac{s}{\sqrt{10.000}}\right]$
Entries				
Diesel fuel	304,49803	Kg/h	0,499681	0,0049968
Outputs				
Benzene	0,456959	g/h	0,000049	0,0000005
Methane	15,66720	g/h	0,000502	0,0000050
Carbon monoxide	1877,005	g/h	0,495673	0,0049567
Particulate matter (<2,5 μm)	306,002	g/h	0,504966	0,0050497

(*continued*)

Table 3. (*continued*)

Inventory item	Value	Unit	Standard deviation [s]	Uncertainty $\left[\frac{s}{\sqrt{10.000}}\right]$
Nitrous oxide	45,007	g/h	0,500641	0,0050064
Non-methane volatile organic compounds	637,006	g/h	0,503045	0,0050305
Nitrogen oxides	12749,998	g/h	0,503447	0,0050345
Sulphur dioxide	121,794	g/h	0,503794	0,0050379
Toluene	0,065279	g/h	0,000050	0,0000005
Xylene	5,7446399	g/h	0,000050	0,0000005

Source: prepared by the author

Treatment of the Uncertainties of the Base Inventory Through the Fuzzy Set Theory. Figures 6 and 7 show the results of defuzzification by the center of gravity method for the items Diesel Oil and Benzene from the life cycle inventory.

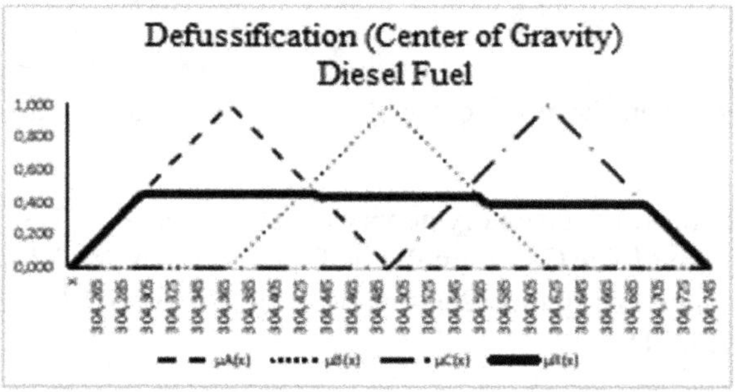

Fig. 6. Results of defuzzification by the center of gravity method for the item Diesel Fuel

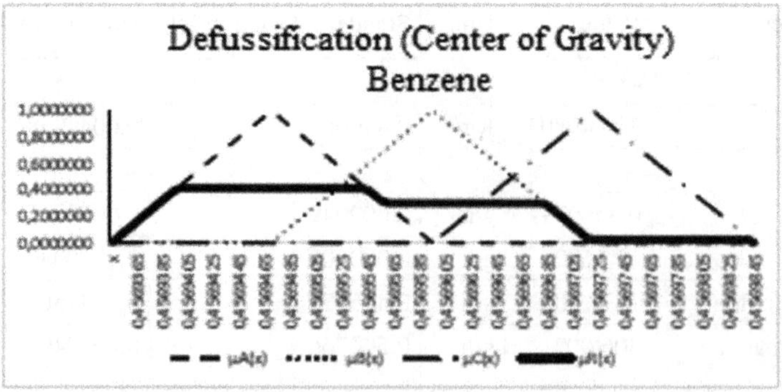

Fig. 7. Results of defuzzification by the center of gravity method for the item Benzene

Simulations for the other inventory items were also calculated. Table 4 shows the values adjusted by the Monte Carlo simulation for the Mining Truck life cycle inventory, consumption mix, technology mix, 220 t payload, 1,700 kW (GLO) - https://nexus.openlca.org/database/ELCD.

Table 4. Base inventory adjusted by Fuzzy Set Theory

Inventory item	Value	Unit	Semi-amplitude [a]	Uncertainty $\left[\frac{a}{\sqrt{6}}\right]$
Entries				
Diesel fuel	304,49439	Kg/h	0,5	0,20412
Outputs				
Benzene	0,456953	g/h	0,456953	0,00002
Methane	15,66714	g/h	0,0005	0,0002
Carbon monoxide	1876,997	g/h	0,5	0,20412
Particulate matter (<2,5 μm)	305,988	g/h	0,5	0,20412
Nitrous oxide	45,008	g/h	0,5	0,20412
Non-methane volatile organic compounds	636,957	g/h	0,5	0,20412
Nitrogen oxides	12749,992	g/h	0,5	0,20412
Sulphur dioxide	200,011	g/h	0,5	0,20412
Toluene	0,065280	g/h	0,00005	0,00002
Xylene	5,744639	g/h	0,00005	0,00002

Source: prepared by the author

Comparative Analysis Between Base Inventory and Modified Inventories. Applying the hypothesis test in the three life cycle inventories, it is verified the hypothesis that the mean values obtained in the different treatments do not vary within the significance limit of 1%.

Thus, the treatments allow the treatment of uncertainties aiming at their quantification without changing the pre-established values in the original inventory, obtained through consensus among specialists and references to important secondary sources, such as technical literature of manufacturers and academic papers.

8 Life Cycle Impact Assessment (LCA) Results

8.1 Impact Assessment Results Considering Baseline Inventory

The baseline inventory data were used to conduct a Life Cycle Impact Assessment (LCA) study based on the following service parameters:

- Engine load2factor: 60%
- Off-road truck productivity: 300 m^3/h
- Iron ore density: 1.8 ton/m^3
- Fuel with sulphur content of 200 ppm (Diesel B23)

Table 5 presents the results of the Life Cycle Impact Assessment (LcA) considering the mining truck base inventory, consumption mix, technology mix, 220 t payload, 1,700 kW (GLO) - https://nexus.openlca.org/database/ELCD.

Table 5. Life Cycle Impact Assessment results - baseline inventory

Impact category	Reference unit	Result	[%]
Ecosystems - climate change	Species, year[a]	0,000000039400000	99,53
Ecosystems - terrestrial acidification	Species, year	0,000000000182000	0,46
Ecosystems - terrestrial ecotoxicity	Species, year	0,000000000002520	0,01
Ecosystems - eutrophication of fresh water	Species, year	0,000000000000699	0,00
Ecosystems - marine ecotoxicity	Species, year	0,000000000000245	0,00
Ecosystems - freshwater ecotoxicity	Species, year	0,000000000000100	0,00
Ecosystems - occupation of agricultural land	Species, year	–	0,00
Ecosystems - transformation of the natural land	Species, year	–	0,00
Ecosystems - occupation of urban land	Species, year	–	0,00
Ecosystems - total	**Species, year**	**0,0000000396000**	**100**
Human health - climate change	DALY[b]	0,0000069600000	67,12
Human health - human toxicity	DALY	0,0000000125000	0,12
Human health - ionizing radiation	DALY	0,0000000001990	0,00
Human health - depletion of the ozone layer	DALY	0,0000000000176	0,00
Human health - formation of particulate matter	DALY	0,0000033900000	32,76
Human health - formation of photochemical oxidants	DALY	0,0000000021400	0,02
Human Health - total	**DALY**	**0,0000104000000**	**100**
Resources - fossil depletion	$	0,2730000000000	99,97
Resources - metal depletion	$	0,0000000396000	0,03
Resources - total	**$**	**0,0000069600000**	**100**

Source: prepared by the author

[a]Uma maneira de medir a taxa de extinção. Estudos mostram que há aproximadamente uma extinção estimada por milhão de espécies-anos. Isso significa que se houver um milhão de espécies na Terra, uma seria extinta a cada ano, enquanto se houvesse apenas uma espécie, seria extinta em um milhão de anos (Goedkoop et al. [25]).

[b]DALY - Disability-Adjusted Life Year (Ano de Vida Ajustado por Incapacidade). Um DALY pode ser considerado como um ano perdido de vida "saudável". A soma desses DALYs em toda a população, ou a carga da doença, pode ser pensada como uma medida da lacuna entre o estado de saúde atual e uma situação de saúde ideal, onde toda a população vive até uma idade avançada, livre de doenças e incapacidades (Goedkoop et. al. [25]).

Figure 8 shows the contribution of each of the processes in relation to the three categories of end-of-process impact.

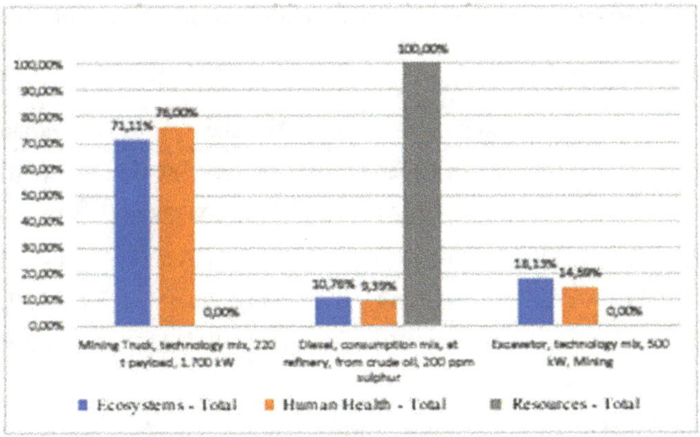

Fig. 8. Contribution [%] in the process impact categories (prepared by the author)

The iron ore transport process is primarily responsible for the impacts related to the ecosystem and human health (72.11% and 76%, respectively), while the supply process of trucks and excavators is responsible for resource consumption (100%).

This result is consistent with the reality of the process, since the impacts on the ecosystem and human health stem from the action of emissions resulting from the burning of fuel in transport operations. In relation to resource depletion, the burned fossil fuel represents non-renewable energy, being totally consumed in the process. On the other hand, transport equipment is susceptible to several end-of-life treatments that allow its reintegration in almost all of the same chain or in parallel chains, such as reuse and recycling, especially metal and polymeric materials.

8.2 Impact Assessment Results Considering the Base Post-treatment Inventory

As described in the methodology, the items in the base inventory and the items of the treated inventories were purchased through single factor variance analysis. For the purpose of representation, the treatments are named as follows:

- Base - Original inventory
- MCNORM - original inventory handled by Monte Carlo Simulation with normal distribution function
- MCUNIF - original inventory handled by Monte Carlo Simulation with uniform distribution function
- FTRI - original inventory treated by Fuzzy Set Theory with triangular distribution function
- FTRAP - original inventory treated by Fuzzy Set Theory with trapezoidal distribution function

The results of the ANOVA are represented in Table 6. As the P-value is greater than 0.05, we can conclude that the mean values of the different treatments are equal to the significance level(α) to 1%.

Table 6. Variance analysis for inventory items in relation to treatment factor

Summary				
Group	Count	Sum	Average	Variance
BASE	11	16.141,4340	1.467,403098	14.303.406,75
MCNORM	11	16.141,4531	1.467,404828	14.303.398,9
MCUNIF	11	14.452,9346	1.313,903152	14.424.923,85
FTRI	11	16.141,3814	1.467,398309	14.303.394,65
FTRAP	11	16.141,4840	1.467,407643	14.303.415,83

ANOVA					
Variation source	SQ	gl	MQ	F	P-value
Between groups	207. 348,6573	4	51. 837,16432	0,003617966	0,999973
Within the groups	716.385.399,8	50	14.327.708		
Total	716.592.748,4	54			

Source: prepared by the author

The same procedure was used to compare the mean values of uncertainties. Table 7 presents the results for ANOVA. As a P-value is less than 0.05, we can conclude that the mean values of uncertainties generated in the different treatments are different from the significance level(α) to 1%.

Table 7. Analysis of variance for uncertainties in relation to the uncertainty factor

Summary				
Group	Count	Sum	Average	Variance
BASE	11	2,021101	0,183736455	0,021198294
MCNORM	11	0,0351145	0,003192227	0,00000640
MCUNIF	11	0,866201	0,078745545	0,0056746
FTRI	11	1,24539	0,113217273	0,01094068
FTRAP	11	1,24962	0,113601818	0,011017719

ANOVA					
Variation source	SQ	gl	MQ	F	P-value
Between groups	0,189020	4	0,047255247	4,837988863	0,002236
Within the groups	0,488376	50	0,009767539		
Total	0,677397	54			

To identify the differences between the different treatments for the uncertainty value, the Turkey Test was applied.

Using the data above, with $q = 4.01$ (5 treatments, 50 degrees of freedom) we found a DMS of 0,119492492.

Table 8 presents the comparisons between the means of the pairs of treatments and the DMS value.

Table 8. Tukey Test

Treatment	Average of the first term [m1]	Average of the second term [m1]	$\|m1 - m2\|$	Comparison
BASE-NORMAL	0,183736	0,00319	0,180544227	$\|m1 - m2\| > DMS$
BASE-UNIFORM	0,183736	0,07875	0,104990909	$\|m1 - m2\| < DMS$
BASE-TRI	0,183736	0,11322	0,070519182	$\|m1 - m2\| < DMS$
BASE-TRAP	0,183736	0,1136	0,070134636	$\|m1 - m2\| < DMS$
NORMAL-UNIFORM	0,003192	0,07875	0,075553318	$\|m1 - m2\| < DMS$
NORMAL-TRI	0,003192	0,11322	0,110025045	$\|m1 - m2\| < DMS$
NORMAL-TRAP	0,003192	0,1136	0,110409591	$\|m1 - m2\| < DMS$
UNIFORM-TRI	0,078746	0.1132	0,034471727	$\|m1 - m2\| < DMS$
UNIFORM-TRAP	0,078746	0,1136	0,034856273	$\|m1 - m2\| < DMS$
TRI-TRAP	0,113217	0,1136	0,000384545	$\|m1 - m2\| < DMS$

Source: prepared by the author

The Tukey Test recommends that the pairs compared have a significant difference in the level of significance chosen (in our case, 1%) when the difference between the mean values is greater than the DMS value.

Therefore, we can conclude that there is a significant difference between the uncertainties obtained in the treatment by the Monte Carlo Simulation and the base inventory. The other pairs do not present significant differences.

9 Final Considerations

The results of the research allow us to consider that the Monte Carlo simulation approaches and application of fuzzy set theory in the treatment of uncertainties in life cycle inventories (ICV) are effective, since such approaches allow the quantification of these uncertainties without significantly altering the pre-established values in the original inventory; values obtained through consensus among specialists and with consultations with important secondary sources, such as the technical literature of manufacturers and academic papers.

In several situations, the Life Cycle Inventory is presented to the community of practitioners and researchers in Life Cycle Assessment for global use, without the quantitative expression of uncertainty. In these cases, it is considered that the values are subject to the international standards of expression of measurement uncertainty, especially the Measurement Uncertainty Guide (ISO GUM 2008). The treatment of life cycle inventory data through Monte Carlo simulation and Fuzzy Set Theory resulted in the possibility of expression of ICV results with a lower degree of uncertainty than that used to express uncertainty under the standard.

The data treatment of the values of the life cycle inventory through the Monte Carlo Simulation with normal probability distribution presented the lowest values of uncertainty expression, with significant difference in relation to the base inventory, at the significance level of 1%. The Tukey Test results reinforce this fact.

Analyzing the sensitivity of the data treatments of the original inventory with the variation of the probability distribution, it is observed that there is no significant difference in the application of Monte Carlo simulation with normal distribution in relation to Monte Carlo simulation with uniform distribution. Regarding the uncertainties addressed by fuzzy set theory, there are also no significant differences between triangular and trapezoidal distributions. This leads us to the conclusion that there is no direct correlation between the calculation of uncertainties and the choice of probability distribution.

Therefore, as LCA studies have in most cases comparative character, aiming at the selection between conflicting choices, such post-treatment data are capable of generating inventories that lead to more consistent studies, allowing decisions making with a higher level of safety.

It should be emphasized that the low operational cost, the low computational effort and, above all, the non-requirement of programming skills by users are favorable factors to the use of spreadsheets in the treatment of data, both through Monte Carlo simulation and fuzzy set theory.

Future work may deepen the findings of the present study, applying them to other productive systems and even gradually expanding approaches to cradle-to-grave multifunctional and consequential approaches.

It will also be possible to apply the same approach to other phases of life cycle assessment, especially the Life Cycle Impact Assessment (LCA) phase and the interpretation of life cycle assessment results in order to compare the results.

From the point of view of spreadsheets, they can be improved through the application of macros and scripting language, but always considering the need for the human-machine interface to be as user friendly as possible and does not require more in-depth knowledge of programming by users.

References

1. ABNT NBR ISO 14040:2009. Gestão Ambiental – Avaliação do Ciclo de Vida – Princípios e Estrutura
2. Kloepffer, W.: Life cycle sustainability assessment of products. Int. J. Life Cycle Assess. **13**, 89–95 (2008)
3. Qin, Y., Suh, S.: What distribution function do life cycle inventories follow? Int. J. Life Cycle Assess. **22**, 1138–1145 (2017)
4. Althaus, H., et al.: Ecoinvent report No 1: Overview and methodology. Swiss centre for life cycle inventories, Dübendorf (2004)
5. Weidema, B.P., et al.: Reducing uncertainty in LCI: developing a data collection strategy. Environmental project No. 862. Danish environmental protection agency, Danish Ministry of the Environment, Ramboll, Denmark (2003)
6. Heijungs, R., Huijbregts, M.A.J.: A review of approaches to treat uncertainty in LCA (2004). http://www.iemss.org/iemss2004/pdf/lca/heijarev.pdf. Accessed 19 October 2017

7. Pizzol, M., et al.: Normalisation and weighting in Life Cycle Assessment: quo vadis? Int. J. Life Cycle Assess. **22**, 853–866 (2017)
8. Hung, M.-L., Ma, H.: Quantifying system uncertainty of Life Cycle Assessment based on Monte Carlo simulation. Int. J. Life Cycle Assess. **14**, 19–27 (2009)
9. Groen, E.A., Heijungs, R., Bokkeres, E.A.M., Boer, I.J.M.: Methods for uncertainty propagation in Life Cycle Assessment. Environ. Model Softw. **62**, 316–319 (2014)
10. Baustert, P., Benetto, E.: Uncertainty analysis in agente-based modelling and consequential life cycle assessment coupled model: a critical review. J. Clean. Prod. **156**, 378–394 (2017)
11. International Organization for Standardization. Avaliação de Dados de Medição: Guia para a expressão de incerteza de medição, 1ª edição brasileira (2008)
12. International Reference Life Cycle Data System (ILCD). General Guide for Life Cycle Assessment: detailed guidance. Luxemburgo, 398 p. (2010)
13. Huijbregts, M.A.J.: Application of uncertainty and variability in LCA. Part I: a general framework for the analysis of uncertainty and variability in the Life Cycle Assessment. Int. J. Life Cycle Assess. **5**, 273–280 (1998)
14. Groen, E.A., Heijungs, R.: Ignoring correlation in uncertainty and sensitivity analysis in life cycle assessment: quat the risk? Environ. Impact Assess. Rev. **62**, 98–109 (2017)
15. Baek, C., Park, K., Tahara, K., Chun, Y.: Data quality assessment of the uncertainty analysis applied to the greenhouse gas emissions of a dairy cowsystem. J. Sustain. **9**, 1676–1693 (2017)
16. Benini, L., Sala, S.: Uncertainty and sensitivity analysis of normalization factors to methodological assumptions. Int. J. Cycle Assess. **21**, 224–236 (2016)
17. Lo, S.-C., Ma, H., Lo, S.-L.: Quantifying and reducing uncertainty in life cycle assessment using the Bayesian Monte Carlo method. Sci. Total Environ. **340**, 23–33 (2005)
18. Ewertowska, A., Pozo, C., Gavaldà, J., Jiménez, L., Guillén-Gosálbez, G.: Combined use of life cycle assessment, data envelopment analysis and Monte Carlo simulation for quantifying environmental efficiencies under uncertainty. J. Clean. Prod. 166, 771–783 (2017)
19. Lanzuela, N.E., Sanchis, F.J.R., Señer, A.R., Polo, G.C., Vidal, A.P., Pollicer, N.S.: Uncertainty analysis in the environmental assessment of na integrated management system for restaurant and catering waste in Spain. Int. J. Life Cycle Assess. **20**, 244–262 (2015)
20. Geisler, G., Hellweg, S., Hungerbühler, K.: Uncertainty Analysis in Life Cycle Assessment (LCA): case study on plant-protection products and implications for decision making. Int. J. Life Cycle Anal. **10**, 184–192 (2005)
21. Kim, Y.-J.: Monte Carlo vs. Fuzzy Monte Carlo Simulation for uncertainty and global sensitivity analysis. Sustainability **9**, 539–552 (2017)
22. Suhariyanto, T.T., Wahab, D.A., Rahman, M.N.: Multi-life cycle assessment for sustainable products: a systematic review. J. Clean. Prod. **165**, 677–696 (2017)
23. Hongxiang, C., Wei, C.: Uncertainty analysis by Monte Carlo simulation in a life cycle assessment of water-saving project in green buildings. Inf. Technol. J. **12**, 2593–2598 (2013)
24. Buckley, J.J., Jowers, L.J.: Monte Carlo Methods in Fuzzy Optimization, p. 260. Springer, Berlin (2008)
25. Goedkoop, M., Heijungs, R., Huijbregts, M., Schryver, A., Struijs, J., Zelm, R.: Recipe 2008. A life cycle impact assessment method which comprises harmonised category indicators at the midpoint and the endpoint level (2009)
26. Ferreira, H., Leite, M.G.P.: A life cycle assessment study of iron ore mining. J. Clean. Prod. **108**, 1081–1091 (2015). https://doi.org/10.1016/j.jclepro.2015.05.140

Manufacturing Variability of 3D Printed Broadband Multi-frequency Metastructure

Adriano T. Fabro[1(✉)], Han Meng[2], and Dimitrios Chronopoulos[2]

[1] Department of Mechanical Engineering, University of Brasilia, Brasilia 709010-900, Brazil
fabro@unb.br

[2] Institute for Aerospace Technology and The Composites Group, University of Nottingham, Nottingham NG8 1BB, UK

Abstract. Additive manufacturing has been used to propose several designs of phononic crystals and metamaterials due to the low cost to produce complex geometrical features. However, like any other manufacturing process, it can introduce material and geometrical variability in the nominal design and therefore can affect the structural dynamic performance. Locally resonant metamaterials are typically designed such that the distributed resonators have the same natural frequency or, in the case of rainbow metastructures, a well-defined spatial profile. In this work, manufacturing tolerances of beam samples produced from a Selective Laser Sintering process are assessed and variability levels are used to investigate the vibration suppression performance of broadband multi-frequency metastructures. Evenly spaced non-symmetric resonators are attached to a beam with U-shaped cross-section and partitioned by parallel baffled plates. An analytical model based on a transfer matrix approach is used to calculate transfer receptance due to a point time harmonic force. Moreover, a random field model is assumed based in previous experimental results and the effects of the correlation length, a measure of the spatial fluctuation, are also investigated for individual. The obtained results are expected to be useful for further robust design in mass produced industrial applications.

Keywords: Vibration attenuation · Rainbow metamaterial · Additive manufacturing · Uncertainties

1 Introduction

Additive manufacturing (AM) has been used to propose several designs of phononic crystals and metamaterials due to the low cost to produce complex geometrical features and has the potential to make them feasible in several industrial applications [1] along with the use of smart structures (e.g. [2]). However, like any other manufacturing process, AM can introduce material and geometrical variability in the nominal design [3, 4] and therefore affect the structural dynamic performance. Currently, there is few experimental work available in the literature investigating metamaterial performance for vibration attenuation applications (e.g. [5–10]) and most of them do not consider the effects of spatially correlated disorder on the band gap.

J. E. S. De Cursi (Ed.): Uncertainties 2020, LNME, pp. 198–208, 2021.
https://doi.org/10.1007/978-3-030-53669-5_15

It has been shown that the break of periodicity introduced by the spatial variability of material and geometrical properties of metastructures creates a resonator mistuning which can induce wave trapping and thus greatly affecting the vibration attenuation performance by either band gap annihilation or attenuation bandwidth widening, depending on thc imposed spatial profile [6]. Overall, there is a need to investigate the effects of the manufacturing variability on the performance of metastructures [1], which can also affect the coupling of the metastructures to the vibration source and receiver. Near-periodic structures are referred herein as systems where material or geometric properties vary spatially following a deterministic function or a random field.

In this paper, the results presented by Fabro et al. [11] are further explored with additional numerical samples. The effects of the break of periodicity caused by additive manufacturing variability on the attenuation performance of a multi-frequency metastructure is investigated. Numerical results are presented for some highlighted individual samples numerically generated from the proposed random field model considering only randomness on the resonators tip masses. The effects of individual samples of the metastructures are analysed and the main physical consequences of the break in the periodicity are discussed.

2 Metamaterial Beam with Non-symmetric Resonators

The proposed metastructure and an analytical model for beam with two non-symmetric resonators, shown in Fig. 1, is briefly presented, such that the frequency response function can be efficiently predicted. The metastructure is composed of a total of 17 unit cells with two non-symmetric cantilever-mass attached and acting like local resonators and the presented analytical model was previously proposed and experimentally validated by Meng et al. [12, 13]. It assumes a Euler-Bernoulli beam theory for both the baseline host structure and the cantilever-mass resonator. Each resonator is modelled as a beam point attached to the baseline structure at one end and a lumped mass with negligible inertia at the other. The resonators are modelled as point attachments displacements at the n-th segment and the periodic plate insertions are modelled as added masses to the Π-shaped beam at the ends of each segment. A detailed description of the analytical model can be found in [12].

For a periodic structure, the transfer matrix $\mathbf{T}_n$ is identical for every segment. Consequently, according to the Bloch theorem [14, 15], a wavenumber k_n can be defined from the propagation constant $\mu_n = \exp(-ik_nL_d)$, which is given from the transfer matrix of the periodic metamaterial such that

$$|\mathbf{T}_n - \mu_n\mathbf{I}| = 0. \tag{1}$$

However, this assumption does not hold when variability effects are taken into account thus breaking the periodicity of the metastructure. In this sense, an equivalent wavenumber k_T is also defined from the propagation constant $\mu_T = \exp(-ik_TL)$ of the finite length metamaterial with spatially varying properties, i.e. near-periodic, as

$$|\boldsymbol{\Xi} - \mu_T\mathbf{I}| = 0, \tag{2}$$

The equivalent wavenumber k_T can be also interpreted as the wavenumber when the finite rainbow metamaterial is a periodic unit of an infinite complex beam and it represents the total phase change and attenuation of a travelling wave over m unit cells. It is used as a tool to understand the wave-like behaviour of non-periodic structures. Note that the total phase and attenuation change $k_T L$, which is a complex number for the case of metastructure, is not generally equal to the contribution of the phase and attenuation change at each segment in the near periodic metastructure, i.e., $k_T L \neq \sum_{n=1}^{m} k_n L_d$. This is because of the additional scattering created due to the differences in neighboring segments. However, it can be shown that $k_T L \approx \sum_{n=1}^{m} k_n L_d$ for the case of slowly varying properties in which this internal scattering is negligible [16].

3 Manufacturing Variability on the Periodic Design

In this section, the material and geometrical variability introduced by the additive manufacturing process in the nominal design is briefly discussed based on the experimental results from a 3D printer. Several manufacturing parameters in have to be taken in to account [4, 17, 18] but for comprehensive UQSM on structural dynamics applications, focus should be given in macro or meso scale mechanical parameters [19]. In this sense, Machado et al. [20] estimated spatially correlated variability of the Young's modulus, modelled a Gaussian random field using a model updating approach. Beli et al. [6] proposed a random field identification approach based on sides blocks printed along the metastructure. They have shown that spatial variability plays a major role on the vibration response of locally resonating metastructures. Most importantly, both works have shown that the available analytical models for spatial correlation functions can accurately represent the spatial uncertainty of polyamide beams produced by Selective Laser Sintering (SLS).

An experimental investigation is carried out with a total of 15 test samples produced by SLS. The test samples are beams with prismatic cross-section with 10 mm $\times$ 4 mm $\times$ 80 mm. The obtained mean value, standard deviation and coefficient of variation COV, i.e., the standard deviation over mean value, are presented in Table 1 for the mass, density and Young's modulus. The measured dimensions of the test samples presented negligible variability, which shows that the SLS is much more accurate for geometric than for material properties. More details are presented in Fabro et al. [11]. This result is similar to the obtained by Beli et al. [6]. The considered metastructure is 490 mm length with 17 unit cells, therefore the test samples are approximately 2.5 larger than the unity cell. It means that the Young's modulus and mass density spatial variability of the test sample is averaged out more significantly than for the unit cell, thus reducing the overall dispersion. This effect is similar to the local average process as described by Vanmarke [21], which does not affect the mean value but can reduce the estimated variance.

Table 1. The mean value, standard deviation and coefficient of variation COV of the Young's modulus, density and mass of the 15 test samples.

	E, MPa	ρ, kg/m^3	m, g
Mean	1621.7	948.9	3.16
Standard deviation	49.9	7.4	0.03
COV, %	3.07	0.78	0.95

Although the obtained results are very specific to the printer and the printing process used to manufacture the metastructure, the analytical model presented in the previous section requires the characterization of a few parameters, such as the Young's modulus and mass density. Typically, this information be made available from the printer's manufacturer or other reliable source. In addition to the dispersion and correlation between the Young's modulus and mass density, a random field model has to be introduced to properly represent the mistuning of the resonators in the metastructure, i.e., the spatial changes also introduced by manufacturing variability.

3.1 Resonators Mistuning and Near-Periodicity

One of the most important effects for the vibration attenuation performance is the mistuning of the resonators [5, 6], which is caused by the spatial profile on the material properties created due to the manufacturing variability. In this section, the effects of the mistuning of resonators are modelled. It is assumed that the material variability can be represented by a change on the resonators mass, given by a lumped parameter in the analytical model, and by the Young's modulus, assumed constant within each unity cell.

To include the random spatial variability on the analytical model, a random field model is assumed based on available experimental results in 3D printed structures [6]. Random fields are multidimensional random processes and can be used to model spatially distributed variability using a probability measure [21]. The random field is Gaussian if its random variables are Gaussian and it can be completely defined by its mean value, standard deviation and correlation function $C(\tau)$, where τ is the lag, i.e. the distance between any two points in the homogeneous random field. The manufacturing variability is assumed to be given by a Gaussian Homogeneous random field $H_1(x)$ for each mass, i.e. the both masses at each unity cell have the same statistical properties, and $H_2(x)$ for the Young's modulus, both with same correlation function $C(\tau) = \exp(-\tau/c_l)$, where c_l is the correlation length, i.e., the level of statistical fluctuation of the spatial variability. In other words, this parameter controls the smoothness of the spatial variation, given the spatial profile of the tip masses of the resonators along the beam.

Due to the discrete nature of the resonators tip mass, the random fields $H_{1,2}(x)$ are discretized over the metamaterial length such that it is approximated by the random vector $\boldsymbol{\xi}_{\mathrm{m}}$. The discretization process also results in a correlation matrix $\mathbf{C}$ from $C(\tau)$ [22]. Similarly, the Young's modulus that is assumed constant within each unit cell can be represented by a random vector $\boldsymbol{\xi}_{\mathrm{E}}$. This approach is known as the midpoint method, first introduced by Der Kiureguian [23] for random field discretization. Moreover, experimental results presented by Fabro et al. [11] suggests that both random fields are correlated, with correlation coefficient $\rho_{E,m} = 0.6359$.

A realization of the random field given by both tip masses and Young's modulus at each unit cell are respectively given by

$$\mathbf{m}_{1,2} = m_{0_{1,2}}\left(1 + \sigma_m \boldsymbol{\xi}_{\mathrm{m}}\right), \mathbf{E} = E_0\left(1 + \sigma_E \boldsymbol{\xi}_{\mathrm{E}}\right), \tag{3}$$

where $m_{0_{1,2}}$ and E_0 are the nominal values of the resonators tip masses and Young's modulus in the periodic design, 1 is a vector $m \times 1$ vector filled with 1 and $\sigma_{E,m}$ controls the level of statistical dispersion of the random vector. From this expression, both random fields are subsequently correlated using the estimated correlation coefficient $\rho_{E,m}$.

4 Numerical Results

In this section, a numerical analysis is carried out considering the mean value of density and Young's modulus given in Table 1, Poisson ratio $\nu = 0.3$ and structural damping $\eta = 0.02$. The baseline or host beam presents a Π-shaped cross-section with height $H_d = 10$ mm and width $w_d = 51$ mm, side wall thickness $t_d = 2$ mm, bottom plate thickness $b_d = 5$ mm and lateral plate thickness $t_w = 2$ mm distanced by $L_d = 15$ mm, with a total of 17 unit cells and total beam length $L_T = 257$ mm. The full description of the metamaterial beam in found in [11]. The tip mass in both resonators at each unit cell is assumed to have a mean value of 1 g. Note that, in this case, the difference in length of the beams produces the differences in the resonance frequencies and therefore in the band gaps.

Figure 1 (a) presents the amplitude of the transfer receptance of proposed metastructure considering the nominal periodic design from 0 Hz to 350 Hz at steps of 0.5 Hz and Fig. 1 (b) presents the real and imaginary parts of the normalized equivalent wavenumber from a unit cell, as described in Eq. (1). Note that $|\mathrm{Re}\{k_n L_d/\pi\}| \leq 1$ due to the periodic nature of the metastructure, where L_d is the length of periodic unit cell. For a lossless waveguide, real wavenumber indicates purely propagating waves, imaginary wavenumber indicates evanescent waves while complex wavenumber indicates oscillating and decaying waves.

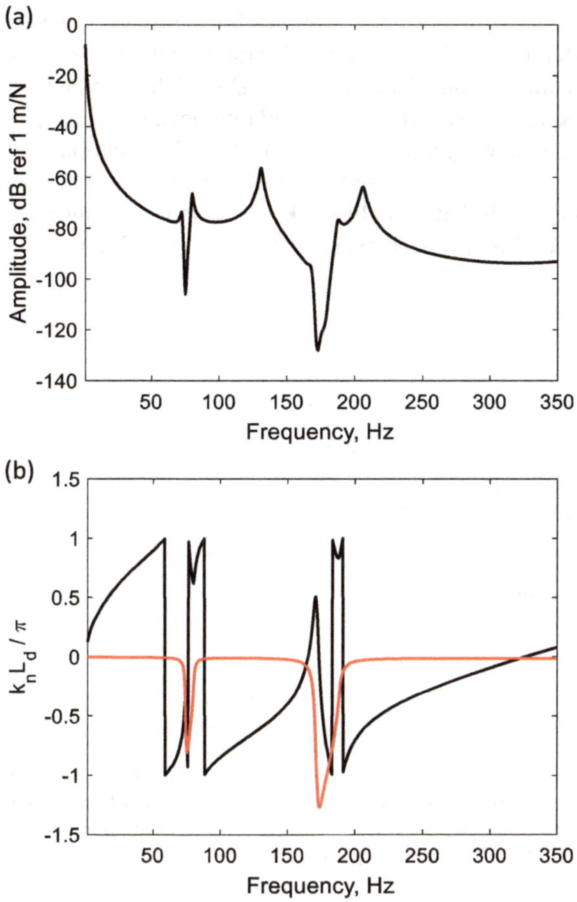

Fig. 1. (a) Amplitude of the transfer receptance of proposed metastructure considering the nominal periodic design and (b) real (black) and imaginary (red) parts of the normalized equivalent wavenumber Adapted from [11].

4.1 Effects of the Manufacturing Variability

In this section, the effects of manufacturing uncertainty are considered on the tip mass of both resonators and on the Young's modulus of the metastructure, represented by $\mathbf{m}_1$, and $\mathbf{m}_2$ as described in Sect. 3.1. Note that the additional variability of the Young's modulus is not taken into account in the results presented in this paper but its effects can be found in [11]. It is considered that the tip masses of both resonators at each unit cell are identical, which is equivalent to assume that $\mathbf{m}_1 = \mathbf{m}_2$, and the dispersion parameters $\sigma_m = 0.10$, which is significantly higher than the values shown in Table 1. Additionally, a correlation length is assumed varying as $c_l = 0.25\,L$, $c_l = 0.50\,L$, $c_l = 0.75\,L$ and $c_l = 1.00\,L$. This is to further understand the effects of the correlation length in the band gap performance. The effects of individual samples of the metastructures are analysed and the main physical consequences of the break in the periodicity are discussed.

Figure 2 presents the random field samples for the tip masses $\mathbf{m}_1 = \mathbf{m}_2$ with same dispersion but different correlation length. The random fields are numerically generated by a Cholesky decomposition of the kind $\mathbf{C} = \Sigma\Sigma^{\mathrm{T}}$. Then, the lower triangular matrix $\boldsymbol{\Sigma}$ can be used to generate a vector of correlated random number by $\boldsymbol{\xi} = \boldsymbol{\Sigma}\boldsymbol{\zeta}$, in which $\boldsymbol{\zeta}$ is a vector of uncorrelated independent and identically (i.i.d.) random numbers. In practice, the vector $\boldsymbol{\zeta}$ can be created from a random number generator. The samples presented in Fig. 2 are generated such that $\boldsymbol{\zeta}$ is always kept the same and then only the correlation matrix $\mathbf{C}$ is updated. This approach excludes the statistical fluctuation from the analysis.

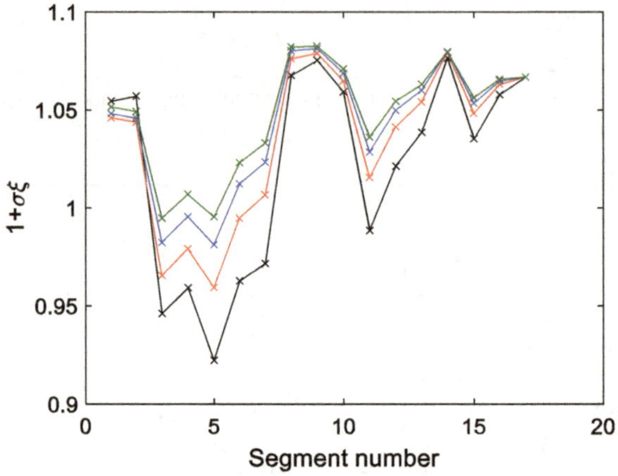

Fig. 2. Highlighted cases of independent realizations of the tip masses (a) $\mathbf{m}_1$ and (b) $\mathbf{m}_2$ assuming the same dispersion $\sigma_m = 0.10$ but different correlation lengths c_l and with dispersion $c_l = 0.25\,L$ (red line), $c_l = 0.50\,L$ (blue line), $c_l = 0.75\,L$, (green line) and $c_l = 1.00\,L$ (black line). Both masses are normalized by the nominal values m_0.

Figure 3 presents the corresponding amplitude of the transfer receptance of the metastructure for the highlighted cases and the imaginary part of the normalized equivalent wavenumber for each of the samples. Note that each isolation band in the vibration response corresponds to a dip in the imaginary part of the equivalent wavenumber. Moreover, the shorter correlation length induces a slightly smaller maximum attenuation for each corresponding band gap region but with a slightly wider attenuation band. It is not shown here, but other random samples present worse or better attenuation performances. Figure 4 presents the amplitude of the receptance along the metastructure for each of the highlighted cases. Note that for the shorter correlation length, the band gap region is divided in two with a critical zone, also known as a turning point, from which the wave is propagating, consequently with significant vibration amplitude, to a non-propagating region, with negligible vibration amplitude. This is known as wave trapping and explain the jagged profile in the band gap frequency band of the vibration response. This effect shows that the correlation length of random spatially varying resonators can potentially be explored for improving the attenuation performance of metastructures.

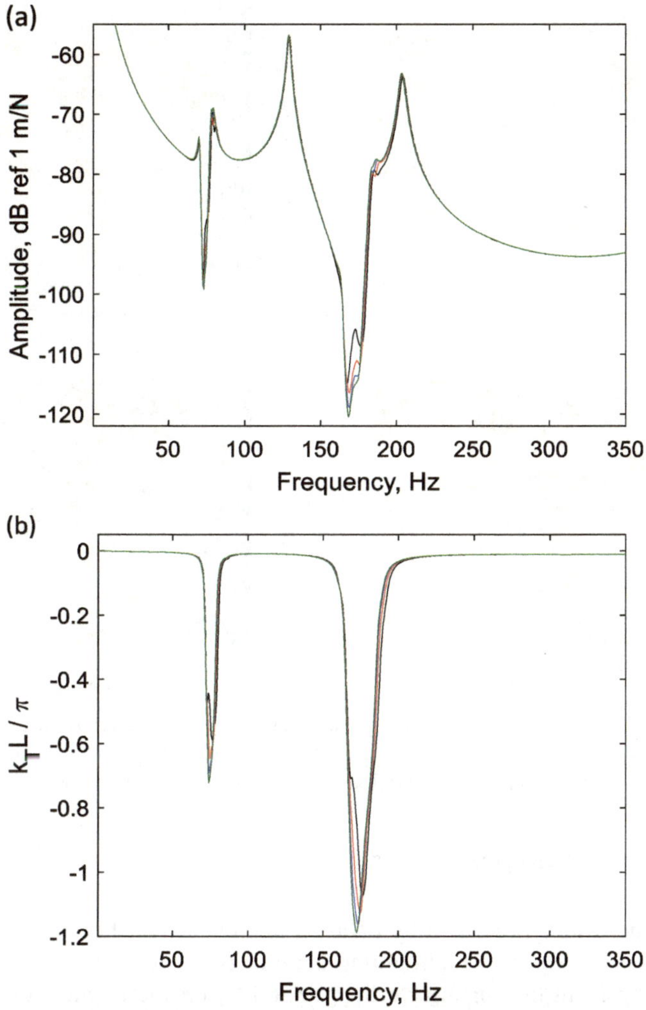

Fig. 3. (a) Amplitude of the transfer receptance of the metastructure for the highlighted cases and (b) the corresponding imaginary part of the normalized equivalent wavenumber. Independent realizations of $\mathbf{m}_1 = \mathbf{m}_2$ are considered with the same dispersion $\sigma_m = 0.10$ and with correlation length $c_l = 0.25$ (black line), $c_l = 0.50$ (red line), $c_l = 0.75$ (blue line) and $c_l = 1.00$ (green line).

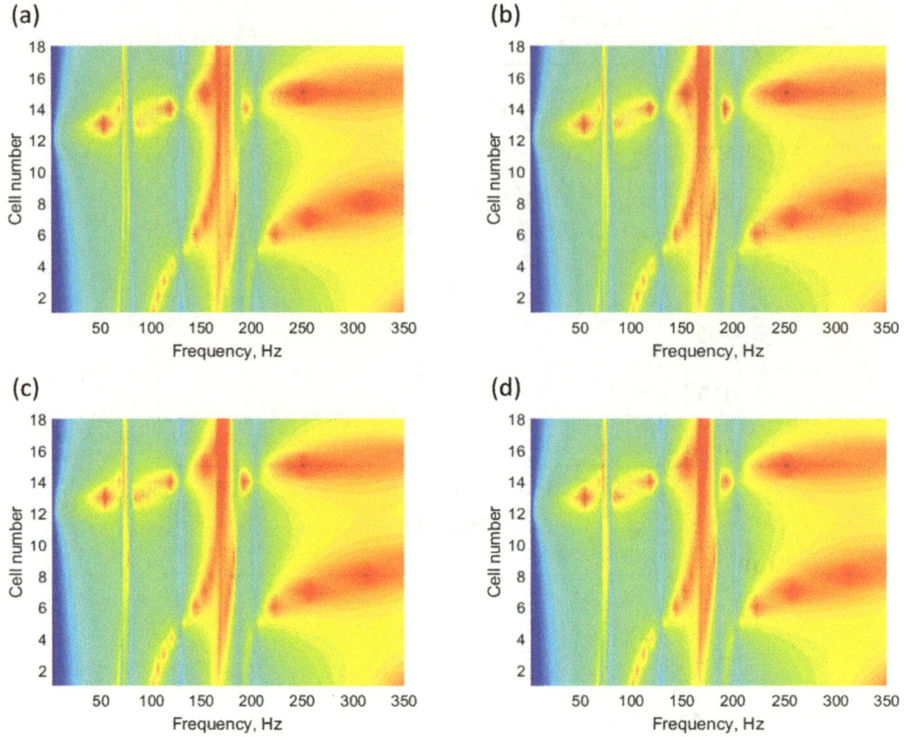

Fig. 4. Amplitude of the receptance along the metastructure and excitation at cell number 1 assuming dispersion $\sigma_m = 0.10$ (a) $c_l = 0.25$, (b) $c_l = 0.50$, (c) $c_l = 0.75$, (d) $c_l = 1.00$. Red colours indicate low amplitude and blue colours indicate high amplitude levels.

5 Concluding Remarks

In this paper the results presented by Fabro et al. [11] are further explored with some additional numerical samples highlighting the effects of the spatial correlation length, typically arising from the manufacturing variability, on the attenuation performance of the locally resonant metastructures. It is shown that shorter correlation lengths can induce wider band gaps but with smaller attenuation performance. The presented results highlight that the random spatial correlation in the frequency of the resonators along the metastructure can be potentially explored to further improve or optimize the vibration response of metastructures. The obtained results are expected to be useful for further robust design in mass produced industrial applications.

Acknowledgments. The authors would like to acknowledge the support acquired by the H2020 DiaMoND project (Grant Agreement ID: 785859), Royal Society Grant: PURSUIT, the Brazilian National Council of Research CNPq (Grant Agreement ID: 420304/2018-5) and the Federal District Research Foundation FAPDF (Grant Agreement ID: 0193.001507/2017).

References

1. Bertoldi, K., Vitelli, V., Christensen, J., van Hecke, M.: Flexible mechanical metamaterials. Nat. Rev. Mater. **2**, 17066 (2017). https://doi.org/10.1038/natrevmats.2017.66
2. Fan, Y., Collet, M., Ichchou, M., Li, L., Bareille, O., Dimitrijevic, Z.: A wave-based design of semi-active piezoelectric composites for broadband vibration control. Smart Mater. Struct. **25**, 055032 (2016). https://doi.org/10.1088/0964-1726/25/5/055032
3. Hague, R., Campbell, I., Dickens, P.: Implications on design of rapid manufacturing. Proc. Inst. Mech. Eng. Part C J. Mech. Eng. Sci. **217**, 25–30 (2003). https://doi.org/10.1243/095440603762554587
4. Goodridge, R.D., Tuck, C.J., Hague, R.J.M.: Laser sintering of polyamides and other polymers. Prog. Mater Sci. **57**, 229–267 (2012). https://doi.org/10.1016/j.pmatsci.2011.04.001
5. Sugino, C., Xia, Y., Leadenham, S., Ruzzene, M., Erturk, A.: A general theory for bandgap estimation in locally resonant metastructures. J. Sound Vib. **406**, 104–123 (2017). https://doi.org/10.1016/j.jsv.2017.06.004
6. Beli, D., Fabro, A.T., Ruzzene, M., Arruda, J.R.F.: Wave attenuation and trapping in 3D printed cantilever-in-mass metamaterials with spatially correlated variability. Sci. Rep. **9**, 5617 (2019). https://doi.org/10.1038/s41598-019-41999-0
7. Celli, P., Yousefzadeh, B., Daraio, C., Gonella, S.: Bandgap widening by disorder in rainbow metamaterials. Appl. Phys. Lett. **114**, 091903 (2019). https://doi.org/10.1063/1.5081916
8. Van Belle, L., Claeys, C., Deckers, E., Desmet, W.: Measurement of dispersion curves for locally resonant metamaterials with damping. In: Proceedings of the 27th International Conference on Noise and Vibration Engineering (ISMA 2016) (2016)
9. Droz, C., Robin, O., Ichchou, M., Atalla, N.: Improving sound transmission loss at ring frequency of a curved panel using tunable 3D-printed small-scale resonators. J. Acoust. Soc. Am. **145**, EL72–EL78 (2019)
10. Jung, J., Kim, H.-G., Goo, S., Chang, K.-J., Wang, S.: Realisation of a locally resonant metamaterial on the automobile panel structure to reduce noise radiation. Mech. Syst. Signal Process. **122**, 206–231 (2019). https://doi.org/10.1016/j.ymssp.2018.11.050
11. Fabro, A.T., Meng, H., Chronopoulos, D.: Uncertainties in the attenuation performance of a multi-frequency metastructure from additive manufacturing. Mech. Syst. Signal Process. **138**, 106557 (2020). https://doi.org/10.1016/j.ymssp.2019.106557
12. Meng, H., Chronopoulos, D., Fabro, A.T., Elmadih, W., Maskery, I.: Rainbow metamaterials for broadband multi-frequency vibration attenuation: numerical analysis and experimental validation. J. Sound Vib., 115005 (2019). https://doi.org/10.1016/j.jsv.2019.115005
13. Meng, H., Chronopoulos, D., Fabro, A.T., Maskery, I., Chen, Y.: Optimal design of rainbow elastic metamaterials. Int. J. Mech. Sci., 105185 (2019). https://doi.org/10.1016/j.ijmecsci.2019.105185
14. Bloch, F.: Über die Quantenmechanik der Elektronen in Kristallgittern. Z. Für Phys. **52**, 555–600 (1929). https://doi.org/10.1007/BF01339455
15. Zhang, H., Xiao, Y., Wen, J., Yu, D., Wen, X.: Flexural wave band gaps in metamaterial beams with membrane-type resonators: theory and experiment. J. Phys. Appl. Phys. **48**, 435305 (2015). https://doi.org/10.1088/0022-3727/48/43/435305
16. Fabro, A.T., Ferguson, N.S., Jain, T., Halkyard, R., Mace, B.R.: Wave propagation in one-dimensional waveguides with slowly varying random spatially correlated variability. J. Sound Vib. **343**, 20–48 (2015). https://doi.org/10.1016/j.jsv.2015.01.013
17. Huang, Y., Leu, M.C., Mazumder, J., Donmez, A.: Additive manufacturing: current state, future potential, gaps and needs, and recommendations. J. Manuf. Sci. Eng. **137**, 014001-014001–10 (2015). https://doi.org/10.1115/1.4028725

18. Moges, T., Ameta, G., Witherell, P.: A review of model inaccuracy and parameter uncertainty in laser powder bed fusion models and simulations. J. Manuf. Sci. Eng. **141**, 040801-040801–14 (2019). https://doi.org/10.1115/1.4042789
19. Wang, Z., Liu, P., Ji, Y., Mahadevan, S., Horstemeyer, M.F., Hu, Z., Chen, L., Chen, L.-Q.: Uncertainty quantification in metallic additive manufacturing through physics-informed data-driven modeling. JOM **71**, 2625–2634 (2019). https://doi.org/10.1007/s11837-019-03555-z
20. Machado, M.R., Adhikari, S., Dos Santos, J.M.C., Arruda, J.R.F.: Estimation of beam material random field properties via sensitivity-based model updating using experimental frequency response functions. Mech. Syst. Signal Process. **102**, 180–197 (2018). https://doi.org/10.1016/j.ymssp.2017.08.039
21. Vanmarcke, E.: Random Field: Analysis and Synthesis. World Scientific, Cambridge (2010)
22. Ghanem, R., Spanos, P.D.: Stochastic Finite Elements: A Spectral Approach. Dover Publications, Minneola (2012)
23. Der Kiureghian, A., Ke, J.-B.: The stochastic finite element method in structural reliability. Probabilist. Eng. Mech. **3**, 83–91 (1988)

Data Analysis and Identification

UAV Autonomous Navigation by Image Processing with Uncertainty Trajectory Estimation

Gerson da Penha Neto[1], Haroldo Fraga de Campos Velho[1(✉)], and Elcio Hideiti Shiguemori[2]

[1] National Institute for Space Research, São José dos Campos, SP, Brazil
{gerson.penha,haroldo.camposvelho}@inpe.br
[2] Department of Aerospace Science and Technology, São José dos Campos, SP, Brazil
elcio@ieav.cta.br

Abstract. Unmanned Aerial Vehicles (UAV) is a technology under strong development, with application on several fields. For the UAV autonomous navigation, a standard scheme is to use signal from a Global Navigation System by Satellite (GNSS) onboard. However, such signal can suffer natural or human interference. Our approach applies image processing procedure for the UAV positioning: image edge extraction and correlation between drone image and georeferenced satellite image. A data fusion is also applied, for combining the inertial sensor data and positioning by image. The data fusion is performed by using neural network. The output from the data fusion neural network is the correction for the UAV trajectory. Here, the variance of the trajectory error is also predicted to quantify the uncertainty.

Keywords: Unmanned Aerial Vehicles · Autonomous navigation · Image processing · Self-configuring neural network · Uncertainty quantification

1 Introduction

Unmanned Aerial Vehicles (UAV) application has been on the rise in recent years. UAV can be used in various situations such as rescue missions or to supply resources, environmental and climate monitoring, agriculture, tourism and entertainment industries, and for situations where sending people can be very risky [9,10,12,16]. UAVs are also known as *drones*. Usually, the UAV navigation is done by a remote pilot or specialized software, requiring a communication with a ground station. Many UAV systems use IEEE 802.11-based wireless technologies and these technologies have limited range operation [11,14].

Another way of UAV navigation is the autonomous navigation [8,20]. For this type of navigation, the aircraft is controlled by an onboard computer. Therefore, the autonomous navigation does not require a human pilot, ground station, or radio control. The UAV computer uses information from onboard sensors, i.e.,

J. E. S. De Cursi (Ed.): Uncertainties 2020, LNME, pp. 211–221, 2021.
https://doi.org/10.1007/978-3-030-53669-5_16

from the inertial navigation system (INS), global navigation satellite system (GNSS), barometer, and other sensors with information about the aircraft attitude. Indeed, GNSS signal suffers impact from ionospheric irregularities [3,19].

When the GNSS signal fails or missing, alternative approaches must be employed for the autonomous navigation. Our approach combines inertial sensor data and UAV positioning estimated by computer vision using a neural network. The data fusion is performed by a *second* neural network (NN). The architecture for both neural networks is determined by an optimization problem procedure. UAV trajectories found by the neural fuser were compared with measurements made by a real-time kinematic GNSS (RTK) device.

Here, the uncertainty quantification is not computed from a Monte Carlo type method [5]. Different from our previous results [17,18], the neural data fuser is designed with double outputs: a new (correction) UAV positioning, and the variance associated to the uncertainty of the UAV position estimation.

The results show a very good trajectory estimation by the neural fuser, and it was possible to calculate the confidence interval for the estimated UAV trajectory from the predicted variance.

2 Multi-sensor Data Fusion, Image Matching, Selfing-Configured Artificial Neural Network

One solution already presented for multi-sensor data fusion for UAV autonomous navigation is to use a data fuser based on an extended Kalman Filter [15]. For the cited data fusion, the UAV position is estimated by coupling computer vision technique and a signal from an INS sensor [16]. Figure 1 shows a scheme for estimating aircraft position using a data fusion and image matching.

Here, the data fuser is based on a selfing-configured artificial neural network (ANN), emulating the Kalman filter for data fusion. The employed neural network is a multi-layer perceptron (MLP-NN), with the learning phase executed by the back-propagation algorithm.

The computer vision technique uses edge detection from a georeferenced image, herein called *reference image*, and from an image obtained by the UAV camera. Another selfing-configured artificial neural network has been defined for edge detection [4,6,13]. The maximum correlation between the reference image and the image obtained by the drone indicates the UAV geographical position. Figure 2 shows the scheme used for image matching.

The architecture for the both cited ANNs is computed as an optimization problem. The objective is to determine optimal values for the neural network parameters. For determining the optimal topology for the neural network, the metaheuristic named multiple particle collision algorithm (MPCA) [15] was applied – see also the MPCA applications [1,2] and [7]. The cost function $J(\mathbf{Q})$ to be minimized is a combination of two factors: a metric for the ANN complexity – the first term for the right hand side in the below equation, and the quadratic difference between the desired values and the ANN output:

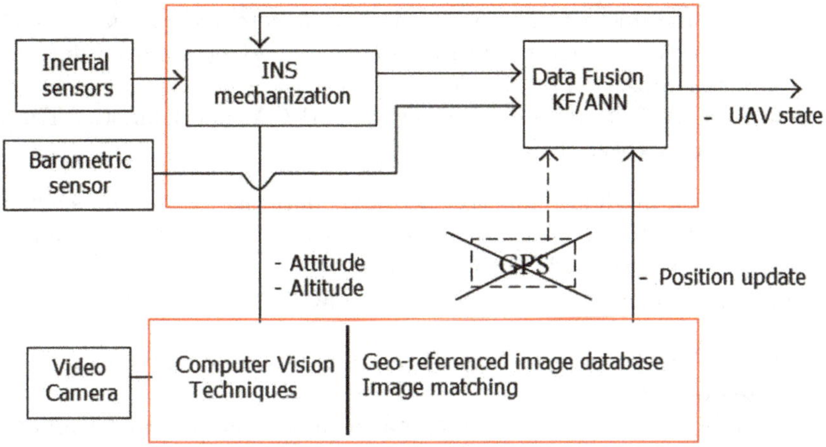

Fig. 1. Scheme for estimating aircraft position; Using a data fusion and image matching.

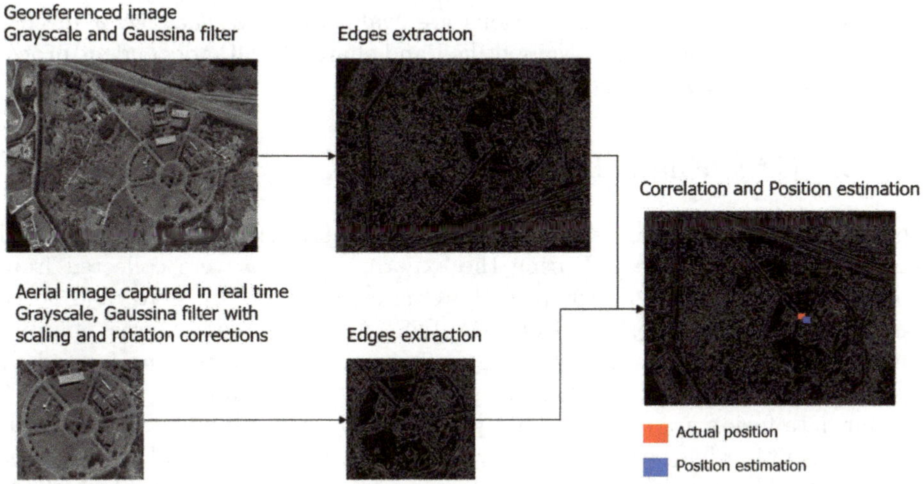

Fig. 2. Image matching scheme.

$$J(\mathbf{Q}) = \Phi(x,y) \times \left[\frac{\rho_1\, E_{\text{train}}(\mathbf{Q}) + \rho_2\, E_{\text{gen}}(\mathbf{Q})}{\rho_1 + \rho_2}\right] \tag{1}$$

$$\Phi(x,y) = \theta_1\,[e^{x^2}] + \theta_2\,[y] + 1 \tag{2}$$

where $\mathbf{Q}$ is the unknown vector; E_{train} and E_{gen} are respectively the training and generalization errors, ρ_1 and ρ_2 relates to the balance between training (E_{train}) and generalization (E_{gen}) errors; $\Phi(x,y)$ is a measurement on neural network *complexity*, with x the number of neurons and y corresponds to the number of epochs until the convergence during the training phase, θ_1 and θ_2

are adjustment parameters. This way, the neural network optimal topology is looking for the simplest architecture with the smallest errors from datasets for training and generalization.

The functional $J(\mathbf{Q})$ is minimized by the MPCA metaheuristic. The algorithm can be summarized as following [15]:

1. Starting from a candidate solution $\mathbf{Q}_m$, the operator *exploration* is activated, where a new candidate $\mathbf{Q}_{k+1}$ is randomly selected inside from a small sphere around the previous candidate $\mathbf{Q}_m$.
2. If $J(\mathbf{Q}_{k+1}) < J(\mathbf{Q}_k)$, the new candidate is accepted, and the convergence is verified.
3. If $J(\mathbf{Q}_{k+1}) \geq J(\mathbf{Q}_k)$, a propability is computed to decide if the new candidate is accepted or not.
4. If the candidate $\mathbf{Q}_{k+1}$ is not accepted, a new solution will be found in a different region of the search space, where the operator *scattering* is activated to find a new candidate with the convergence being checked again.
5. Many particles (candidate solutions) are part of the process, where each particle follows steps (1)–(4), independently, up to a value of iterations is reached, when different sub-optimal solutions are evaluated by a *black-board* mechanism, and the best solution is identified and shared to all independent procedures of searching by *particle collisions*.

3 The UAV Flight Experiment

The flight experiment was carried out in Southeastern Brazil. The aircraft used was the Matrice 600 UAV. During the experiment, data were collected from embedded sensors, such as compass, barometer, GNSS, INS, gyro, and servo motor position, among others. Here, the INS (Accelerometers, Gyros), barometer, and video camera were used. Figure 3 shows the trajectory performed by the UAV. The Quickbird satellite image was used as a reference.

Our data fusion approach was not performed by a Kalman filter – see Fig. 1. The Kalman filter has been replaced by an selfing-configured neural network: the neural fuser. The best architecture for the neural fuser was defined by the MPCA. The neural fuser was implemented on dedicated hardware by using FPGA (Field Programmable Gate Array). As mentioned before, the ANN is configured with as fewer neurons as possible. Such feature is also important to implement the neural processing saving energy consumption and for taking into account the use of FPGA as a co-processor [17].

The MPCA algorithm was employed to determine the simplest architecture with the best performance. The performance means the difference between the position estimation by the neural fuser and the position obtained by the high precision GNSS device. A total of 13,880 temporal data were used for self-configured ANN. The dataset was separated as follows: 70% were used for training phase, 10% for the generalization phase, and 20% for the validation. Each partition of

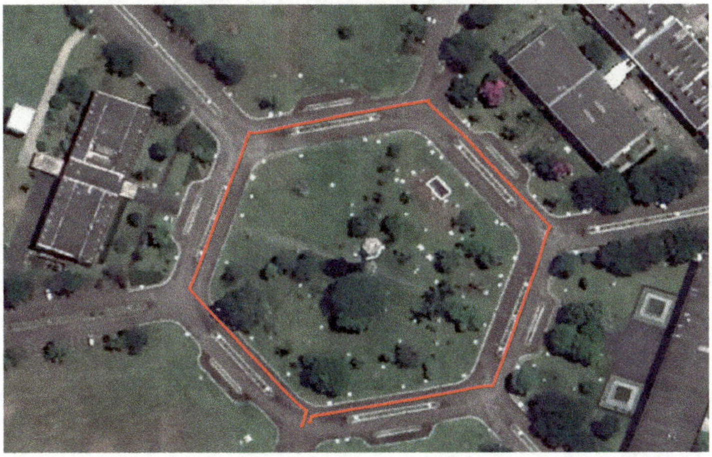

Fig. 3. UAV trajectory

the dataset defines a vector with values of estimated (by image processing) latitude and longitude, inertial sensor data (Accelerometers, Gyros), and altitude obtained by a barometric sensor.

For the neural fuser, the best architecture determined by the MPCA has two hidden layers with ten neurons per hidden layer. Table 1 presents a summary of the architecture obtained by the MPCA.

Table 1. ANN architecture to neural fuser

Parameter	Value computed by MPCA
Neurons in the input layer	12
Number of hidden layers	2
Neurons in the hidden layer	10
Neurons in the output layer	2
Learning rate	0.7
Rate of momentum	0.52
Activation Function	Sigmoid

4 Results

The neural fuser combines data from onboard sensors and the position estimation from the image matching. Figure 4 shows the path taken by the aircraft and the position estimation carried out by the neural fuser. Estimates were made for the latitude, longitude, and attitudes of the aircraft – the Euller, roll, pitch, and yaw angles.

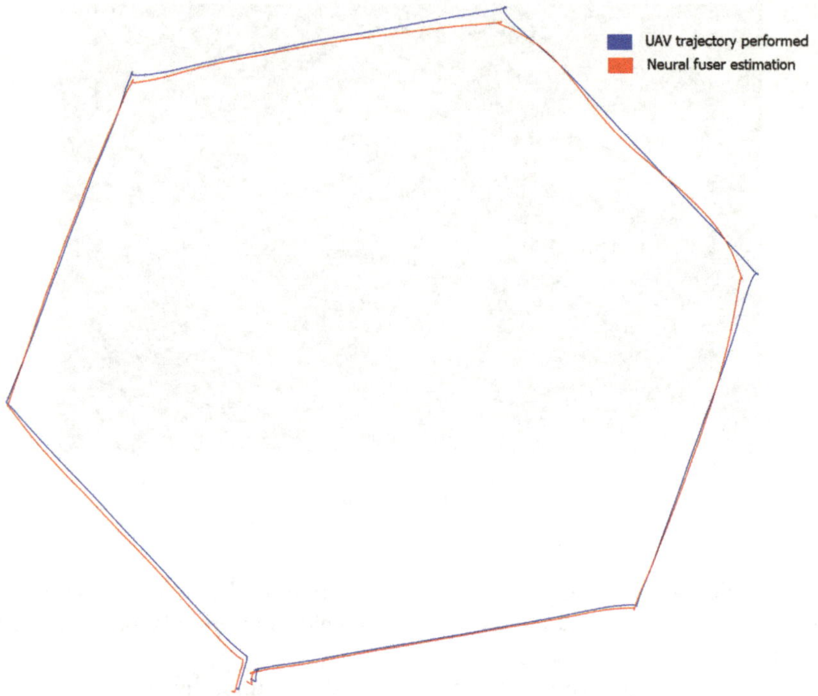

Fig. 4. Position estimation performed by neural fuser.

The GNSS device was used for the ANN configuration and training and to evaluate the quality of the neural UAV position estimation. During the positioning estimation, the GNSS device information is neglected. The neural positioning estimation is close to the value obtained from the high precision GNSS equipment. There are errors noted by the neural fuser for the UAV positioning. The errors are expressed in centimeters. Latitude and longitude errors are values of the order $O(10^{-5})$. Table 2 presents the errors for longitude and latitude respectively. Figures 5 and 6 show the estimation made by the neural fuser for latitude and longitude. All used data were normalized for the ANN self-configured process and for the training phase.

Table 2. Errors, calculated in centimeters.

Latitude error	$9.975169144372613 \times 10^{-5}$ cm
Longitude error	$4.690294186740925 \times 10^{-5}$ cm

The algorithmic complexity for the neural fuser is simpler than the Kalman filter. In addition, for applying the neural fuser for UAV position identification, the procedure does not require a mathematical model for aircraft flight dynamics. The neural fuser is just a non-linear mapping between the inputs (data from sensors) and desired outputs (UAV position), and the process is fully formulated by data driven framework.

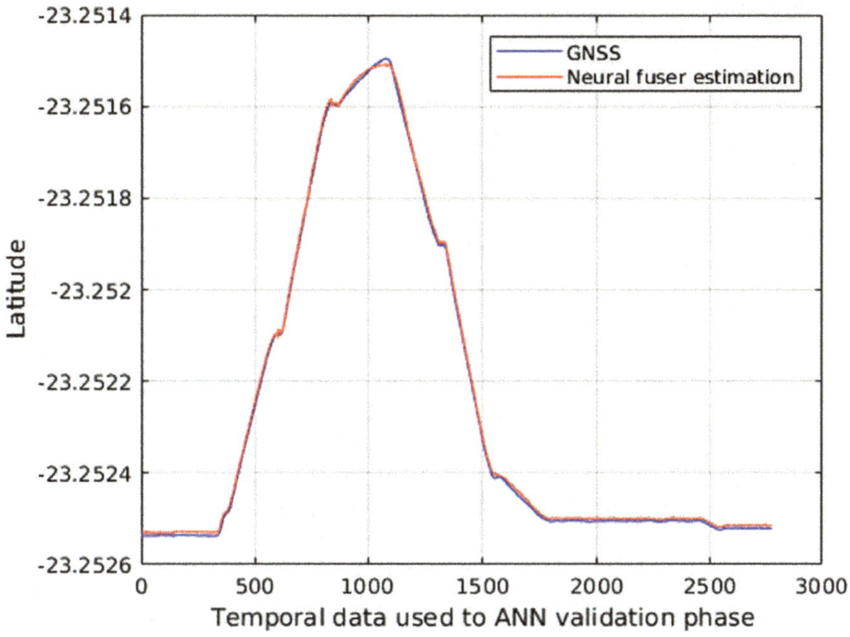

Fig. 5. Estimated latitude by neural fuser.

An important topic of this work is to calculate the uncertainties related to the estimation made by the neural fuser. From the set of estimated trajectories, the mean and variance of the position difference were calculated (in centimeters) between the trajectory measured by the high-precision GNSS device and the neural fuser. Once the variance is predicted, the confidence interval can be calculated for the estimated trajectories. Figures 7 and 8 show the confidence interval for the latitude and longitude estimations.

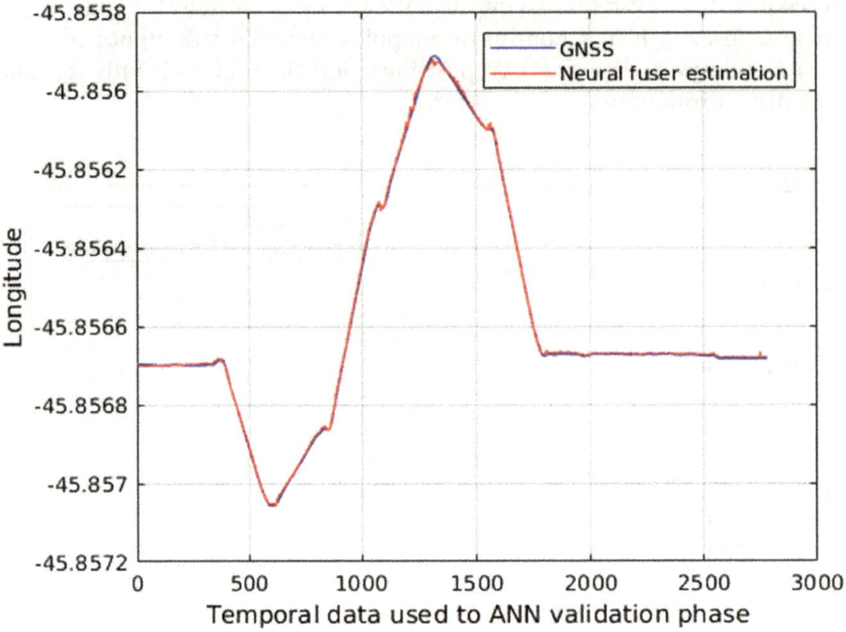

Fig. 6. Estimated longitude by neural fuser.

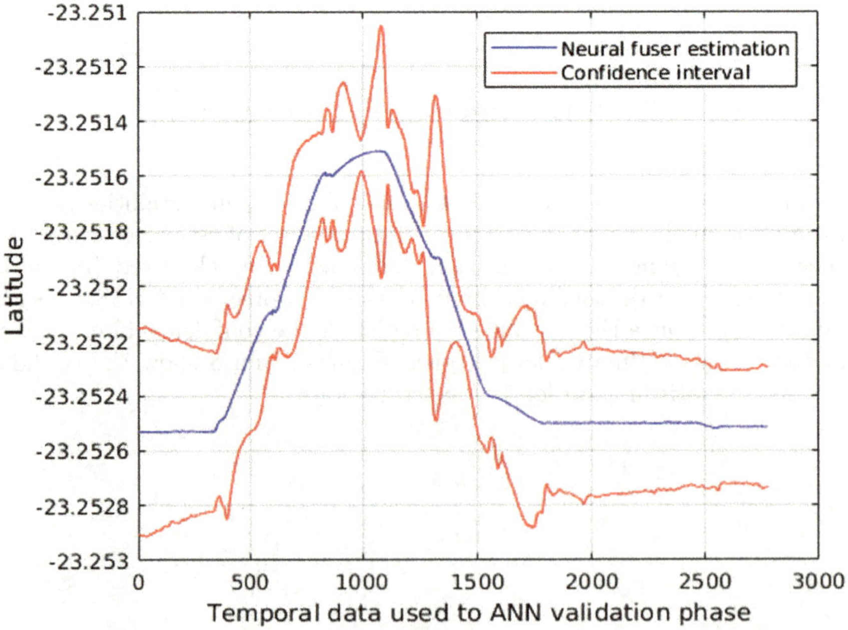

Fig. 7. Confidence interval for latitude.

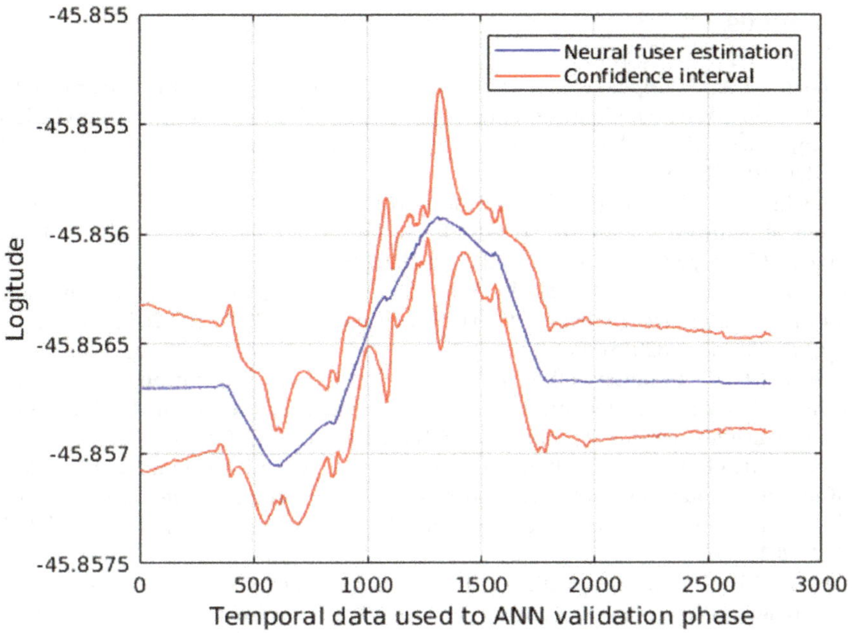

Fig. 8. Confidence interval for longitude.

5 Conclusions

The paper reports the use of data fusion procedure to estimate in-flight positioning of a drone, without the use of a GNSS signal. Different from other approaches, we have applied an artificial neural network as a data fusion operator. The positioning process is computed by a data fusion from the inertial sensor and by neural image processing. The image processing has several steps, with image edges extraction by a neural network and to find the maximum correlation point between reference and UAV segmented images. Both neural networks used here had their architectures designed as optimal solutions calculated from the MPCA metaheuristic. We have named the process as *self-configuring neural network*.

The *optimal* neural fuser was an effective as a UAV positioning system, as can be seen in Fig. 4 for the planned flight (blue color) and for the estimated trajectory by the neural fuser (red color). In other words, the neural positioning system is fully competitive with one based on GNSS signal – see Figs. 5 and 6.

Finally, uncertainty quantification was addressed by a new output from the data fuser: the UAV position estimation variance. From this consideration, the confidence interval can be obtained, as shown in Figs. 7 and 8.

Acknowledgments. Authors thank to the Brazilian research support agencies: CNPq and FAPESP.

References

1. Anochi, J.A., Campos Velho, H. F.: Optimization of feedforward neural network by Multiple Particle Collision Algorithm. In: IEEE Symposium on Foundations of Computational Intelligence (FOCI), pp 128–134 (2014) https://doi.org/10.1109/FOCI.2014.7007817
2. Anochi, J.A., Velho, H.F.C., Furtado, H.C.M., Luz, E.P.: Self-configuring two types of neural networks by MPCA. J. Mechan. Eng. Automat. **5**, 112–120 (2015). https://doi.org/10.17265/2159-5275/2015.02.008
3. Balan, N., Liu, L.-B., Le, H.-J.: A brief review of equatorial ionization anomaly and ionospheric irregularities. Earth Planetary Phys. **2**, 1–19 (2018). https://doi.org/10.26464/epp2018025
4. Braga, J.R.G., Campos Velho, H.F., Shiguemori, E.H.: Estimation of UAV position using LiDAR images for autonomous navigation over the ocean. In: International Conference on Sensing Technology (ICST), pp. 811–816 (2015)
5. Braga, J.R.G., Campos Velho, H.F., Shiguemori, E.H.: Determining the trajectory of unmanned aerial vehicles by a novel approach for the particle filter. Uncertainty Quantification and Stochastic Modeling, Mecánica Computacional XXXVI, 683–692 (2018)
6. Braga, J.R.G., Campos Velho, H.F., Conte, G., Doherty, P., Shiguemori, E.H.: An image matching system for autonomous UAV navigation based on neural network. In: International Conference on Control, Automation, Robotics and Vision (ICARCV), pp. 1–6 (2016)
7. Cintra, R.S.C., Cocke, S., Campos Velho, H.F.: Data assimilation by neural networks with ensemble prediction. In: Proceedings of the International Symposium on Uncertainty Modeling and Analysis. Florianópolis (SC), Brazil, pp. 35-43 (2018)
8. Conte, G., Doherty, P.: Vision-based unmanned aerial vehicle navigation using georeferenced information. J. Adv. Signal Process. (EURASIP) **1**, 308–387 (2009)
9. Dash, J.P., Watt, M.S., Pearse, G.D., Heaphy, M., Dungey, H.S.: Assessing very high resolution UAV imagery for monitoring forest health during a simulated disease outbreak. J. Photogrammetry Remote Sens. (ISPRS) **131**, 1–14 (2017)
10. Eltner, A., Baumgart, P., Maas, H., Faust, D.: Multi-temporal UAV data for automatic measurement of rill and interrill erosion on loess soil. Earth Surface Process. Landforms **40**, 741–755 (2014)
11. Fraden, J.: Handbook of Modern Sensors. Springer, New York (2010)
12. Goncalves, J., Renato, H.: UAV photogrammetry for topographic monitoring of coastal areas. J. Photogrammetry Remote Sens. (ISPRS) **104**, 101–111 (2015)
13. Goltz, G.A.M., Shiguemori, E.H., Campos Velho, H. F.: Position estimation of UAV by image processing with neural networks. In: X Congresso Brasileiro de Inteligencia Computacional (CBIC-2011), pp. 9–17 (2011)
14. Groves, P.D.: Principles of GNSS, inertial, and multisensor integrated navigation systems (2nd edition) – book review. IEEE Aerospace and Electronic Systems Magazine **30**, pp. 26–27 (2015)
15. Luz, E.F.P., Becceneri, J.C., Velho, H.F.C.: A new multi-particle collision algorithm for optimization in a high performance environment. J. Comput. Interdisciplinary Sci. **1**, 3–10 (2008)
16. James, M.S.R., d'Oleire, S.U.N.: Optimising UAV topographic surveys processed with structure-from-motion: Ground control quality quantity and bundle adjustment. Geomorphology **280**, 51–66 (2017)

17. Penha Neto, G., Campos Velho, H.F., Shiguemori, E.H.: UAV autonomous navigation by data fusion and FPGA. Mecânica Computacional, v. XXXVII, pp. 609–618 (2019)
18. Penha Neto, G., Campos Velho, H.F., Shiguemori, E.H.: Autonomous navigation by data fusion with optimized architecture artificial neural networks. In: Proceedings of the XL Ibero-Latin-American Congress on Computational Methods in Engineering (CILAMCE 2019). Natal (RN), Brazil, vol. 1. pp. 1–10 (2019)
19. Wernik, A.W., Alfonsi, L.M.M.: Ionospheric irregularities, scintillation and its effects on systems. Acta Geophisica Polonica **52**, 237–249 (2004)
20. Zhilenkov, A.A., Epifantsev, I.R.: System of autonomous navigation of the drone in difficult conditions of the forest trails. In: IEEE Conference of Russian Young Researchers in Electrical and Electronic Engineering, pp. 1036–1039 (2018)

Uncertainty Quantification in Data Fitting Neural and Hilbert Networks

Leila Khalij and Eduardo Souza de Cursi(✉)

LMN, Normandy University, INSA Rouen, 76801 Saint-Etienne du Rouvray CEDEX, France
souza@insa-rouen.fr

Abstract. We analyze the uncertainties in Neural and Hilbert networks. The first source of uncertainty is the variability of partition of the data in subsets for training, testing and validating. The analysis is made on the variability of the performance and of the weights of the networks. The effects of additive and multiplicative noises in the data are studied. The results of Neural and Hilbert Networks are compared. It appears that Hilbert Networks are more robust but imply a higher computational cost. The distributions of the outputs are studied and it appears that their means and modes may be used to improve the estimates furnished by the nets. The extension of Hilbert Networks to Element Based Networks is considered.

Keywords: Neural networks · Hilbert Networks · Uncertainty Quantification

1 Introduction

Uncertainties in Artificial Intelligence and Data Mining is a topic that is becoming increasingly important in the literature. Indeed, many works are devoted to the study of the impacts of noise in classification (see, for instance, [1–11]). In the particular case of Artificial Neural Networks (ANN), the effects of noisy on data were considered from the point of view of inverse problems and led to the use of regularization and margin techniques, such as, for instance, Tikhonov's regularization (see, for instance, [12–15]). Recent works suggested the addition of noise in the training process as a form of regularization and improvement in the determination of the weights (see, for instance, [16–19]. The procedure may be compared to the stochastic perturbations in optimization [20] or the Uncertainty Quantification (UQ) approach in Optimization [21]). Nevertheless, noise is not the only source of uncertainty in ANN: in general, the separation of the datasets in training set, validation set and testing set is random; the initialization of the network may be random; the algorithms for the training may introduce random variables or noise; the data may be corrupted by noise; the architecture of the net - namely the number of units in each layer and the number of layers – and the choice of the activation functions and of the aggregation operators may influence the result. Thus, many sources of uncertainty may be involved in ANN. In despite of these sources of uncertainty, the applications of UQ approaches to the analysis of ANN remain rare. In this work, we examine ANN from the standpoint of UQ and we apply UQ procedures to characterize the uncertainties in fitting ANN. The results are compared to Hilbert

J. E. S. De Cursi (Ed.): Uncertainties 2020, LNME, pp. 222–241, 2021.
https://doi.org/10.1007/978-3-030-53669-5_17

Approximations by Hilbert Networks, which are flexible ways for the implementation of Hilbertian approximations, id est, approximations based on Hilbert spaces. Hilbert Networks are described in Sect. 2.1

2 Variability of the Training, Validation, and Testing Sets

As observed in the Introduction, a first source of uncertainty is the choice of the subsets for training, validation and testing. Let us analyze this variability by using UQ tools: to apply the UQ approach, we start by generating a sample. We consider initially a first unnoisy dataset $\mathcal{D} = \{(x_1, y_1), \ldots, (x_n, y_n)\}$ corresponding to $n = 21$ equally spaced points on $(-2\pi, 2\pi)$ and the values $y_i = \sin(x_i)$. We consider an ANN having one hidden layer with 10 units. The transfer function is *tansig* and the aggregation operator is the weighted sum with a bias. The net is trained by Bayesian Regularization with a maximum of $1e4$ epochs. The training set corresponds to 70% of the dataset, the validation set is formed by 15% of the dataset and the testing set is formed by 15% of the dataset. These subsets are chosen randomly. The performance function is the mean square error. We generate a sample of $1e3$ ANN solving this problem. The results furnish a sample of the weights and the bias, which is analyzed by the UQ approach: for each weight w_i, we generate an approximation F_i of its Marginal Cumulative Distribution Function (CDF) by using the data and 21 points equally spaced in the interval $(\min\{w_i\}, \max\{w_i\})$. The resulting data is interpolated by using the particle method ([22]) to generate a good representation of the CDF. Analogously, the derivative f_i of F_i is evaluated by the particle method ([22]) and the result is interpolated by a particle method to generate a good representation of the Marginal Probability Density Function (PDF). We illustrate below the distribution of the weight w_1. As shown in the Fig. 1 below, the distribution is bimodal.

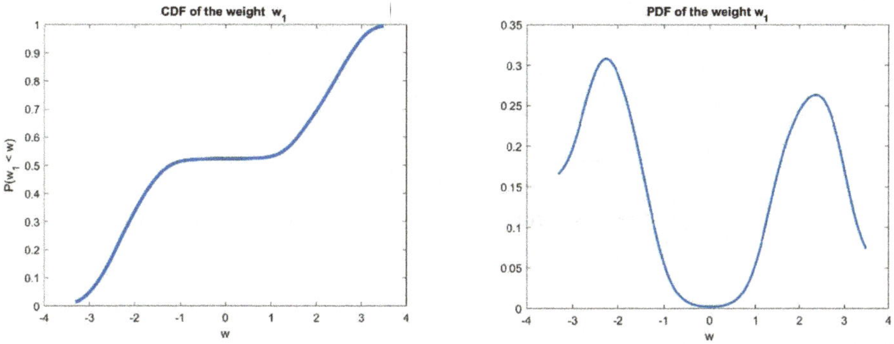

Fig. 1. Empirical Marginal Cumulative Function and Probability Density Functions of w_1.

The marginal distribution of the other weights is analogous. The performance of the network is also affected by variability: in Fig. 2, we show the distribution of the negative natural logarithm of the performance – the grater is the value, the smaller is the mean squared error. We observe that the mode of the distribution appears a little bit inferior to 20, corresponding to mean square error of about $1e-8$. This value may be considered as the most probable one when running the algorithm. The results appear as relatively good on the dataset (see Fig. 3)

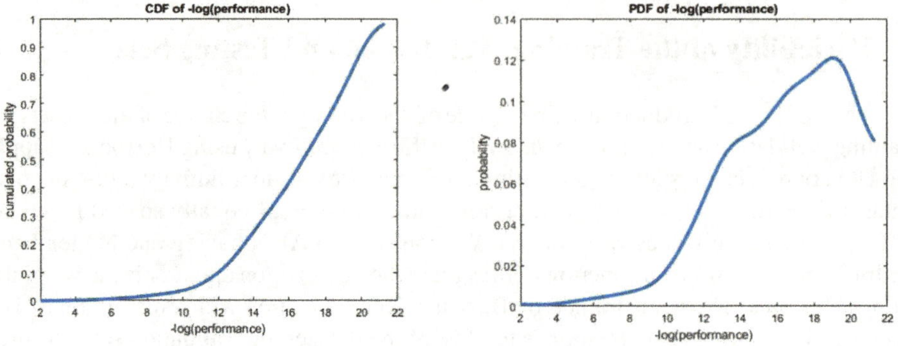

Fig. 2. Empirical Cumulative Function and Probability Density Functions of the negative natural logarithm of the performance.

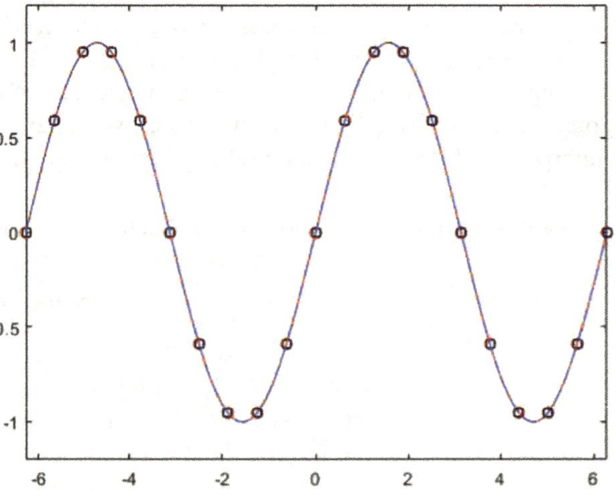

Fig. 3. Typical results corresponding to the mode of the performance.

2.1 Hilbert Networks

Hilbertian approximations may be determined by considering a convenient Hilbert spaces V and an associated Hilbert basis $\mathcal{B} = \{\varphi_i : i \in \mathbb{N}\}$. In practice, we look for a finite linear combination.

$$y \approx Py = \sum_{i=0}^{k} w_i \varphi_i(x)$$

In the framework of UQ, the basic procedures for the determination of the approximation are collocation, variational approximation, moments matching, adaptation of numerical methods [23]. To keep the idea of ANN, we may use a Hilbert Network (HN): indeed, the basic operation $w_i \varphi_i(x)$ corresponds to the unbiased unity $H_i(\mathcal{B})$ shown in the Figure below.

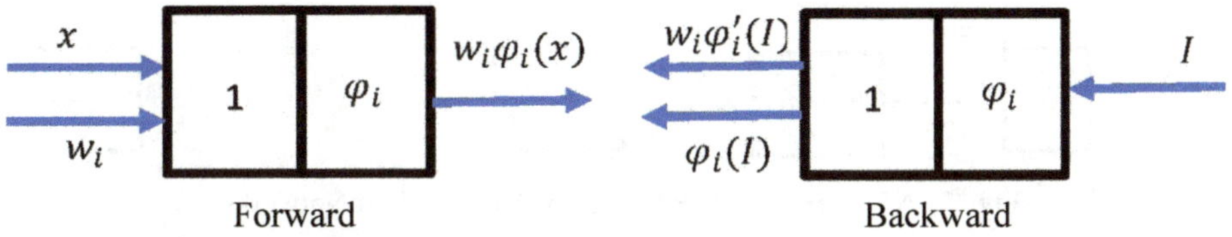

Fig. 4. The basic unit $H_i(\mathcal{B})$ of a Hilbert Network

Then, the evaluation of Py may be carried out by a network $H(\mathcal{B}, k)$ having the following architecture (Fig. 5):

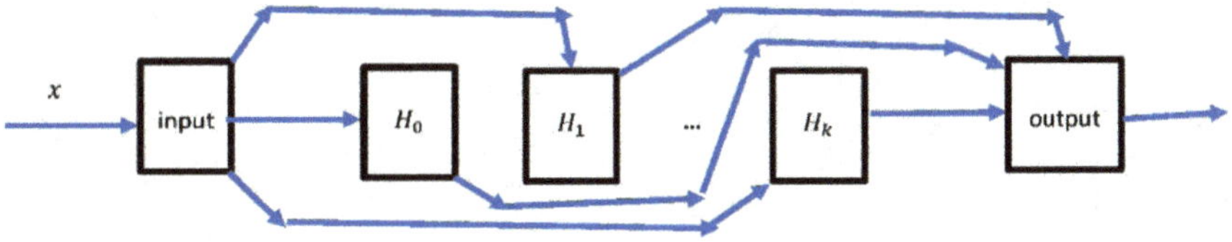

Fig. 5. A basic Hilbert Network $H(\mathcal{B}, k)$.

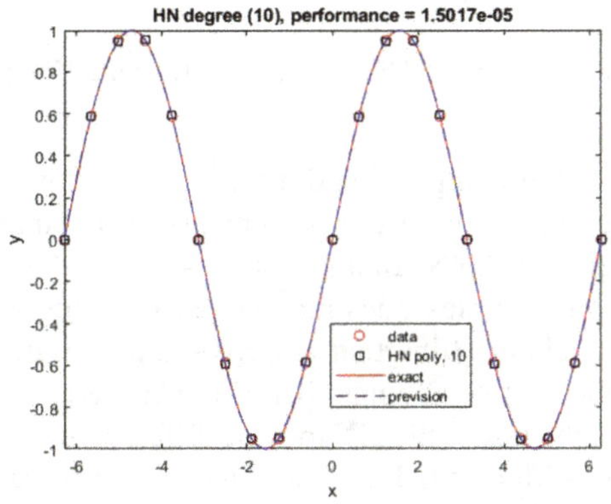

Fig. 6. Results furnished by a simple Hilbert Network (polynomial basis, $k = 10$)

Thus, the Hilbert approximation may be considered as a particular ANN. As an example, let us consider the unnoisy dataset $\mathcal{D}$ and an approximation using a polynomial basis $\varphi_i(x) = x^i$, with $k = 9$. The weights are determined by a hybrid algorithm: firstly, the net is trained with a maximum 500 epochs using Bayesian regularization. Then, 50 stochastic perturbations of the weights try to improve the weights. The resulting weights are used as initial point to the minimization of the mean square error by Nelder-Meade. This procedure is iterated 20 times. The results are shown in Fig. 6.

In practice, the network may be replicated, and we may consider networks as shown in the Fig. 7:

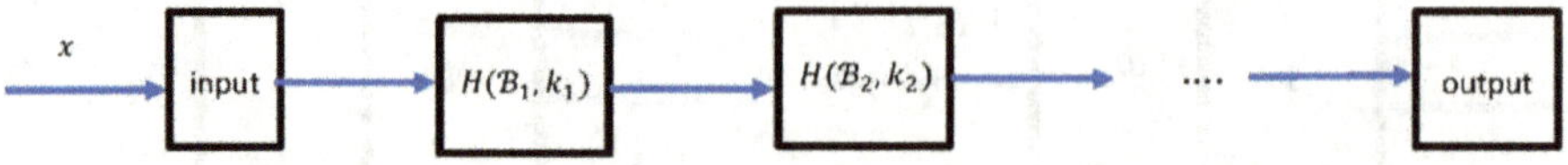

Fig. 7. A Hilbert Network formed by basic Hilbert Networks

As an example, let us consider again the unnoisy dataset $\mathcal{D}$ and the polynomial basis $\varphi_i(x) = x^i$. We use 2 and 3 HN with the polynomial basis of the same degree $k = 3$. The results are shown in the Fig. 8.

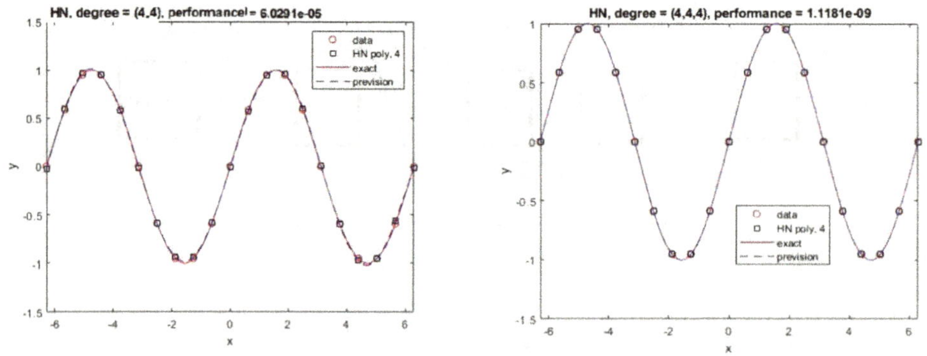

Fig. 8. Results furnished by an Iterated Hilbert Network (2 and 3 HN, polynomial basis, $k = 4$)

The formulation of Hilbert Approximations using HN allows the use of ANN training methods for the determination of the coefficients of the approximation, which are assimilated to the weights of ANN. In addition, instead of the sequential architecture shown in the Fig. 4, more complex ones may be used, analogously to classical ANN: the mathematical results about Hilbertian Approximations state that a single network $H(\mathcal{B}_1, k_1)$ is able to approximate any member of the Hilbert space V under consideration, but the numerical behavior of the HN may be different according to its architecture – for instance, the use of different Hilbert basis in each subnet $H(\mathcal{B}_i, k_i)$ may be useful from the numerical standpoint.

2.2 Effects of Uncertainty in the Division in Subsets in HN Approximation

Analogously to ANN, HN is affected by uncertainty due to randomness in the choice of the subsets for training, validating and testing. As previously observed, HN may be considered as a particular ANN, so that we may use the same approach: we generate a sample of $1e3$ HN solving this problem. The iterated HN used involve 2 HN using a polynomial basis and $k = 5$. The results furnish a sample of the weights: as in the preceding analysis, the sample is interpolated by the particle method ([22]) to generate the CDF. Analogously, the derivative of the CDF is evaluated by the particle method and interpolated to generate a good representation of the Marginal Probability Density Function (PDF). We illustrate below the distribution of the weight w_1. As shown in the Fig. 9 below, the distribution is unimodal.

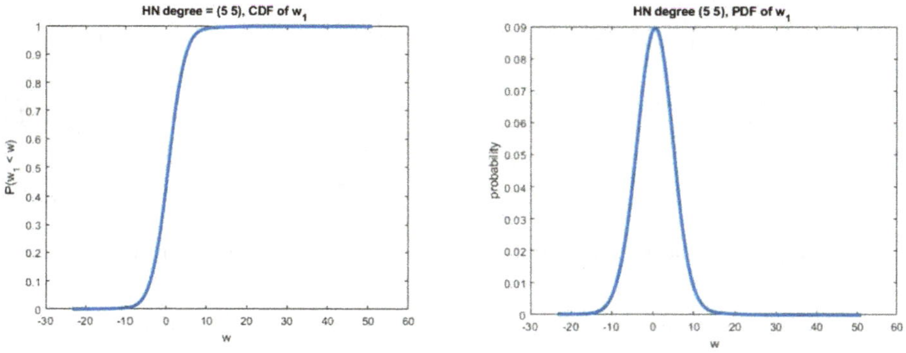

Fig. 9. Empirical Marginal Cumulative Function and Probability Density Functions of w_1.

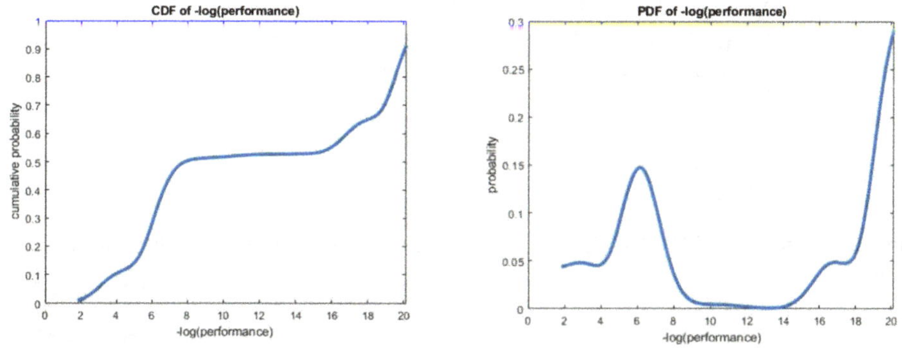

Fig. 10. Empirical Cumulative Function and Probability Density Functions of the negative natural logarithm of the performance.

In this case too, the marginal distribution of the other weights is analogous. The distribution of the performance of the network is shown in Fig. 10 – again, we use the distribution of the negative natural logarithm of the performance. We observe that there are two modes: about 6 and 20, corresponding to mean square errors of about $2e-3$ and $2e-9$, respectively. These values may be considered as the most probable ones when running the algorithm. The bimodal behavior is connected to the fact that training HN is more difficult than training ANN: the number of results having of inferior quality increases. Nevertheless, the results appear as relatively good in both the cases (see Fig. 11).

If we are interested in extrapolation of the data outside the dataset, a well-trained HN produces results of best quality (Fig. 12).

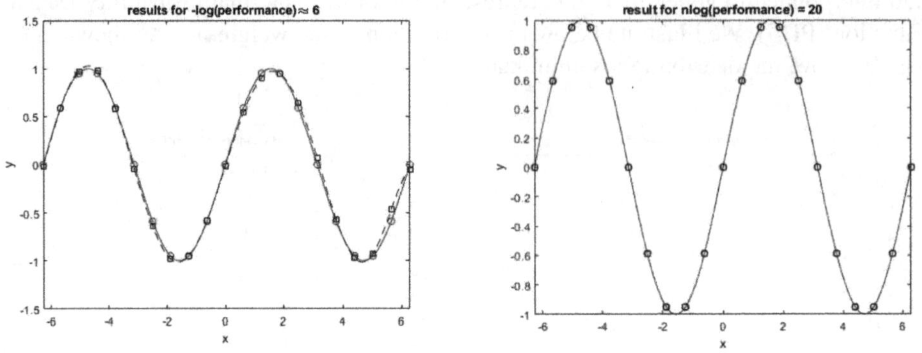

Fig. 11. Typical results corresponding to the modes of the performance.

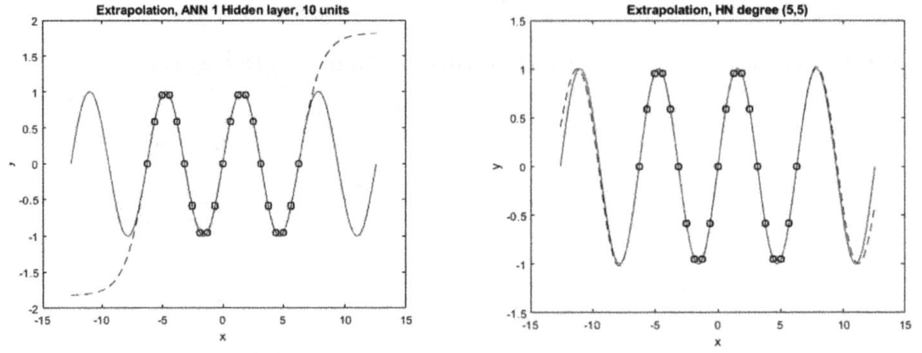

Fig. 12. Comparison ANN/HN for extrapolation: a well-trained HN extrapolates with a better quality.

3 Effect of an Addictive Noise in the Target Data

Noisy data introduces uncertainties. For instance, let us consider a first noisy dataset where the inputs x are unnoisy, but the targets y are affected by additive gaussian noise corresponding to a signal-noise ratio (SNR) equal to 40.

$$y_i = \sin(x_i) + \varepsilon_i, SNR = \frac{E\left(\sin^2(x)\right)}{E\left(\varepsilon^2\right)} = 40 \tag{1}$$

We repeat the experiments presented in the preceding section: on the one hand, we consider a shallow ANN with one hidden layer formed by 10 units with the same parameters given in Sect. 2. As in the preceding section, we generate a sample of $1e3$ ANN solving this problem for this dataset. For comparison, we consider an iterated HN formed by 2 HN with a polynomial basis with $k = 5$ – we generate $1e3$ HN solutions. The results are used to generate the CDF and the PDF of the negative natural logarithm of the performance, analogously to the preceding section.

In a first test, we fixed the subsets of training, testing and validation. The results are shown in Fig. 13: we observe that both HN and ANN have a mode about 3, what corresponds to a performance of about 5E-2. The distribution furnished by HN appears as more concentrated, with a smaller standard deviation.

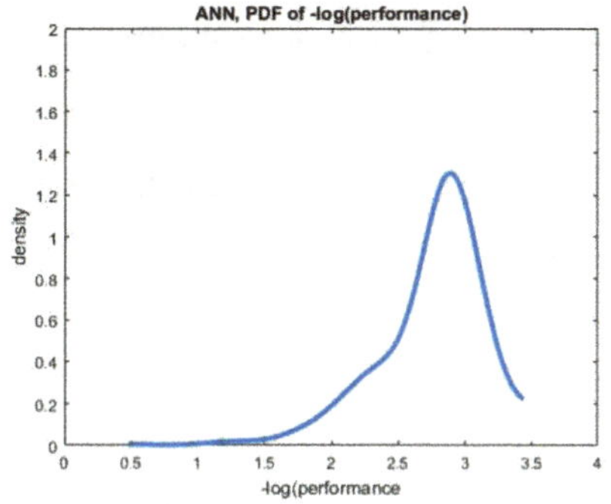

Fig. 13. Empirical Probability Density Functions of the negative natural logarithm of the performance. At left, ANN results. At right, HN results. Test with fixed subsets

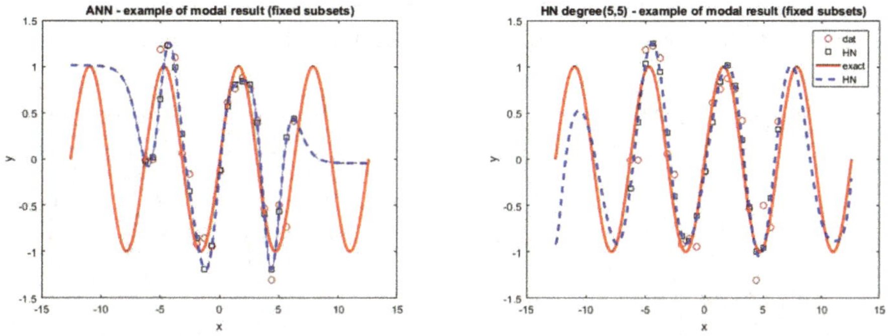

Fig. 14. Comparison ANN/HN for extrapolation: HN extrapolates better. Test with fixed subsets

The modal results are shown in Fig. 14. Analogously to the preceding situation, the results are comparable on the dataset, but the extrapolation gives better results when using HN. As in the preceding the distribution of the coefficients is unimodal for HN and bimodal for ANN (Fig. 15).

In a second test, we used random subsets of training, testing and validation. For this test, the results are exhibited in Figs. 16 to 17; the results are analogous to those obtained with fixed training sets (Fig. 18).

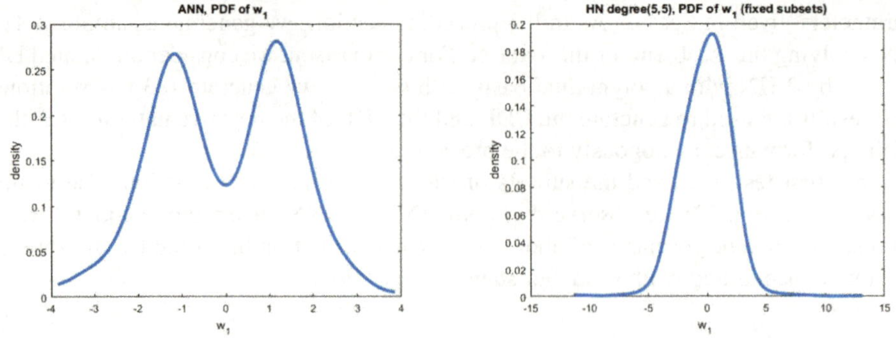

Fig. 15. PDF of w_1. At left, the bimodal distribution furnished by ANN, while the results of HN are unimodal. Test with fixed subsets

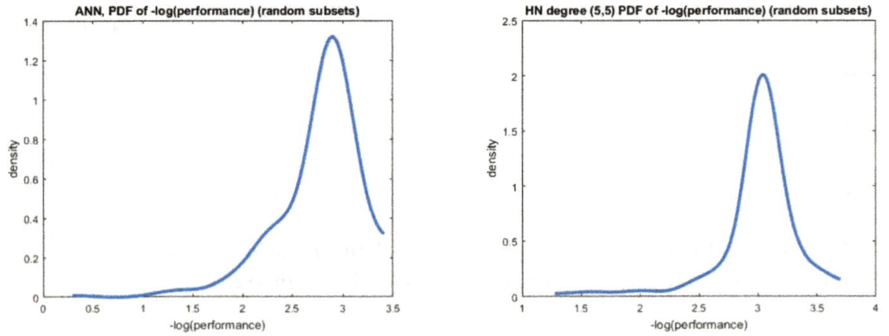

Fig. 16. Empirical Probability Density Functions of the negative natural logarithm of the performance. At left, ANN results. At right, HN results. Test with random subsets

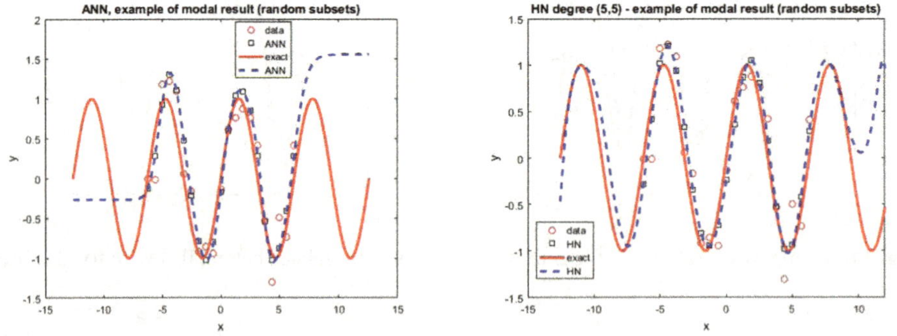

Fig. 17. Comparison ANN/HN for extrapolation: HN extrapolates better. Test with random subsets

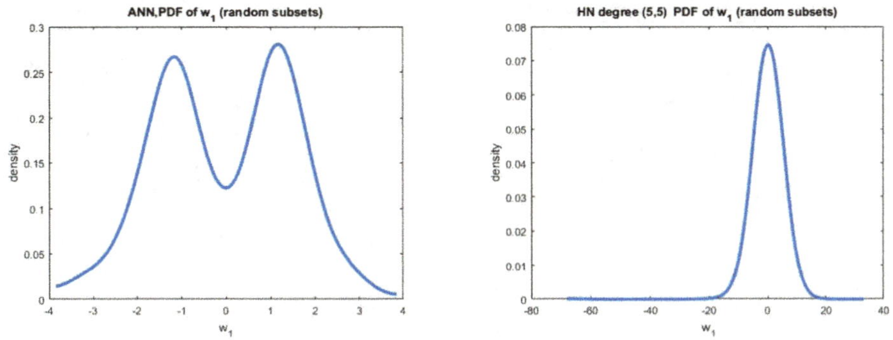

Fig. 18. PDF of w_1. At left, the bimodal distribution furnished by ANN, while the results of HN are unimodal. Test with random subsets

4 Effect of a Multiplicative Noise in the Target Data

Let us consider a second noisy dataset where the inputs x are unnoisy, but the targets y are affected by multiplicative uniform noise corresponding to 10% of the signal:

$$y_i = (1 + \varepsilon_i)\sin(x_i), \varepsilon \sim U(-0.1, 0.1) \tag{2}$$

We repeat the experiments presented in the preceding section, with the same parameters (Fig. 19). The modal results are shown in Fig. 20. Analogously to the preceding situation, the results are comparable on the dataset, but the extrapolation gives better results when using HN. As in the preceding the distribution of the coefficients is unimodal for HN and bimodal for ANN (Figs. 21, 22, 23, 24 and 25).

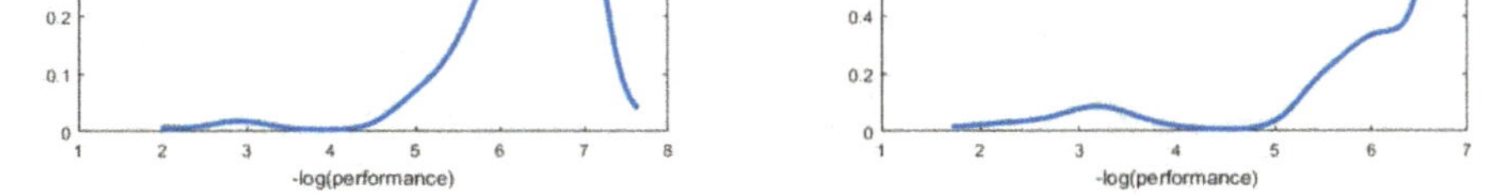

Fig. 19. Empirical Probability Density Functions of the negative natural logarithm of the performance. At left, ANN results. At right, HN results

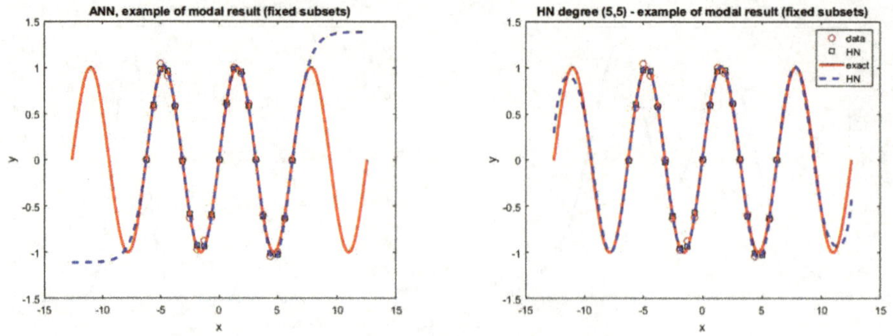

Fig. 20. Comparison ANN/HN for extrapolation: HN extrapolates with a better quality.

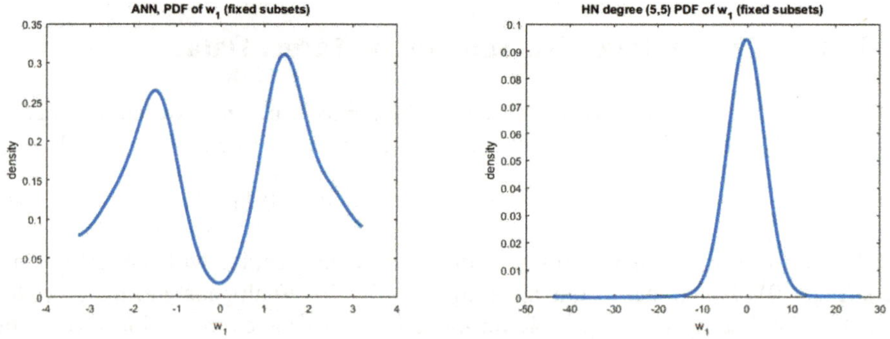

Fig. 21. PDF of w_1. At left, the bimodal distribution furnished by ANN, while the results of HN are unimodal.

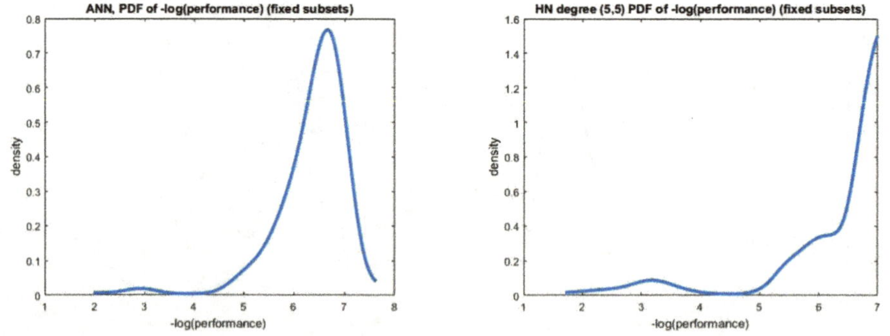

Fig. 22. Empirical Probability Density Functions of the negative natural logarithm of the performance. At left, ANN results. At right, HN results

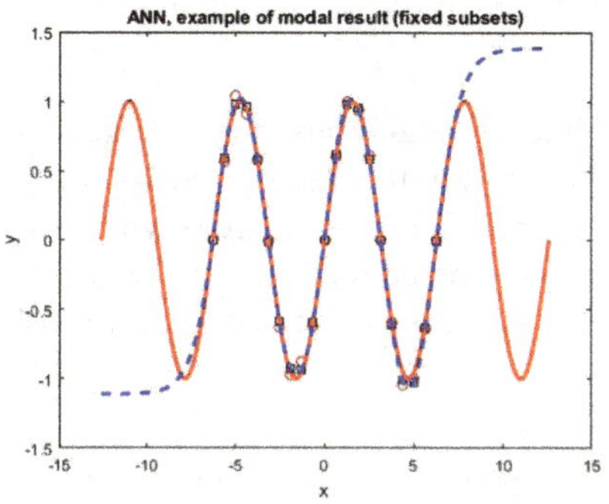

Fig. 23. Comparison ANN/HN for extrapolation: HN extrapolates with a better quality.

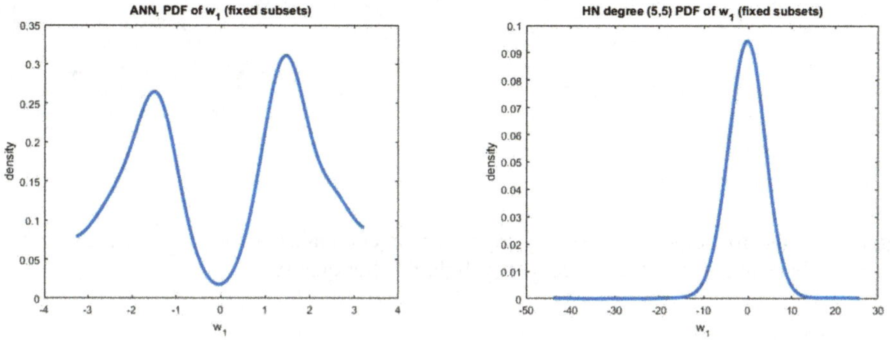

Fig. 24. PDF of w_1. At left, the bimodal distribution furnished by ANN, while the results of HN

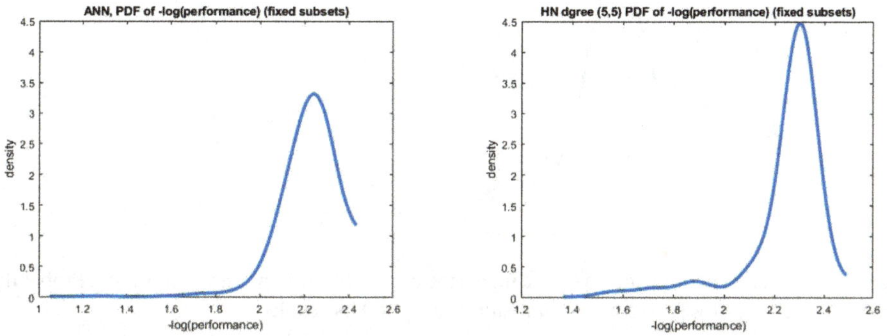

Fig. 25. Results for additive noise on the input (the target remains unnoisy): Empirical Probability Density Functions of the negative natural logarithm of the performance. At left, ANN results. At right, HN results are unimodal.

5 Effect of Noise in the Input

We studied the effect of the noise in the input into an analogous way: we generated additive noise and multiplicative noise on the input x as in the Sects. 3 and 4, with the same parameters. The target values were the exact ones. The results are sown below.

For the additive noise, the results are similar to the preceding ones, but the extrapolation by HN losses in quality – but it remains better than the extrapolation furnished by ANN (Figs. 26 and 27).

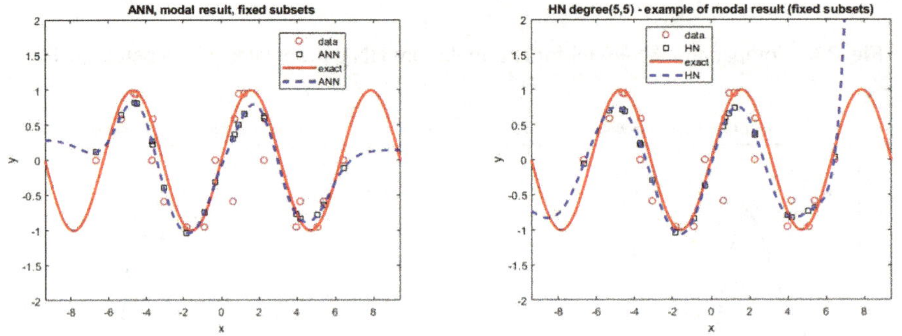

Fig. 26. Results for additive noise on the input (the target remains unnoisy). At left, ANN results. At right, HN results. HN extrapolates with a slightly better quality.

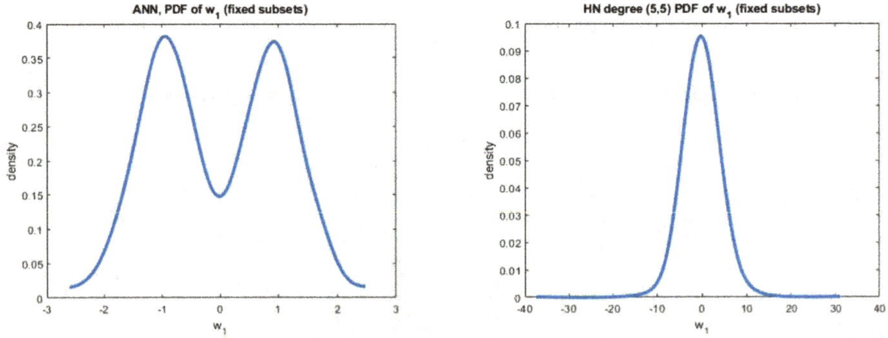

Fig. 27. Results for additive noise on the input (the target remains unnoisy): Empirical Probability Density Functions of w_1. At left, ANN results. At right, HN results

Now, let us consider the results for the multiplicative noise. The results are similar to the preceding ones. Again, the extrapolation by HN losses in quality, but remains better than the extrapolation furnished by ANN. In addition, the distribution of the negative logarithm of the performance appears as bimodal (Figs. 28, 29 and 30).

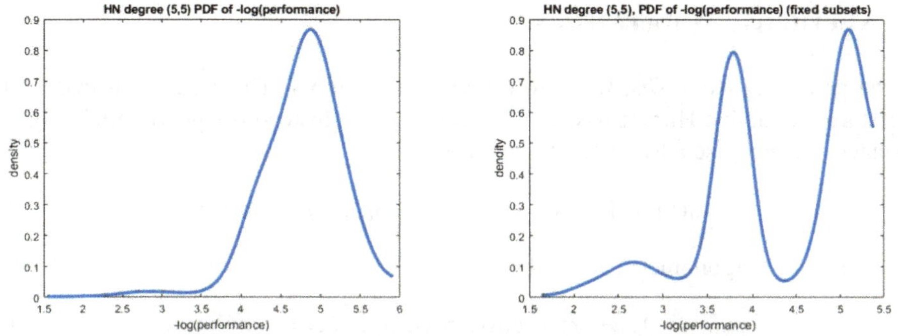

Fig. 28. Results for multiplicative noise on the input (the target remains unnoisy): Empirical Probability Density Functions of the negative natural logarithm of the performance. At left, ANN results. At right, HN results

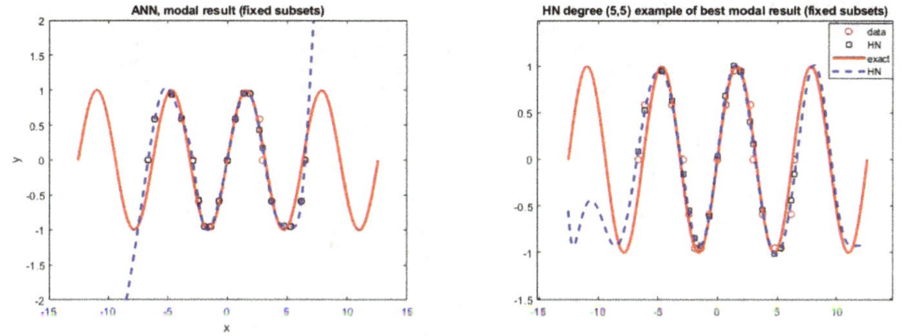

Fig. 29. Results for multiplicative noise on the input (the target remains unnoisy): At left, ANN results. At right, HN results. HN extrapolates with a better quality.

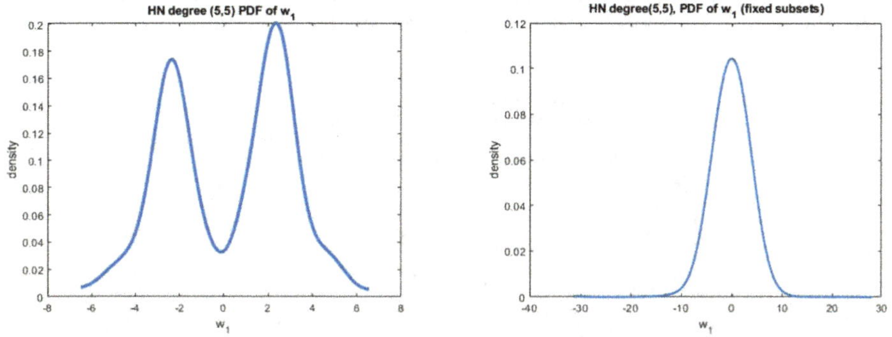

Fig. 30. Results for multiplicative noise on the input (the target remains unnoisy): Empirical Probability Density Functions of w_1. At left, ANN results. At right, HN results

6 Alternative Hilbert Basis

In the preceding examples, we used a polynomial family. The reader can find in the literature alternative Hilbert basis, which can be used instead the polynomial one. For instance, we may use a trigonometrical basis

$$\varphi_0(x) = 1, \varphi_{2i}(x) = \cos(ix), \varphi_{2i+1}(x) = \sin(ix). \tag{3}$$

or the periodic trigonometrical family.

$$\varphi_0(x) = 1, \varphi_{2i}(x) = \cos(2i\pi x), \varphi_{2i+1}(x) = \sin(2i\pi x). \tag{4}$$

We illustrate the use of alternative basis by using these bases. Figures 31, 32 and 33 exhibit results furnished by the trigonometrical family. Figure 31 concerns unnoisy data. Figures 32–33 concern multiplicative noise. In all the cases, we observe the HN extrapolates better than RN.

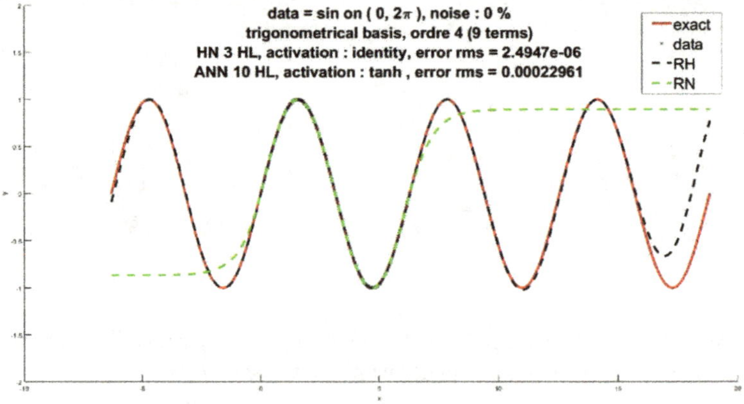

Fig. 31. Results for unnoisy data. The black curve is furnished by HN using trigonometrical basis with 9 terms (4 sines, 4 cosines, constant term) and an iteration of 3 HN.

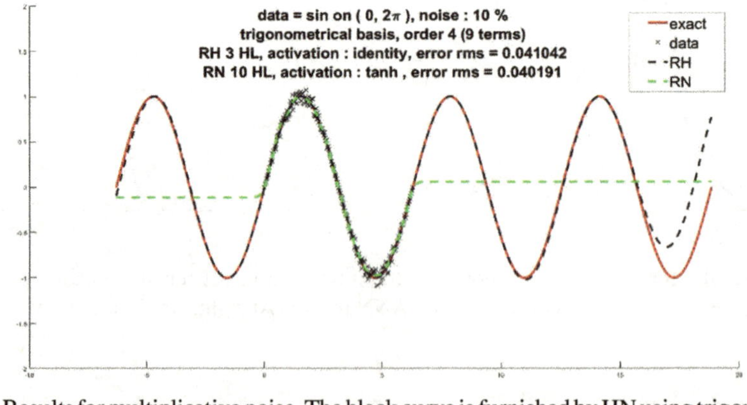

Fig. 32. Results for multiplicative noise. The black curve is furnished by HN using trigonometrical basis with 9 terms (4 sines, 4 cosines, constant term) and an iteration of 3 HN.

Figure 33 results are obtained by realizing a careful optimization of the weights of the HN – notice that this result is expensive in terms of computational cost.

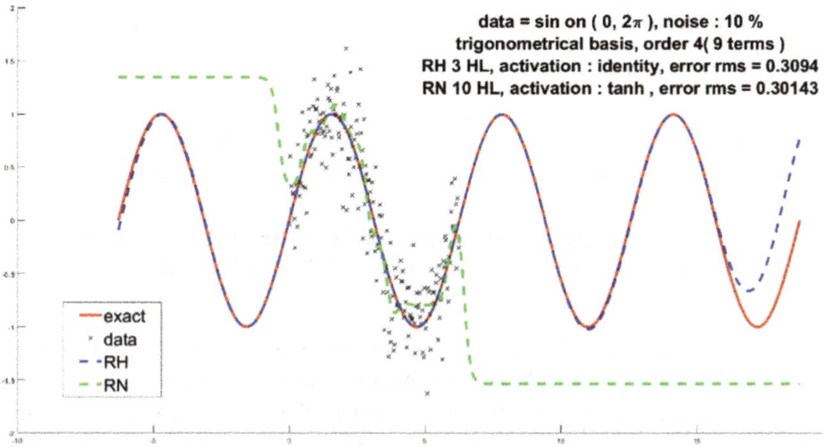

Fig. 33. Results obtained for additive noise with the trigonometrical Hilbert basis. In this case, a careful optimization of the weights was carried on with an expensive computational cost.

7 Determination Probability Distributions of Outputs

UQ techniques may be used to determine the probability distribution of the outputs of ANN or HN: we may generate a sample from the outputs at a given point x and use UQ techniques to determine its distribution. Examples are given in the Figs. 34 and 35 below. We used a sample of 1000 values to determine the distributions.

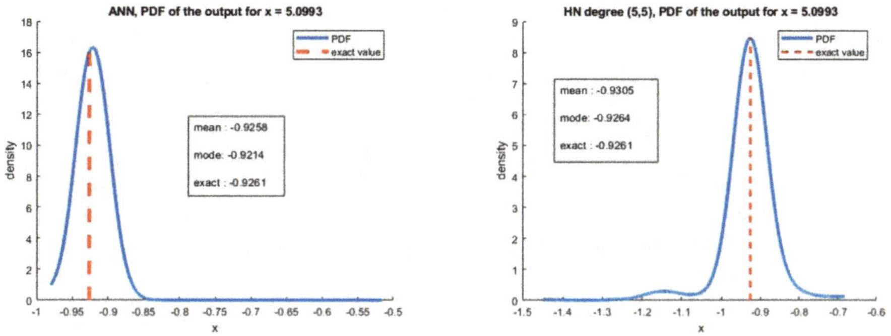

Fig. 34. An example of determination of the PDF of the output. At left, ANN results. At right, HN results. In both the cases, the mean and the mode are close to the exact value.

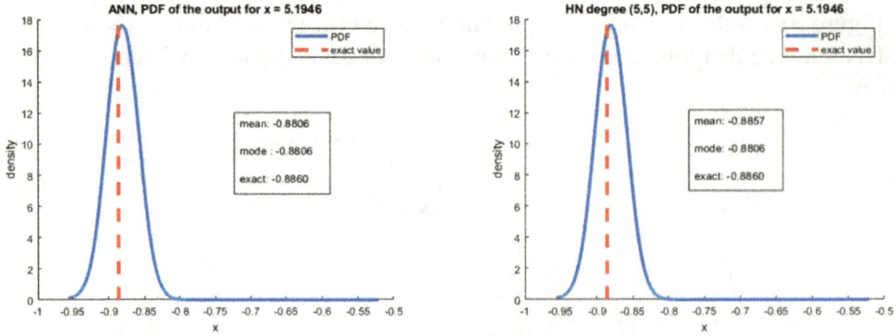

Fig. 35. An example of determination of the PDF of the output. At left, ANN results. At right, HN results. In both the cases, the mean and the mode are close to the exact value.

UQ techniques may also be used to improve the quality of the input or target data. For instance, let us consider a dataset $\mathcal{D}$. Assume that the data is noisy in the target value y, but a sample is available for each point in the dataset: for each $(x_i, \blacksquare) \in \mathcal{D}$, values $(x_i, y_{i1}), (x_i, y_{i2}), \ldots, \left(x_i, y_{in_i}\right)$ are available. In this situation, we may consider the sample $\mathcal{Y} = \left(y_{i1}, y_2, \ldots, y_{in_i}\right)$ and use UQ techniques to determine the distribution of y. Then, the mean of the distribution may be determined. Assuming that the noise has a null mean, the estimation of the mean furnishes an estimation for the real value of y. Let us illustrate this approach with 5 samples of the dataset generated by Eq. (2) (multiplicative noise). The results are shown in the Fig. 35 and 36.

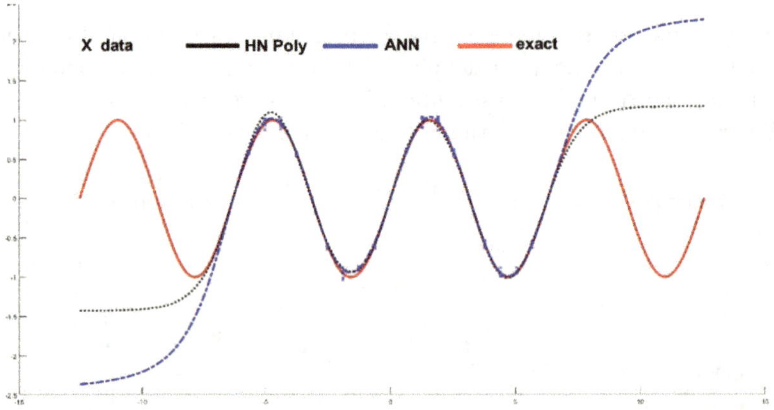

Fig. 36. Results obtained using 5 samples of the data

8 Element Based Networks

In the framework of Hilbert spaces, Hilbert basis are generalized by total families and dense enumerable reunions of finite families (see, for instance, [24]) – which include

Finite Elements and Splines. An analogous generalization may be implemented in Hilbert Networks: we may replace $\mathcal{B}$ by a total family or a convenient member of a dense reunion of enumerable families. Nevertheless, in such a situation, it appears as adequate to take into account the specific properties of the members of $\mathcal{B}$. For instance, Splines or Finite Element families are *element-based*, id est, *their support is a bounded subset*: in other words, for each member φ_i of the family, there is a closed bounded subset $supp(\varphi_i)$ such that $\varphi_i = 0$ outside $supp(\varphi_i)$. Thus, using such a family to approximate data is equivalent to partitioning the space of the inputs in elements - analogously to convolutional or capsule networks, the elements of the network are assigned to specific regions or specific tasks. We illustrate this approach in Fig. 37, where the space of inputs is partitioned in 10 elements, each one corresponding to a polynomial HN (3, 3). The global solution corresponds to a polynomial HN(5, 5, 5).

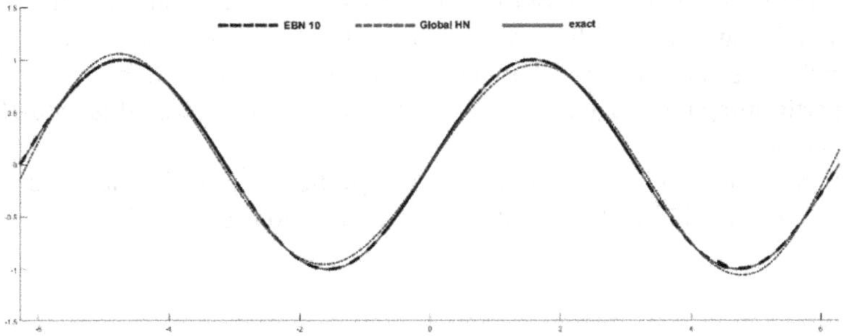

Fig. 37. Element based network (10 elements) compared to a global network.

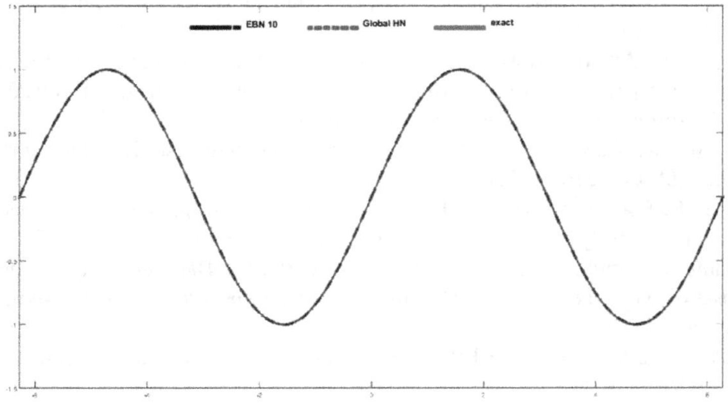

Fig. 38. Element based network (20 elements) compared to a global network.

In Fig. 38, the space of inputs is partitioned in 20 elements, each one corresponding to a polynomial HN (3, 3). The global solution corresponds to a polynomial HN(5, 5).

9 Concluding Remarks

We considered some sources of uncertainty in fitting ANN and HN and performed an analysis of their impact on the results furnished by the network. The first source of uncertainty considered was the random choice of the subsets for training, testing and validation. The experiments with data furnished by a sine showed a bimodal distribution for the weights of ANN and a unimodal distribution for HN. Performance of HN tends to be better than the performance of ANN, but with a more expensive computational cost and may lead, sometimes, to hard optimization problems.

We analyzed the performance of both the networks when data is noisy. It appeared that noise on the inputs produces worse results when compared to noise in the targets. The results show that the determination of the mean and/or of the mode of inputs and targets may improve the performance of the network. Analogously, the determination of the mean and/or the mode of the outputs furnishes a better estimation of the real value.

HN extends naturally by the introduction of total families, namely element-based families such as splines or finite elements. Using element-based families may be interpreted as partitioning the space of inputs. The first experiments with this kind of families are encouraging.

Although the number of realizations was significant in each test presented, other datasets must be considered to consolidate the conclusions above.

References

1. Talmon, J.L., McNair, P.: The effect of noise and biases on the performance of machine learning algorithms. Int. J. Biomed. Comput. **31**, 45–57 (1992)
2. Napolitano, A.: Classification techniques for noisy and unbalanced data. Ph. D. Thesis. Florida Atlantic University (2009)
3. Nettleton, D.F., Albert Orriols-Puig, A., Fornells, A.: A study of the effect of different types of noise on the precision of supervised learning techniques. Artif. Intell. Rev. **33**, 275–306 (2010). https://doi.org/10.1007/s10462-010-9156-z
4. Zhu, X., Wu, X.: Class noise vs. Attribute noise: a quantitative study of their impacts. Artif. Intell. Rev, **22**, 177–210 (2004)
5. García, S., Luengo, J., Herrera, F.: Dealing with noisy data. In: Data Preprocessing in Data Mining. Intelligent Systems Reference Library, vol 72. Springer, Cham (2015)
6. Saseendran, A., Setia, L., Chhabria, V., Chakraborty, D., Barman, R.A.: Impact of Noise in Dataset on Machine Learning Algorithms (2019). https://doi.org/10.13140/rg.2.2.25669.91369
7. Sessions, V., Valtorta, M.: The Effects of Data Quality on Machine Learning Algorithms, pp. 485–498 (2006)
8. Raychev, V., Bielik, P., Vechev, M., Krause, A.: Learning programs from noisy data. IEEE Trans. Inf. Theory **37**(4), 1085–1094 (1991)
9. Evans, R., Grefenstette, E.: Learning explanatory rules from noisy data. In: Proceedings of the Twenty-Seventh International Joint Conference on Artificial Intelligence (IJCAI 2018) (2018). https://www.ijcai.org/Proceedings/2018/0792.pdf. Accessed 22 Jan 2020
10. Song, S., Chaudhuri, K., Sarwate, A.D.: Learning from data with heterogeneous noise using SGD. In: Proceedings of the 18th International Conference on Artificial Intelligence and Statistics (AISTATS), San Diego, CA, USA. JMLR: W&CP, vol. 38 (2015)

11. Blechner, M.D.: Behavior of Various Machine Learning Models in the Face of Noisy Data. https://ocw.mit.edu/courses/health-sciences-and-technology/hst-951j-medical-decision-support-fall-2005/projects/full_text.pdf. Accessed 22 Jan 2020
12. Meechan-Maddon, A.: The effect of noise in the training of convolutional neural networks for text summarization. Master's Thesis, Uppsala University (2019)
13. Al-Mashouq, K.A.: Prediction of neural net tolerance to noise. Neural Process. Lett. **5**, 25–34 (1997)
14. Khardon, R., Wachman, G.: Noise tolerant variants of the perceptron algorithm. J. Mach. Learn. Res. **8**, 227–224 (2007)
15. Burger, Martin, Engl, Heinz W.: Training neural networks with noisy data as an ill-posed problem. Adv. Comput. Math. **13**, 335–354 (2000)
16. Noh, H., You, T., Mun, J., Han, B.: Regularizing deep neural networks by noise: its interpretation and optimization. In: 31st Conference on Neural Information Processing Systems (NIPS 2017), Long Beach, CA, USA (2017). https://arxiv.org/pdf/1710.05179.pdf. Accessed 22 Jan 2020
17. Khalfaoui, B., Boyd, J., Vert, J.-P.: Adaptive structured noise injection for shallow and deep neural networks (2019). https://arxiv.org/pdf/1909.09819.pdf. Accessed 22 Jan 2020
18. Bishop, C.M.: Training with noise is equivalent to tikhonov regularization. Neural Comput. **7**(1), 108–116 (1995)
19. Zur, R.M., Jiang, Y., Pesce, L.L., Drukker, K.: Noise injection for training artificial neural networks: a comparison with weight decay and early stopping. Med. Physics **36**(10), 4810–4818 (2009). https://doi.org/10.1118/1.3213517
20. Pogu, M., de Cursi, J.E.S.: Global optimization by random perturbation of the gradient method with a fixed parameter. J. Glob. Optim. **5**, 159–180 (1994). https://doi.org/10.1007/BF01100691
21. Lopez, R., de Cursi, E.S., Lemosse, D.: Approximating the probability density function of the optimal point of an optimization problem. Eng. Optimizat. **43**, 281–303 (2011). https://doi.org/10.1080/0305215x.2010.489607
22. de Cursi, E.S.: Variational methods for engineers with Matlab. ISTE, Ltd., Wiley, London, Hoboken (2015)
23. de Cursi, E.S., Sampaio, R.: Uncertainty Quantification and Stochastic Modeling with Matlab. Elsevier Science Publishers B.V., NLD (2015)
24. de Cursi, E.S., Sampaio, R.: Modeling and convexity. ISTE, Ltd., Wiley, London, Hoboken (2010)

Climate Precipitation Prediction with Uncertainty Quantification by Self-configuring Neural Network

Juliana A. Anochi[1], Reynier Hernández Torres[2], and Haroldo F. Campos Velho[3](✉)

[1] National Institute for Space Research (INPE), Center for Weather Forecasting and Climate Studies (CPTEC), Cachoeira Paulista, SP, Brazil
juliana.anochi@inpe.br
[2] National Institute for Space Research (INPE), Earth Observation General Coordination (OBT), São José dos Campos, SP, Brazil
reynier.torres@inpe.br
[3] National Institute for Space Research (INPE), Laboratory for Computing and Applied Mathematics (LABAC), São José dos Campos, SP, Brazil
haroldo.camposvelho@inpe.br

Abstract. Artificial neural networks have been employed on many applications. Good results have been obtained by using neural network for the precipitation climate prediction to the Brazil. The input are some meteorological variables, as wind components for several levels, air temperature, and former precipitation. The neural network is automatically configured, by solving an optimization problem with Multi-Particle Collision Algorithm (MPCA) metaheuristic. However, it is necessary to address, beyond the prediction the uncertainty associated to the prediction. This paper is focused on two-fold. Firstly, to produce a monthly prediction for precipitation by neural network. Secondly, the neural network output is also designed to estimate the uncertainty related to neural prediction.

Keywords: Neural network · Precipitation climate prediction · MPCA metaheuristic

1 Introduction

The climate forecast is a key factor for many social and economic sectors, such as civil defense for disasters, energy production, agricultural, transportion systems, insurance price. Brazil is a country with a great territorial extension, with different precipitation and temperature regimes. The CPTEC-INPE (CPTEC: Center for Weather Prediction and Climate Studies, INPE: National Institute for Space Research) has global and regional models to perform the weather and climate prediction based on numerical time integration by partial differential equations. These are complex and sophisticated computer models executed by a supercomputer.

J. E. S. De Cursi (Ed.): Uncertainties 2020, LNME, pp. 242–253, 2021.
https://doi.org/10.1007/978-3-030-53669-5_18

Models for carrying out prediction and climate monitoring using artificial intelligence are already employed in research related to climate precipitation by Anochi et al. [2], hydrology by Sousa et al. [22], severe weather by Ruivo et al. [18], Baboo et al. [7], among others.

Brazil has several climate regimes: equatorial, tropical, and subtropical patterns, due to different latitude ranges within the Brazilian territory. In addition, there are great variations in the rainy and dry seasons, with some regions registering long periods of drought – a semi-arid region from the Brazilian Northeast.

Anochi et al. [4] have developed a climate prediction model of the precipitation based on neural networks (NN) approach. This model was adapted for monthly and seasonal forecasting.

Different from our previous studies, where the NN were designed for climate prediction associated at each Brazilian regions, here the NN forecaster is applied for the Northeast of Brazil. In addition, the NN is designed to estimated the prediction variance, in order to find the uncertainty of the prediction. Some approaches use ensemble prediction [8,13,17] or bred-vector [20] to quantify the uncertainty. Our formulation follows an alternative methodology, where the uncertainty is estimated as an output from the neural climate predictor.

2 Theoretical Foundations

2.1 Neural Networks

Artificial neural network (ANN) is one of the main branches of Artificial Intelligence, characterized as a learning machine, and has been used in a variety of applications with great success. Inspired by the functioning of the neural structure of intelligent organisms, neural networks are able to acquire knowledge through experience. ANNs are efficient machines and can provide robust solutions for areas involving classification, forecasting, optimization, pattern recognition, and approximation of functions.

Neural networks can be found with very simple architectures, even very complex configurations. The choice of architecture to be used is made by specialists in artificial intelligence, based on experience and empirical analysis.

Neural networks are distributed parallel systems, composed of neurons or processing units, which compute certain mathematical functions, usually non-linear. Processing neurons can be distributed in one or more layers and interconnected by a large number of connections (synaptic weights), which store the knowledge represented in the model.

Mathematically, we can describe a neuron k by writing the following pair of equations [12]:

$$input : v_k = \sum_{j=1}^{n} w_{kj}\, x_j, \tag{1}$$

$$output : y_k = \varphi(v_k + b_k), \tag{2}$$

where x_n are the inputs; w_{kj} are the connection weights; b_k is the bias; φ is the activation function; and y_k is the output.

The different architectures of neural networks can be formed by the combination of artificial neurons and are defined by the type of connection between networks. Each neuron transfers its signal to the neurons that are in one of the subsequent layers.

In this work, the Multi Layer Perceptron (MLP) network was used. The MLP has emerged as an alternative solution for non-linearly separable problems and has been successfully applied to solve complex problems through its supervised training using the error backpropagation algorithm based on the learning rule for correction of error [12].

The architecture of the MLP network consists of the topological arrangement of the processing units of the neurons with the respective values of weights associated with the connections. The synaptic weights are adjusted by delta rule. The MLP network has an input layer, at least one intermediate layer, and an output layer.

Although a NN model has great potential, its performance depends on the definition of its parameters, since the definition of the topology can significantly influence the phase of the training process.

2.2 MPCA for Identify the Best NN Topology

The performance of neural networks depends heavily on the data analyzed. For example, researchers who are familiar with the data, do not necessarily have the experience of designing ANN architectures and vice versa. To this end, there is a great demand for the development of algorithms that allow researchers without any knowledge of machine learning to obtain models to describe their data in an automatic way.

The task of optimizing neural networks is very computationally expensive, compared to other optimization problem solving tasks. It is therefore necessary to develop effective methods of evolutionary computing that significantly accelerate the assessment of the suitability of ANNs.

In practice, the NN topology is usually selected by using empirical or statistical methods that are used to find the best parameters. Here, we use the Multiple Particle Collision Algorithm (MPCA) metaheuristic for configuration the topology of the MLP network.

This strategy was developed by Anochi et al. [3] can be considered as an optimization problem, where each point in the search space represents a NN with different topologies.

The MPCA metaheuristic was introduced by Luz et al. [14], and is an extension of the canonical Particle Collision Algorithm (PCA) [19]. The proposed structure for the MPCA algorithm uses a set of n particles, independently exploring but collaboratively, the same space. The introduction of n particles leads to the need to implement an indirect communication mechanism between particles.

The MPCA starts with a selection of an initial solution, it is modified by a stochastic perturbation conducing to the construction of a new solution. The new solution is compared and this solution can or cannot be accepted. If the new solution is not accepted, the particle can be send to a different location of the search space. If a new solution is better than the new solution is absorbed.

The MPCA has been used successfully in several optimization problems, such as: diagnosis of failures [11], identification of atmospheric temperature profile [21], climate prediction [2], solution of a radiative inverse problem [23], and the other applications.

3 Methodology

The proposal is focused on the application of neural networks in meteorology. The methodological novelty consists in the definition of an optimal neural network. The optimal network identification is formulated as an optimization problem, solved by the meta-heuristic MPCA – this methodology is called the self-configuring network.

Despite the development of numerical models and computer systems, the prognosis of atmospheric conditions still has uncertainties [15]. Indeed, there are a lot of uncertainties for the numerical weather prediction (NWP), when a set of partial differential equation are employed: calculation of initial condition[1], representation and calibration of physical processes[2]. For the NWP, uncertainties are enveloped by the *ensemble prediction* procedure [9,16], allowing to compute a forecasting confidence interval. In this context, our goal is to present a decision support approach, providing the seasonal forecasting to the precipitation field, as well as the uncertainty associate to this climate forecast.

One advantage of the presented methodology is to apply it without the support of a neural network expert: the user provides the data set and the system defines the best multilayer perceptron neural network for the application.

3.1 Meteorological Data

For the climate prediction of precipitation using neural networks, different data sets were collected, from reanalysis data from the National Centers for Environmental Prediction (NCEP/NCAR) provided by the NOAA/OAR/ESRL PSD, Boulder, Colorado, USA, from the website: https://www.esrl.noaa.gov/psd/, and the Global Precipitation Climatology Project monthly precipitation dataset (GPCP) NOAA/OAR/ESRL PSD, Colorado, USA [6].

In our previous results for predicting climate precipitation [1–3], the neural network input/output are described below. The same values are used in the neural network here, but the variance associated to the previous forecasting is another input feature, and there is one more output: the variance of the predicted precipitation.

---

1 The process to compute the initial condition combining observations and the background data from the mathematical model is called *data assimilation* [13].

2 Radiation, turbulence, surface representation, cloud formation, rainfall processes.

- **Input variables**: The variables used as inputs to the neural network were selected from the NCEP Reanalysis data: zonal and meridional winds at three levels (500 hPa, 850 hPa), air temperature at 850 hPA, air temperature (surface), specific humidity at 850 hPa, and pressure surface.
- **Output variable**: The variable used as an output to the NN was selected from the monthly precipitation dataset, this dataset consistent analysis of global precipitation from an integration of various satellite data sets over land and ocean and a calibration analysis over land. – GPCP.

3.2 Study Area

The study region was Northeast Brazil. The rainy season in this region is characterized by low rainfall levels, restricted to a few months, featuring a semi-arid climate and presenting high climatic predictability.

The precipitation regime in northeastern Brazil is characterized by scarcity, spatio-temporal irregularity and long periods of drought, where most of the precipitation usually occurs in three months, with an annual average of less than 800 mm. These characteristics result in the frequent occurrence of days without precipitation. The variability of precipitation may be associated with variations in Sea Surface Temperature (TSM) patterns over tropical oceans, which affect the position and intensity of the Intertropical Convergence Zone (ZCIT) over the Atlantic Ocean, as well as anomalies observed in the Pacific Ocean, resulting in years with La Niña and/or El Niño [10].

The climate prediction problem was applied in Northeast region with coordinates (Lat 17S, 0) to (Lon 47W, 35W). The considered region is illustrated in Fig. 1.

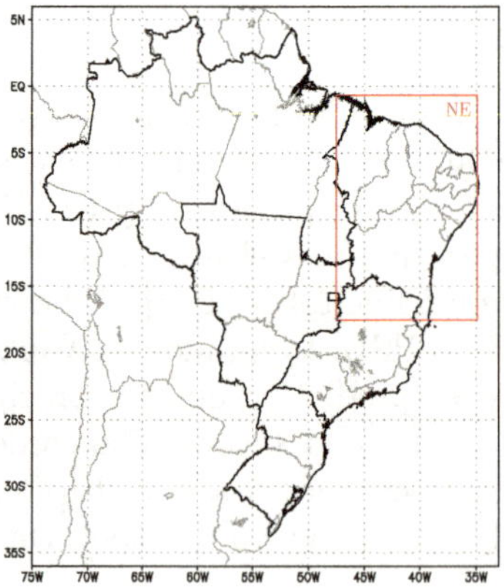

Fig. 1. Analyzed area from the Northeast (NE) region of Brazil

4 Experimental Settings

The strategy for optimizing the parameters of the NN topology is considered as a mono-objective optimization problem. Four mixed-variable domain parameters will be optimized: two continuous variables (the learning rate parameter η, and the momentum constant α) and two discrete variables (the number of neurons in the hidden layer, and the type of activation function.

For the neural network learning phase, a well known procedure is the Delta rule, where a correction $\Delta w_{ji}(n)$ is applied to the synaptic weight, for minimizing the square difference between the network output and the target values $\varepsilon(n)$. Following the rule delta, the synaptic weight are found by [12]:

$$\Delta w_{ji}(n) = \alpha \Delta w_{ji}(n-1) - \eta \frac{\partial \varepsilon(n)}{\partial w_{ji}}, \tag{3}$$

where $\Delta w_{ji}(n) \equiv w_{ji}(n) - w_{ji}(n-1)$.

The objective function used in the MPCA metaheuristic optimization process was implemented to weight the average training E_{train}) and generalization error E_{gen}), and a penalty term. The function is given by [5]:

$$J = penalty \times \left(\frac{\rho_1 \times E_{\text{train}} + \rho_2 \times E_{\text{gen}}}{\rho_1 + \rho_2} \right), \tag{4}$$

where $\rho_1 = 1$ and $\rho_2 = 0.1$, the same values proposed by Carvalho et al. [5].

4.1 Training Strategy ANN

The data was sub-divided, in which we created a partition in the previous data (training set). From this partition, we create a time series, with the values of difference, mean and variance for each partition (sub-interval). From this new data series (variance), a new neural network was created to estimate the variance. This paper is focused on two-fold. Firstly, to produce a monthly prediction for precipitation. Secondly, to design the neural network to has an output to estimate the uncertainty related to the neural prediction.

For our experiments, monthly averages for meteorological variables are considered for the period January-1990 up to December-2018.

- The training data subset was formed with data from January-1990 up to December-2001;
- The period from January-2002 up to December-2017 was used for the test;
- The generalization (prediction) subset corresponds to the period from January-2018 up to December-2018.

The "Mean Squared Error" (MSE, Eq. 5) metric is adopted to quantify learning performance and the ability of the neural network to identify and respond to similar patterns.

$$\text{MSE} = \frac{1}{N} \sum_{k=1}^{N} (d_k - y_k)^2, \tag{5}$$

where N is the number of training patterns, d_k is the target values and y_k is the output produced by the neural network.

This work presents the climate precipitation prediction for the Northeast Region using neural networks as a predictor model. However, in addition to prediction, it is necessary to address the uncertainty associated to the prediction. For this, a neural network was designed to have an output to estimate the uncertainty related to the neural precipitation prediction.

This uncertainty was calculated from the variance of the average error obtained from the difference between the network output and the target value, is given by Eq. 6.

$$\sigma_j^2 = \frac{1}{n_j - 1} \sum_{k=1}^{n_j} (e_{k,j} - \overline{e}_j)^2 \tag{6}$$

where σ_j^2 is the variance of the subset j; the number of points in the subset j is denoted by n_j. The difference between the desired response and the prediction obtained with the neural network is given by $e_{k,j} - \overline{e}_j$.

In this experiment, several realizations are executed to find representative solutions. The configuration of the NN topology for the climate prediction experiment is described below – the NN is applied for each grid point:

- Nine inputs (meteorological variables).
- Two output nodes for the results: one to estimate the uncertainty related to the neural model forecast and the other to the precipitation prediction.
- One hidden layer with 12 neurons.
- The logistic sigmoid is the activation function.
- Learning rate η and momentum value α produce the best fitness (MLP-MPCA) for the following numerical values: $\eta = 0.57$ and $\alpha = 0.65$.
- The iteration for the training phase stops when the error reaches the value 10^{-6}.
- For determining the NN configuration by the MPCA, 25 experiments are performed to find the best fitness to NN.

Figure 2 shows an example of a neural network model designed with an input layer, an intermediate layer (hidden), and two outputs: one to estimate the uncertainty related to the neural forecast and the other to the precipitation prediction. The term vm^k is k-th input variable meteorological, the p_k is the precipitation predicted by the NN, and $\sigma^2{}_k$ is the uncertainty related to the neural prediction.

5 Results

The results presented in this section show the behavior of the neural network as a predictor model of the precipitation variable.

The numerical experiments carried out in this study show the advantage of using the MPCA tool. The procedure identifies the best parameters for the NN

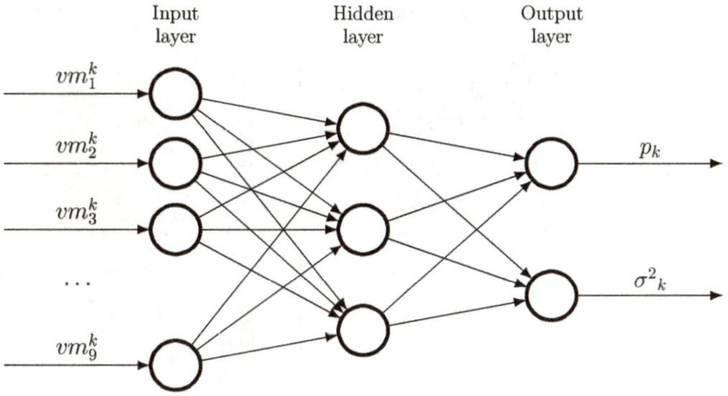

Fig. 2. Neural network model for climate precipitation prediction.

application, finding its topology by minimizing the a cost functional The MPCA approach does not require a specialist to the task for NN configuration. The neural model can be implemented for both applications, for operational use and in research institutions.

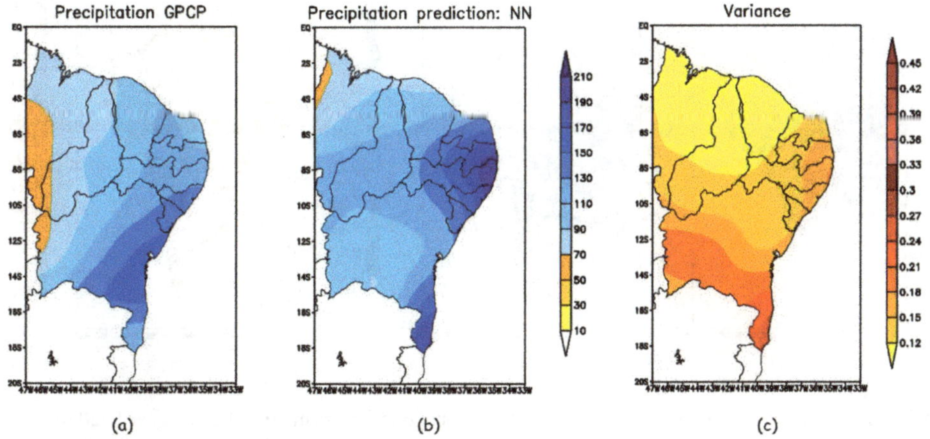

Fig. 3. Precipitation prediction for the Northeast region in the January of 2018. (a) observed precipitation by GPCP; (b) the result of the precipitation prediction estimated by the NN model; (c) estimated variance.

Figures 3, 4, 5, and 6 show the results of the month forecast for January, March, July, and December, respectively. In general, the NN was able to catch the patterns for precipitation field to the region under analysis. From the qualitative point of view, Fig. 3 looks to have the worse result, in comparison with other months. However, the lower error on the North part of the studied region was noted at January (Fig. 3) – see the variance maps Figs. 3c, 4c, 5c, and 6c.

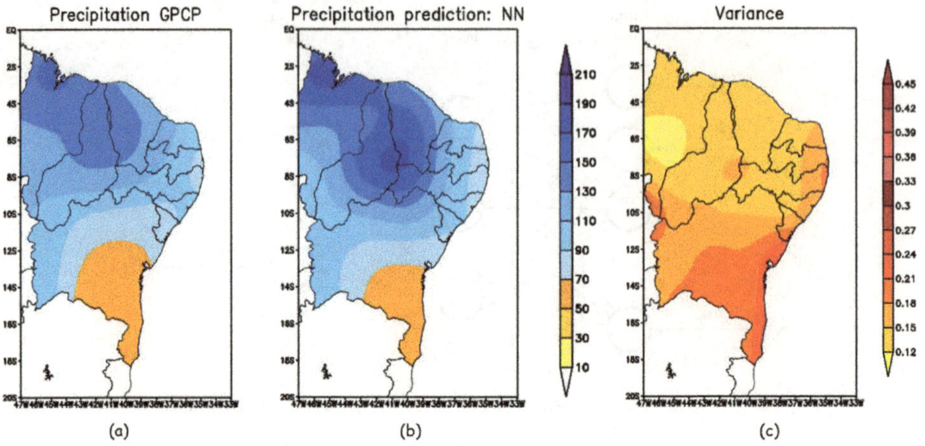

Fig. 4. Precipitation prediction for the Northeast region in the March of 2018. (a) observed precipitation by GPCP; (b) the precipitation prediction calculated by the NN model; (c) estimated variance.

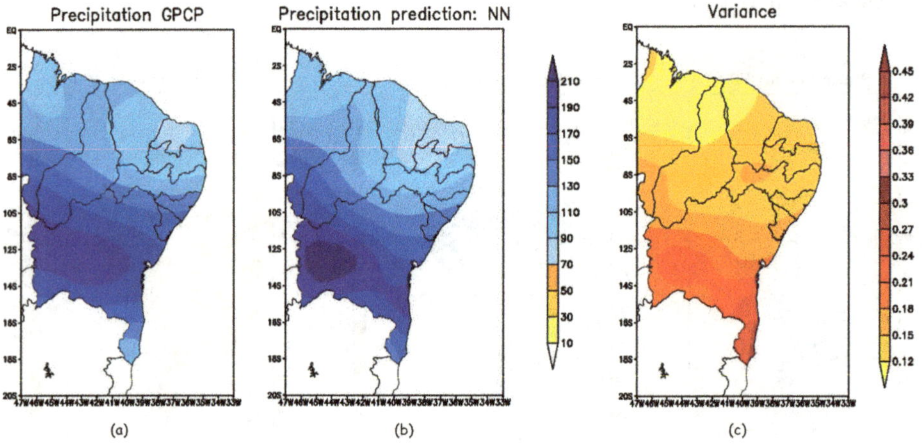

Fig. 5. Precipitation prediction for the Northeast region in the July of 2018. (a) observed precipitation by GPCP; (b) the result of the precipitation prediction estimated by the NN model; (c) estimated variance.

The poorest prediction was verified to the south of the region – see the variance maps. More studies need to be carried out for a better understanding of such behaviour is a bias from the neural network, and/or additional information (inputs) is necessary to include in the system.

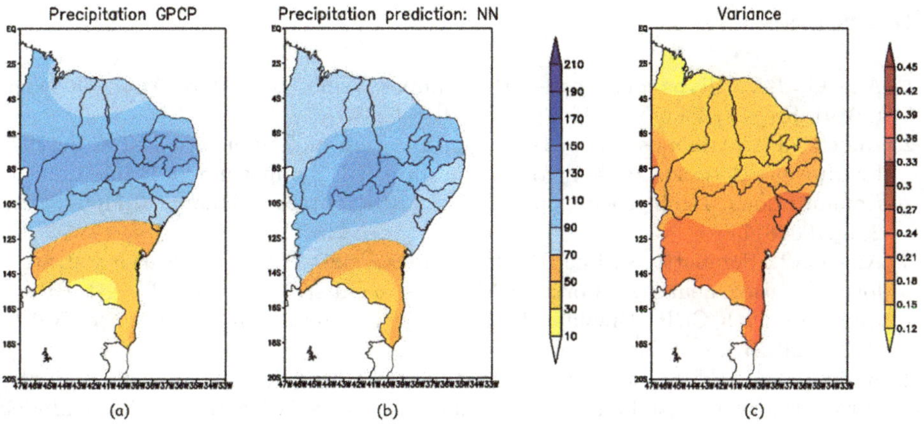

Fig. 6. Precipitation prediction for the Northeast region in the December of 2018. (a) observed precipitation by GPCP; (b) the precipitation prediction estimated by the NN model; (c) estimated variance.

6 Conclusions

Brazil is a country with different climate conditions: equatorial, tropical, and subtropical climatic zones. The variable precipitation has a high degree of difficulty to be predicted.

Precipitation is the most difficult meteorological variable to be predicted. Brazil is a country with different climate conditions, with equatorial, tropical, and sub-tropical climate regions. The present paper applied a *data driven* framework to perform a monthly climate prediction to the precipitation for the Brazilian territory. An optimal Multi-Layer Perceptron (MLP), one of the most used supervised Neural Network (NN), is employed as a neural climate forecaster. An automatic configuration for the MLP-NN is obtained by using the Multi-Particle Collision Algorithm (MPCA) to minimize a function to identify the best topology for the neural network.

It is a hard task to develop models to predict the precipitation. There are many reasons for such difficulties: local and synoptic patterns has different influences, and different variables and/or conditions are associated with the rain-fall depending on the season of the year. In addition, the precipitation has high variability in space and time. All these factors has strongly influence in the behavior of the precipitation.

Neural network is a prestigious area of Artificial Intelligence and has shown their efficiency in various application areas, and the meteorology is one of them. Predictive models based on NNs are easy to use, and do not require very powerful computers.

The results present in this paper are consistent with observation data (GPCP), becoming a good tool to support month climate prediction. Our approach is also applied to identify where the prediction is most reliable in the domain.

References

1. Anochi, J.A., Hernández Torres, R., Campos Velho, H.F. Pure Appl. Geophys. (2019). https://doi.org/10.1007/s00024-019-02386-y
2. Anochi, J.A., Campos Velho, H.F.: Optimization of feedforward neural network by Multiple Particle Collision Algorithm. In: IEEE Symposium on Foundations of Computational Intelligence , pp. 128–134 (2014). https://doi.org/10.1109/FOCI.2014.7007817
3. Anochi, J., Sambatti, S., Luz, E., de Campos Velho, H.F.: New learning strategy for supervised neural network: MPCA meta-heuristic approach. In: 1st BRICS Countries 11th CBIC Brazilian Congress on Computational Intelligence, Recife, Brasil (2013)
4. Anochi, J.A., Silva, J.D.: Uso de redes neurais artificiais e teoria de conjuntos aproximativos no estudo de padrões climáticos sazonais. Learn. Nonlinear Models **7**, 83–91 (2009)
5. Carvalho, A., Ramos, M., Chaves, A.: Metaheuristics for the feedforward artificial neural network (ANN) architecture optimization problem. Neural Comput. Appl. **20**, 1273–1284 (2011)
6. Adler, R.F., Huffman, G.J., Chang, A., Ferraro, R., Xie, P., Janowiak, J., Rudolf, B., Schneider, U., Curtis, S., Bolvin, D., Gruber, A., Susskind, J., Arkin, P.: The version 2 global precipitation climatology project (GPCP) monthly precipitation analysis (1979-Present). J. Hydrometeor. **4**, 1147–1167 (2003)
7. Baboo, S.S., Shereef, I.K.: An efficient weather forecasting system using artificial neural network. Int. J. Environ. Sci. Dev. **1**, 321–326 (2010)
8. Cintra, R.S., Cocke, S. Campos Velho, H.F.: Data assimilation by neural networks with ensemble prediction. Proceedings of the joint ICVRAM ISUMA UNCERTAINTIES Conference, Florianopolis, SC, Brazil, 8–11 April 2018 (2018). http://icvramisuma2018.org/cd/web/PDF/ICVRAMISUMA2018-0095.PDF
9. Cintra, R.S.C., Cocke, S., Campos Velho, H.F.: Data assimilation by neural networks with ensemble prediction. In Proceedings of the Joint ICVRAM ISUMA Uncertainties Conference. ICVRAM ISUMA Uncertainties, Florianapolis, SC, Brazil, pp. 1–8 (2018)
10. Cavalcanti, I.F.: Tempo e clima no Brasil. Oficina de textos (2016)
11. Echevarría, L.C., Santiago, O.L., Silva Neto, A.J.: Aplicación de los algoritmos Evolución Diferencial y Colisión de Partículas al diagnóstico de fallos en sistemas industriales. Investigación Operacional **33**, 160–173 (2012)
12. Haykin, S.: Neural Networks: A Comprehensive Foundation. Prentice Hall, New Jersey (1994)
13. Kalnay, E.: Atmospheric Modeling. Cambridge University Press, Data Assimilation and Predictability (2002)
14. Luz, E.F., Becceneri, J., Campos Velho, H.F.: A new multi-particle collision algorithm for otimization in a high-performance environment. J. Comput. Interdisciplinary Sci. **1**, 3–10 (2008)
15. Neal, R.A., Boyle, P., Grahame, N., Mylne, K., Sharpe, M.: Ensemble based first guess support towards a risk-based severe weather warning service. Meteorol. Appl. **21**(3), 563–577 (2014)
16. Palmer, T.: The ECMWF ensemble prediction system: looking back (morethan) 25 years and projecting forward 25 years. Quarterly J. Royal Meteorol. Soc. **145**(S1), 12–24 (2018). https://doi.org/10.1002/qj.3383

17. Reich, S., Cotter, C.: Probabilistic Forecasting and Bayesian Data Assimilation. Cambridge University Press, Cambridge (2015)
18. Ruivo, H.M., Campos Velho, H.F., Sampaio, G., Ramos, F.M.: Analysis of extreme precipitation events using a novel data mining approach. Am. J. Environ. Eng. **5**, 96–105 (2015)
19. Sacco, W., Oliveira, C.: A new stochastic optimization algorithm based on a particle collision metaheuristic. In: Proceedings of 6th World Congress of Structural and Multidisciplinary Optimization (WCSMO) (2005)
20. Salgueiro Romero, L.F., Sandri, A.A., Campos Velho, H.F.: Analysis of the breeding technique applied to the CPTEC-AGCM model. In: Conference of Computational Interdisciplinary Science (CCIS 2016), São José dos Campos (SP), Brazil (2016). http://www.epacis.net/ccis2016/papers/paper_57.pdf
21. Sambatti, S.B.M., Anochi, J.A., Luz, E.F.P., Carvalho, A.R., Shiguemori, E.H., Campos Velho, H.F.: Automatic configuration for neural network applied to atmospheric temperature profile identification. In: 3rd International Conference on International Conference on Engineering Optimization, pp. 1–9 (2012)
22. Sousa, W., Sousa, F.: Rede neural artificial aplicada a previsão de vazão da bacia hidrográfica do rio piancó. Revista Brasileira de Engenharia Agrícola e Ambiental **14**, 173–180 (2010)
23. Torres, T.R., Luz, E.F., Campos Velho, H.F.: Multi-particle collision algorithm for solving an inverse radiative problem. In: Constanda, C., Kirsch, A. (eds.) Integral Methods in Science and Engineering, Springer, Cham, pp. 275–284 (2015) https://doi.org/10.1007/978-3-319-16727-5_26

Uncertainty Quantification in Risk Modeling: The Case of Customs Supply Chain

Lamia Hammadi[1,2](✉), Eduardo Souza de Cursi[1], and Vlad Stefan Barbu[3]

[1] LMN, National Institute of Applied Sciences INSA of Rouen, Saint-Étienne-du-Rouvray, France
[2] Laboratory of Engineering Sciences for Energy, National School of Applied Sciences-ENSAJ, UCD, El Jadida, Morocco
hammadi.l@ucd.ac.ma
[3] LMRS, University of Rouen - Normandy, Mont-Saint-Aignan, France

Abstract. In this paper, Uncertainty Quantification (UQ) is used in risk modeling to de-scribe the time series describing the risks in customs supply chain. We start by introducing the Hilbertian approach related to representation of Random Variables and addressing these approximations and their applications in UQ. Then an extension where these models are applied to the time series in order to handle the seasonal components of risks in customs supply chain. The models are fitted to time series describing the seized quantities of the smuggling of drugs on two sites using Moment Matching Method. the results furnish a good description of important properties of the data, namely, the polynomial expansion and Cumulative Distribution Function (CDF).

Keywords: Uncertainty quantification · Hilbert expansion · Risk time series · Moment matching method · Cumulative distribution function · Customs supply chain

1 Introduction

In recent times, uncertainty quantification has gained a massive attention from many researchers in many scientific fields: Engineering, Physics, Applied Mathematics, Management sciences and Economics. Uncertainties are seen not only as a governing element in human activities, namely, environmental and social phenomenon, industrial processes, risk analysis and economies of governments, but also as an effective way of generating a better accuracy. However, some difficulties are still with us, namely the approximation of a random variable by another one without the previous knowledge of their joint probability distribution.

Many studies were conducted to handle the uncertainties and researchers developed various methods in order to improve the quality and the efficiency of UQ. Among these methods, Monte Carlo Simulation (MCS) may be considered as being the basic UQ approach to be used when no other information is known. The MCS provides successive resolutions of a deterministic system incorporating uncertain parameters modeled by

J. E. S. De Cursi (Ed.): Uncertainties 2020, LNME, pp. 254–270, 2021.
https://doi.org/10.1007/978-3-030-53669-5_19

random variables. Historically, this technique was developed for situations where the lack of information about the problem to be solved makes that only generation of variates is available. It generates, for uncertain parameters and according to their probability distributions and their correlations, random draws. For each draw, a set of parameters is obtained and a deterministic calculation, following analytical or numerical models well defined, is made. The statistics generated by this method are poor in accuracy [1] and significant results require a large number of samples which makes the MCS prohibitive in terms of computational cost [2]. Improvements may be obtained by using the Latin Hypercube Sampling (LHS). Such a sampling approach gives a better accuracy [1]. However, it has some limitations announced by [3]: "*The error estimates may not be improved even if the number of samples increases*".

In this regard, an appropriate approach has been emerged, namely, Polynomial Chaos Expansion (PCE), where random variables are represented by multivariate series of Gaussian variables [4–7]. When considering random variables, the terminology representation is generally used in the sense of Hilbert basis. Approximation results from the truncation of the representation at a given level. This approach has proved to be less expensive and computationally more efficient than the MCS and LHS, but it was limited by the use of Gaussian variables and the need of a particular statistical information about the distributions under consideration: when approximating a random variable **Y** by a random Gaussian vector **U**, the knowledge of the joint distribution of the pair $(\boldsymbol{Y}, \boldsymbol{U})$ is necessary- what introduces some difficulty. Indeed, the use of independent variables leads to poor quality approximations [8].

Customs supply chain is introduced briefly as an application context of risk modeling in Sect. 2. UQ based-models are presented in Sect. 3. Then in Sect. 4, we first block the time series by season, in order to remove the seasonal components, and fit a separate UQ based- model each season. These seasonal models are validated by checking their ability to generate Cumulative Density Functions CDFs per season per site. The results are satisfactory in the both sites. Finally, we conclude in Sect. 5.

2 Literature Review

2.1 Customs Supply Chain

The proposed definition of "customs supply chain" is developed based on a work with experts from the Moroccan customs administration. In this work, we gathered, on the one hand, a list of definitions and a categorization available on literature, and on the other hand, suggestions made by supply chain managers and customs experts. The proposed definition is the following: "*Customs supply chain is a set of all aspects that incorporate the moving of cargo and information from the exporter through the transport process, the logistics operations, customs crossing and financial process to the final importer. The customs supply chain is no longer contained within a country's borders, but encompasses all nations, whether they are exporters, importers or manufacturers*" [9, 10].

As a result, our supply chain structure is composed of five blocks. The proposed building blocks of our definition, which come from the existing operations management within supply chain, emphasize customs operations [11]:

i. **Customs operators:** the parties that receive and send goods: an importer in the receiving country and an exporter in the sending country. Importation and exportation are the main financial transactions of international trade.
ii. **Transport process:** multimodal transport, which covers at least two modes of transport, with the main one being sea transport (CTBL: combined transportation bill of lading).
iii. **Customs crossing:** includes any point authorized by customs authorities for the crossing of external borders. It covers declaration processing, customs clearance, data analysis, risk assessment, document checking, scanning, physical inspection, etc. Accordingly, customs crossing is the main block for securing the supply chain, due to customs intervention in all stages along with the routing of cargo. It includes the border checks both for goods entering and leaving the country.
iv. **Logistics operations:** encompasses all activities associated with the flow and transportation of goods as well as the associated information. Such activities follow three steps: "organize" (organize the activities for each function of the supply chain to deliver results); "carry out" (implementing and controlling what was planned); and "monitor" (which denotes checks and measurement of the functions and results against customs policies, objectives and requirements). It comprises activities such as warehousing, inventory and materials management, order fulfillment, supply/demand planning, and management of third-party logistics service providers.
v. **Financial process:** supports financial transactions between the actors in the supply chain and facilitates the monetary flows.

These elements are presented using generic descriptions [12]; i.e., the characterization of the building blocks of our customs supply chain are described in a general way without specification because we aimed to provide a simplified view of this concept that can be used to guide future researchers and practitioners to gain an overview or understanding of this particular field. The definition anchors its explanation of the customs supply chain with the aid of a newly proposed framework that illustrates the different components within the customs environment as well as the relationships between them.

2.2 Risk in Customs Context

It is a commonly held belief that "risk is a strategic prevention and response to potential threats" [9]. "Risk" in the customs context can be defined as the "likelihood that something will prevent the application of customs union measures and/or national measures concerning the customs treatment of goods" [13]. In the same regard, Widdowson stated that the concept of risk in the customs context refers "to the possibility of events and activities occurring that may prevent the customs from achieving their objectives" (Widdowson, 2004). Risks facing customs are related to "the potential for noncompliance with customs laws, regulatory requirements and international standards such as restricted and prohibited goods, rules of origin, duty exemption regimes, security and safety regulation, sanitary, environmental and technical standards, intellectual property, transnational crime, commercial fraud, and illicit traffics, as well as the organizational risks" [10].

In practical terms, risk in the customs context refers to illicit activities related to global supply chains. Those illicit activities can be regarded as a process involving several tasks or conduct in which, the ultimate goal is the delivery of goods to consumers in violation of applicable laws. Those goods may be intrinsically legal (for example, genuine tobacco products and alcoholic beverages, legitimately produced nuclear material, etc.); however, they may be transported and traded in contravention of tax laws, safety and security regulations, and so forth. Instead, other goods may be intrinsically.

3 Uncertainty Quantification in Risk Time Series Modeling

This paper develops stochastic models for risk time series. A particular incitation for this study was the need to generate realistic risk time series of the most critical illicit traffic: smuggling of drugs [9, 10]; in two risk detection sites, with the aim of analyzing the risk evolution over time.

3.1 Model Description

A model-based on UQ is a discrete process used the approximation involve the evaluation of expressions that depend on the joint distribution of the couple $(\boldsymbol{Y}, \boldsymbol{U})$, where: the uncertainties on a system is modelled by a random vector $\boldsymbol{U}$ and the system's response is denoted by $\boldsymbol{Y}$.

In this case, we assume that **Y** is a function of **U**, $\boldsymbol{Y} = \boldsymbol{Y}(\boldsymbol{U})$ and that statistical information is available on a sample of the couple $(\boldsymbol{Y}, \boldsymbol{U})$. In the following, this condition is not always guaranteed and that convenient approximations will be generated when this information is unknown. Since $\boldsymbol{Y} = \boldsymbol{Y}(\boldsymbol{U})$, we may consider a Hilbertian approach [2]: admitting that $\boldsymbol{Y} = \boldsymbol{Y}(\boldsymbol{U}) \in \boldsymbol{W}$, where $\boldsymbol{W}$ is a separable Hilbert space for the scalar product, we can choose an adequate Hilbert basis $\{\varphi_i\}_{i \in N^*}$ and look for a representation

$$\boldsymbol{Y} = \sum_{i=1}^{+\infty} y_i \varphi_i(\boldsymbol{U}) \tag{1}$$

In practice: Let $\boldsymbol{Y}$ and $\boldsymbol{U}$ be two random vectors (eventually having different dimensions) on the probability space (Ω, P). Let $W =$ be a subspace of functions of U. Determine an element **PY** from W such that **PY** is the best approximation of **Y** on W, i.e. It is expected that $\mathbf{PY} \to \mathbf{Y}$ for $\mathbf{n} \to +\infty$. The main purpose of a representation is to use $\boldsymbol{PY}$ instead of $\boldsymbol{Y}$ in the evaluation of the probability of crucial events, in a way that $\boldsymbol{PY}$ is a convenient approximation of $\boldsymbol{Y}$, in the sense that the probabilities are close to the real values when using $\boldsymbol{PY}$ instead of $\boldsymbol{Y}$. In general, this aim cannot be achieved and we look for an approximation such that $\boldsymbol{PY}$ may replace $\boldsymbol{Y}$ in practice, i.e. such that the difference $\boldsymbol{Y} - \boldsymbol{PY}$ is small enough in a sense corresponding to our goals. If $\{\mathrm{\varphi_i}\}_{\mathrm{i \in N^*}}$ is not orthonormal, a sequence of orthogonal projections are requested

$$\boldsymbol{Y} \approx \boldsymbol{PY} = \sum_{i=1}^{k} y_i \varphi_i(\boldsymbol{U}) \in \boldsymbol{W}; y = (y_1, y_2, \ldots, y_k)^t \tag{2}$$

where $\boldsymbol{y} \in \boldsymbol{R}^k$ denotes the unknown coefficients of the expression of $\boldsymbol{PY}$ which describes the evolution of the response of the system and $\{\varphi_i\}_{i \in N^*}$ is an Hilbert basis.

In our particular application:

- $\boldsymbol{U}$ is a random vector representing the uncertainties on a risk process, and
- $\boldsymbol{Y}$ denotes the monthly quantities seized of the illicit traffic by customs authorities in two sites (**S1**: site characterized by a high frequency of risk and **S2**: site characterized by a low frequency of risk).

$\boldsymbol{Y}$ is a time series with positive values. Risk times series in customs context are generally not stationary, with typically important seasonal and monthly components but also interannual variability. Interannual components are generally neglected. Notice that the risk's seasonality is an empirical observation, linked to divers factors such as: the number of passengers, illegal markets, consumption and holidays, etc. Seasonality is one of the rarely used statistical behaviors to analyze the evolution of the time series over time. Among methods available in the literature for modelling seasonal components, the differentiation or scaling with mean and variance functions which evolve periodically e.g. [14, 15].

Our approach consists in approximating seasonal components of $\boldsymbol{Y}$ by a linear combination of orthogonal polynomials of $\boldsymbol{U}$. This model consists in blocking the data by short time periods, typically per season depending on the conditional distribution of data available, then looking for a representation per season considering a polynomial basis: in a first time we make varying both the probability distribution function (PDF) of the random variable $\boldsymbol{U}$ and of the data $\boldsymbol{Y}$, in a second time we estimate these quantities from samples. In the sequel, we propose to use such models based-on uncertainty quantification for each risk per season per site and replace (2) by (3):

$$\boldsymbol{Y} \approx \boldsymbol{PY} = \sum_{i=1}^{k} \boldsymbol{y}_i^{(t)} \varphi_i(\boldsymbol{U}) \in \boldsymbol{W}; \boldsymbol{y}^{(t)} = \left(\boldsymbol{y}_1^{(t)}, \boldsymbol{y}_2^{(t)}, \ldots, \boldsymbol{y}_k^{(t)}\right)^t \tag{3}$$

with the unknown parameters $\boldsymbol{y}^{(t)} \in \boldsymbol{R}^k$ and $\{\varphi_i\}_{i \in N}$ is a polynomial basis. To follow the seasonal evolution of the risks, the states are numbered $\boldsymbol{y}_i^{(t)}$ according to their season of the time series Y. In practice, the exact distribution of $\boldsymbol{U}$ is unknown. It often assumes that $\boldsymbol{U}$ is not known and only samples from $\boldsymbol{Y}$ are available. In such a situation, a convenient solution consists in considering an artificial random variable $\boldsymbol{V}$ and replace (3) by (4):

$$\boldsymbol{Y} \approx \boldsymbol{PY} = \sum_{i=1}^{k} \boldsymbol{y}_i^{(t)} \varphi_i(\boldsymbol{V}) \in \boldsymbol{W}; \boldsymbol{y}^{(t)} = \left(\boldsymbol{y}_1^{(t)}, \boldsymbol{y}_2^{(t)}, \ldots, \boldsymbol{y}_k^{(t)}\right)^t \tag{4}$$

For the application considered in this paper, the observations are per month in such a way that Eq. (4) defines a seasonal function with period: a combination of three seasons; namely; high season, medium season and low season. Some observations of risk have a combination of more than three season. In other words, the observed data are generated by several distinct underlying causes of seasonality that have some unknown relationship between them. Each cause contributes independently to the generation process but we can only see the observation sequence as a final seasonal function, without knowing how much and in what way each cause contributes to the resulted observation sequence. This phenomenon is of great interest in customs risk time series analysis.

Hereafter, we focus on those particular time series of risks with a combination of three season within customs detected in two sites. These time series, which is shown in Figs. 1, are generated by simulation. Such a simulation consists of 10 years of data, with monthly quantities seized; quantities in kilograms (kg). It has to consider a long time series in this study in order to gather significant results for modeling seasonal variability, but good results were also obtained with the methodology described in this paper on shorter risk time series at various risk sites which permit to save computational time.

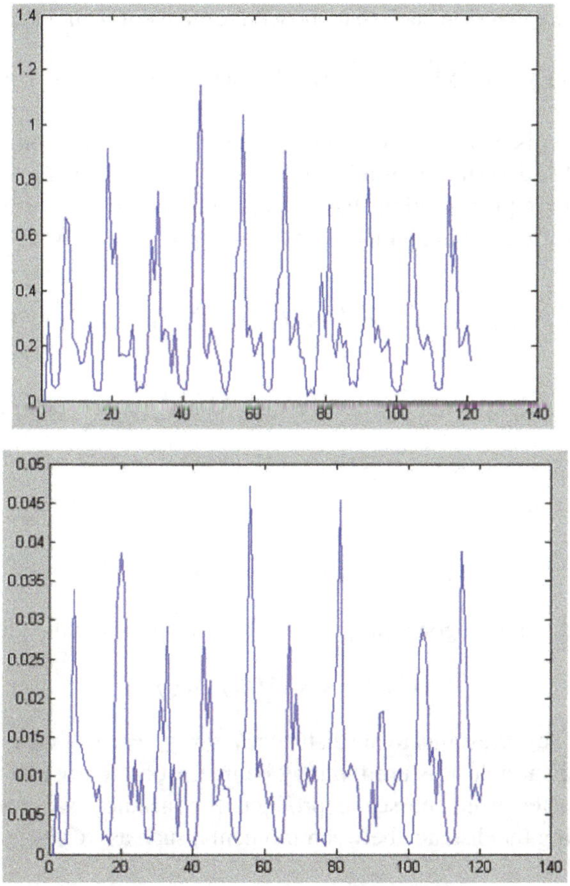

Fig. 1. Seized quantity of smuggling of drugs in (10^4) kg (y-axis) at site **S1** (upper) and at site **S2** (lower) between 2006 and 2016

3.2 Statistical Inference

Fitting the underlying model, for given distribution of the couple $(\boldsymbol{Y}, \boldsymbol{V})$, consists of using the approximation. This approximation involve the evaluation of expressions depend on the joint distribution of the pair $(\boldsymbol{Y}, \boldsymbol{V})$. In many situations, the approximation uses quantities having unknown nor difficult to determinate distributions and we have only a sample of the pair. In such a situation, it may be practical to use an approach that does not imply the evaluation of quantities depending upon the joint distribution of the pair and estimates all the necessary statistical properties directly from a sample.

In the following, we may determine an element $\boldsymbol{PY} \in \boldsymbol{W}$ such that some statistics of $\boldsymbol{Y}$ coincide with those of $\boldsymbol{PY}$. Among these approaches, the moment matching method may be considered as the most known method of approximation based on statistical properties. It looks for a finite expansion such that chosen moments of Y and **PY** coincide:

$$\boldsymbol{M_i(Y)} = \boldsymbol{E}\left(\boldsymbol{Y^i}\right) = \boldsymbol{E}\left((\boldsymbol{PY})^{\boldsymbol{i}}\right) = \boldsymbol{M_i(PY)};\ \ \boldsymbol{for}\ 1 \leq \boldsymbol{i} \leq \boldsymbol{n} \tag{5}$$

This approximation is connected with Levy's theorem; namely, convergence in distribution. The detailed description of this method can be found in [2]. This is an iterative series of steps, looking for solutions based on a minimization procedure as below

From numerical viewpoint, let us consider a map $\mathbf{F}\colon \mathbb{R}^k \to W$ given by:

$$\boldsymbol{x} = (\boldsymbol{x}_1, \boldsymbol{x}_2, \ldots, \boldsymbol{x}_k)^t \to \boldsymbol{F}(\boldsymbol{x}) = \sum_{i=1}^{k} \boldsymbol{x}_i^{(t)} \boldsymbol{\varphi}_i(\boldsymbol{V}) \in W \tag{6}$$

where $\{\varphi_i\}_{i\in N}$ is a polynomial basis. And a map $\mathbf{M} : \mathrm{R}^k \to \mathrm{R}^n$ such that:

$$\boldsymbol{x} = (\boldsymbol{x}_1, \boldsymbol{x}_2, \ldots, \boldsymbol{x}_k)^t \to \boldsymbol{M}(\boldsymbol{x}) = (M_1(\boldsymbol{F}(\boldsymbol{x})), \ldots, M_n(\boldsymbol{F}(\boldsymbol{x})))^t \in \mathbb{R}^n \tag{7}$$

And

$$\boldsymbol{M_Y} = (M_1(\boldsymbol{Y}), \ldots, M_n(\boldsymbol{Y}))^t \in \mathbb{R}^{\mathbf{n}} \tag{8}$$

We notice that the orthogonal projection $\boldsymbol{PY}$ corresponds to the global minimum of

$$C(\boldsymbol{x}) = dist(\boldsymbol{M}(\boldsymbol{x}), \boldsymbol{M_Y}) \tag{9}$$

The equations above correspond to the distance $dist(\boldsymbol{x}, \boldsymbol{y}) = \|\boldsymbol{x} - \boldsymbol{y}\|$. For our application where an artificial variable V is used, it may be interesting to use a distance or pseudo-distance disconnected from the scalar product: for instance, we may consider other distances measuring the distance between moments, such as, for instance

$$dist = \sum_{i=1}^{k} |M_i(\boldsymbol{F}(\boldsymbol{x})) - M_i(\boldsymbol{Y})|, \tag{10}$$

or the pseudo-distances such as, for instance,

$$dist = \sum_{i=1}^{k} \left| \frac{M_i(\boldsymbol{F}(\boldsymbol{x})) - M_i(\boldsymbol{Y})}{M_i(\boldsymbol{Y})} \right|. \tag{11}$$

Then

$$y = argmin(\boldsymbol{C}). \tag{12}$$

This kind of approximation based on minimization procedure has some limitations. It may converge to a non-interesting local minimum of the objective function which leads to a poor approximation. In practice, it means that a careful choice of the starting points has to be made; this is further set in the next section. For our particular implementation we used usual quasi-Newton algorithm which is known as an efficient algorithm with a quick rate of convergence near the minima. Another important problem in practice, which has received lots of attention in the last few years, is the problem of model selection which aims at finding the "optimal" value of of the degree of expansion $\boldsymbol{d}$ and the number of the moments $\boldsymbol{n}$. there is not a criterion from theoretical and/or computational point of view to select parsimonious model which fit the data well.

4 UQ-Based Models with Seasonal Components

4.1 Data and Model Parameters Setting

The models described in the previous section are tested on an artificially generated data set. The data is generated for each Risk per site by a function defined as

$$Y_t^{(i)} = \begin{cases} m_{h_i} + \left(\sigma_{h_i}\right)^2 \varepsilon_t \\ m_{m_i} + \left(\sigma_{m_i}\right)^2 \varepsilon_t;\ i = 1, 2 \\ m_{l_i} + \left(\sigma_{l_i}\right)^2 \varepsilon_t \end{cases} \tag{13}$$

where $\boldsymbol{\varepsilon_t}$ is a sequence of independent and identically distributed Gaussian variable with zero mean and unit variance. $\boldsymbol{m_h}$, $(\boldsymbol{\sigma_h})^2$ denote respectively mean and variance of high season of risk seizures, $\boldsymbol{m_m}$, $(\boldsymbol{\sigma_m})^2$ are mean and variance of medium season of seizures and $\boldsymbol{m_l}$, $(\boldsymbol{\sigma_l})^2$ present mean and variance of low season of seizures. The annual risk averages per site are computed based on illicit trade reports of World Customs Organization (WCO) [16] since 2011 until 2016 and WCO customs and Tobacco reports [17] since 2008 until 2011. Then the annual averages are converted to monthly averages which are weighted depends on seasonality of risks.

In Eq. (13), the models are assumed to vary with time. Loosely speaking, it permits to model that the mean and the variance of the seized quantities exhibits monthly variation. This seems to be realistic for the particular time series considered in this work. This models could be obviously generalized by assuming that seasonal components are periodic functions, but the various attempts that we have done in this regards for our particular data set are to consider a combination of three Gaussian process depending on the period of the year. Indeed, the observation $\boldsymbol{Y_t^{(i)}}$ is related to the extent of the monthly variation in site i (i.e., two sites of risk detection).

The degree of expansion and the number of the moments are involved in the quality of the approximation. The degree of expansion and the number of the moments should be neither too large nor too small. An expansion with a high degree and/or with a many number of moments will increase the cost and time of computing. On the other hand, if the number of moments and/or the degree of expansion are too small, risk of miss-approximation will increase. We have launched experiments with different degrees of expansion and moments numbers. Based on our experience, the best results are obtained for an expansion of degree **d = 6** and **n = 10** first empirical moments. And the initial components of the orthogonal projection ***PY*** are assigned a set of random values of 6 entries.

4.2 Results

We now come to the core part of the problem: making the approximation and determining the Cumulative Distribution Function (CDF) of each Risk time series per site: we apply our approach by considering a polynomial basis $\{\varphi_i\}_{i \in N}$, then we look for a finite expansion based on Moment matching method. In the following, we focus on the case of one risk time series since the approach applied and outputs obtained for this risk are the same quantities for all other risks. For instance, let us consider the case of smuggling of drugs given by:

$$Y_t^{(i)} = \begin{cases} m_{h_i} + \left(\sigma_{h_i}\right)^2 \varepsilon_t; & t = 6, 7, 8 + 12n; \quad n \in \mathbb{N} \\ m_{m_i} + \left(\sigma_{h_i}\right)^2 \varepsilon_t; & t = 1, 5, 9, 10, 11, 12 + 12n; \quad i = 1, 2 \\ m_{l_l} + \left(\sigma_{h_i}\right)^2 \varepsilon_t; & t = 2, 3, 4 + 12n; \quad n \in \mathbb{N} \end{cases} \tag{14}$$

$\boldsymbol{m_h}$, $(\boldsymbol{\sigma_h})^2$ denote respectively the mean and the variance of high season of seizures which correspond to the months of June, July and August, $\boldsymbol{m_m}$, $(\boldsymbol{\sigma_m})^2$ are the mean and the variance of medium season which correspond to the months of January, May, September, October, November and December and $\boldsymbol{m_l}$, $(\boldsymbol{\sigma_l})^2$ present the mean and the variance of low season of seizures which correspond to the months of February, March, April [16, 17]. The results furnished by the uncertainty quantification model with the pseudo-distance (11) are shown in the Figures below (Notice that for a reason of simplification, we divided the mean and the standard deviation of risk by $(\mathbf{10^4})$ per the two sites).

The results show a good quality of approximations for an expansion of degree d = 6 and n = 10 first empirical moments as well as the Cumulative Distribution Functions (CDF) are correct. Except for low season of site S2, as, the cumulative distribution functions are correct –even the quality of approximations are relatively good. these approximations have been accepted as the final distances between the moments are negligible (of the order of 1E-4). The quality of the approximation depends on the quality of the optimization – if the algorithm converges to a non-interesting global optimum of C, a poor quality of approximation is obtained. For our application, the approximations are conducted-based on Moment Matching which is connected to Levy's theorem. Such a theorem, ensures the convergence in distribution, which is a weak convergence involving approximation of the cumulative distribution function of the variable, but not the approximation of the variables themselves [2]. In practice, it means that the moment matching is not anticipated to furnish a good approximation of the variables, but only of their distributions (Figs. 2, 3, 4, 5, 6 and 7).

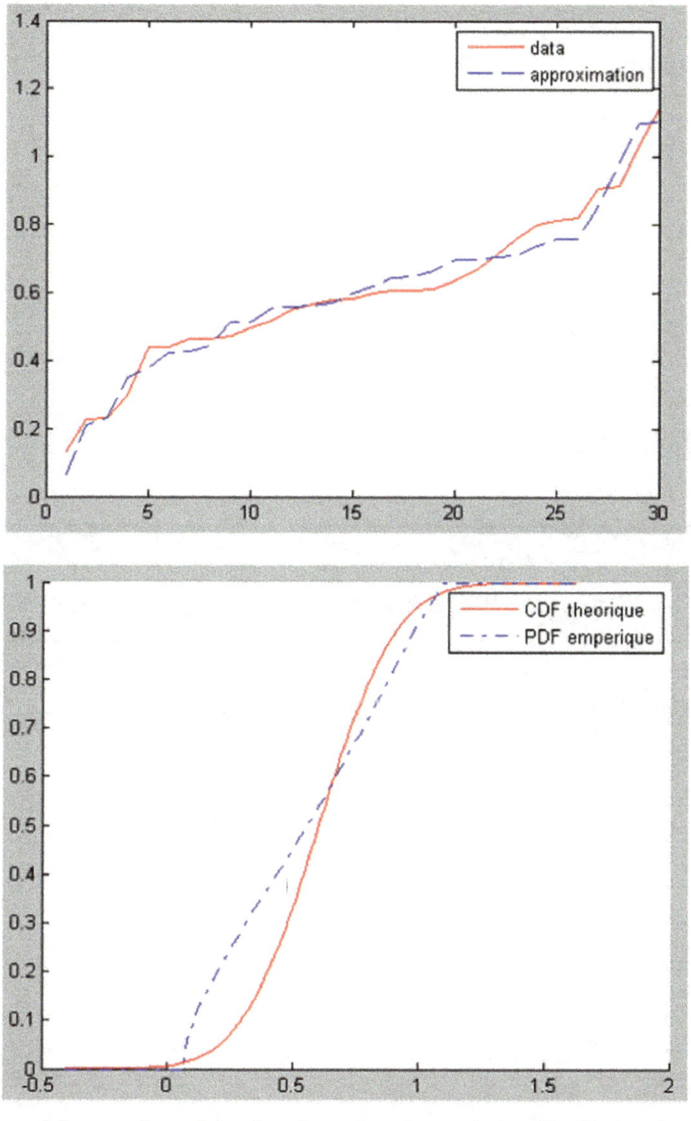

Fig. 2. Polynomial expansion of the date (upper) and cumulative distribution function (lower) (theoretical & empirical) (y-axis) of smuggling of drugs in site **S1** approximated using moment matching: **A-1:** high season.

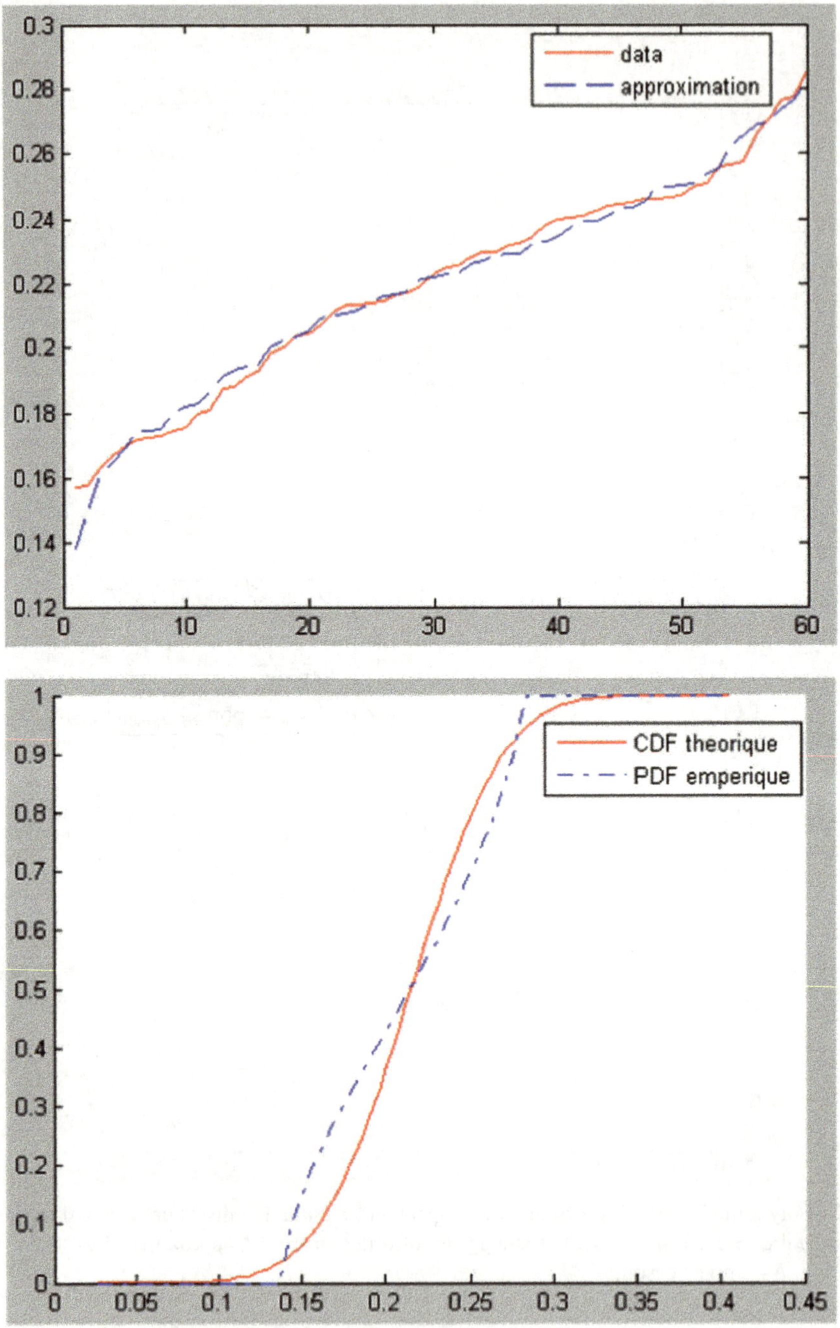

Fig. 3. Polynomial expansion of the date (upper) and cumulative distribution function (lower) (theoretical & empirical) (y-axis) of smuggling of drugs in site **S1** approximated using moment matching: **A-2**: medium season.

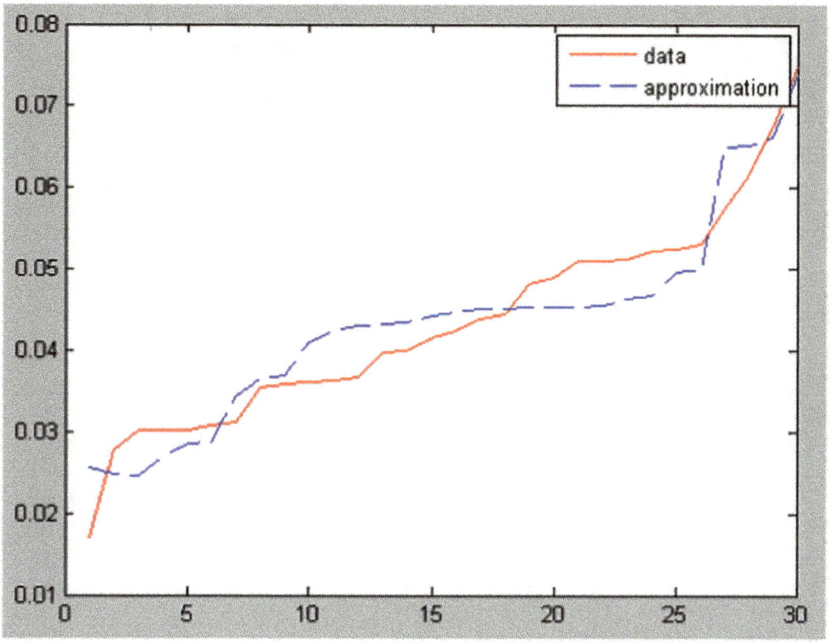

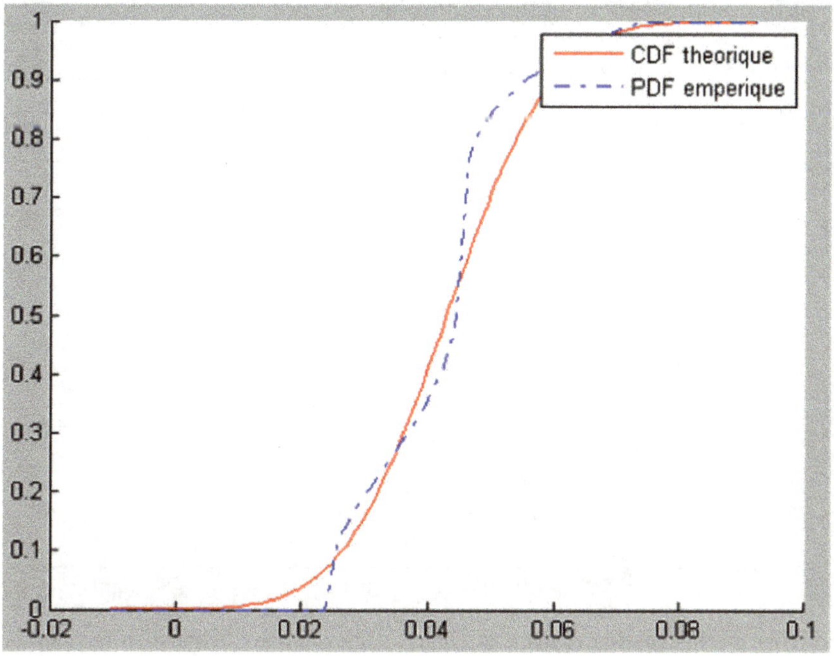

Fig. 4. Polynomial expansion of the date (upper) and cumulative distribution function (lower) (theoretical & empirical) (y-axis) of smuggling of drugs in site **S1** approximated using moment matching: **A-3**: low season.

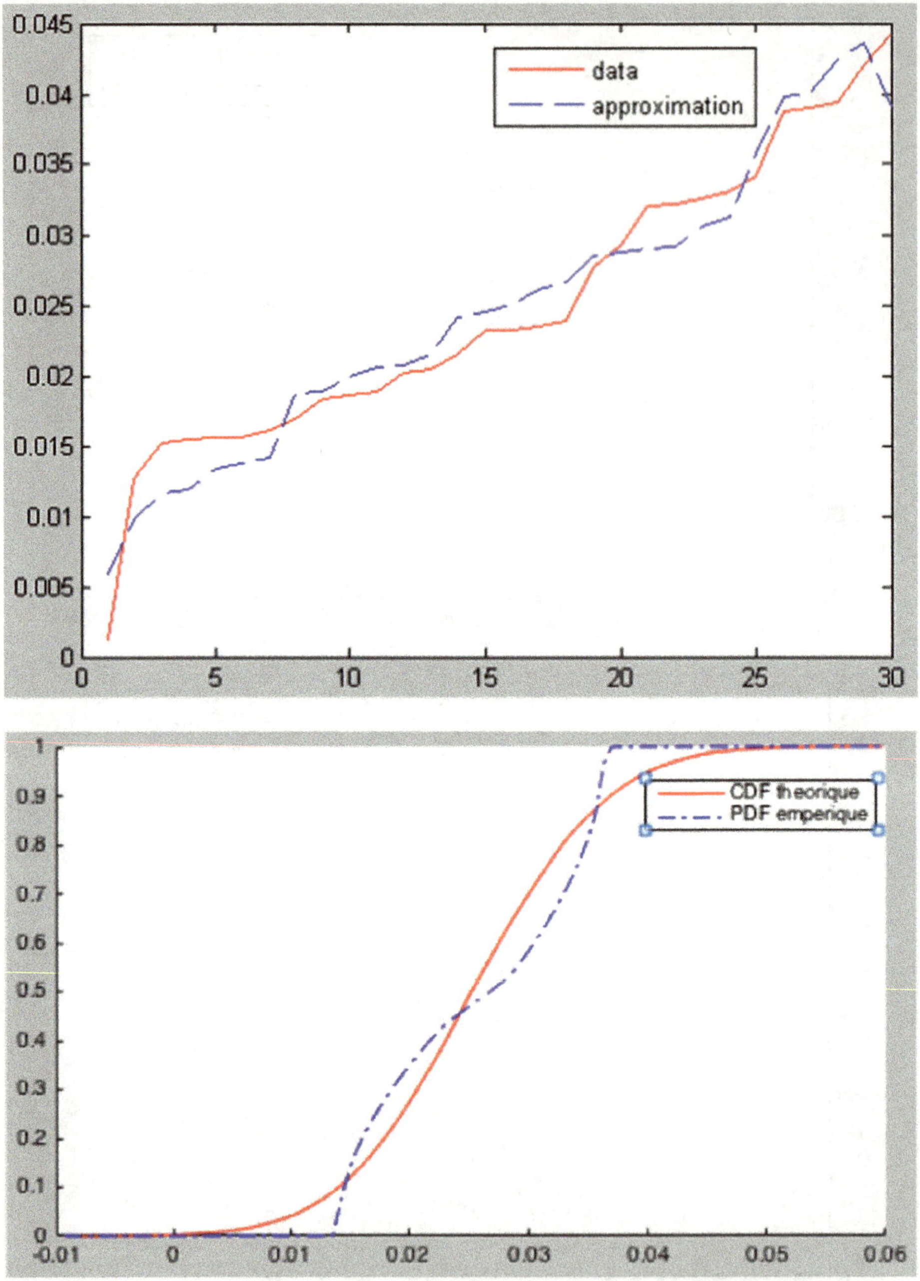

Fig. 5. Polynomial expansion of the date (upper) and cumulative distribution function (lower) (theoretical & empirical) (y-axis) of smuggling of drugs in site **S2** approximated using moment matching: **D-1**: high season.

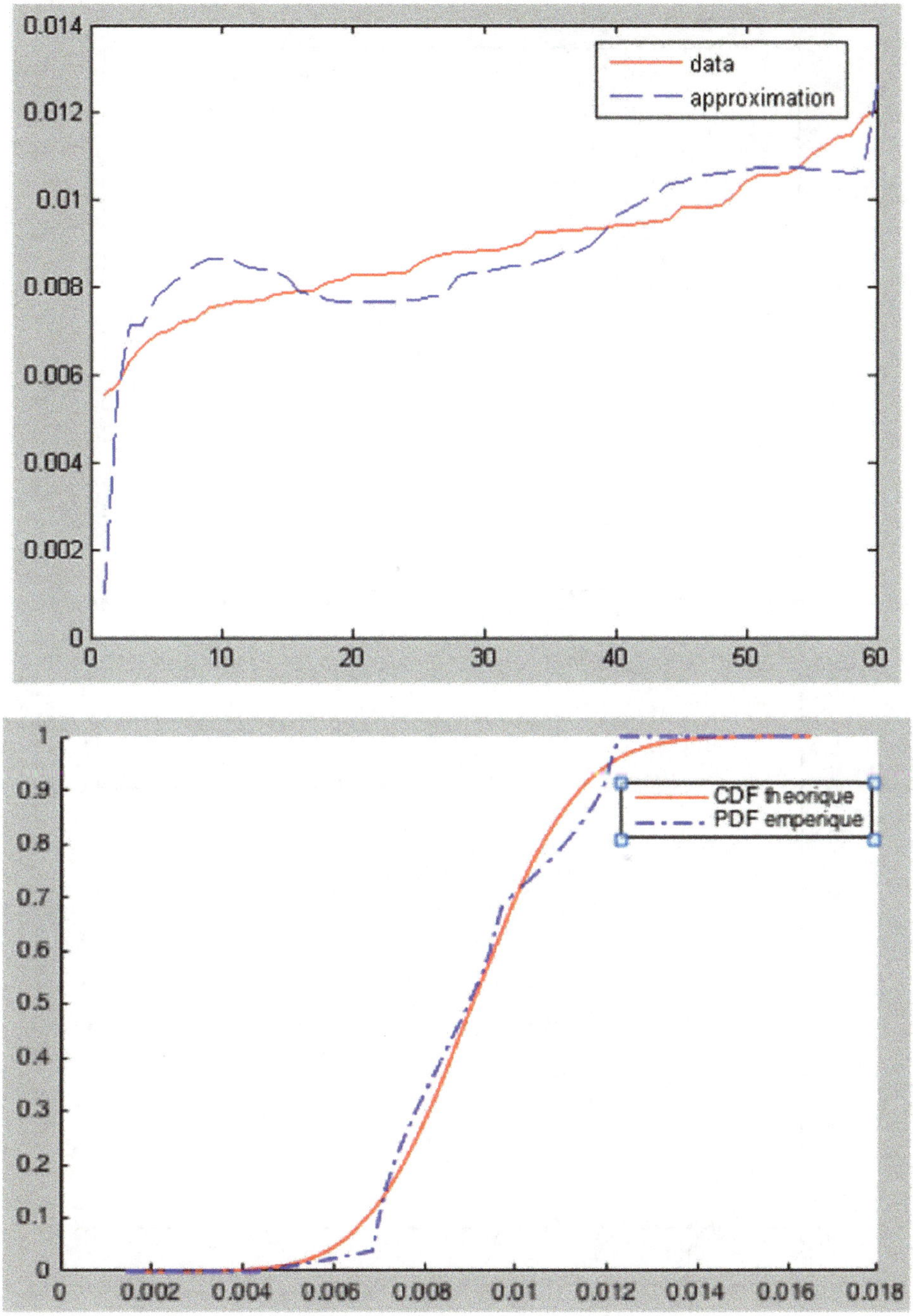

Fig. 6. Polynomial expansion of the date (upper) and cumulative distribution function (lower) (theoretical & empirical) (y-axis) of smuggling of drugs in site **S2** approximated using moment matching: **D-2**: medium season.

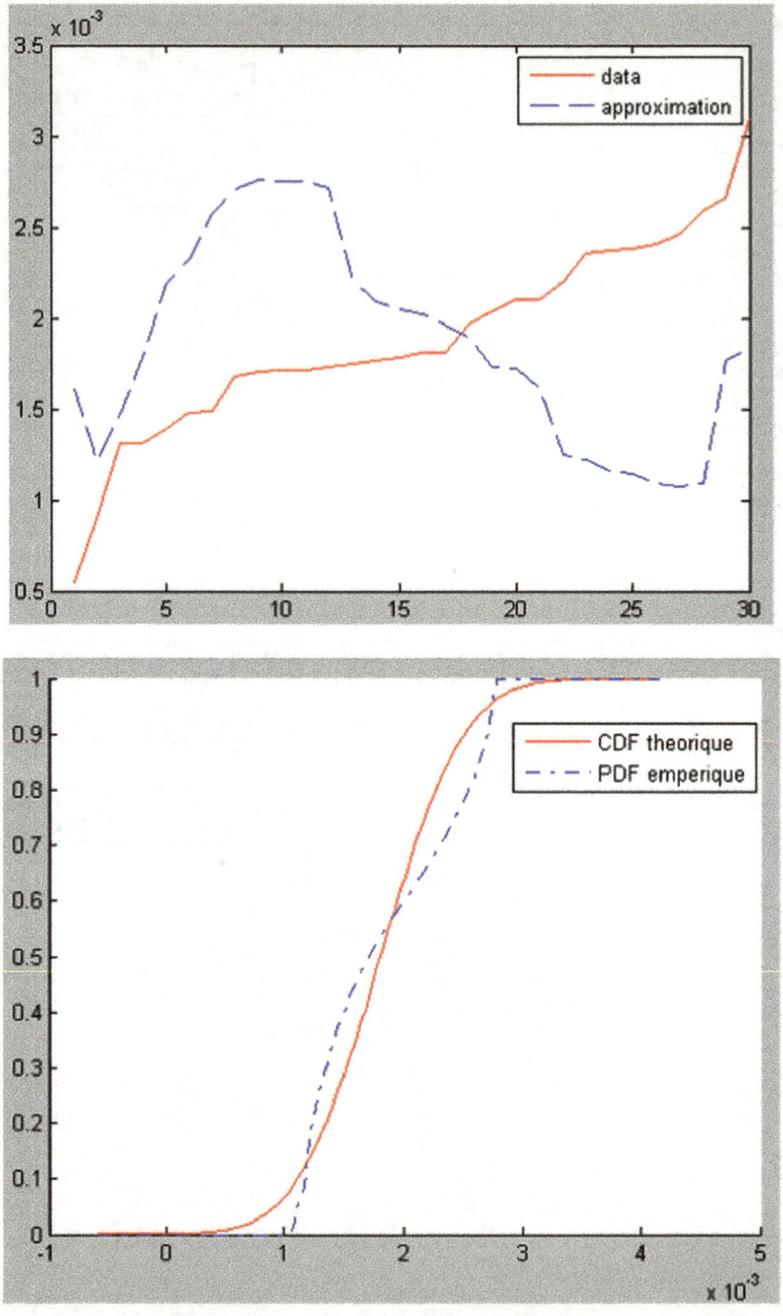

Fig. 7. Polynomial expansion of the date (upper) and cumulative distribution function (lower) (theoretical & empirical) (y-axis) of smuggling of drugs in site **S2** approximated using moment matching: **D-3**: low season.

5 Conclusion and Future Work

In this study, models-based on uncertainty quantification is developed to describe risk time series, especially the case of smuggling of drugs and it is shown that these models have several benefits. First, thanks to their distributional versatility, they are able to describe seasonal behavior of the time series and thus pre-processing the data to obtain stationarity, is not needed. Then, these models have the ability to model diverse time scales which are present in risk time series in customs supply chain context, which is an important aspect for many applications of these models. Finally, their interpretability which provides divers form of outputs such as polynomial approximation of risk and CDFs. In this work, this is used to include seasonality with monthly time scale. Other related risk modeling issues which remain to investigate is using various time scale, from daily to interannual components, into the model and the generalization to analyze the customs risk behavior over time and space in the same model (multivariate model).

References

1. Poles, S., Lovison, A.: A Polynomial Chaos Approach to Robust Multiobjective Optimization, Dagstuhl Seminar Proceedings 09041. Hybrid and Robust Approaches to Multiobjective Optimization (2009)
2. de Cursi, J.E.S., Sampaio, R.: Uncertainty Quantification and Stochastic Modeling with Matlab. Elsevier, 1st Edition (2015)
3. Lebon, J., Le Quilliec, G., Breitkopf, P., Filomeno Coelho, R., Villon, P.: Fast Latin Hypercube Sampling and efficient Sparse Polynomial Chaos Expansion for uncertainty propagation on finite precision models (2014)
4. Wiener, N.: The homogeneous chaos. Am. J. Math. **60**(4), 897–936 (1938)
5. Xiu, D., Karniadakis, G.E.: The Wiener-As key polynomial chaos for stochastic differential equations. SIAM J. Sci. Comput. **24**(2), 619–644 (2002)
6. Sudret, B.: Global sensitivity analysis using polynomial chaos expansions. Reliabil. Eng. Syst. Saf. **93**(7), 964–979 (2008)
7. Branicki, M., Majda, A.J.: Fundamental limitations of polynomial chaos for uncertainty quantification in systems with intermittent instabilities. Commun. Math. Sci. **11**(1), 55–103 (2013)
8. Holdorf Lopez, R.: Optimisation en présence d'incertitudes. Ph. D. Thesis, INSA de Rouen, Rouen, France (2010)
9. Hammadi, L., Ouahman, A.A., De Cursi, E.S., Ibourk, A.: An approach based on FMECA methodology for a decision support tool for managing risk in customs supply chain: a case study. Int. J. Manuf. Technol. Manage. **32**(2), 102–123 (2018)
10. Hammadi. L., Ouahman A.A., Souza De Cursi. E., Ibourk. A.: Risk Management and Security Practice in Customs Supply Chain: Application of Cross ABC Method to the Moroccan Customs. International Journal of Social, Behavioral, Educational, Economic, Business and Industrial Engineering 8 (5), 1586–1596 (2014)
11. Hammadi, L., De Cursi, E.S., Ouahman, A.A., Barbu, V.S., Ibourk, A.: Customs supply chain process design: an application of SCOR model. In: IEEE 7th International Conference on Advanced Logistics and Transport (IEEE ICALT 2019), 14–16 June, Marrakech, Morocco (2019)
12. Hammadi, L, Souza, D.C.E, Barbu, V.S., Ait Ouahman, A., Ibourk, A.A.: SCOR model for customs supply chain process design. World Customs J. **12**(2), 95–106 (2018)

13. Hammadi, L.: Customs supply chain Engineering: modeling and risk management: application to the customs. Ph. D. Thesis, INSA de Rouen, Rouen, France (2018)
14. Brockwell, P.J., Davis, R.A.: Introduction to Time Series and Forecasting, 2nd edn. Springer, New York (2002)
15. Brown, B.G., Katz, R.W., Murphy, A.H.: Time series models to simulate and forecast wind speed and wind power. J. Climate Appl. Meteorol. **23**, 1184–1195 (1984)
16. WCO Publications. 2012, 2013, 2014, 2015 & 2016. Illicit Trade Report
17. WCO Publications. 2008, 2009 2010 & 2011, WCO customs and TOBACCO report

Uncertainty Analysis and Estimation

PLS Application to Optimize the Formulation of an Eco-Geo-Material Based on a Multivariate Response

Saber Imanzadeh[1,2(✉)], Armelle Jarno[2], and Said Taibi[2]

[1] Laboratoire de Mécanique de Normandie, Normandie Univ., INSA Rouen Normandie, 76801 Saint-Etienne du Rouvray, France
saber.imanzadeh@insa-rouen.fr
[2] Normandie Univ., Université Le Havre Normandie, CNRS, LOMC, 76600 Le Havre, France
{jarnoa,Said.Taibi}@univ-lehavre.fr

Abstract. The ecological and environmental issues today encourage the use of eco-geo-materials that consume less «grey energy» , such as raw earth building material. This raw earth material is inexpensive and is generally available on the construction site. The compressive strength and ductility have a major role on its mechanical behavior. However, few studies are effectively dedicated to estimate both the compressive strength and the ductility property of raw earth materials. Treated with a very low percentage of binders, the raw earth is transformed into an eco-material suitable for use in the construction. However, the experimental study of the raw earth concrete mechanical properties presents uncertainties related to the bio-sourced nature of its components, such as plant fibers and raw earth. In this study, the raw earth samples of 25 different formulations were casted and tested for 90 days of curing time. The samples were tested for unconfined compressive strength through which ductility index were inferred. In order to optimize the formulation of this raw earth material, a «Design of Experiments» was conducted to study the effect of the various components on these two mechanical properties. A multivariate statistical regression technique of PLS, Partial Least Square, was performed to evaluate the design. This PLS technique was selected because of the complicated experimental design data along with different constraints on model based on the two responses. The obtained results show this technique could be a helpful tool to improve and optimize a raw earth concrete formulation.

Keywords: Eco-geo-materials · Unconfined compressive strength · Ductility · Design of Experiments · PLS technique · Optimization · Uncertainty

1 Introduction

Raw earth materials have been used as a main building construction material since the earliest civilizations settled on the continent. Raw earth materials are of attracting renewed interest due to their green characteristics along with very small energy for their transformation into building construction material compared to traditional building

J. E. S. De Cursi (Ed.): Uncertainties 2020, LNME, pp. 273–284, 2021.
https://doi.org/10.1007/978-3-030-53669-5_20

construction materials such as concrete. They are recyclable and need just simple technology. Building with raw earth materials, which are the local materials, are a solution to prevent energy-intensive transport [1].

The main properties of raw earth materials are the mechanical strength, the shrinkage and swelling, the hygrothermal properties and the cracking process [2]. Nevertheless, shrinkage in these raw earth-based materials is one of their weak point. It is the most common cause of the development of cracks. To decrease this impact, their plastic behavior requires to be improved to avoid brittle failures and thus limit cracking [3]. A suitable ductility of the raw earth material will avoid the sudden failure of the structure. Either the raw earth material is suitably ductile enough or a component, like plant and animal aggregates, fibers or biopolymers can be added.

It is not commonly adequate to reach the performance required for a building construction material and different stabilizers may be used to improve its properties [4]. Among them, when the mechanical strength of a raw earth material is the desired property, stabilizers such as lime, gypsum and cement can be utilized. Some experimental studies were performed to study the effect of these binders on the raw earth properties to make an improved construction material [5]. For instance, Zak et al. 2016 [1] studied the influence of reinforcement by natural fibers, gypsum and cement on compressive strength of unfired earth bricks materials. They pointed out that the mixing of earth material with gypsum has no favorable effect on the compressive strength. Furthermore, they found that the hemp and flax fibers have rather low effect on the compressive strength of earth bricks, but they have high effect on the breaking behavior [1].

A new concrete made of raw earth was recently developed by a French firm from Normandy region called CEMATERRE, collaborating with Le Havre Normandie University. Its advantage and originality are its ability to be cast in place as a traditional concrete. This raw earth concrete as an eco-geo-material is made of five components: fiber, lime, cement, water and silt. The main aim of this research paper is to study simultaneously the two responses: unconfined compressive strength (UCS) and ductility index *i'*.

The Design of experiments, DOE, can be a suitable method to study the two responses simultaneously varying the proportion of mixture constituents [6]. To fulfill this aim, combinations of five-constituent mixtures were formulated by a D-optimal mixture design. The experimental region was defined through different constraints. Several sets of laboratory tests were carried out to establish model formulations targeting the sought unconfined compressive strength and ductility index simultaneously for 90 days of curing time. Afterwards, the derived model was verified. A multivariate statistical regression method of PLS, Partial Least Square projections to latent structures, was selected to assess the design [7, 8]. This PLS technique was selected to study the two responses simultaneously with the complicated experimental design data due to different constraints on model. The Partial Least Square projections on latent structures were determined to calculate the coefficients of the model. Lastly, results were analyzed to improve the formulation of raw earth concrete material thanks to the loading plot of PLS technique.

2 Design of Experiment Using Partial Least Square Technique

Design of experiments, DOE, is a practical approach for exploring complex problems. It is the design of any task that aims to explain the variation of information under conditions to reflect the variation. The term is commonly related with experiments in which the design introduces conditions that directly affect the variation [9]. Relied on statistics analysis, it gives the maximum amount of relevant information with a selected number of experimentations. Different types of DoE exist like fractional, full factorial, composite and D-optimal designs [10].

Several classical methods of statistics exist to study the design of experiment. An example is Multiple Linear Regression; MLR, [11] an often used regression tool. As a predictive analysis, the Multiple Linear Regression is performed to explain the relationship between one continuous dependent variable and two or more independent variables fitting a linear equation to measured data. Nevertheless, there are some disadvantages with MLR technique: i) It has problems in respect to missing values. If some data element is lacking that row and that column must be deleted from the modeling. ii) This technique is not adapted to complicated experimental data design with different constraints on model. Considering these disadvantages, a multivariate statistical regression technique of PLS, Partial Least Square projections to latent structures can be an alternative choice to MLR technique.

The PLS technique was initially presented in chemometrics and economics [12]. The pioneering work with PLS technique was done by Wold in the field of econometrics [7]. This technique has been successfully used in other scientific fields, like bioinformatics, medicine, computer vision and civil engineering. PLS model gives approximations of an underlying complicated reality. The goal of PLS model is to estimate a set of response variables from a set of predictor variables using latent variables through an iterative process. PLS model has two main advantages. PLS generates low-rank approximations of the values that are aligned with the response and then uses low-rank approximations of both the input values and response values to predict the final regression model. The PLS model is divided in two terms: linear and non-linear PLS. A main limitation of linear PLS is that some problems are often displaying nonlinear characteristics; then applying the nonlinear PLS is suggested [7].

Related with PLS model, some tools like scores and loadings can be used to describe the model. Within the PLS modelling framework, a major attention is given on the plotting of model parameters such as scores and loadings, because such plots are useful and constructive for model verification and interpretation [13].

The analysis with PLS model gives some advantages [8]: i) there is no limitation by the number of experiments and the degree of freedom of the model. ii) if there are several response variables, their covariance is taken into account in the model. iii) furthermore the results given by the impact or response surface analysis, PLS model provides tools to identify outliers. As each experiment (except the center one) is commonly not replicated, it is hard and difficult to locate an outlier result in classical method.

This PLS model has been adequately described in the literature [14] and its application to mixture data was explained by Kettaneh-Wold [15].

3 Raw Earth-Based Materials

Eco-geo-materials studied in this paper, are composed of five constituents: fiber, lime, cement water and silt (Fig. 1). Natural silt used for a raw earth concrete. Silt is granular material of a size between sand and clay, whose mineral origin is quartz and feldspar [16]. It was chosen because locally available material in abundance on the construction site. Some properties of this natural silt are: the effective diameter of 32 μm, the Hazen uniformity factor of 4.37 and the curvature factor of 0.94. The Atterberg limits are respectively 20% for the liquid limit and 6% for the plasticity index [17]. Portland cement and lime as two binders are used to make a raw earth concrete. The used cement is CEM I 52.5 N, according to the NF EN197-1, NF P15-318 and NF EN196-10 standards. It is a Portland cement, compounded of 95 to 100% of clinker, with an unconfined compressive strength measured at 28 days of 52.5 MPa (smallest value), with an ordinary short-term strength class (2 or 7 days). The used lime comes from the Proviacal® DD range. More information concerning these two binders' properties can be found in Eid 2017 [18]. The used flax fibers are extracted locally from the Normandy region. These used fibers are natural fibers without any chemical surface treatment. Their diameters are between 10 and 15 μm [19]. Regarding preparation of samples, it should be mentioned that incorporating fibers without favoring any fiber orientation and with a final homogeneous distribution in the material is a difficult task. Then, fiber content was limited in the range between 0.3% and 0.45% in mass: 0.3% was taken as a small level and 0.45% as a high level of fiber content in samples. A constant quantity of additive of 5 ml/m^3 has been added for each specimen preparation. A potable tap water in the laboratory was used for preparing the raw earth concrete.

The mixing procedure was carried out in a laboratory mixer with a capacity of four liters. Thereafter, the molds of 100 mm in height and 50 mm in diameter were filled by vibration for two minutes with a vibrating table. Then, the samples were stored for 90 days of curing-time in controlled laboratory environment. Laboratory prepared raw earth concrete samples with different mix proportion were used to estimate unconfined compressive strength (UCS) and ductility index *i*'. The samples were sheared on unconfined compressive strength path according to NF P94–420 and NF P94-425 French standards. The stress-strain curve was analyzed through its maximum value (UCS) and the ductility index *i*' estimated from the post-peak region of the stress-strain curve, representative of the post peak behavior of the material [20]. The *i*' index is defined as $i' = e_3/e_2$ where e_3 refers to the axial strain when UCS drops down to 0.85 times of UCS_{max} and e_2 refers to the axial strain for UCS_{max}.

Fig. 1. Materials for a raw earth concrete

4 Mixture Design Concept for Raw Earth Concrete

Quadratic polynomial model was selected for five-constituents to obtain the model that fulfill the best fit to the experimental measurement data then to make predictions of the two responses simultaneously: Unconfined Compressive Strength (UCS) and ductility index *i'* of a raw earth concrete whatever the mixture of constituents. The quadratic model takes into account binary interaction effect for all possible pairs of constituents as expressed by Eq. 1.

$$\text{Response} = a_0 + \sum_{i=1}^{n} a_i x_i + \sum_{i,j=1}^{i,j=n} a_{ij} x_i x_j \qquad \text{(Quadratic model)} \qquad (1)$$

In Eq. 1, a_0 is a constant. n is the number of components. x_i *and* x_j are the amount of i and j components in weight percent. a_i the regression coefficients for the linear terms and a_{ij} are the regression coefficients for the binary interaction parameters.

The Analysis Of Variance (ANOVA) includes several diagnostic tools to verify the validity of the models [18, 21]. The two first diagnostic tools are the regression coefficients, R^2 and R^2_{adj}, presenting the information on the ability of the model to fit the measured data. In addition, Q^2 coefficient verifies the model validity that means its ability to estimate new data. Thereafter, the first F-Test and the second F-Test should be verified. The F-test is used to assess the regression model significance test. It compares regression variance to residual variance. For the second F-Test, residual error is divided in two terms: lack of fit due to imperfection of the model and pure error calculated from replicates data error [22].

Three constraints should be taken into account:

i) Fundamental constraint where the sum of the constituents of the mixture is come to 100% in weight for all the mixes of the design,
ii) Economic and ecological mixture constraints: the raw earth concrete should be designed to provide the suitable mechanical properties to be used in the construction of buildings and houses. Moreover, it should be less energy intensive. In consequence, the quantity of these two binders, cement and lime, should be limited. Then, the maximum quantities of cement and lime are respectively limited to 16% and 12% and a condition of maximum binder proportion was determined: Cement % + Lime % < 16%.
iii) Workability constraint: workability has a significant role on the mechanical properties. Some properties such as cohesion, plasticity and consistency can affect the workability of a raw earth material. In this research study, a S3 level of consistency calibrated by standard slump test was carried out to verify a fluidity near to a very plastic concrete according to the standard NF EN 206-1.

In consonance with above-mentioned constraints, the mixing range determined for each of the components, is given in Table 1.

Table 1. Mixing range

x_i	Lower limit (%)	Upper limit (%)
x_1: Fiber	0.3%	0.45%
x_2: Lime	0	12%
x_3: Cement	4%	16%
x_4: Water	20%	25%
x_5: Silt	47%	75%

In this research paper, the experimental domain is constrained by the three above-mentioned conditions to an irregular 3D-polyhedron. After that, D-optimal design based on a computer-aided is adapted to generate the set of experiments [6]. The 27 formulations were tested in the random run order suggested by MODDE© software along with three repetitions including three center points (Table 2).

Once tests were completed, a multivariate statistical regression method of PLS, Partial Least Square projections to latent structures, was performed to analyze the design. PLS technique is appropriated to study the two responses simultaneously with complicated experimental design data with different constraints on model. The Partial Least Square projections on latent structures were estimated to calculate the coefficients of the model.

Table 2. Experimental design of the D-optimal design for a raw earth concrete (x_i is the proportion of the different mixture constituents)

Formulation	Fiber (x_1)	Lime (x_2)	Cement (x_3)	Water (x_4)	Silt (x_5)
1	0.0030	0.0000	0.1600	0.2239	0.6131
2	0.0030	0.0647	0.0400	0.2500	0.6423
3	0.0030	0.0000	0.0400	0.2290	0.7280
4	0.0045	0.0000	0.0400	0.2287	0.7268
5	0.0030	0.0000	0.0400	0.2345	0.7225
6	0.0045	0.0000	0.0400	0.2342	0.7213
7	0.0030	0.0000	0.1600	0.2295	0.6075
8	0.0045	0.0000	0.1600	0.2292	0.6063
9	0.0045	0.0719	0.0881	0.2500	0.5855
10	0.0030	0.0560	0.1040	0.2500	0.5870
11	0.0045	0.0568	0.1032	0.2500	0.5855
12	0.0045	0.0000	0.1200	0.2253	0.6502
13	0.0045	0.0000	0.1600	0.2255	0.6100
14	0.0040	0.0000	0.1600	0.2237	0.6123
15	0.0030	0.0237	0.1363	0.2326	0.6044
16	0.0045	0.0543	0.0400	0.2500	0.6512
17	0.0045	0.0438	0.0400	0.2429	0.6688
18	0.0045	0.0479	0.1121	0.2412	0.5943
19	0.0035	0.0480	0.0400	0.2500	0.6585
20	0.0035	0.0713	0.0887	0.2500	0.5865
21	0.0038	0.0000	0.1000	0.2291	0.6672
22	0.0030	0.0299	0.0841	0.2396	0.6433
23	0.0038	0.0261	0.0859	0.2409	0.6433
24	0.0038	0.0320	0.1280	0.2383	0.5980
25	0.0038	0.0301	0.0840	0.2395	0.6426
26	0.0038	0.0301	0.0840	0.2395	0.6426
27	0.0038	0.0301	0.0840	0.2395	0.6426

5 Statistical Model Validation

Quadratic polynomial model was first fit to the measured experimental data following a detailed examination of R^2, R^2_{adj} and Q^2 to obtain the most adequate model representing the measured experimental values for the two responses: unconfined compressive strength and ductility index *i'* for 90-day curing time. For unconfined compressive strength, the obtained values of R^2, R^2_{adj} and Q^2 are respectively 0.973, 0.937 and 0.939. For ductility index *i'* the obtained values of R^2, R^2_{adj} and Q^2 are respectively 0.936,

0.775 and 0.807. For this 90 days of curing time, the values of R^2, R^2_{adj} can be considered as good. In addition, this model illustrates a good predictive relevance ($Q^2 > 0.8$). Moreover, it was found that the chosen model verified the first F-Test and the second F-Test. Then, chosen model can be validated. Once the best-fitting model was selected, an equation describing the predictions of unconfined compressive strength and ductility index *i'* was determined for raw earth concrete formulations.

6 Results and Discussion

6.1 Loading Plot of PLS Technique

Scores and loading plots are derived from the model fitted with PLS regression. Three PLS components were used in this research study. In the following, the loading plot was used to study the influence of the different terms simultaneously on the two responses: UCS and ductility index. UCS and ductility index are defined from unconfined compressive stress versus strain curve. The plot of X (constituents) and Y (response) weights (w and c, respectively) shows how the X variable influences the Y variable, and the correlation structure between X and Y variables. It gives a representation of how the X and Y variables combine in the projections, and finally inform on the relation between X and Y variables.

This plot simplifies the analysis of how the responses vary in relation to constituents and the existing interactions between them. Moreover, this plot provides a direct indication of which constituent or existing interaction between constituents contribute significantly to the two responses simultaneously: UCS and ductility index *i'*.

It explains how the five constituents affect simultaneously these two responses. For the UCS response, first, a base line is drawn between UCS and the plot origin (0, 0). Second, all constituents and the existing interactions between constituents are projected orthogonally onto this line. The relative position of the projections to the origin gives information on the significance of the different terms in the model and on their positive or negative correlation with the studied responses. The more the orthogonal projection of the terms deviates from the origin (0, 0), the more they affect the unconfined compressive strength. The same work was performed for the second response (*i'*). Figure 2 presents both constituents weights (w) and response weight (c), for the first and the second significant PLS components (wc [1], wc [2]). For unconfined compressive strength, this figure illustrates that cement, lime-silt, water-silt, fibre-water, fibre-lime and fibre-silt interaction terms are positively associated to UCS since they are projected on the same quadrant than UCS. It is found that the distance of the orthogonal projection to origin for cement constituent is larger than for these interaction terms. Cement is, as expected, the major factor that contributes to increase the unconfined compressive strength. In fact, when the quantity of cement increases, the unconfined compressive strength increases too. The positive influence of the lime-silt interaction on UCS is slightly larger than the water-silt interaction. The positive influence of the fibre-water, fibre-lime and fibre-silt interactions on the unconfined compressive strength are almost the same and less than those ones for lime-silt and water-silt interactions. On the contrary, fibre, lime, water and silt constituents and fibre-cement, cement-water and lime-cement interactions act with a negative effect on the unconfined compressive strength. The negative influence of fibre, lime, water and silt constituents are respectively increased on UCS. Fiber has a smallest negative influence and silt has a biggest negative influence. The negative influence of fibre-cement interaction is smaller than cement-water and lime-cement interactions. The negative influence of cement-water and lime-cement interactions are nearly same on UCS.

An inverse behavior can be observed for ductility index *i'* response compared to UCS response (Fig. 2). For ductility index *i'*, this figure shows that fibre, lime, water and silt constituents and fibre-cement, cement-water and lime-cement interactions act with a positive effect on the ductility index *i'*, since they are projected on the same quadrant than ductility index *i'*. The positive effect of fibre, lime, water and silt constituents are respectively increased on the ductility index *i'*. Fiber has a smallest positive impact and silt has a biggest positive impact. Silt is the major factor that contributes to increase the ductility index *i'*. In fact, silt shows an elasto-plastic behavior and is naturally ductile; it can tolerate high strains during loading. The positive influence of fibre-cement interaction is smaller than cement-water and lime-cement interactions. The positive influence of cement-water and lime-cement interactions are nearly same on the ductility index *i'*. On the contrary, this figure illustrates that cement, lime-silt, water-silt, fibre-water, fibre-lime and fibre-silt interaction terms are negatively associated to the ductility index *i'*. Cement is, as expected, the major factor that contributes to decrease the ductility index *i'*. In fact, when the quantity of cement increases, the ductility index decreases. Indeed, this negative effect on ductility index is caused by pozzolanic chemical reactions including cement that generate a very hard bond between the soil particles [20]. The negative influence of the lime-silt interaction on ductility index *i'* is slightly larger than the water-silt interaction. The negative influence of the fibre-water, fibre-lime and fibre-silt interactions on the ductility index *i'* are almost the same and less than those ones for lime-silt and water-silt interactions.

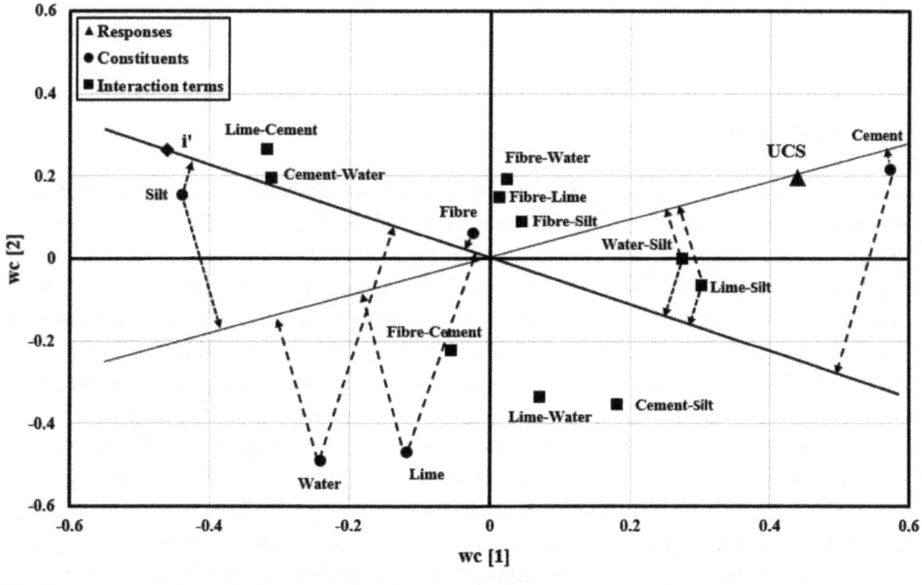

Fig. 2. Interpretation of the unconfined compressive strength (UCS) and ductility index *i'* responses simultaneously by a PLS loading plot

Finally, lime and fibre components, cement-silt, lime-water, fibre-cement, fibre-water, fibre-silt and fibre-lime interactions have a little impact on the two responses simultaneously: UCS and ductility index *i'*. They could be less important due to their low values of Variable Importance for Projection (VIP) smaller than 0.8 (Fig. 3, [7]).

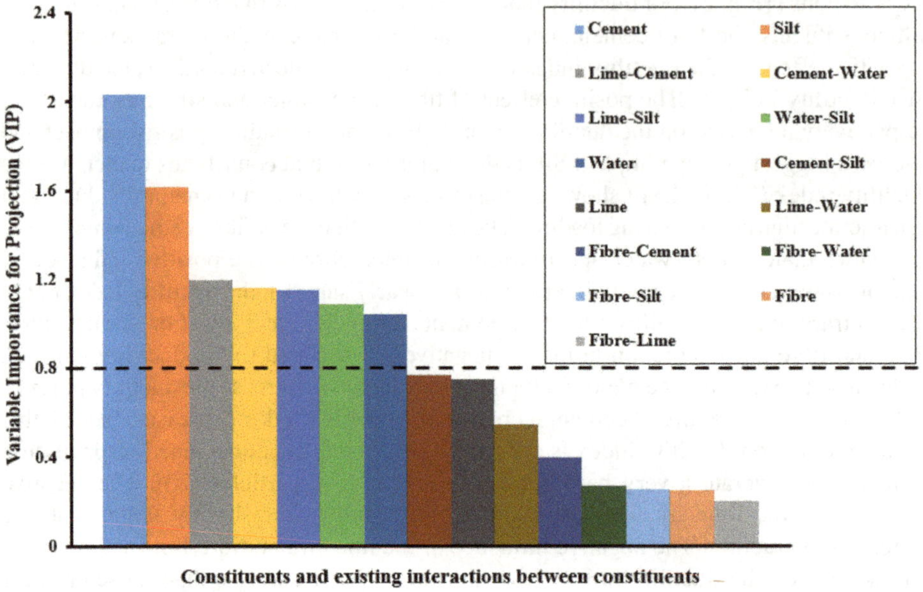

Fig. 3. Interpretation of Variable Importance for Projection (VIP) versus constituents and existing interactions between constituents for the two responses simultaneously: Unconfined Compressive Strength (UCS) and ductility index *i'*

7 Conclusions

In this research paper, a building eco-material based on raw earth stabilized with binders and vegetal fibers was presented. It was demonstrated that Design of Experiments method is an optimization tool adapted to assess to the two responses simultaneously: unconfined compressive strength and ductility index *i'* of a raw earth material. The design tests were chosen by taking into account three different constraints: fundamental constraints, economical and ecological constraints, and workability. A multivariate statistical regression technique of PLS was chosen to study the design. This PLS technique was used to assess the two responses simultaneously along with the complicated experimental design values together with different constraints on model. Through the loading plot of PLS technique, the positive or negative influences of each component and existing interactions between components were discussed. As an example, the model has shown fiber has a smallest negative effect and silt has a biggest negative effect on UCS response. The negative effect of fiber-cement interaction is smaller than cement-water and lime-cement interactions. An inverse behavior can be found for ductility index *i'* response compared to

UCS response. For example, the main positive effect of silt on ductility index *i'* response was found. Indeed, silt shows an elasto-plastic behavior and is naturally ductile; it can tolerate large strains in the course of loading.

PLS is a suitable technique to examine the two responses simultaneously in the presence of complicated experimental design values along with different constraints on model.

References

1. Zak, P., Ashour, T., Korjenic, A., Korjenic, S., Wu, W.: The influence of natural reinforcement fibers, gypsum and cement on compressive strength of earth bricks materials. Constr. Build. Mater. **106**, 179–188 (2016)
2. Ashour, T., Korjenic, A., Korjenic, S., Wu, W.: Thermal conductivity of unfired earth bricks reinforced by agricultural wastes with cement and gypsum. Energy Build. **104**, 139–146 (2015)
3. Araldi, E., Vincens, E., Fabbri, A., Plassiard, J.P.: Identification of the mechanical behavior of rammed earth including water content influence. Mater. Struct. **51**, 88 (2018)
4. Earth building in Spain: Carmen Jimenez Delgado, M., Canas Guerrero, I. Constr. Build. Mater. **20**, 679–690 (2006)
5. Al-Mukhtar, M., Lasledj, A., Alcover, J.F.: Lime consumption of different clayey soils. Appl. Clay Sci. **95**, 133–145 (2014)
6. Herrero, A., Ortiz, M.C., Sarabia, L.A.: D-optimal experimental design coupled with parallel factor analysis 2 decomposition a tool in determination of triazines in oranges by programmed temperature vaporization-gas chromatography-mass spectrometry when using dispersive-solid phase extraction. J. Chromatogr. A **1288**, 111–126 (2013)
7. Wold, S., Kettaneh-Wold, N., Skagerberg, B.: Nonlinear PLS modelling. Chemometrics Intell. Lab. Syst. **7**, 53–65 (1989)
8. Eriksson, L., Byrne, T., Johansson, E., Trygg, J., Vikstrom, C.: Multi - and megavariate data analysis, basic principal and applications. 3rd edn, Umetrics Academy (2013)
9. Myers, R.H., Montgomery, D.C., Anderson-Cook, C.M.: Response surface methodology: process and product optimization using designed experiments, 4th edn. Wiley, New York (2016)
10. Eriksson, L., Johansson, E., Kettaneh-Wold, N., Wikström, C., Wold, S.: Design of Experiments: Principles and Applications. Umetrics AB, Umeå Learnways AB, Stockholm (2000)
11. Geladi, P., Kowaiski, B.R.: Partial least-squares regression: a tutorial. Anal. Chim. Acta **185**, 1–17 (1986)
12. Wold, H.: Path Models with Latent Variables: The NIPALS Approach. Academic Press, New York (1975)
13. Hoskuldsson, A.: Prediction Methods in Science and Technology. Thor Publishing, Denmark (1996)
14. Li, B., Morris, A.J., Martin, E.B.: Generalized partial least squares regression based on the penalized minimum norm projection. Chemom. Intell. Lab. Syst. **72**, 21–26 (2004)
15. Kettaneh-Wold, N.: Analysis of mixture data with partial least squares. Chemometr. Intell. Lab. Syst. **14**, 57–69 (1992)
16. Assallay, A.M., Rogers, C.F., Smalley, I.J., Jefferson, I.F.: Silt. Earth-Sci. Rev. **45**, 20–30 (1998)
17. Imanzadeh, S., Hibouche, A., Jarno, A., Taibi, S.: Formulating and optimizing the compressive strength of a raw earth concrete by mixture design. Constr. Build. Mater. **163**, 149–159 (2018)

18. Eid, J.: New construction material based on raw earth: cracking mechanisms, corrosion phenomena and physico-chemical interactions. Eur. J. Environ. Civ. Eng. **8189**, 1–16 (2017)
19. Abbar, B., Alem, A., Pantet, A., Marcotte, S., Ahfir, N.D., Duriatti, D.: Experimental investigation on removal of suspended particles from water using flax fibre geotextiles. Environmental Technology, Taylor & Francis Group, pp. 1–15 (2017)
20. Imanzadeh, S., Jarno, A., Hibouche, A., Bouarar, A., Taibi, S.: Ductility analysis of vegetal-fiber reinforced raw earth concrete by mixture design. Constr. Build. Mater. **239**, 1–15 (2020)
21. Fisher, R.A.: The design of experiments. Hafner: Libraries Australia, New York (1971)
22. Goupy, J.: Plans D'expériences: Les Mélanges. DUNOD, Paris (2001)

Noise, Channel and Message Identification on MIMO Channels with General Noise

Lucas Nogueira Ribeiro[4], João César Moura Mota[1], Didier Le Ruyet[2], and Eduardo Souza de Cursi[3](✉)

[1] GTEL, UFC, Campus du Pici, Fortaleza, CE, Brazil
[2] CEDRIC, CNAM, 292 rue Saint Martin, 75141 Paris Cedex 03, France
[3] LMN, Normandy University, INSA Rouen, 76801 St-Etienne du Rouvray Cedex, France
souza@insa-rouen.fr
[4] CRL, Ilmenau University of Technology, 98684 Ilmenau, Germany

Abstract. This work presents an application of Uncertainty Quantification (UQ) approaches to a Multiple-Input Multiple-Output (MIMO) system. The use of UQ techniques in this field is new and is one of the originalities of the paper. A second originality of this work is to furnish an efficient method to deal with non-gaussian noise, with non-zero mean. A third originality is that the method proposed allows the estimation of the noise, of the channel and the decoding of messages - the approach makes all the chain: determination of the distribution of the noise, identification of the Channel Matrix and detection of the symbols transmitted. The first step is made by using UQ techniques, which furnish the distribution of the noise, without assumption of Gaussian distribution or a mean equal to zero. A new application of UQ techniques furnishes the channel matrix. Then, the detection of symbols is performed by a method based on the determination of a selection matrix formed by lines of the identity matrix. Numerical examples show that the proposed approach is practical and efficient.

Keywords: Uncertainty Quantification · Identification · MIMO systems

1 Introduction

Multiple-input multiple-output (MIMO) systems involve a transmitter entity with multiple inputs which communicates with a receiver entity with multiple outputs. MIMO systems are typically adopted to describe many communication systems, including wireless [1], acoustic [2], and underwater systems [3–5]. These different communication scenarios are often deployed in uncontrolled and changing environments, hereafter referred to as communication channels. Therefore, the messages sent and received involve noise, measurement errors, and modeling errors that compromise proper decoding.

A widely adopted channel model in wireless communications is the Rayleigh fading [1, Chapter 3], wherein the received signals can be described in terms of a channel matrix with Gaussian-distributed elements and an additive noise component. To ensure proper message detection, the receiver needs to correctly identify the channel matrix and

J. E. S. De Cursi (Ed.): Uncertainties 2020, LNME, pp. 285–305, 2021.
https://doi.org/10.1007/978-3-030-53669-5_21

the additive noise distribution by means of an adapted procedure. In many application scenarios, it is usually assumed that the noise has a Gaussian distribution and that its statistics are known a priori [6]. Scenarios with non-Gaussian additive noise [4, 5] are little explored in the literature, as well as scenarios where the noise is multiplicative so that the performance of MIMO symbol detectors is not fully understood in these scenarios yet.

In recent years, Uncertainty Quantification (UQ) proposed methods and tools to deal with general noise distributions, including non-Gaussian and multiplicative (see, for instance, [7–10]). These developments remain little exploited in the framework of Telecommunications. This work explores the use of UQ approaches to deal with noise in MIMO, namely for Channel and Noise Identification, then for Message decoding.

In the following, we present a novel pilot-based channel estimation method for a general additive noise, eventually non-Gaussian. As an example, we consider uniform and Laplace distributions, what shows that the approach is effective to calculate in non-Gaussian situations. These examples are chosen as an illustration, without loss of generality – other noise distributions can be assumed for the approach proposed. The UQ approach resumes as follows:

- In a first step, we determine the distribution of the noise, using a few measurements in "silent mode", where the transmitter does not transmit anything. UQ approach furnishes a fine estimation of the noise distribution from a small number of measurements, so that we obtain good estimates of the statistics of the noise from a few measurements.
- In a second step, we identify the channel matrix using a pilot method: the transmitter sends convenient pilot symbols that allows the receiver to identify successively each column of the matrix channel individually.
- Finally, we use the results of these steps to decode messages and we analyze the error rate in some situations, with different alphabets.

2 System Model

Let us consider a single-user multiple-input multiple-output (MIMO) system with N_t transmission and N_r reception antennas. Under the flat fading assumption, the discrete-time received signal vector $\boldsymbol{y} \in \mathbb{C}^{N_r}$ is given by

$$\boldsymbol{y} = \boldsymbol{H}\boldsymbol{x} + \boldsymbol{n}, \tag{1}$$

where $\boldsymbol{x} \in \mathbb{C}^{N_t}$ is the vector of the transmitted signal, $\boldsymbol{H} \in \mathbb{C}^{N_r \times N_t}$ is the channel matrix, and $\boldsymbol{n} \in \mathbb{C}^{N_r \times 1}$ is a vector of additive noise. Assuming independent and identically distributed (i.i.d.) Rayleigh fading, the elements of $\boldsymbol{H}$ are i.i.d. complex Gaussian random variables with zero mean and unit variance. Each transmitted signal $\boldsymbol{x_n}$, $n = 1, \ldots, N_t$, consists of a digitally-modulated symbol in the constellation alphabet $\boldsymbol{a}$, of order M : $\boldsymbol{a} = (a_1, \ldots, a_M) \in \mathbb{C}^M$.

If convenient, the complex-valued signal model (1) may be rewritten in terms of its real and imaginary parts as follows:

$$\bar{\boldsymbol{y}} = \bar{\boldsymbol{H}}\bar{\boldsymbol{x}} + \bar{\boldsymbol{n}} \tag{2}$$

with

$$\bar{y} = \begin{pmatrix} y_{re} \\ y_{im} \end{pmatrix}, \ \bar{H} = \begin{pmatrix} H_{re} & -H_{im} \\ H_{im} & H_{re} \end{pmatrix}, \ \bar{x} = \begin{pmatrix} x_{re} \\ x_{im} \end{pmatrix}, \ \bar{n} = \begin{pmatrix} n_{re} \\ n_{im} \end{pmatrix} \tag{3}$$

Here, we denote $\xi_{re} = Re(\xi)$ and $\xi_{im} = Im(\xi)$. Thus, $\bar{y} \in \mathbb{R}^{2N_r}$, $\bar{H} \in \mathbb{R}^{2N_r \times 2N_t}$, $\bar{x} \in \mathbb{R}^{2N_t}, \bar{n} \in \mathbb{R}^{2N_r}$, which are real-valued vectors corresponding to the received signal, the channel matrix, the transmitted signal, the additive noise, respectively.

3 Determination of the Distribution of the Noise

As previously observed, the determination of the distribution of the noise may be performed in a "silent mode", where the transmitter does not transmit anything. In this case, the signal x in Eq. (1) becomes $x = 0$, so that $\bar{x} = 0$ in Eq. (2). The received message is $y = n, \bar{y} = \bar{n}$. Taking s measurements, we generate a sample $\mathcal{N} = \{n^1, \ldots, n^s\}$ of the noise, which is brought to a matrix $N = (n^1, \ldots, n^s) \in \mathbb{C}^{N_r \times s}$ having as columns the measurements for each realization: $N_{ij} = n^i_j$. The noise measurements are also expanded into real and imaginary parts:

$$\bar{N} = \begin{pmatrix} Re(n^1) \cdots Re(n^s) \\ Im(n^1) \cdots Im(n^s) \end{pmatrix} \in \mathbb{R}^{2N_r \times s} \tag{4}$$

Using the UQ approach, we consider a Hilbert basis $\Phi = \{\varphi_n : n \in \mathbb{N}\}$ and a random variable U (for instance, uniform or Gaussian). Assuming that the noise has a finite second moment (i.e., that the noise is square summable), we have

$$\bar{N}_i \approx \sum_{j=0}^{+\infty} c_{ij}\varphi_j(U) \tag{5}$$

In practice, we approximate the noise using a finite sum, truncated at order k:

$$\bar{N}_i \approx \sum_{j=0}^{\mathrm{k}} c_{ij}\varphi_j(U) \tag{6}$$

Let us introduce

$$C = \left(c_{ij} : 1 \le i \le N_r, 1 \le j \le k\right), \quad \Phi(\mathrm{U}) = \begin{pmatrix} \varphi_0(\mathrm{U}) \\ \vdots \\ \varphi_k(\mathrm{U}) \end{pmatrix}.$$

Then

$$\bar{N}_i = (C\Phi(U))_i \tag{7}$$

Notice that, a priori, the random variable U is different for each index i. The coefficients C must be determined by one of the following UQ methods: collocation, variational approximation, moments matching . (See [10] for a detailed description). As an example, let us assume that $\bar{N}_1$ is uniformly distributed on $(-1, 1)$ and we have $s = 20$ measurements of $\bar{N}_1$: $\{\bar{N}_1^1, \ldots, \bar{N}_1^{20}\}$ (Table 1):

Table 1. Measurements of $\bar{N}_1$

−0.0924	−0.0827	−0.0715	−0.0621	−0.0566	−0.0529	
−0.0158	−0.0133	0.0032	0.0083	0.0151	0.0193	0.0285
0.0291	0.0297	0.0581	0.0807	0.0812	0.0828	0.0938

Let U_1 be Gaussian $N(0, 1)$. We consider a sample from U_1: (Table 2)

Table 2. A sample of 20 variates from the Gaussian distribution (randomly generated)

−1.5932	−1.0216	−0.7483	−0.6147	−0.2585	0.2510	
0.0180	0.1335	0.1954	0.2398	0.3041	0.5680	0.6214
0.6324	0.7506	0.8446	1.1717	1.2247	1.2325	1.5017

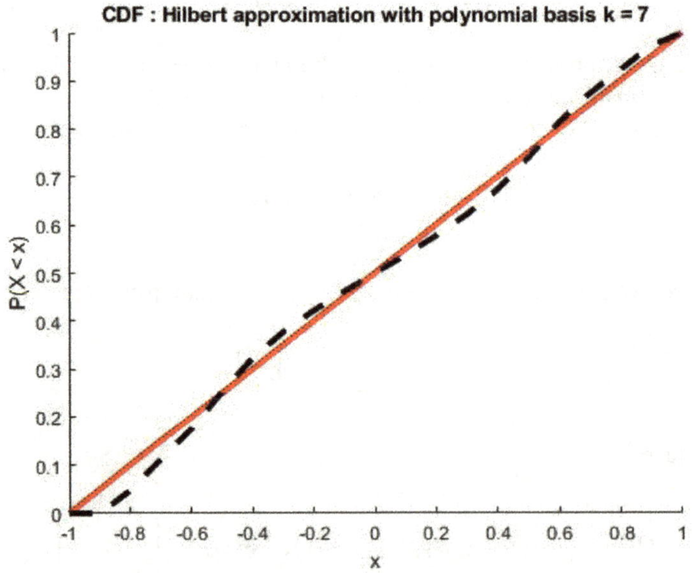

Fig. 1. Hilbert approximation of a noise. The noise is uniformly distributed, but we consider a Gaussian variable in the expansions (5)–(6). The results are obtained by Moments Matching for the first 8 moments [4].

We apply the UQ approach with a polynomial basis

$$\varphi_j(U) = \left(\frac{U-a}{b-a}\right)^{\mathrm{i}-1} \tag{8}$$

where $a = \min(U)$ and $b = \max(U)$. We take $k = 9$ (thus, the degree of the polynomial is 9). By using the Moments Matching Approach [4], with 8 moments, we obtain the coefficients given in Table 3:

Table 3. Coefficients of the polynomial

$i = 0$	$i = 1$	$i = 2$	$i = 3$	$i = 4$
−0.0927	9.416	−181.460	1449.778	−6246.36
$i = 5$	$i = 6$	$i = 7$	$i = 8$	$i = 9$
15937.485	−24760.923	22981.886	−11704.300	2514.663

As soon as the coefficients are obtained, we use the polynomial to generate a large sample formed by 1e4 variaties from $\bar{N}_1$. This large sample is used to generate an empirical CDF of $\bar{N}_1$. The resulting CDF is shown in Fig. 1 above. Notice that, despite the use of a Normal Variable in the expansion (5)–(6), the resulting distribution is close to a Uniform Variable.

As an example, let us consider a $8 \times 8-$ MIMO System ($N_r = N_t = 8$) such that the noise is additive and distributed as follows:

- in the real part, it is uniformly distributed with mean −0.03 and variance 0.1;
- in the imaginary part follows a Laplace's distribution having a mean 0.02 and a variance 0.1;
- The components of the noise are mutually independent.

These assumptions do not imply a loss of generality: indeed, the approach applies to arbitrary noises of finite variance, including dependent components. It is also possible to introduce a given covariance matrix as constraint to be verified.

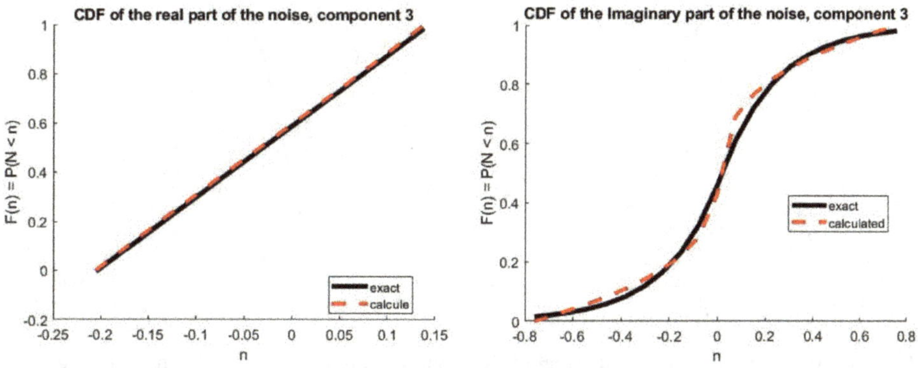

Fig. 2. Determination of the distribution of the noise: a typical result. Results for component 3: at left, the CDF of the real part; at right the CDF of the imaginary part

Let us consider 250 measurements of the noise. We use the approach by collocation [4] with a polynomial basis involving $k = 3$ (thus, a polynomial of degree 3) and U uniformly distributed on (0, 1). By this way, we determine the coefficients $(c_{in} : 1 \le i \le 16,\ 0 \le n \le 3)$ of the approximation, which is, then, used to generate a large sample of $1e4$ variates from each component of the noise. The large sample is used to generate the empirical distribution of the noise. The results obtained correspond to an error of 3.5% in the CDF of the noise. We show above (Fig. 2) a typical result – which are analogous for the other components.

The results may be improved by using, on the one hand, a larger number of measurements and, on the other hand, by using the approach by moment matching. For instance, for 1000 measurements, the error in the CDF is of 1.6%. The mean of the real part of the noise is estimated by the empirical mean of the large sample. We obtain the results shown in Table 4. As we may observe, the results are close to the real means.

Table 4. Empirical means for each component (in ascending order: at left, the first component; at right, the 8th. one)

Real	−0.032	−0.030	−0.028	−0.027	−0.023	−0.032	−0.030	−0.023
Imaginary	0.034	0.027	0.019	0.021	0.033	0.043	0.029	0.012

To get some statistics about the behavior of the proposed method, we realized 1000 approximations using different realizations of the noise and we analyzed the distributions of the errors. We determined the empirical CDF by the same way as the one upper described: a Hilbert approximation by a polynomial of degree 4, then the generation of a sample of $1e4$ variates. The sample is used to determine the CDF. The PDF is found by particle derivation of the CDF (see [11]).

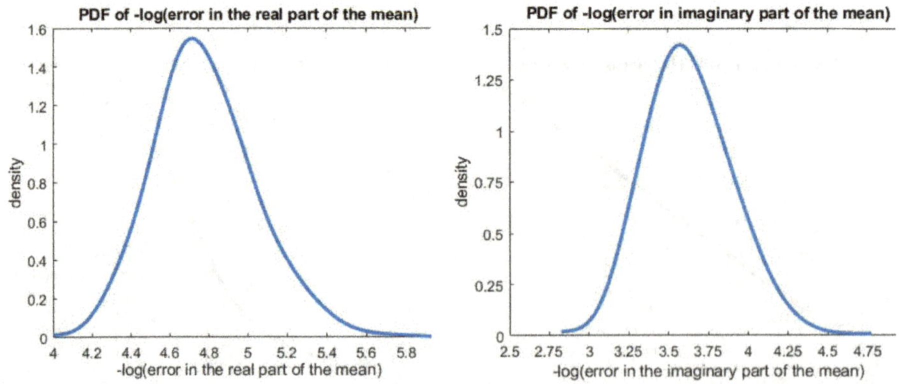

Fig. 3. PDF of $-\log(error)$ for the real (at left) and the imaginary (at right) parts of the mean (natural logarithm)

In Fig. 3 above, we exhibit the PDF obtained for the negative natural logarithms of the errors in the real and the imaginary parts of the mean of the noise (recall that it is not equal to zero): notice that the larger is the abscissa, the smaller is the error. We observe that both the distributions are unimodal. The mode for the real part is about 4.8 and the mode for the imaginary part is about 3.6. Indeed, the statistics of the errors in the original sample of 1000 realizations are given in the Table 5 below.

Table 5. Statistics of the errors in the estimation of the mean of the noise. These values correspond to about 3.6% of error in the real part and 11% in the imaginary part.

	Mean	Standard Deviation
Real part	0.0087	0.0021
Imaginary part	0.0273	0.0070

We observe that the errors are larger in the imaginary part: this is connected to the original probability distribution: indeed, the estimation of the Laplace distribution mean appears to be harder. Improvements may be obtained by using other approaches, such as other Hilbert basis leading to approximations that are more accurate.

4 Channel Estimation

The standard procedure for the estimation of the channel matrix involves the exchange of *pilot signals* – i.e., reference signals known both by the transmitter and receiver: the transmitter sends a N_p-length sequence of reference signals $\boldsymbol{p}^1, \ldots, \boldsymbol{p}^{N_p}$ (the *pilots*). Then, the comparison with the received signals $\boldsymbol{y}^i = \boldsymbol{H}\boldsymbol{p}^i + \boldsymbol{n}^i$, $i = 1, \ldots, N_p$ furnishes information about the channel matrix $\boldsymbol{H}$.

One of the limitations of this procedure is the necessary knowledge of the distribution of the noise $\boldsymbol{n}$: indeed, let $\boldsymbol{P} = \left(\boldsymbol{p}^1, \ldots, \boldsymbol{p}^{N_p}\right) \in \mathbb{C}^{N_t \times N_p}$ denote the *pilot matrix*, having the pilots as columns. Then, the signals received may be collected in a matrix $\boldsymbol{Y} = \left(y^1, \ldots, y^{N_p}\right) \in \mathbb{C}^{N_r \times N_p}$ given by

$$\boldsymbol{Y} = \boldsymbol{H}\boldsymbol{P} + \boldsymbol{N} \tag{9}$$

where $\boldsymbol{N} = \left(\boldsymbol{n}^1, \ldots, \boldsymbol{n}^{N_p}\right) \in \mathbb{C}^{N_r \times N_p}$ is the additive noise matrix, having as columns the noise corresponding to each pilot. The least squares (LS) channel estimation based on Eq. (4) is given by

$$\hat{\boldsymbol{H}}_{LS} = \boldsymbol{Y}\boldsymbol{P}^{\dagger} \tag{10}$$

where $\boldsymbol{P}^{\dagger}$ is the generalized inverse of $\boldsymbol{P}$. We have $\boldsymbol{P}^{\dagger} = \boldsymbol{P}^{*}\left(\boldsymbol{P}\boldsymbol{P}^{*}\right)^{-1} \in \mathbb{C}^{N_p \times N_t}$, where $\boldsymbol{P}^{*} \in \mathbb{C}^{N_p \times N_t}$ is the Hermitian adjoint matrix associated to $\boldsymbol{P}$. Notice that the existence of $\boldsymbol{P}^{\dagger}$ assumes that $\boldsymbol{P}\boldsymbol{P}^{*} \in \mathbb{C}^{N_t \times N_t}$ is invertible – thus, $\boldsymbol{P}$ has full row rank N_t and $(N_p \geq N_t)$.

$$\hat{\boldsymbol{H}}_{LS} = \boldsymbol{H} + \boldsymbol{P}^{\dagger}\boldsymbol{N}, \tag{11}$$

so that the result contains noise. To eliminate – or simply reduce – the effect of the noise, some information about its distribution is required. For instance, a minimum mean square error (MMSE) estimator can be devised if convenient statistics of the signals and noise are available.

In many application scenarios, it is reasonable to assume that the additive noise has a complex-valued Gaussian distribution. Moreover, it is usual to assume that the noise statistics are known a priori. Scenarios where the additive noise is not Gaussian-distributed are little explored in the literature and the performance of MIMO symbol detectors is not fully understood yet.

In the recent years, UQ proposed methods and tools to deal with general distributions of probability: for instance – as previously remarked - the approach presented in Sect. 3 deals with any distribution having a finite variance. The UQ approach may be employed to improve the identification of the Channel Matrix by two basic procedures, described and illustrated in the sequel.

In the first one, the distribution of $\boldsymbol{n}$ is determined as shown in Sect. 3. Then, the information obtained is used to improve the estimation given in Eq. (9): denoting by $E(\blacksquare)$the mean, we have

$$E\left(\hat{\boldsymbol{H}}_{LS}\right) = \boldsymbol{H} + \boldsymbol{P}^{\dagger}\boldsymbol{E}(\boldsymbol{N}) \tag{12}$$

Thus, we may generate a sample $\boldsymbol{Y}^1, \ldots \boldsymbol{Y}^{N_s}$ from $\boldsymbol{Y}$ to estimate

$$E\left(\hat{\boldsymbol{H}}_{LS}\right) = E(\boldsymbol{Y})\boldsymbol{P}^{\dagger} \approx \left(\frac{1}{\mathrm{N_s}}\sum_{j=1}^{\mathrm{N_s}}\mathbf{Y}^{\mathbf{j}}\right)\boldsymbol{P}^{\dagger} \tag{13}$$

and

$$E(N_i) \approx \frac{1}{N_u}\sum_{r=1}^{\boldsymbol{N_u}}(\boldsymbol{C\Phi}(U_r))_i = \frac{1}{N_u}\sum_{r=1}^{N_u}\sum_{j=0}^{\mathrm{k}} c_{ij}\varphi_j(U_r) \tag{14}$$

may be estimated using a large sample of N_U variates from U. Then, we obtain $\boldsymbol{H}$ from Eq. (12).

The second approach consists in applying the UQ procedures directly to $\boldsymbol{Y}$: we may consider the expansion

$$\mathrm{Y}_i = \sum_{j=0}^{+\infty} y_{ij}\varphi_j(U) \approx \sum_{j=0}^{\mathrm{k}} y_{ij}\varphi_j(U) \tag{15}$$

Then, we may estimate

$$E\left(\hat{\boldsymbol{H}}_{LS}\right) = \boldsymbol{E}(\boldsymbol{Y})\boldsymbol{P}^{\dagger} \tag{16}$$

and get $\boldsymbol{H}$ from Eq. (12).

As an example, let us consider the situation where the Channel Matrix has real and imaginary parts given in Tables 6 and 7

Table 6. Real part $\boldsymbol{H_{re}}$ of the test channel matrix

0.3802	2.5303	−0.0878	0.3457	−0.7558	−0.0723	0.7731	−0.5442
1.2968	1.9583	1.0534	0.7316	−0.5724	−0.1707	0.7844	0.2626
−1.5972	−0.9545	0.9963	0.5140	−2.0819	0.2257	−0.6107	−0.1595
0.6096	2.1460	1.0021	−0.2146	1.0171	0.2212	0.0547	0.7901
0.2254	0.5129	0.4748	0.2078	0.2299	−0.6116	−0.8585	−0.7701
−0.9247	−0.0446	−0.8538	−0.5567	−0.5338	−0.0212	−0.7874	0.0230
−0.3066	0.5054	0.5072	0.6282	0.9689	−0.1166	−0.0048	0.3907
0.2423	−0.1449	1.1528	−0.8111	−1.2102	0.4439	1.0837	0.7782

Table 7. Imaginary part $\boldsymbol{H_{im}}$ of the test channel matrix

1.0919	−0.1361	1.0036	0.1525	−0.0583	0.2146	0.9578	2.0563
0.0608	0.6283	0.2062	−0.8244	−1.3669	−0.4245	−0.7581	0.5835
−1.0547	−0.5408	0.1399	−0.8117	−0.3104	0.3465	0.6795	0.9751
−0.5249	−0.9916	1.1227	0.0742	−1.2690	0.5228	0.0877	−0.7482
−0.7507	−1.0058	−0.5688	0.5107	0.5942	1.2105	1.0159	−0.3314
1.6620	0.3452	0.4926	1.8282	−0.6279	−0.1373	−1.3866	−0.1927
−0.4353	−0.1254	−0.5905	−0.4716	0.0708	−1.5120	−0.1398	0.7767
0.5290	−0.1386	−0.1723	0.1325	−0.3850	−0.5937	−0.8541	−0.1965

To simplify the procedure, we consider the successive identification of the columns of $\boldsymbol{H}$ – there is no loss of generality, since the UQ procedures apply to the complete matrix $\boldsymbol{H}$. To perform the identification of the first column, we consider as pilot

$$\boldsymbol{P} = \left(\boldsymbol{p}^1, \ldots, \boldsymbol{p}^{N_p}\right), \quad \boldsymbol{p}^i = \begin{pmatrix} 1 \\ 0 \\ \ldots \\ 0 \end{pmatrix}$$

Let us denote $\boldsymbol{H}^1$ the first column of $\boldsymbol{H}$. Then,

$$\boldsymbol{Y} = \boldsymbol{HP} + \boldsymbol{N} = \left(\boldsymbol{H}^1 + \boldsymbol{n}^1 \ \ldots \ \boldsymbol{H}^1 + \boldsymbol{n}^{N_p}\right),$$

so that we obtain a noisy sample of N_p variates from $\boldsymbol{Y}^1 = \boldsymbol{H}^1 + \boldsymbol{n}$. Let us apply the first procedure defined above: we have

$$\boldsymbol{H}^1 \approx \frac{1}{N_p} \sum_{r=1}^{N_p} \boldsymbol{Y}^r - \boldsymbol{E}(\boldsymbol{N}),$$

where $\boldsymbol{E}(\boldsymbol{N})$ is estimated by using Eq. (14). The procedure may be iterated for all the columns of the matrix $\boldsymbol{H}$, leading to the complete determination of the Channel matrix. The total number of messages requested for the identification is $N_t \times N_p$ (N_p messages for each column).

Let us consider again the same $8 \times 8-$ MIMO System introduced in Sect. 3, with additive noise uniformly distributed in the real part of $\boldsymbol{H}$ and Laplace distributed in the imaginary part of $\boldsymbol{H}$. We use 1000 measurements to determine the distribution of the noise and $N_p = 100$. To obtain some statistics about the behavior of the proposed method, we realized 1000 identifications using different realizations of the noise and we analyzed the distributions of the errors. In general, the first procedure furnishes a result with errors of about 2.5%. Typical results are shown in Tables 8 and 9.

Table 8. A typical result for the identification of $\boldsymbol{H}_{re}$: the relative error is 2.5%. In this case, we used a polynomial of degree 3 ($k = 3$)

0.3197	2.5331	−0.1123	0.3573	−0.7387	−0.0688	0.7639	−0.5622
1.3180	1.9880	1.0252	0.7476	−0.5982	−0.1660	0.8447	0.2547
−1.5747	−0.9262	0.9858	0.5271	−2.0471	0.2200	−0.6355	−0.1403
0.6090	2.1529	1.0666	−0.2134	1.0019	0.2164	0.0477	0.7759
0.2145	0.4843	0.4695	0.2636	0.2107	−0.6655	−0.8945	−0.7468
−0.9401	−0.0512	−0.8667	−0.5595	−0.5474	−0.0765	−0.7875	0.0107
−0.2738	0.5069	0.5372	0.6183	0.9725	−0.1592	−0.0099	0.3689
0.2243	−0.1180	1.1575	−0.8363	−1.2034	0.4597	1.1054	0.7495

Table 9. A typical result for the identification of $\boldsymbol{H}_{im}$: the relative error is 2.7%. In this case, we used a polynomial of degree 5 ($k = 5$)

1.1223	−0.1371	1.0276	0.1651	−0.0996	0.2413	0.9847	2.0552
0.0900	0.5813	0.2079	−0.8344	−1.4037	−0.3948	−0.8001	0.5528
−1.0403	−0.5373	0.1504	−0.7784	−0.3145	0.3370	0.6600	1.0037
−0.5472	−1.0040	1.1498	0.0735	−1.3057	0.4959	0.0892	−0.7480
−0.7535	−1.0056	−0.6029	0.5181	0.5898	1.1942	1.0119	−0.2977
1.6606	0.3181	0.4542	1.7691	−0.6316	−0.1724	−1.3636	−0.2176
−0.4341	−0.1186	0.5958	−0.4734	0.1117	−1.5063	−0.1536	0.7360
0.5280	−0.1820	−0.1652	0.1276	−0.3488	−0.5760	−0.8474	−0.1550

Analogously to the noise, we generated statistics about the behavior of the proposed method: 1000 approximations using different realizations of the noise were used to determine the empirical CDF by the same way as preceding: a Hilbert approximation by a polynomial of degree 4, then the generation of a sample of $1e4$ variates. As in the study of the noise, the sample is used to determine the CDF and the PDF is found by particle derivation of the CDF [11].

In Fig. 4 below, we exhibit the PDF obtained for the relative errors in the real and the imaginary parts of the channel matrix. Again, both the distributions are unimodal. The mode for the real part is about 2.5% and the mode for the imaginary part is about 3.5%. Indeed, the statistics of the errors in the original sample of 1000 realizations are given in the Table 10 below

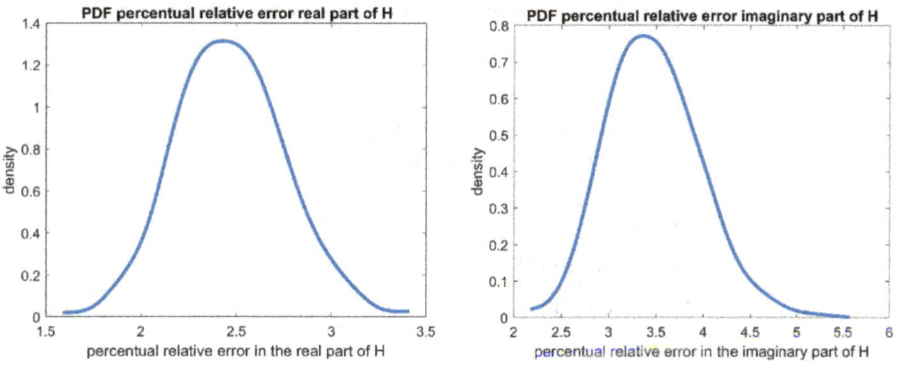

Fig. 4. PDF of the relative errors for the real and the imaginary parts of H (first approach)

Table 10. Statistics of the relative errors in the estimation of the channel matrix. These values correspond to about 2.5% of error in the real part and 3.5% in the imaginary part (first approach)

Quantity	Mean	Standard Deviation
Percent error in $\boldsymbol{H}_{re}$	2.4666	0.2804
Percent error in $\boldsymbol{H}_{im}$	3.4772	0.4861

Again, the errors are larger in the imaginary part, due to the difficulty introduced by the Laplace distribution. Here yet, improvements may be obtained by using other approaches, such as other Hilbert basis leading to more accurate approximations (Fig. 5).

Now, let us consider the second approach: as previously mentioned, the UQ procedures are applied directly to $\boldsymbol{Y}$. The results are given in Tables 11, 12, 13 below. We observe that the results are close to the preceding ones.

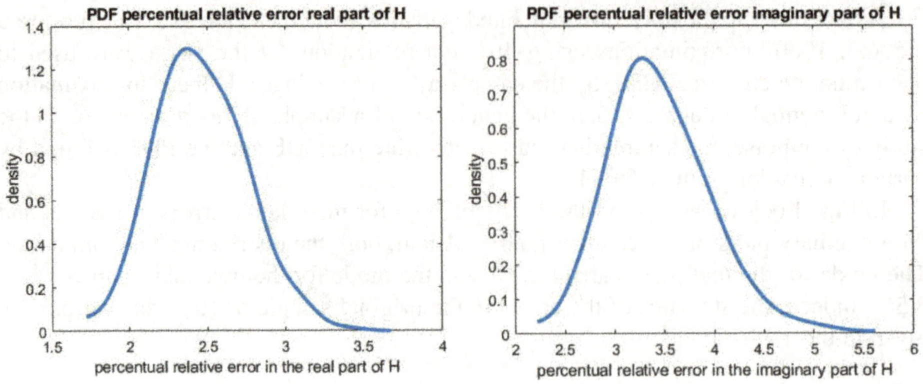

Fig. 5. PDF of the relative errors for the real and the imaginary parts of H (second approach)

Table 11. Statistics of the errors in the estimation of the channel matrix. These values correspond to about 2.5% of error in the real part and 3.5% in the imaginary part (second approach)

Quantity	Mean	Standard deviation
Percent error in $\boldsymbol{H}_{re}$	2.4632	0.2861
Percent error in $\boldsymbol{H}_{im}$	3.4593	0.5047

Table 12. A typical result for the identification of $\boldsymbol{H}_{re}$: the relative error is 2.7%. In this case, we used a polynomial of degree 3 ($k = 3$)

0.3680	2.5028	−0.0990	0.3692	−0.7568	−0.1403	0.7332	−0.4937
1.2847	1.9414	1.0555	0.7532	−0.6005	−0.2234	0.7769	0.2865
−1.5988	−0.9406	0.9933	0.5163	−2.0528	0.2602	−0.6184	−0.1595
0.5886	2.1526	1.0214	−0.2446	1.0243	0.2035	0.0250	0.7791
0.2338	0.4664	0.4736	0.1865	0.2833	−0.6027	−0.8558	−0.7494
−0.9310	−0.0365	−0.8368	−0.5052	−0.5354	0.0047	−0.7993	0.0442
−0.2799	0.5414	0.5156	0.6768	0.9260	−0.1125	0.0092	0.3585
0.2214	−0.1848	1.1550	−0.8090	−1.2271	0.4246	1.0873	0.7406

Table 13. A typical result for the identification of $\boldsymbol{H_{im}}$: the relative error is 3.04%. In this case, we used a polynomial of degree 5 ($k = 5$)

1.0586	−0.1284	1.0096	0.1549	−0.0503	0.2289	0.9755	2.0566
0.0730	0.6450	0.2133	−0.8185	−1.3692	−0.4237	−0.7618	0.6284
−1.0485	−0.5586	0.1311	−0.8217	−0.2576	0.3661	0.6772	1.0161
−0.5068	−0.9969	1.1592	0.0935	−1.2508	0.5323	0.0611	−0.7377
−0.7666	−1.0275	−0.5537	0.5166	0.5941	1.2245	0.9634	−0.3507
1.6436	0.3618	0.4758	1.8125	−0.5872	−0.1457	−1.4069	−0.1555
−0.4201	−0.1268	0.6263	−0.5105	0.0407	−1.5517	−0.1747	0.7578
0.4866	−0.1457	−0.1615	0.0918	−0.3937	−0.6077	−0.8930	−0.2004

5 Symbol Detection

At this stage, we obtained information about the distribution of the noise and about the channel matrix. The next step is the detection of symbols. Let us assume that both the real and imaginary parts of the symbols are chosen in an alphabet $\boldsymbol{a}$ corresponding to the so-called Quadrature Amplitude Modulation (QAM) of order $M = 2^d$: $\boldsymbol{a} = (a_1, \ldots, a_M) \in \mathbb{C}^M$, with

$$a_i = -2^d + 2i - 1 : 1 \le i \le 2^d.$$

For $d = 1$, we have $\boldsymbol{a} = (-1, 1)$. For $d = 2$, $\boldsymbol{a} = (-3, -1, 1, 3)$. In the general situation:

$$\boldsymbol{a} = \left(-2^d + 1, -2^d + 3, \ldots, -1, 1, 3, \ldots, 2^d - 3, , 2^d - 1\right).$$

Let us consider the situation where each transmitter sends a symbol: for the $8 \times 8-$ MIMO System introduced in Sect. 3, the transmitters send 8 symbols and the receivers receive 8 values, modified by the channel matrix and the noise. The aim is to determine the symbols sent by the transmitters using the values detected by the receivers, the estimation of the channel matrix and our knowledge about the distribution of the noise.

In the sequel, we consider two methods of detection. The first approach associates each symbol to a vector of length M, corresponding to lines of the $M \times M$ identity matrix:

$$\boldsymbol{a_i} = (s_{i1}, \ldots, s_{iM}),\ s_{ij} \in \{0, 1\},\ s_{ii} = 1,\ s_{ij} = 0\ if\ i \neq j.$$

Thus, an unknown symbol a corresponds to an unknown line of the $M \times M$ identity matrix, i.e., to a vector

$$\boldsymbol{a} = (z_1, \ldots, z_M), z_i \in \{0, 1\}, z_1 + \ldots + z_M = 1.$$

According to these observations, N_t unknown real symbols are equivalent to N_t unknown lines of the $M \times M$ identity matrix, i.e., to N_t unknown vectors of length

M. Thus, we have $N_t \times M$ unknowns to be determined. These unknowns are collected into a $N_t \times M$ matrix $\mathbf{Z} = \left(z_{ij}: 1 \leq i \leq N_t, 1 \leq j \leq M\right)$, having as lines the vectors corresponding to the unknown symbols. Analogously to Eqs. (2)–(3), we may consider separately the real and imaginary parts of the symbols, so that the message transmitted, and the data received read as

$$\bar{\boldsymbol{x}} = \bar{\boldsymbol{Z}}\bar{\boldsymbol{a}}; \quad \bar{\boldsymbol{y}} = \bar{\boldsymbol{H}}\bar{\boldsymbol{Z}}\bar{\boldsymbol{a}} + \bar{\boldsymbol{n}}$$

Each line of $\bar{Z}$ is a line of the $M \times M$ identity matrix, so that

$$\forall i: \quad \bar{z}_{ij} \in \{0, 1\}, \ \forall j; \quad \bar{z}_{i1} + \ldots + \bar{z}_{iM} = 1.$$

These restrictions are equivalent to

$$\forall i \ : \ \bar{z}_{ij}\left(1 - \bar{z}_{ij}\right) \leq 0, \ \bar{z}_{ij} \geq 0, \bar{z}_{ij} \leq 1, \ \forall j; \quad \bar{z}_{i1} + \ldots + \bar{z}_{iM} = 1. \tag{17}$$

Let us introduce $\overline{\boldsymbol{\mathcal{H}}} = \left(\bar{\mathcal{H}}_{ijk}\right)$ given by

$$\bar{\mathcal{H}}_{ijk} = \bar{H}_{ij}\bar{a}_k.$$

Then,

$$\left(\bar{\boldsymbol{H}}\bar{\boldsymbol{Z}}\bar{\boldsymbol{a}}\right)_i = \sum_{j,k} \bar{H}_{ij}\bar{Z}_{jk}\bar{a}_k = \sum_{j,k} \bar{\mathcal{H}}_{ijk}\bar{Z}_{jk},$$

and

$$\bar{\boldsymbol{H}}\bar{\boldsymbol{Z}}\bar{\boldsymbol{a}} = \overline{\boldsymbol{\mathcal{H}}}\bar{\boldsymbol{Z}}$$

Notice that $\overline{\boldsymbol{\mathcal{H}}}$ may be transformed into a matrix $\bar{\mathbb{H}}$ and $\bar{\boldsymbol{Z}}$ may be transformed into a vector $\bar{\mathbb{Z}}$ by using the correspondence $(j, k) \to \ell = j + 2(k-1)N_t$, $\bar{\mathbb{H}}_{i\ell} = \bar{\mathcal{H}}_{ijk}$, $\bar{\mathbb{Z}}_\ell = \bar{Z}_{jk}$. Then,

$$\bar{\boldsymbol{y}} = \overline{\boldsymbol{\mathcal{H}}}\bar{\boldsymbol{Z}} + \bar{\boldsymbol{n}} = \bar{\mathbb{H}}\bar{\mathbb{Z}} + \bar{\boldsymbol{n}} \tag{18}$$

Equation (18) is analogous to Eq. (9), but restrictions (17) must be taken into account. Notice that Eqs. (17)–(18) may be solved for $\left(\bar{\boldsymbol{Z}}, \bar{\boldsymbol{n}}\right)$. For instance, we may use a constrained least squares method may be used. We may also use algebraic equation methods: notice that Eq. (17) is equivalent to

$$\forall i \quad : \ \bar{z}_{ij}\left(1 - \bar{z}_{ij}\right) = 0, \ \forall j \ ; \quad \bar{z}_{i1} + \ldots + \bar{z}_{iM} = 1. \tag{19}$$

Thus, Eqs. (18)–(19) form a overdetermined nonlinear system which may be solved by adapted methods – eventually by a least squares approach. The number of unknowns is $2N_tM + 2N_r$, while the number of equations is $2N_t(M+1) + 2N_t$. A simplified approach may use the approximation

$$\bar{\boldsymbol{y}} \approx \boldsymbol{E}(\bar{\boldsymbol{y}}) = \overline{\boldsymbol{\mathcal{H}}}\bar{\boldsymbol{Z}} + \boldsymbol{E}(\bar{\boldsymbol{n}}) = \bar{\mathbb{H}}\bar{\boldsymbol{Z}} + \boldsymbol{E}(\bar{\boldsymbol{n}}) \tag{20}$$

Since $\boldsymbol{E}(\bar{\boldsymbol{n}})$ was previously estimated, Eq. (20) may be used in combination with Eq. (17) or Eq. (19) to furnish an estimation of $\bar{\boldsymbol{Z}}$. The results may be corrected by replacing x_i by the closest member of $\boldsymbol{a}$:

$$x_i = a_{j(i)},\ j(i) = \arg\min\{|x_i - a(j)| : 1 \le j \le M\} \tag{21}$$

A second approach, more simplified yet, consists in using the equation

$$\boldsymbol{H}^*\boldsymbol{E}(\boldsymbol{y}) = \boldsymbol{H}^*\boldsymbol{H}\boldsymbol{x} + \boldsymbol{H}^*\boldsymbol{E}(\boldsymbol{n}),$$

we have

$$\boldsymbol{x} = \left(\boldsymbol{H}^*\boldsymbol{H}\right)^{-1}\boldsymbol{H}^*(\boldsymbol{E}(\boldsymbol{y}) - \boldsymbol{E}(\boldsymbol{n})) \approx \left(\boldsymbol{H}^*\boldsymbol{H}\right)^{-1}\boldsymbol{H}^*(\boldsymbol{y} - \boldsymbol{E}(\boldsymbol{n})). \tag{22}$$

The estimation furnished by Eq. (22) is corrected as indicated in Eq. (21).

Let us illustrate these approaches. Initially we consider $M = 2\ (QAM4), \boldsymbol{a} = \{-1, 1\}$ and the message

$$\bar{\boldsymbol{x}} = (1, -1, 1, -1, 1, -1, 1, -1, 1, 1, -1, -1, 1, 1, -1, -1)^t.$$

sThe noise is the same as in the preceding sections (uniformly distributed for the real part and Laplace distributed for the imaginary part). Initially, we determine the noise characteristics by the method exposed in Sect. 3, using 1e4 measurements: this step furnishes also an estimation of $\boldsymbol{E}(\bar{\boldsymbol{n}})$. Then, we identify the channel matrix as indicated in Sect. 4, using 1e3 measurements. We perform 1e4 estimations of the message using a different realization of the noise each time. The performance of the methods is shown in the Fig. 6. The first approach produced the exact result in 100% of the runs, while the second one produced the exact result in 90.4% of the runs.

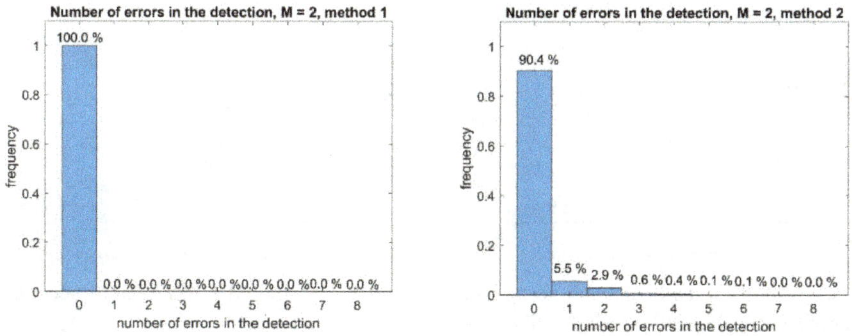

Fig. 6. Results for 1E4 runs. The alphabet has length 2 (QAM4). Method 1 has 100% of success (no erroneous symbol). Method 2 finds all the exact symbols in 90.4% of the runs.

Now, let us consider $M = 8(QAM64, d = 3), \boldsymbol{a} = \{-7, -5, -3, -1, 1, 3, 5, 7\}$ and the message (randomly generated)

$$\bar{\boldsymbol{x}} = (-3, 3, 3, 7, 1, -7, 7, 1, -7, 1, 1, -3, 5, -7, 5, 7)^t.$$

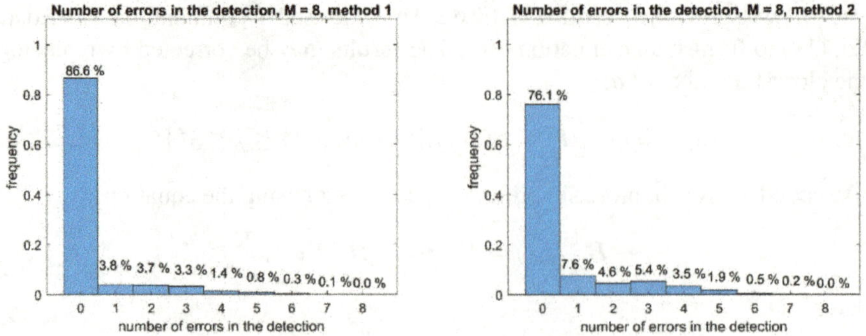

Fig. 7. Results for 1E4 runs, with an alphabet of length 8 (QAM64). Method 1 has no errors in 87% of the runs. Method 2 in 76%.

Again, we realize 1e4 runs using distinct variates of the noise each time. The results are shown in the Fig. 7. The first approach produced the exact result in 87% of the runs, while the second one produced the exact result in 76% of the runs.

Then, let us consider $M = 32\,(d = 5), \boldsymbol{a} = \{-31, -29, \ldots, -1, 1, \ldots, 29, 31\}$ and the message (randomly generated)

$$\bar{\boldsymbol{x}} = (7, 1, 15, 11, 15, -7, 21, 23, -19, -7, -9, 3, -23, -1, -29, 25)^t.$$

The performance of the methods is shown in the Fig. 8. The first approach produced the exact result in 87% of the runs, while the second one generally fails.

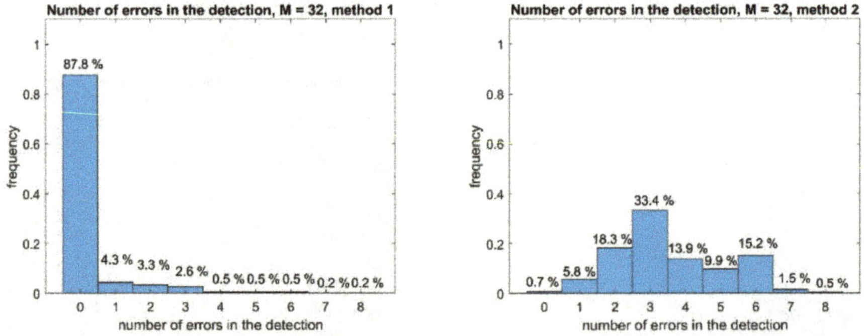

Fig. 8. Results for 1E4 runs and an alphabet of length 32. Method 1 has 88% of success (no erroneous symbol). Method 2 generally fails.

Finally, let us consider $= \{-127, -125, \ldots, -1, 1, \ldots, 125, 129\}$, $M = 128\,(d = 7)$, and the message (randomly generated)

$$\bar{\boldsymbol{x}} = (-117, 65, -1, -27, -95, 33, -43, 59, 13, -113, 75, 31, 47, -95, -77, 77)^t.$$

The performance of the method proposed is shown in the Fig. 9: it produced the exact result in 87% of the runs.

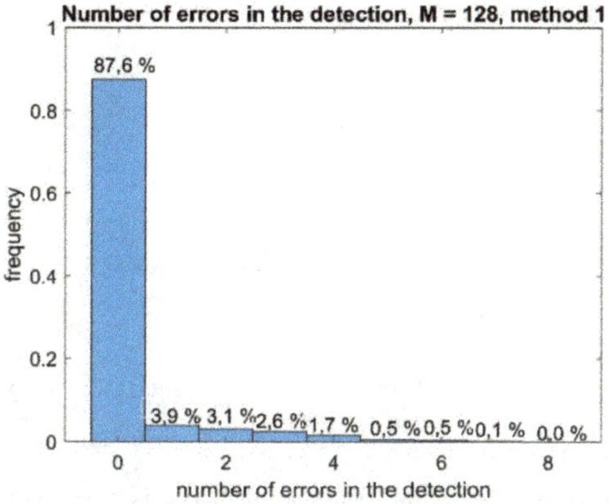

Fig. 9. Results for 1E4 runs. The alphabet has length 128 and spread is 2. Method 1 has 87.6% of success (no erroneous symbol).

The performance of the methods depends on the spread of the alphabet, id est, of the distance between consecutive symbols: for instance, the alphabets considered up to this point have a spread of 2: the difference between consecutive symbols is 2 – what is the actual standard in telecommunications. As a prospective analysis, let us consider $= \{-15, -11, -7, -3, 1, 5, 9, 13\}$, which corresponds to $M = 8$ with a spread of 4. Notice that the alphabet becomes asymmetric. We consider the message

$$\bar{x} = (9, 5, 13, 13, -7, 1, 5, -3, 9, -11, -15, -15, -11, 1, 1, -3)^t.$$

This message corresponds to symbols in the same positions as in the test performed for the alphabet of length 8 in the preceding. The results are shown in Fig. 10 below: we observe that the performance of both the methods increase significantly: the method 1 has found all the correct symbols in 99.9% of the runs.

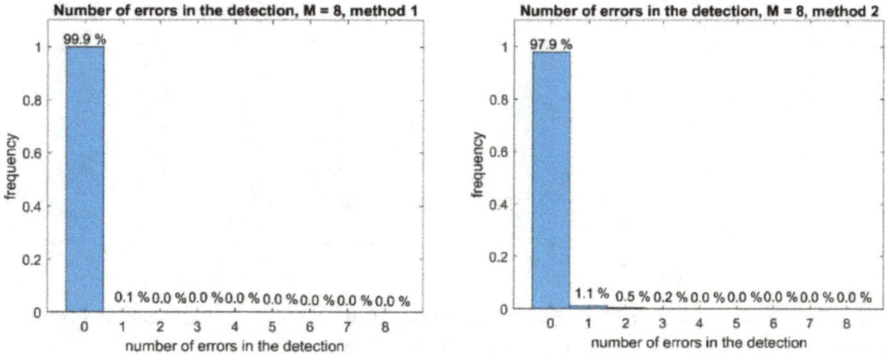

Fig. 10. Results for 1E4 runs of the method with an asymmetric alphabet spread out. The alphabet has length 8, and the spread is 4 (the double of the one initially considered). Method 1 has 99.9% of success (no erroneous symbol) and method 1 has 97.9% of success.

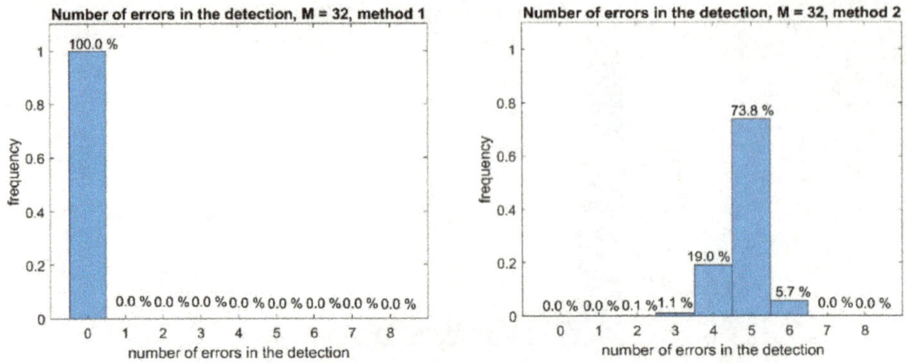

Fig. 11. Results for 1E4 runs of the method. The alphabet is asymmetric, of length 32, with a spread of 8. Method 1 has 100% of success (no erroneous symbol). Method 2 generally fails.

A second example of the influence of the spread is given in Fig. 11 above. We consider $M = 32$ with an spread of 8: $\boldsymbol{a} = \{-127, -119, \ldots, -7, 1, \ldots, 113, 121\}$ and the message corresponding to symbols at the same positions as the used in the test previously presented for this alphabet:

$$\bar{\boldsymbol{x}} = (25, 1, 57, 41, 57, -71, -31, 81, 89, -79, -39, 9, -95, -7, -119, 97)^t.$$

The first approach produced the exact result in 100% of the runs, while the second one fails.

It is also possible to consider symmetric alphabets with larger spread. For instance, $a = \{-14, -10, -6, -2, 2, 6, 10, 14\}$ is a symmetric alphabet of length 8, having a spread of 4. The effects of the spread are illustrated in the Fig. 12 below.

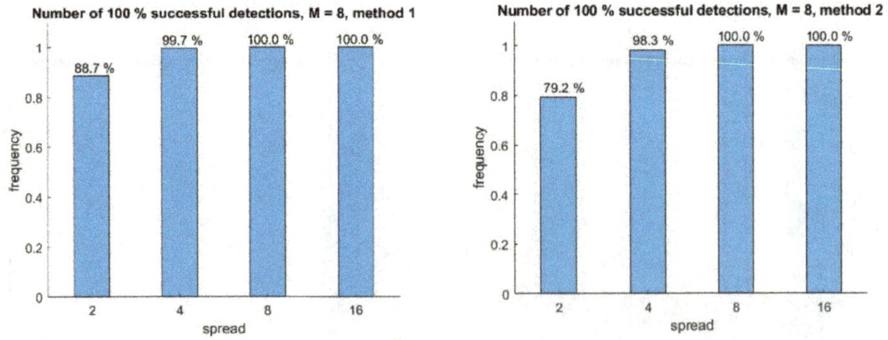

Fig. 12. Influence of the spread: results for 1E4 runs of the method. The alphabet is symmetric, of length 8. The rate of success increases for both the methods.

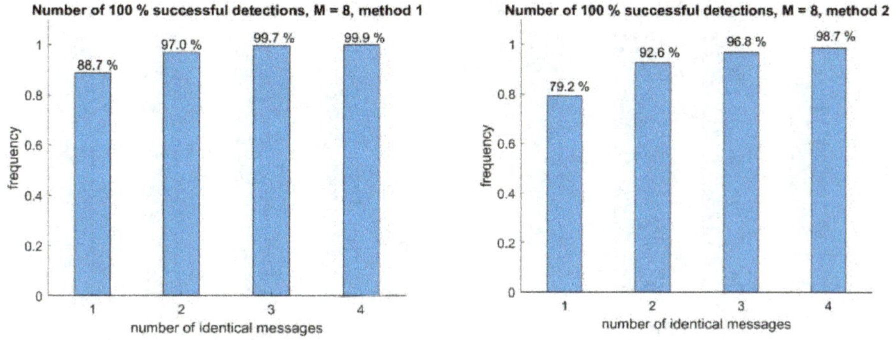

Fig. 13. Influence of the redundancy: results for 1E4 runs. The alphabet is symmetric, of length 8, spread 2. The rate of success increases with redundancy for both the methods.

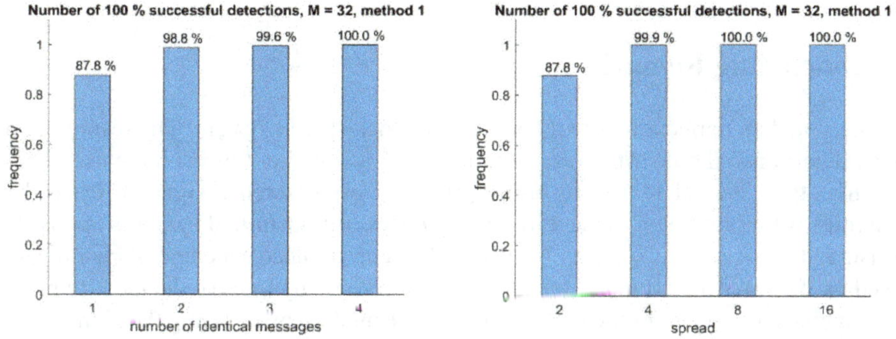

Fig. 14. Influence of the redundancy and spread: results for 1E4 runs. The alphabet is symmetric, of length 32. Redundancy analysis uses a fixed spread equal to 2. Spread analysis is made with no redundancy (a single sending). Method 2 fails in this case

The quality of the detection depends also upon the redundancy of the message. For instance, we may consider a simple redundancy implemented by sending r times the same message. The influence of r is illustrated in the Figs. 13 (above), 14 (above) and 15 (below).

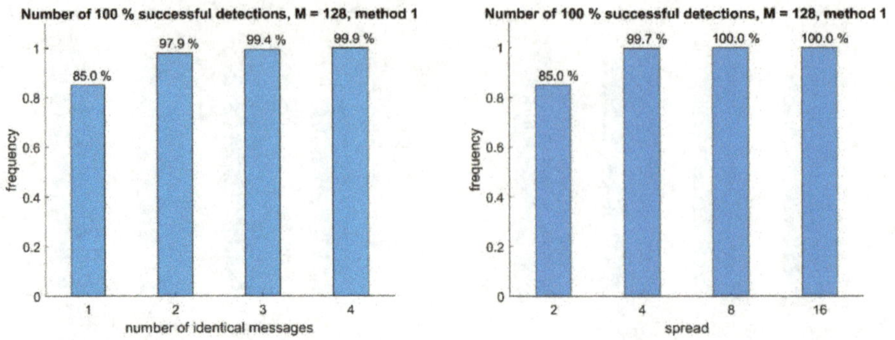

Fig. 15. Results for a symmetric alphabet of 128 symbols, 1E4 runs of the method. For the analysis of the effect of redundancy, the spread is fixed to 2. For the analysis of the effect of the spread, there is no redundancy (a single message without any repetition). For this alphabet, Method 2 fails.

6 Concluding Remarks

We presented an application of UQ approaches to a MIMO system. The resulting approach determines the distribution of the noise by using silent measurements. Then, it identifies the Channel Matrix by using pilots (id est, reference signals). Once these quantities were identified, the lecture of sequences of transmitted symbols is possible, by using identification techniques. In this work, we introduced a method of identification based on the determination of a selection matrix formed by lines of the identity matrix. We established statistics of the performance of the method by realizing the identification of the same sequence of symbols ten thousand times and drawing up the histograms of the number of errors in each identification. No correcting code was used in the tests, which show crude results, but we examined the effects of redundancy by considering repeated messages. We examine also the effects of the distance between symbols: the larger is the spread and the larger is the number of repetitions, the smaller are the errors: for convenient values, the rate of error is almost null – this is a prospective analysis, since the usual spread in telecommunications is 2. Two methods were considered: the faster one does not involve selection matrix but fails for alphabets using more than 8 symbols. The second method uses a selection matrix and appears as more robust. Several variations of the parameters were considered – for instance, changes in the number of measurements (divided by 10), using the rough estimation $\boldsymbol{E}(\boldsymbol{n}) \approx \mathbf{0}$, using different methods of optimization, ... - and led to analogous results. The preliminary results of this work may be considered as promising and encourage further research in the application of UQ approaches to MIMO systems.

References

1. Heath, R.W., Angel, L.: Foundations of MIMO Communication. Cambridge University Press, Cambridge (2018)
2. Huang, Y., Benesty, J., Chen, J.: Identification of acoustic MIMO systems: challenges and opportunities. Signal Process. **86**(6), 1278–1295 (2006)

3. Li, B., et al.: MIMO-OFDM for high-rate underwater acoustic communications. IEEE J. Ocean. Eng. **34**, 4 (2009)
4. Zhang, L., Ming, L., Li, G.: Symbol estimation for MIMO underwater acoustic communication based on multiplicative noise model. In: 2014 IEEE International Conference on Communication Problem-Solving, IEEE (2014)
5. Zhang, L., et al.: Symbol estimation algorithm for MIMO underwater acoustic communication system based on multiplicative noise model. Math. Prob. Eng. **2015**, 7 (2015)
6. Barhumi, I., Leus, G., Moonen, M.: Optimal training design for MIMO OFDM systems in mobile wireless channels. IEEE Trans. Signal Process. **51**(6), 1615–1624 (2003)
7. Lopez, R.H, de Cursi, J.E.S., Lemosse, D.: Approximating the probability density function of the optimal point of an optimization problem, Eng. Optimizat. **43**(3), 281–303 (2011). https://doi.org/10.1080/0305215x.2010.489607
8. Lopez, R.H., Miguel, L., de Cursi, E.S.: Uncertainty quantification for algebraic systems of equations. Comput. Struct. **128**, 189–202 (2013). https://doi.org/10.1016/j.compstruc.2013.06.016
9. Lopez, R.H., Cursi, JESouza, Carlon, A.G.: A state estimation approach based on stochastic expansions. Comput. Appl. Math. **37**(3), 3399–3430 (2017). https://doi.org/10.1007/s40314-017-0515-0
10. de Cursi, E.S., Sampaio, R.: Uncertainty Quantification and Stochastic Modeling with Matlab. Elsevier Science Publishers B.V., NLD (2015)
11. de Cursi, E.S.: Variational methods for engineers with Matlab. ISTE, Ltd., Wiley, Hoboken, London, UK (2015)

Estimation of the Extreme Response Probability Distribution of Offshore Structures Due to Current and Turbulence

Oscar Sanchez Jimenez[1(✉)], Emmanuel Pagnacco[1], Eduardo Souza de Cursi[1], and Rubens Sampaio[2]

[1] LMN, INSA de Rouen-Normandie,
BP8, 76801 St. Etienne du Rouvray Cedex, France
oscar.sanchez_jimenez@insa-rouen.fr

[2] PUC-Rio, Mechanical Engineering Department, Rua Marques de Sao Vicente,
Rio de Janeiro, RJ 225 22451-900, Brazil

Abstract. For a model of offshore structure standing in a fluid, we derived the cumulative distribution function of the extreme base shear and the extreme bending moment, that is, the extreme value produced by the weighted sum of a spatially continuous non-Gaussian stochastic horizontal force field produced by current and turbulence, in a given time interval it is assumed to be stationary. In the proposed methodology, the theory of translated processes is used to solve the problem of finding the extreme of non-Gaussian stochastic processes. The application is carried out on a thick column whereas the stochastic field of turbulence is supposed to be Gaussian but not necessarily narrow-band. The cumulative distribution functions obtained are in good agreement with statistical results derived from simulations in the time domain.

Keywords: Offshore structure · Uncertainty quantification · Non-gaussian process · Random field · Extreme value distribution · Translated gaussian process

1 Introduction

Immersed offshore structures are subject to the effects of oceanic current and their inherent turbulence. These phenomena are modelled through stochastic processes of various classes. Since one of the main considerations when designing this kind of structure is its capacity to withstand the maximum possible effort induced by these fluid forces, a key element of said design is a reliable estimation of the maximum value of the effects that these hydrodynamic forces produce over the structure. The stochasticity of the velocity field induces a non Gaussian fluid force stochastic field onto the structure, which, under reasonable simplifying assumptions, results in a structural response expressed in terms of a

J. E. S. De Cursi (Ed.): Uncertainties 2020, LNME, pp. 306–322, 2021.
https://doi.org/10.1007/978-3-030-53669-5_22

bending moment and a shear force, that happen to be themselves stochastic processes. The problem of seeking the maximum value of these quantities is in fact the search for the probability distribution of their maximum values (a random variable).

The degree of complexity to model the current and turbulent flow of the ocean can vary. In one end of the spectrum, the approach that has been followed for many years ([1] for instance) is to use a scalar stochastic process for the fluid particle velocity and then adjusting the effects by means of an empirical drag coefficient per unit of length of the structure. In the other end of the spectrum, the more general theoretical models assume that the fluid particle velocity is in fact a stochastic field, thus at each point of the height of the structure there is a corresponding stochastic process associated. The study of the Extreme Value Distribution (EVD) of the first type of model, that is, the distribution of maximum values of shear force and bending moment produced by a single non-Gaussian fluid force has been extensively covered, with analytical solutions that allow the determination of the EVD of the responses as a function of the Gaussian velocity process. In the case of models that include a stochastic field to model these hydrodynamic forces, the bibliography shows a much more restricted success. This latter problem necessitates the determination of the EVD of an arbitrary sum of non Gaussian, stationary processes. We propose a methodology to estimate the EVD of said problem, which is based on the theory of translated Gaussian processes.

Early works in offshore design focused on the analysis of the forces produced by the velocity flow, the so-called wave force (a hydrodynamic force). Reference [2] focuses on the extreme values of the wave forces from Morison-type efforts, comparing what was at the time the common procedure to calculate the response. The extreme values of the wave forces were studied through linearization, whereas the proposed probability distributions are nonlinear. It is shown that depending on the dominant term, drag or inertial, the linearization underestimates the expected occurence of extreme values. In the case of drag dominated situations, the linearization underestimates the probability of occurrence of extremes. The limiting details of this study is first in the assumption of an exclusively narrow band process of the particle velocity, and secondly on the assumption that the response is linear in the forces and their extreme values will behave similar to the force. Finally, in contrast with our study, it focuses on the stresses and it doesn't take into account, for instance, the possible EVD of the bending moment on the pile. Reference [1] uses the Gaussian translated processes to obtain the EVD of a Morison-type force by means of the mean upcrossing rate. Reference [3] gives expressions for the mean upcrossing rate for the wave loading, parting from Morison equation and retaining both terms, drag and inertia. Also, reference [3] decomposes the resulting stochastic process into the sum of two component processes, so in a sense it addresses the problem of the EVD of the sum of processes for $N = 2$. Reference [4] studies the peak distribution of Morison-type forces at a given point on a pile using an analytical approach employing some simplifications like the deep level assumption and truncating higher-than-resonance frequencies from the sea-state power density. Reference [5] models the probability distribution of wave forces as a bi-parametric (Pierse-Holmes) distribution and, based on linearity considerations between the excitation and the response, presents a discretized model for the responses at nodes. Since the nodal response is expressed in terms of the expected values of the excitation, and since no analytical closed-form solution for these is known, it proceeds to propose a series of simplifications to reduce the burden of numerical integration, including polynomial approximations of the nonlinear terms in the particle velocity. It is to note that the EVD is not the main focus of the study. Reference [6] presents an analytical method for the determination of the Probability Density Functions (PDFs) of the base shear of a cylindrical element subject to a Morison force. Their results account for non-linearities in the response, but are limited to narrow band velocity processes. The key points of their results rest on the linearization of the wave theory modelling the sea level evolution and in a proposed transformation of variables.

In more recent efforts, reference [7] studies the structural response (displacements) of slender, fixed elements subject to Morison-type forces. The focus is on the determination of the cumulants, particularly the kurtosis and skewness, of the reported non Gaussian response by means of the Volterra series. The aim is to estimate these cumulants to then approximate the EVD of the response. It highlights the limitations of time-domain, Monte-carlo based simulations on the basis of the computational effort required. It is to note that their development accounts for nonlinear wave modelling and for non Gaussian sea characterization. In references [8,9], the author proposes the use of reliability methods, First Order Reliability Method and Second Order Reliability Method, to reduce the size of time simulations that serve as basis of the determination of the EVD in the response of the structure subject to a random wave. This methodology considers the dynamics effects, *i.e.*, it is not limited to the quasi-static response. The mean number of upcrossings is expressed in terms of reliability parameters and these are obtained through a Monte-Carlo simulation.

Finally, with respect to the formulation of the problem that leads to a response that is expressed as a sum of processes, in reference [10], the author provides a study of a response expressed as the sum of two components, focusing in particular on the Square Root of Sum of Squares (SRSS) method. The focus is to obtain the EVD of the response process. One of the main conclusion is the fact that SRSS, an heuristic implementation, gives accurate prediction of the EVD when the component processes of the response are Gaussian and independent. Dependency between the components leads to an important overestimation. For the problem under study this makes the method ineffective, since the wave force processes are generated by the same velocity field and in general will have some local correlation.

The organization of this paper is as follows: in Sect. 2 the problem to be solved is described along with the relevant definitions and notation pertaining to the physical, mechanical and stochastic aspects; in Sect. 3, essential results from the theory of transformation of stochastic processes are presented, notably integral transformations and a special case of transformation called translation of Gaussian processes. The first type of transformation is useful to obtain necessary information to the determination of the EVD of the response from known information of the input process. The second type of transformation allows to generate the partially characterized response process from Gaussian processes via the translation, which in turn facilitates the determination of the EVD of the response. Section 4 is a concise description of the proposed methodology. The results of numerical simulations of the proposed methodology are presented in Sect. 5, along with comparisons with other approaches as well as descriptions of interesting aspects of the models and assumptions. Additional discussion is provided in Sect. 6, as well as some guidelines for future work.

2 Problem Formulation

2.1 Characterization of the Velocity Field

Let $\mathcal{V}(z,t)$ be the Gaussian stochastic field describing the velocity of a turbulent tidal flow as a function of depth. We assume the flow is unidirectional. Furthermore, we will limit the study of $\mathcal{V}(z,t)$ to a narrow time interval $[0,T]$ in which it is reasonable to assume it is stationary. Associated with this process there is a mean velocity function of depth (or height) noted $\mu_{\mathcal{V}}(z) = \mathrm{E}[\mathcal{V}(z,t)]$ and a variance we denote by $\sigma^2_{\mathcal{V}}(z) = \mathrm{E}[(\mathcal{V}(z,t) - \mu_{\mathcal{V}}(z))^2]$, $\mathrm{E}[\bullet]$ being the expectation operator.

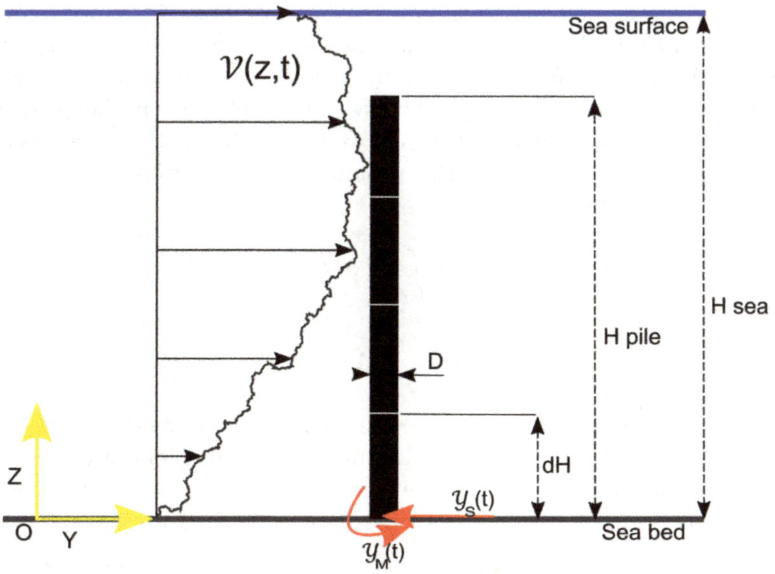

Fig. 1. Sketch of the problem of interest

To complete the description of $\mathcal{V}(z,t)$, we introduce the covariance function:

$$\Sigma_{\mathcal{V}(z_1)\mathcal{V}(z_2)}(\tau) = \rho_{\mathcal{V}(z_1)\mathcal{V}(z_2)}(\tau)\sqrt{\sigma^2_{\mathcal{V}}(z_1)\sigma^2_{\mathcal{V}}(z_2)} \tag{1}$$

with $\tau = t_2 - t_1$ and $\rho_{\mathcal{V}(z_1)\mathcal{V}(z_2)}(\tau)$ is the covariance coefficient for two arbitraty times t_1 and t_2 and heights z_1 and z_2 of the stochastic field. The covariance coefficient is further decomposed into a spatial $\rho^{\text{space}}(z_1, z_2)$ and a temporal function $\rho^{\text{time}}(\tau)$ as:

$$\rho_{\mathcal{V}(z_1)\mathcal{V}(z_2)}(\tau) = \rho^{\text{space}}(z_1, z_2)\rho^{\text{time}}(\tau) \tag{2}$$

2.2 Description of the Structure and Its Loads

We study the effects of the flow on a simplified model of a supporting pile immersed in it. We assume this pile to be a rigid cylinder with constant cross section with a height H_{pile} and a diameter D_{pile}. The assumption of the rigid body is reasonable if the frequency content of the excitation (the flow) is sufficiently lower than the natural frequency of the pile itself. We also assume that the pile is slender, so diffraction effects are negligible.

Assuming $\mathcal{V}(z,t) > 0\ \forall t$, the fluid load on this pile can be modelled by means of the Morison formula [11]:

$$F_H(z,t) = \rho_{\text{f}} C_m \frac{\pi}{4} D^2_{\text{pile}} \frac{\text{d}\mathcal{V}(z,t)}{\text{d}t} + \frac{1}{2}\rho_{\text{f}} C_d D_{\text{pile}} \mathcal{V}^2(z,t) \tag{3}$$

The first term is said to be inertial and the second describes the drag. For this application, the inertial term turns out to be negligible, so we are concerned with the second term alone, that is, the quasi-static effort produced by the drag component of the wave force induced by a velocity field that is stochastic in nature. From this load we can predict two main effects on the pile: a shear force $\mathcal{Y}_{\text{S}}(t)$ and a bending moment $\mathcal{Y}_{\text{M}}(t)$. Both are stochastic processes and our proposed methodology estimates the probability distribution of their extreme values. We define them as:

$$\mathcal{Y}_{\text{S}}(t) = \int_{\mathbb{V}} K_c \mathcal{V}^2(z,t)\text{d}z \tag{4}$$

$$\mathcal{Y}_{\text{M}}(t) = \int_{\mathbb{V}} K_c \mathcal{V}^2(z,t) z\text{d}z \tag{5}$$

where $K_c = \frac{\pi}{4}\rho_{\text{f}} D_{\text{pile}} C_d$. In these expressions ρ_{f} is the density of the fluid(sea water in our case), C_m and C_d are the inertia and drag empirical coefficients describing the interaction between the fluid and a given body submerged in it, and $\mathbb{V}$ is the domain of the integral describing the range of possible values of height z.

2.3 Extreme Value Distribution and Mean Upcrossing Rate of a Stochastic Process

We are interested in the analytical determination of EVD of the stochastic processes, using information pertaining to their mean upcrossing rate.

The mean number of upcrossings of a level s of one stochastic process $\mathcal{X}(t)$ in the time interval $[0, T]$ is [12]:

$$\mathrm{E}[N_{\mathcal{X}}^{+}(s;[0,T])] = \int_{0}^{T} \nu_{\mathcal{X}}^{+}(x = s, t)\mathrm{d}t \tag{6}$$

where $\nu_{\mathcal{X}}^{+}(x, t)$ is the mean rate of upcrossings of the process, for ergodic stationary processes given by the Rice formula [13]:

$$\nu_{\mathcal{X}}^{+}(x, t) = \nu_{\mathcal{X}}^{+}(x) = \int_{0}^{+\infty} \dot{x} f_{\mathcal{X}\dot{\mathcal{X}}}(x, \dot{x}; t)\mathrm{d}\dot{x} \tag{7}$$

where $f_{\mathcal{X}\dot{\mathcal{X}}}(x, \dot{x}; t)$ is the joint PDF of the process and its time derivative $\frac{\mathrm{d}}{\mathrm{d}t}(\mathcal{X}) = \dot{\mathcal{X}}$ at time t.

For a sufficiently large band, the following relationship holds asymptotically:

$$\mathrm{EVD}[\mathcal{X}(t)]_{[0,T]} = F_{\mathcal{X}_m}(s;[0,T]) = \mathrm{Prob}\left[\mathcal{X}_m([0,T]) < s\right] = F_{\mathcal{X}}(s;0)\exp\left(-\lambda_{\mathcal{X}}^{+}(s;[0,T])\right) \tag{8}$$

where $\mathcal{X}_m$ denotes the extreme value random variable and $F_{\mathcal{X}}(s;0) = \mathrm{Prob}\left[\mathcal{X}(0) < s\right]$, which means the occurence of peaks is a Poisson process. This expression gives the Cumulative Distribution Function (CDF) of the occurrence of the extreme value under the assumptions that the occurrence of the upcrossings are statistically independent and rare, justifiable for sufficiently large values of the upcrossing level and sufficient wideness of the band width of the underlying process. From the above relations, considering the stationarity hypothesis and neglecting $F_{\mathcal{X}}(s;0)$, we obtain the simplified expression:

$$\mathrm{EVD}[\mathcal{X}(t)]_{[0,T]} \simeq \exp\left(-T\nu_{\mathcal{X}}^{+}(s,t)\right) \tag{9}$$

An alternative theoretical framework exists to address narrow band processes, shifting the focus instead to the analysis of the extreme values of the envelope of said process.

Of particular interest to the following study are the known results pertaining to Gaussian, stationary, zero mean processes with unit variance with respect to the mean rate of upcrossings. We have: $\mu_{\mathcal{X}}(t) = \mu_{\mathcal{X}}$, $\sigma_{\dot{\mathcal{X}}}^2(t) = \sigma_{\dot{\mathcal{X}}}^2$, $\mu_{\dot{\mathcal{X}}}(t) = 0$, $\sigma_{\mathcal{X}}^2(t) = \sigma_{\mathcal{X}}^2$ for stationary process and $\sigma_{\mathcal{X}\dot{\mathcal{X}}}(t) = \mathrm{E}[(\mathcal{X}(t) - \mu_{\mathcal{X}})(\dot{\mathcal{X}}(t) - \mu_{\dot{\mathcal{X}}})] = 0$ since $\mathcal{X}(t)$ is independent of $\dot{\mathcal{X}}(t)$, which leads to:

$$\nu_{\mathcal{X}}^{+}(x) = \frac{\sigma_{\dot{\mathcal{X}}}}{\sqrt{2\pi}} f_{\mathcal{X}}(x) \tag{10}$$

where $f_{\mathcal{X}}(x) = \frac{1}{\sqrt{2\pi}\sigma_{\mathcal{X}}}\exp\left(-\frac{(x-\mu_{\mathcal{X}})^2}{2\sigma_{\mathcal{X}}^2}\right)$ is the CDF of the Gaussian process.

3 Transformation of Stochastic Processes and Stochastic Fields

The theory of transformation of stochastic processeses or stochastic fields is based on the determination of the relationships between a base process whose characteristics are known, to those of a new process obtained by certain function. We call transformations that are independent of the time history of the processes to be "without memory". We introduce certain relationships concerning these type of transformation since they allow us to arrive at some important results later.

3.1 Integral Transformation of Squared Gaussian Stochastic Field

Of particular interest is the case of an integral transformation of the type:

$$\mathcal{Y}(t) = \int_{\mathbb{V}} a(z)\, \mathcal{V}^2(z,t)dz \tag{11}$$

with $a \in \mathbb{R}$. This transformation produces the base shear and the bending moment, with the appropriate coefficient, on a immersed profile under a fluid force dominated by the drag component of the Morison equation. As a direct extension of the previous relationships established, it is of interest to present the same type of link between the processes involved in this new transformation.

Some relevant results that will be required for the determination of the EVD of the former are presented now. From expanding Eq. 11, we obtain the mean of $\mathcal{Y}(t)$:

$$\mu_{\mathcal{Y}} = \textstyle\int_{\mathbb{V}} a(z) \left(\sigma^2_{\mathcal{V}(z)} + \mu^2_{\mathcal{V}(z)}\right) \mathrm{d}z, \tag{12}$$

and its covariance function:

$$\Sigma_{\mathcal{Y}\mathcal{Y}}(\tau) = \int_{\mathbb{V}^2} a(z_1)a(z_2) \left(2\Sigma^2_{\mathcal{V}(z_1)\mathcal{V}(z_2)}(\tau) + 4\mu_{\mathcal{V}(z_1)}\mu_{\mathcal{V}(z_2)}\Sigma_{\mathcal{V}(z_1)\mathcal{V}(z_2)}(\tau)\right) \mathrm{d}z_1\mathrm{d}z_2 \tag{13}$$

at which point we have the information, for stationary processes, to apply the Wiener-Khintchine theorem and arrive at the Power Spectral Density (PSD) from:

$$S_{\mathcal{Y}\mathcal{Y}}(f) = \mathrm{FT}\left[\Sigma_{\mathcal{Y}\mathcal{Y}}(\tau)\right] \tag{14}$$

where FT [•] denotes the Fourier Transform operator and f the frequency variable.

In synthesis, knowing the stochastic field describing the velocity of the fluid flow one can characterize some of the probabilistic quantities of the resulting bending moment and shear force with the specific aim of determining the EVD of the response process.

3.2 Translated Gaussian Process

Let $\mathcal{X}_g(t)$ be a Gaussian stationary random process with a null mean and $\hat{\mathcal{Y}}(t) = \mathrm{h}(\mathcal{X}_g(t))$ a non-necessarily Gaussian process, resulting from the transformation h. This is called a *translated Gaussian process*, a particular case of transformation [14]. Then, $\mathcal{X}_g(t)$ is called the Gaussian parent process of the translated process $\hat{\mathcal{Y}}(t)$. We note $F_{\mathcal{X}_g}(x)$ the CDF of the Gaussian process and $F_{\hat{\mathcal{Y}}}(y)$ the CDF of the non Gaussian process. The transformation is defined by:

$$h(x) = \left(F_{\hat{\mathcal{Y}}}^{-1} \circ F_{\mathcal{X}_g}\right)(x) = F_{\hat{\mathcal{Y}}}^{-1}\left(F_{\mathcal{X}_g}(x)\right) \tag{15}$$

The correlation function of the output stochastic process $\mathcal{Y}(t)$ as a function of the Gaussian parent process $\mathcal{X}_g(t)$ is given by:

$$R_{\hat{\mathcal{Y}}\hat{\mathcal{Y}}}(\tau) = \begin{cases} \displaystyle\int_{\mathbb{R}^2} h(x_1)h(x_2)f_{\mathcal{X}_g\mathcal{X}_g}(x_1, x_2; \tau)\mathrm{d}x_1\mathrm{d}x_2 & \text{if } \tau \neq 0 \\ \displaystyle\int_{\mathbb{R}} h^2(x)f_{\mathcal{X}_g}(x)\mathrm{d}x & \text{if } \tau = 0 \end{cases} \tag{16}$$

such that $R_{\hat{\mathcal{Y}}\hat{\mathcal{Y}}}(\tau) = \Sigma_{\hat{\mathcal{Y}}\hat{\mathcal{Y}}}(\tau) + \mu_{\hat{\mathcal{Y}}}^2$, with $f_{\mathcal{X}_g\mathcal{X}_g}(x_1, x_2; \tau)$ the joint PDF:

$$f_{\mathcal{X}_g\mathcal{X}_g}(x_1, x_2; \tau) = \frac{1}{2\pi\sigma_{\mathcal{X}_g}^2\sqrt{1 - \rho_{\mathcal{X}_g\mathcal{X}_g}^2(\tau)}} \exp\left(-\frac{x_1^2 + x_2^2 - 2x_1x_2\rho_{\mathcal{X}_g\mathcal{X}_g}(\tau)}{2\sigma_{\mathcal{X}_g}^2\left(1 - \rho_{\mathcal{X}_g\mathcal{X}_g}^2(\tau)\right)}\right) \tag{17}$$

for $\rho_{\mathcal{X}_g\mathcal{X}_g}(\tau) = \frac{\Sigma_{\mathcal{X}_g\mathcal{X}_g}(\tau)}{\sigma_{\mathcal{X}_g}^2}$ and:

$$f_{\mathcal{X}_g}(x) = \frac{1}{\sqrt{2\pi}\sigma_{\mathcal{X}_g}} \exp\left(-\frac{x^2}{2\sigma_{\mathcal{X}_g}^2}\right) \tag{18}$$

This leads to an evaluation of $S_{\hat{\mathcal{Y}}\hat{\mathcal{Y}}}(f) = \mathrm{FT}\left[\Sigma_{\hat{\mathcal{Y}}\hat{\mathcal{Y}}}(\tau)\right]$.

From another side, the marginal PDF of $\hat{\mathcal{Y}}(t)$ is [14]:

$$f_{\hat{\mathcal{Y}}}(y; t) = \sum_{i=1}^{n} \left|\frac{1}{h'(x_i(y))}\right| f_{\mathcal{X}_g}(x_i(y); t)$$

where $h'(x) = \frac{\partial h(x)}{\partial x}$ and $x_i(y)$ are the n real solutions of $x = h^{-1}(y)$, knowing that $n = 1$ if the transformation is monotone, in which case the above expression simplify to:

$$f_{\hat{\mathcal{Y}}}(y; t) = \left|\frac{1}{h'(h^{-1}(y))}\right| f_{\mathcal{X}_g}(h^{-1}(y); t)$$

Then, if $h(x)$ is monotone, the following result can be established based on the mean number of upcrossings of processes:

$$\nu_{\hat{\mathcal{Y}}}^{+}(s) = \nu_{\mathcal{X}_g}^{+}(s_g) = \frac{\sigma_{\dot{\mathcal{X}}_g}}{\sqrt{2\pi}} f_{\mathcal{X}_g}(s_g) \tag{19}$$

where s is the exceedence level or critical value of the non Gaussian process and $s_g = h^{-1}(s)$ is the level of exceedence of the Gaussian process.

Note that the Gaussian parent process does not, in general, coincide with the physical stochastic process of entry that generates the stochastic response. However, given a non Gaussian process, it is always possible to build the transformation given by Eq. 15, which in turn is useful in view of the result of Eq. 19 since it facilitates the calculation of the mean upcrossing rate and thus of the EVD.

4 Proposed Methodology

The strategy to find the EVD of $\mathcal{Y}_{\mathrm{M}}(t)$ and $\mathcal{Y}_{\mathrm{V}}(t)$ is as follows:

1. The reference CDF and PSD are obtained from the information of $\mathcal{V}$ using discrete results from references [15,16] and from the propagation of the moments through the integral monomial transformation described in Eq. 11: the processes are expressed as quadratic forms of a Gaussian distribution.
2. Solve the inverse problem to determine a Gaussian parent process $\mathcal{X}_g(t)$ which enables to approximate the reference PSD of $\mathcal{Y}_{\mathrm{M}}(t)$ and $\mathcal{Y}_{\mathrm{V}}(t)$ through the relations given in the previous section. This step is iterative in nature, starting from an initial guess for the PSD of $\mathcal{X}_g(t)$. Then, in case of a discrepancy, the PSD of $\mathcal{X}_g(t)$ needs to be adjusted, until $\hat{\mathcal{Y}}_{\mathrm{M}}(t) = \mathcal{Y}_{\mathrm{M}}(t)$ and $\hat{\mathcal{Y}}_{\mathrm{V}}(t) = \mathcal{Y}_{\mathrm{V}}(t)$ at convergence. If we note $\mathcal{Y}_{\bullet}(t)$ a generic process to represent $\mathcal{Y}_{\mathrm{M}}(t)$ and $\mathcal{Y}_{\mathrm{V}}(t)$ in a generic notation, the proposed algorithm is:

   (a) Let $k = 0$ and $S^{(0)}_{\mathcal{X}_g\mathcal{X}_g}(f)$ an initial PSD for $\mathcal{X}^{(0)}_g(t)$;

   (b) Evaluate $h^{(k)}(x) = F^{-1}_{\mathcal{Y}_{\bullet}}\left(F_{\mathcal{X}^{(k)}_g}(x)\right)$;

   (c) Evaluate $S^{(k)}_{\hat{\mathcal{Y}}_{\bullet}\hat{\mathcal{Y}}_{\bullet}}(f)$ from relation (16);

   (d) If $S^{(k)}_{\hat{\mathcal{Y}}_{\bullet}\hat{\mathcal{Y}}_{\bullet}}(f) \neq S_{\mathcal{Y}_{\bullet}\mathcal{Y}_{\bullet}}(f)$, let:

   $$S^{(k+1)}_{\mathcal{X}_g\mathcal{X}_g}(f) = \frac{S_{\mathcal{Y}_{\bullet}\mathcal{Y}_{\bullet}}(f)}{S^{(k)}_{\hat{\mathcal{Y}}_{\bullet}\hat{\mathcal{Y}}_{\bullet}}(f)} S^{(k)}_{\mathcal{X}_g\mathcal{X}_g}(f),$$

   $k \leftarrow k + 1$

   and go to step (b).

   In practice, it appears that this strategy is satisfactory as long as not too tight convergence tolerance for $S^{(k)}_{\hat{\mathcal{Y}}_{\bullet}\hat{\mathcal{Y}}_{\bullet}}(f)$ is requested. When necessary, more ellaborate solving strategies can be used, as long as negative values are not allowed for $S^{(0)}_{\mathcal{X}_g\mathcal{X}_g}(f)$. This problem being non-linear, an iterative solver is carried out and the solution of the above algorithm serves as good starting guess.

3. Using the theory of translated Gaussian processes sketched in Sect. 3.2, and with the CDF of step 1 and frequency-domain information of step 2, we compute the mean upcrossing rate for the corresponding parent Gaussian distribution. From here, only the variance of the velocity of this parent Gaussian process $\mathcal{X}_g(t)$ is necessary to compute the EVD of the targeted quantities (Eq. 9).

5 Numerical Application

In this section we make a numerical implementation of the described methodology. We use the physical, geometrical and discretization parameters collected in Table 1. The mean velocity is expressed as a polynomial of degree 4, to fit experimental data observed in reference [17]:

$$\mu_{\mathcal{V}}(z) = -4.414 \times 10^{-6} z^4 + 3.312 \times 10^{-4} z^3 - 8.350 \times 10^{-3} z^2 + 9.700 \times 10^{-2} z + 9.215 \times 10^{-1} [\text{m/s}] \quad (20)$$

and the variance is expressed as:

$$\sigma_{\mathcal{V}\mathcal{V}}(z) = TI^2 \mu^2{}_{\mathcal{V}}(0) \left(\frac{z}{H_{sea}}\right)^{\frac{1}{\alpha}} \quad (21)$$

where TI is known as the turbulence index of the flow and α is an empirical parameter employed to fit the velocity profile, as in reference [17], given measured data. The polynomial approximation is plotted in Fig. 2, where it should be noted that the reference employed is from the seabed, keeping the conventions of references [17,18].

Table 1. Physical properties and discretization parameters for the numerical simulation

Property	Value	Property	Value
ρ_f	1030 $[kg.m^{-3}]$	H_{sea}	40 $[m]$
D	3.5 $[m]$	H_{pile}	30 $[m]$
C_{hydro}	1.15	α	7.1
A	$\frac{\pi D_{pile}}{2} [m^2]$	TI	$\frac{25}{100}$
dH	7.5 $[m]$	L	$\frac{H_{sea}}{3} [m]$
$N_{processes}$	4	$T_{1,2}$	5 min = 300 $[s]$; 3 h = 10800 $[s]$
$N_{realizations}$	2000	dT	0.5 $[s]$

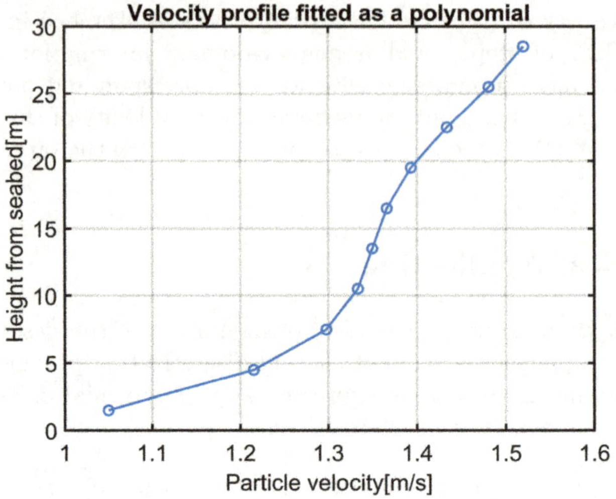

Fig. 2. Velocity profile of flow as function of height from sea

The spatial and temporal components of the correlation are modelled in the following way:

$$\begin{aligned} \rho^{\text{space}}(z_1, z_2) &= (1 - (\tfrac{r}{L})^2) \exp -(\tfrac{r}{L})^2 \quad \text{Spatial correlation} \\ \rho^{\text{time}}(\tau) &= \text{FT}^{-1}[S_{\mathcal{VV}}(f)] \quad \text{Temporal correlation} \end{aligned} \tag{22}$$

where r is the separation between two points of consideration, L is the correlation length. Concerning $S_{\mathcal{VV}}(f)$, we have constructed it such that the frequency content decays with a slope of $m = -\frac{5}{3}$ from the vicinity of $f = \frac{1}{100}$[Hz], so as to fit the measured data presented in reference [19]. The expression reads:

$$S_{\mathcal{VV}}(f) = \frac{1}{f^4} \exp\left(-\frac{A}{f^{\frac{5}{3}}}\right) [\frac{\text{m}^2}{\text{s}^2}\text{H}z^{-1}]; A = 8 \times 10^{-4} \tag{23}$$

The spectrum of the fitted PSD is plotted in Fig. 3 in a logarithmic scale, where the starting frequency and the slope can be seen clearly.

In Table 1, the parameters $N_{\text{processes}}$ and dH are related to the spatial discretization of the problem, the former being the size of the stochastic process vector used to describe the velocity random field and the latter being the height of each discrete section of the pile. This particular discretization choice has been made based on the results of a convergence analysis, from which it is apparent that the statistical properties of the EVD do not change significantly for a higher number of $N_{\text{processes}}$.

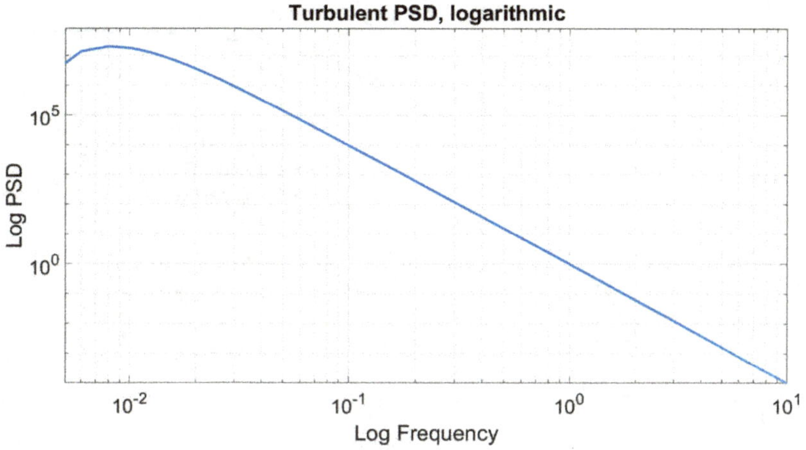

Fig. 3. Power spectral density of turbulent velocity field

First, an initial short-time Monte-Carlo (MC) simulation is carried out and the EVD of resulting bending moment and base shear are compared with the proposed methodology. Reference statistics are computed using a standard empirical distribution function utility from the software *Matlab* (function ECDF), and direct upcrossings are counted from the stochastic process generated. The results of a second simulation for $T = 3$ h (10800 s) are presented too, this range of time being of particular interest since it is a design reference case as it is around the longest duration of time for which it is conventionally assumed that the input stochastic process can be considered stationary.

In Fig. 4 we notice an important discrepancy between the proposed methodology and the MC/ECDF solution for both the bending moment and shear distributions. The method of MC/counting of upcrossings also departs significantly from the MC/ECDF. This discrepancy is associated with the so-called Poisson approximation of the distribution of peaks of the stochastic process, in which both approaches are based. The shortcomings of this model are discussed at length in reference [13], and the main causes for the deviation are the bandwidth and the upcrossing level considered. However, one can infer that the longer the observation time of the process, the more likely it would be for the response process to arrive at higher upcrossing levels, which would result in the conditional terms in Eq. 8 being non neglibible.

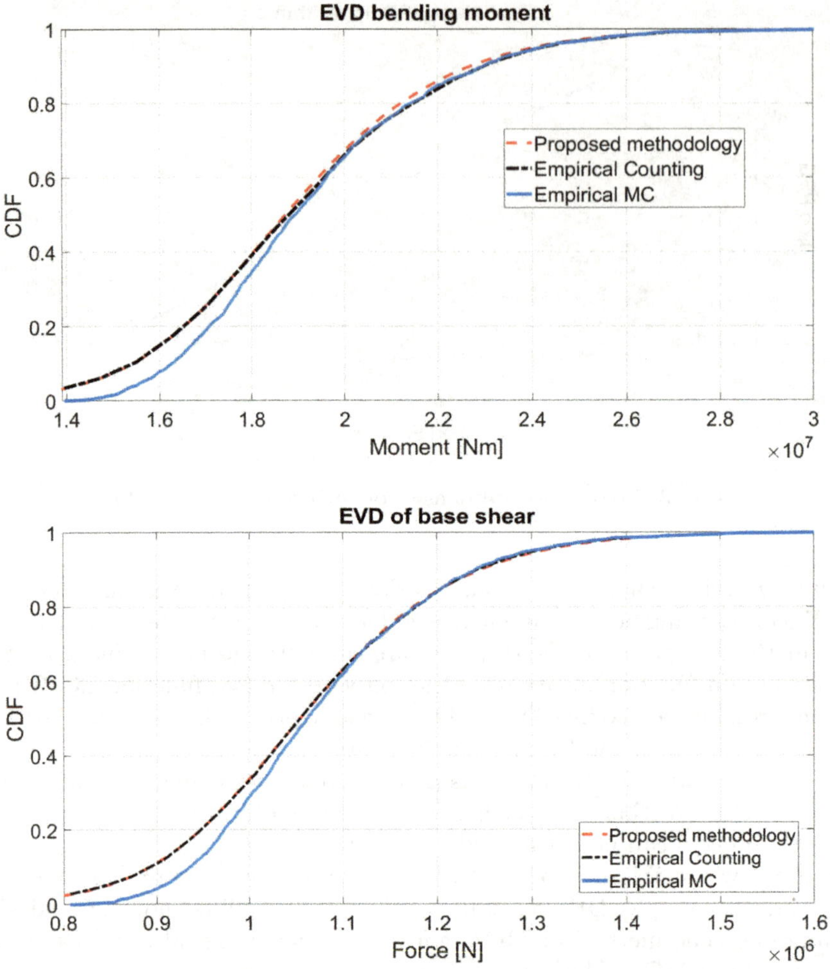

Fig. 4. EVD of bending moment and shear force, $T = 5$ min simulation

The result of the simulation with longer observation time is presented in Fig. 5. For $T = [0, 10800]$ s, the proposed methodology is in good agreement for both output processes.

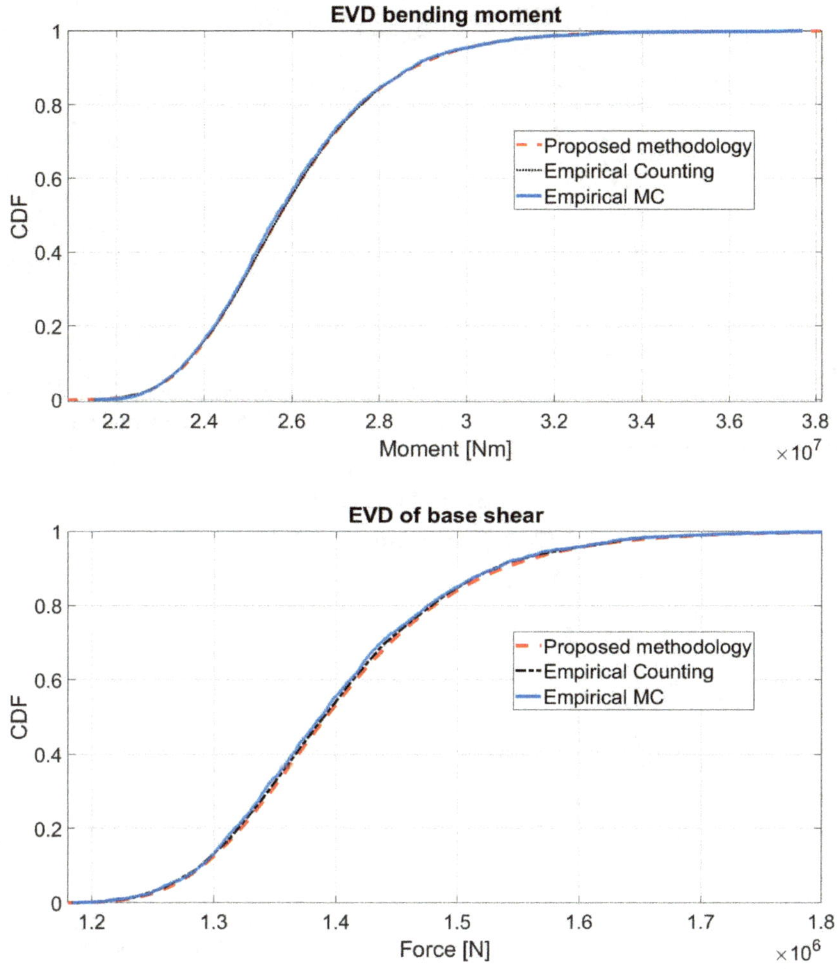

Fig. 5. EVD of bending moment and shear force, $T = 3$ h simulation

To close this section of results, in Fig. 6 we show a comparison of a widely used heuristic method, the SRSS as described in sources like [12], with the results provided by the proposed methodology. As expected, the accuracy of the SRSS for correlated processes is extremely limited.

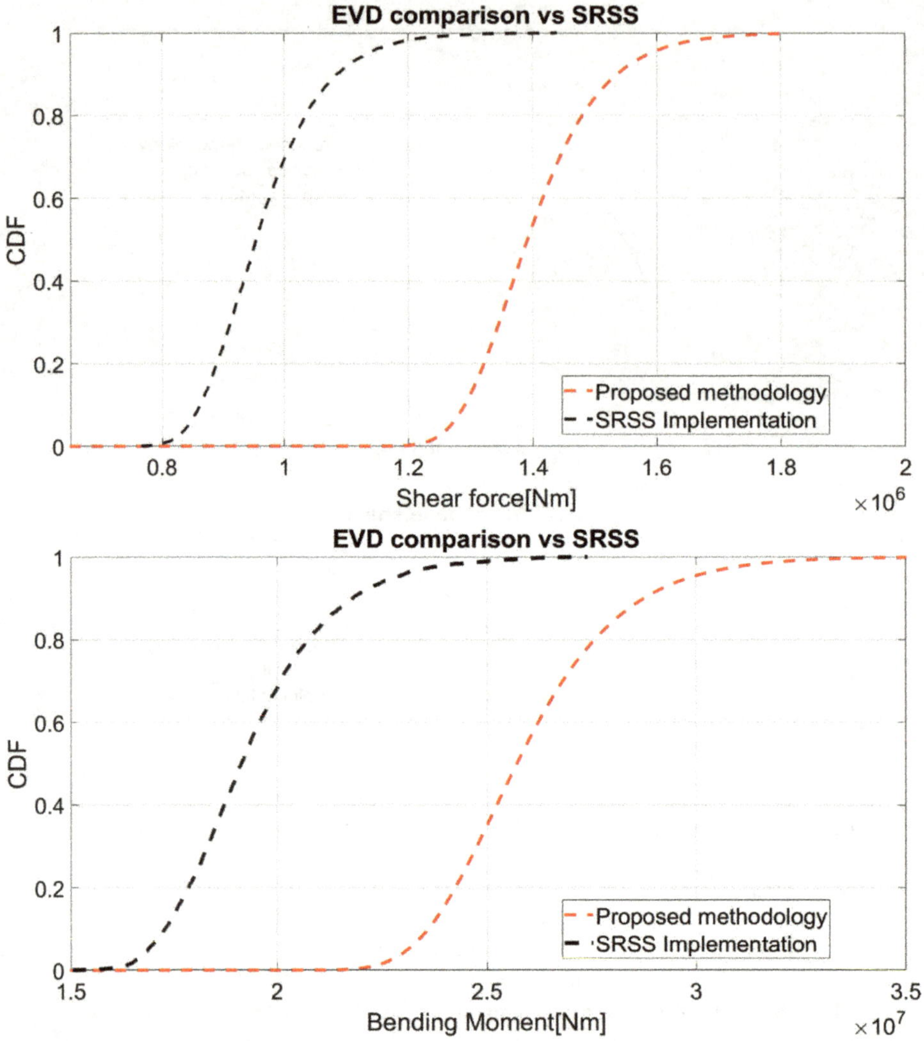

Fig. 6. Comparison of the proposed methodology with an implementation of the SRSS method

6 Conclusions

We propose a methodology to the problem of the determination of the extreme value distribution (EVD) of immerged structure subject to random loading under quasi-static conditions as described by the drag component of the Morison equation. The proposed methodology is analytical, and rests over the theory of translated Gaussian processes and the results of quadratic forms of Gaussian variables from the literature. The methodology has been implemented in the numerical language Matlab, and has been compared to two benchmark approaches based on

Monte-Carlo simulation. When the time scale is chosen to allow good resolution of the frequencies excited on the output stochastic processes, the methodology is in good agreement with the other methods throughout all the regions of the EVD curve. In the case of disparities between the time scale and the excited frequencies, the methodology features some discrepancies with the empirical EVD up to the 2nd quartile of the EVD curve, but the curves converge after this point and thus the proposed methodology produces, even for these conditions, a useful approximation for certain domains pertaining to design and reliability. The source of this discrepancy can be traced to an underlying assumption of independence of the peaks of the process whose EVD is sought, which breaks if the time scale is too short (and presumably if the level of upcrossing is too low).

The problem addressed here can be presented as a limiting case of the EVD of a process that is the weighted sum of n input stochastic processes, with no general limitation to being Gaussian nor imposing limitations on the band-width. Concerning this associated problem, the methods found in the bibliography part from constraints that may render them inapplicable to certain situations and have very limited success. The known Tukstra rule has limited results as exposed by [20], and the heuristic approach known as SRSS (square root of sum of squares) [12] is most effective when the processes that make up the sum are statistically independent. In the mesure that this last condition is not met, the method underestimates the EVD. In this respect, the proposed methodology seems to offer a viable alternative to these approaches. Moreover, the methodology has the advantage of not being limited to narrow-band or wide-band processes exclusively, as many of the methodologies reviewed in the bibliography, and it does not rest over Monte-Carlo methods.

Future work would be focused on extending this methodology to a new model of peak distributions of the output process, the one induced by the inertial component of the Morison equation being the logical continuation. It could also be extended to non Gaussian input processes, which would require to replace the results derived from the study of quadratic forms of Gaussian variable by methods like polynomial chaos. Additionally, a methodology analogous to the one presented here could be applied to study the dynamic response of offshore structures making use of the appropriate transformation, a so-called transformation with memory.

References

1. Grigoriu, M.: Extremes of wave forces. J. Eng. Mech. **110**(12), 1731–1742 (1984)
2. Brouwers, J., Verbeek, P.: Expected fatigue damage and expected extreme response for morison-type wave loading. Appl. Ocean Res. **5**(3), 129–133 (1983)
3. Naess, A.: Prediction of extremes of morison-type loading–an example of a general method. Ocean Eng. **10**(5), 313–324 (1983)
4. Moe, G.: Long-term wave-force statistics for a vertical pile. Coastal Eng. **2**, 297–311 (1978)
5. Najafian, G., Burrows, R.: Probabilistic modelling of quasi-static response of offshore structures subject to nonlinear wave loading: two approximate approaches. Appl. Ocean Res. **16**(4), 205–221 (1994)

6. Tromans, P., Hagemeijer, P., Wassink, H.: The statistics of the extreme response of offshore structures. Ocean Eng. **19**(2), 161–181 (1992)
7. Moan, T., Zheng, X.Y., Quek, S.T.: Frequency-domain analysis of non-linear wave effects on offshore platform responses. Int. J. Non-Linear Mech. **42**(3), 555–565 (2007)
8. Jensen, J.J.: Extreme value predictions and critical wave episodes for marine structures by FORM. Ships Offshore Struct. **3**(4), 325–333 (2008)
9. Jensen, J.J.: Extreme value predictions using monte carlo simulations with artificially increased load spectrum. Probabilistic Eng. Mech. **26**(2), 399–404 (2011)
10. Naess, A.: A study of linear combination of load effects. J. Sound Vibrat. **129**(1), 83–98 (1989)
11. Morison, J., Johnson, J., Schaaf, S.: The force exerted by surface waves on piles. J. Petroleum Technol. **2**(05), 149–154 (1950)
12. Naess, A., Moan, T.: Stochastic Dynamics of Marine Structures. Cambridge University Press, Cambridge (2009)
13. Lutes, L.D., Sarkani, S.: Random Vibrations: Analysis of Structural and Mechanical Systems. Butterworth-Heinemann, Oxford (2004)
14. Grigoriu, M.: Crossings of non-gaussian translation processes. J. Eng. Mech. **110**(4), 610–620 (1984)
15. Kotz, S., Johnson, N.L., Boyd, D.W.: Series representations of distributions of quadratic forms in normal variables. i. central case. Ann. Math. Stat. **38**(3), 823–837 (1967)
16. Kotz, S., Johnson, N.L., Boyd, D.W.: Series representations of distributions of quadratic forms in normal variables II non-central case. Ann. Math. Stat. **38**(3), 838–848 (1967)
17. Lewis, M., Neill, S., Robins, P., Hashemi, M., Ward, S.: Characteristics of the velocity profile at tidal-stream energy sites. Renew. Energy **114**, 258–272 (2017)
18. Jeffcoate, P., Starzmann, R., Elsaesser, B., Scholl, S., Bischoff, S.: Field measurements of a full scale tidal turbine. Int. J. Marine Energy **12**, 3–20 (2015)
19. Thomson, J., Polagye, B., Durgesh, V., Richmond, M.C.: Measurements of turbulence at two tidal energy sites in puget sound WA. IEEE J. Oceanic Eng. **37**(3), 363–374 (2012)
20. Naess, A., Gaidai, O., Batsevych, O.: Extreme value statistics of combined load effect processes. Struct. Safety **31**(4), 298–305 (2009)

Statistical Analysis of Biological Models with Uncertainty

Vicent Bevia[1], Juan-Carlos Cortés[1], Ana Navarro-Quiles[2](✉), and Jose-Vicente Romero[1]

[1] Instituto Universitario de Matemática Multidisciplinar, Universitat Politècnica de València, Camino de Vera s/n, 46022 Valencia, Spain
vibees@posgrado.upv.es,
{jccortes,jvromero}@imm.upv.es

[2] Department of Statistics and Operational Research, Universitat de València, Dr. Moliner 50, 46100 Burjassot, Spain
ana.navarro@uv.es

Abstract. In this contribution relevant biological models, based on random differential equations, are studied. For the sake of generality, we assume that the initial condition and the biological model parameters are dependent random variables with arbitrary probability distributions. We present a general methodology that enables us to provide a full probabilistic description of the solution stochastic process for each stochastic model. The statistical analysis is performed through the calculation of the first probability function by applying the random variable transformation technique. From the first probability density function, we can calculate any one-dimensional moment of the solution, including the mean and the variance as important particular cases. Our theoretical findings are applied to describe the probabilistic dynamics of *Spirulina* sp. biomass production in a particular medium using real data.

Keywords: Uncertainty · Biological models · First probability density function · Random variable transformation method · Real data

1 Introduction and Motivation

Mathematical models based on differential equations aim at describing real situations under variable conditions. In particular, the dynamic behaviour of a certain biological population, whose number of individuals varies over the time, can be modelled via differential equations. Growth models are fundamental to analyse the dynamics of a population, and they permit to construct reliable predictions. In the available literature, we can find many and diverse growth model representations, with a wide range of deterministic curves describing the dynamics followed by the population under study. Originally, growth models intended to describe the evolution of animal populations. The Exponential and the Logistic growth models, proposed by Thomas Malthus and Pierre-François Verhulst,

J. E. S. De Cursi (Ed.): Uncertainties 2020, LNME, pp. 323–336, 2021.
https://doi.org/10.1007/978-3-030-53669-5_23

respectively, are cornerstone models in the realm of Dynamics Population. In [1], Malthus proposed the principle that affirmed that human population grew exponentially while food production did so at arithmetic reason. In this way, he postulated that we would reach a shortage of resources and thus presented the alert about our possible extinction. This behaviour of the population can be modelled by the following initial value problem (IVP)

$$\begin{cases} n'_{\mathrm{E}}(t) &= r n_{\mathrm{E}}(t), \\ n_{\mathrm{E}}(t_0) &= n^0_{\mathrm{E}}, \end{cases} \tag{1}$$

where the growth of population, $n_{\mathrm{E}}(t)$, is determined by the growth rate parameter $r > 0$, which represents the difference between birth and mortality rates. The constant n^0_{E} denotes the initial amount of population at the time instant t_0. The differential equation in IVP (1) is known as the *Malthusian* differential equation and its solution is given by $n_{\mathrm{E}}(t) = n^0_{\mathrm{E}}\, \mathrm{e}^{r(t-t_0)}$. Therefore, Malthus maintained that populations had unbounded growth, whose ratio depended on population density. However, this model is only suitable in ideal conditions since, in nature, any growth tends to stabilize. In fact, in real life situations, populations do not increase indefinitely but during a finite period. When $n_{\mathrm{E}}(t)$ grows, factors causing a brake in the growth rate begin to emerge. This leads to consider alternative and more realistic models whose long-term behaviour is bounded, such as Logistic and Gompertz models. To overcome the above-mentioned drawback of Malthusian model, in 1837 Verhulst proposed the Logistic model [2], where it is intended to mathematically model the effects of competition between individuals for survival taking into account the carrying capacity of the medium. The Logistic differential equation is formulated via the following IVP

$$\begin{cases} n'_{\mathrm{L}}(t) &= a n_{\mathrm{L}}(t) \left(1 - \frac{n_{\mathrm{L}}(t)}{k}\right), \\ n_{\mathrm{L}}(t_0) &= n^0_{\mathrm{L}}, \end{cases} \tag{2}$$

where $a > 0$ denotes the growth input and $k > 0$ represent the limit population, i.e., the maximum number of individuals, also termed the carrying capacity. The solution of IVP (2) is given by

$$n_{\mathrm{L}}(t) = \frac{k n^0_{\mathrm{L}}}{(k - n^0_{\mathrm{L}})\, \mathrm{e}^{-a(t-t_0)} + n^0_{\mathrm{L}}}.$$

Then, the growth begins with an exponential tendency and, from a certain moment, it begins to glimpse a logarithmic tendency (sigmoidal shape). Finally, we introduce the Gompertz growth model that belongs to the family of sigmoidal curves, as the Logistic model. These models are initially concave and from one point they become convex. The Logistic and the Gompertz models have a similar growth behaviour, although there are subtle differences between them. The main one is the location of the inflection point, located between the 35% and 40% growth in Gompertz models and 50% in Verhlust models. Gompertz model, [3], is formulated by the IVP

$$\begin{cases} n'_{\mathrm{G}}(t) &= n_{\mathrm{G}}(t)(c - b\ln(n_{\mathrm{G}}(t))), \\ n_{\mathrm{G}}(t_0) &= n^0_{\mathrm{G}}, \end{cases} \tag{3}$$

where $b > 0$ and $c > 0$ represent the growth rates of the system. The solution of IVP (3) is given by

$$n_{\rm G}(t) = {\rm e}^{\dfrac{c + {\rm e}^{-b(t-t_0)}(b\ln(n_{\rm G}^0) - c)}{b}}.$$

These three models have been extensively applied in numerous problems to describe the dynamics of quantities of interest, see for example [4–6]. As it can be observed, these models depend on specific parameters which need to be set by using sample data, thus they contain uncertainty stemming from measurement errors. Furthermore, there are some external sources that may affect the biological system under study, such as the weather, predators, the shortage of natural resources, etc. These facts make more appropriate to treat model parameters as random variables or stochastic processes rather than deterministic constants or functions, respectively.

The treatment of randomness in the setting of differential equations has been mainly conducted by assuming that uncertainty has specific probabilistic patterns (for instance, in dealing with Itô-type differential equations randomness is assumed to be Gaussian) and then computing the first moments (mean and variance) of the solution stochastic process. Examples of this stochastic approach for the three aforementioned growth models can be found, for instance, in references [7–11]. In all these contributions, apart from obtaining the solution stochastic process, say $N(t,\omega)$, a major goal is to determine its main statistical functions, such as the mean, $\mathbb{E}[N(t,\omega)]$, and the variance, $\mathbb{V}[N(t,\omega)]$. However, a more ambitious target is the computation of its first probability density function (1-PDF), $f(n,t)$. From the 1-PDF, we can be calculated all the one-dimensional statistical moments

$$\mathbb{E}\left[(N(t,\omega))^k\right] = \int_{\mathbb{R}} n^k f(n,t) {\rm d}n, \quad k = 1, 2, \ldots,$$

In particular, the mean and the variance functions can be obtained

$$\mathbb{E}\left[N(t,\omega)\right] = \int_{\mathbb{R}} n f(n,t) {\rm d}n, \quad \mathbb{V}\left[N(t,\omega)\right] = \int_{\mathbb{R}} n^2 f(n,t) {\rm d}n - \left(\mathbb{E}\left[N(t,\omega)\right]\right)^2. \tag{4}$$

In this work, the random variable transformation (RVT) method will be applied to obtain the 1-PDF of the solution of each randomized IVP (1)–(3) (see later (7)–(9), respectively). RVT is a powerful technique that permits the computation of the PDF of a random vector which is obtained after mapping another random vector whose PDF is known, see [12,13]. For the sake of clarity, we state in Theorem 1 the Multidimensional RVT method.

Theorem 1 (Multidimensional Random Variable Transformation method, [14]**).** *Let us consider* $\mathbf{U}(\omega) = (U_1(\omega), \ldots, U_n(\omega))^\top$ *and* $\mathbf{V}(\omega) = (V_1(\omega), \ldots, V_n(\omega))^\top$ *two n-dimensional absolutely continuous random vectors defined on a probability space* $(\Omega, \mathfrak{F}, \mathbb{P})$. *Let* $\mathbf{g} : \mathbb{R}^n \to \mathbb{R}^n$ *be a one-to-one* *deterministic transformation of* $\mathbf{U}$ *onto* $\mathbf{V}$*, i.e.,* $\mathbf{V} = \mathbf{g}(\mathbf{U})$*. Assume that* $\mathbf{g}$ *is continuous in* $\mathbf{U}$ *and has continuous partial derivatives with respect to each* U_i*,* $1 \le i \le n$*. Then, if* $f_{\mathbf{U}}(\mathbf{u})$ *denotes the joint probability density function of random vector* $\mathbf{U}(\omega)$*, and* $\mathbf{h} = \mathbf{g}^{-1} = (h_1(v_1, \ldots, v_n), \ldots, h_n(v_1, \ldots, v_n))^\top$ *represents the inverse mapping of* $\mathbf{g} = (g_1(u_1, \ldots, u_n), \ldots, g_n(u_1, \ldots, u_n))^\top$*, the joint probability density function of vector* $\mathbf{V}(\omega)$ *is given by*

$$f_{\mathbf{V}}(\mathbf{v}) = f_{\mathbf{U}}(\mathbf{h}(\mathbf{v}))\,|J|\,, \tag{5}$$

where $|J|$*, which assumed to be different from zero, is the absolute value of the Jacobian defined by the determinant*

$$J = \det\left(\frac{\partial \mathbf{u}^\top}{\partial \mathbf{v}}\right) = \det\begin{pmatrix} \frac{\partial h_1(v_1,\ldots,v_n)}{\partial v_1} & \cdots & \frac{\partial h_n(v_1,\ldots,v_n)}{\partial v_1} \\ \vdots & \ddots & \vdots \\ \frac{\partial h_1(v_1,\ldots,v_n)}{\partial v_n} & \cdots & \frac{\partial h_n(v_1,\ldots,v_n)}{\partial v_n} \end{pmatrix}. \tag{6}$$

Thus, in this contribution, we consider the following randomized IVPs

$$\textbf{Exponential:} \begin{cases} N_{\mathrm{E}}'(t,\omega) = R(\omega) N_{\mathrm{E}}(t,\omega), \\ N_{\mathrm{E}}(t_0,\omega) = N_{\mathrm{E}}^0(\omega). \end{cases} \tag{7}$$

$$\textbf{Logistic:} \begin{cases} N_{\mathrm{L}}'(t,\omega) = A(\omega) N_{\mathrm{L}}(t,\omega)\left(1 - \frac{N_{\mathrm{L}}(t,\omega)}{K(\omega)}\right), \\ N_{\mathrm{L}}(t_0,\omega) = N_{\mathrm{L}}^0(\omega). \end{cases} \tag{8}$$

$$\textbf{Gompertz:} \begin{cases} N_{\mathrm{G}}'(t,\omega) = N_{\mathrm{G}}(t,\omega)(C(\omega) - B(\omega)\ln(N_{\mathrm{G}}(t,\omega))), \\ N_{\mathrm{G}}(t_0,\omega) = N_{\mathrm{G}}^0(\omega). \end{cases} \tag{9}$$

In IVPs (7)–(9) uncertainty is introduced by randomizing all the inputs parameters and the initial condition of deterministic IVPs (1)–(3), respectively. That is, parameters $N_{\mathrm{E}}^0(\omega)$, $N_{\mathrm{L}}^0(\omega)$, $N_{\mathrm{G}}^0(\omega)$, $R(\omega)$, $A(\omega)$, $K(\omega)$, $C(\omega)$ and $B(\omega)$ are assumed to be absolutely continuous random variables defined in a common complete probability space $(\Omega, \mathcal{F}, \mathbb{P})$. In order to provide as much generality as possible, we do not assume independence between random model parameters involved in each IVP (7)–(9). Then, henceforth $f_{N_{\mathrm{E}}^0,R}(n^0, r)$, $f_{N_{\mathrm{L}}^0,A,K}(n^0, a, k)$ and $f_{N_{\mathrm{G}}^0,C,B}(n^0, c, b)$ will denote the joint PDF of random vector parameter involved in each IVP (7)–(9). For convenience in the notation, throughout this paper the exponential of x will denoted indistinctly by $\exp(x)$ or e^x.

This paper is organized as follows. In Sect. 2, the 1-PDF of the corresponding solution stochastic process to the Exponential, the Logistic and he Gompertz model is determined by applying the RVT method. The applicability and usefulness of this technique is shown in Sect. 3. In Sect. 3, we apply the theoretical results obtained in Sect. 2 to modelling the dynamics of the Spirulina sp. kinetic growth in a particular medium using real data. Finally, conclusions are drawn in Sect. 4.

2 Computing the 1-PDF

In this section, the 1-PDFs, $f_{\mathrm{E}}(n,t)$, $f_{\mathrm{L}}(n,t)$ and $f_{\mathrm{G}}(n,t)$, of the solutions stochastic processes, $N_{\mathrm{E}}(t,\omega)$, $N_{\mathrm{L}}(t,\omega)$ and $N_{\mathrm{G}}(t,\omega)$, respectively, are determined. To this end, the RVT method is applied in each case by choosing an adequate transformation in Theorem 1. For the sake of clarity in the presentation, we follow the same structure in the resolution of each model.

2.1 Exponential Growth Model

The solution of the random IVP (7) is obtained randomizing its deterministic counterpart

$$N_{\mathrm{E}}(t,\omega) = N_{\mathrm{E}}^{0}(\omega)\,\mathrm{e}^{R(\omega)(t-t_0)}.$$

Let $f_{N_{\mathrm{E}}^0,R}(n^0,r)$ be the joint PDF of random vector $(N_{\mathrm{E}}^0(\omega), R(\omega))$, we apply RVT technique to compute the 1-PDF of the solution stochastic process $N_{\mathrm{E}}(t,\omega)$. Now, fixing $t \geq t_0$, we apply Theorem 1 with the following choice

$$\mathbf{U}(\omega) = (N_{\mathrm{E}}^0(\omega), R(\omega))^\top, \quad \mathbf{V}(\omega) = (X(\omega), Y(\omega))^\top,$$
$$\mathbf{g}: \mathbb{R}^2 \to \mathbb{R}^2, \quad \mathbf{g}(n_{\mathrm{E}}^0, r) = (g_1(n_{\mathrm{E}}^0, r), g_2(n_{\mathrm{E}}^0, r))^\top = (x,y)^\top,$$

being

$$x = n_{\mathrm{E}}^0\,\mathrm{e}^{r(t-t_0)}, \quad y = r.$$

Isolating n_{E}^0 and r, the inverse mapping $\mathbf{h}: \mathbb{R}^2 \to \mathbb{R}^2$ is

$$n_{\mathrm{E}}^0 = h_1(x,y) = x\,\mathrm{e}^{-y(t-t_0)}, \quad r = h_2(x,y) = y.$$

The absolute value of the Jacobian of the inverse mapping $\mathbf{h}$ is

$$\begin{aligned} |J| &= \left| \det \begin{pmatrix} \frac{\partial h_1}{\partial x}(x,y) & \frac{\partial h_1}{\partial y}(x,y) \\ \frac{\partial h_2}{\partial x}(x,y) & \frac{\partial h_2}{\partial y}(x,y) \end{pmatrix} \right| \\ &= \left| \det \begin{pmatrix} \mathrm{e}^{-y(t-t_0)} & -x(t-t_0)\mathrm{e}^{-y(t-t_0)} \\ 0 & 1 \end{pmatrix} \right| = \mathrm{e}^{-y(t-t_0)} > 0. \end{aligned}$$

Thus, according to Theorem 1 the joint PDF of the random vector $(X(\omega), Y(\omega))$ is

$$f_{X,Y}(x,y) = f_{N_{\mathrm{E}}^0,R}\left(x\,\mathrm{e}^{-y(t-t_0)}, y\right)\mathrm{e}^{-y(t-t_0)}.$$

Marginalizing with respect to $Y(\omega) = R(\omega)$ and taking $t \geq t_0$ arbitrary, the 1-PDF of the solution $N_{\mathrm{E}}(t,\omega)$ is given by

$$f_{\mathrm{E}}(n,t) = \int_{\mathbb{R}} f_{N_{\mathrm{E}}^0,R}\left(n\,\mathrm{e}^{-r(t-t_0)}, r\right)\mathrm{e}^{-r(t-t_0)}\,\mathrm{d}r. \tag{10}$$

2.2 Logistic Growth Model

The solution of the random IVP (8) is

$$N_{\mathrm{L}}(t,\omega)=\frac{K(\omega)N_{\mathrm{L}}^{0}(\omega)}{(K(\omega)-N_{\mathrm{L}}^{0}(\omega))\,\mathrm{e}^{-A(\omega)(t-t_0)}+N_{\mathrm{L}}^{0}(\omega)},\quad t\geq t_0.$$

Let us assume that the joint PDF of random vector $(N_{\mathrm{L}}^{0}(\omega),A(\omega),K(\omega))$, $f_{N_{\mathrm{L}}^{0},A,K}(n^0,a,k)$ is known. Then, we apply the RVT technique to compute the 1-PDF of the solution $N_{\mathrm{L}}(t,\omega)$ in terms of $f_{N_{\mathrm{L}}^{0},A,K}(n^0,a,k)$. Fixing $t\geq t_0$, we apply Theorem 1 with the following choice

$$\mathbf{U}(\omega)=(N_{\mathrm{L}}^{0}(\omega),A(\omega),K(\omega))^{\top},\quad \mathbf{V}(\omega)=(X(\omega),Y(\omega),Z(\omega))^{\top},$$

$$\mathbf{g}:\mathbb{R}^3\to\mathbb{R}^3,\quad \mathbf{g}(n_{\mathrm{L}}^{0},a,k)=(g_1(n_{\mathrm{L}}^{0},a,k),g_2(n_{\mathrm{L}}^{0},a,k),g_3(n_{\mathrm{L}}^{0},a,k))^{\top}=(x,y,z)^{\top},$$

where

$$x=\frac{kn_{\mathrm{L}}^{0}}{(k-n_{\mathrm{L}}^{0})\,\mathrm{e}^{-a(t-t_0)}+n_{\mathrm{L}}^{0}},\quad y=a,\quad z=k.$$

The inverse mapping $\mathbf{h}:\mathbb{R}^3\to\mathbb{R}^3$ of $\mathbf{g}$ is obtained isolating n_{L}^{0}, a and k

$$n_{\mathrm{L}}^{0}=h_1(x,y,z)=\frac{xz}{x+\mathrm{e}^{y(t-t_0)}(z-x)},\quad a=h_2(x,y,z)=y,\quad k=h_3(x,y,z)=z.$$

Then, the joint PDF of the random vector $\mathbf{V}(\omega)$ is

$$f_{X,Y,Z}(x,y,z)=f_{N_{\mathrm{L}}^{0},A,K}\left(\frac{xz}{x+\mathrm{e}^{y(t-t_0)}(z-x)},y,z\right)\frac{\mathrm{e}^{y(t+t_0)}\,z^2}{(x\,\mathrm{e}^{yt_0}+\mathrm{e}^{yt}(z-x))^2},$$

where $\frac{\mathrm{e}^{y(t+t_0)}\,z^2}{(x\,\mathrm{e}^{yt_0}+\mathrm{e}^{yt}(z-x))^2}>0$ is the absolute value of the Jacobian of mapping $\mathbf{h}$. As the previous development is valid for every $t\geq t_0$, the 1-PDF of the solution $N_{\mathrm{L}}(t,\omega)$ is obtained the $(Y(\omega),Z(\omega))=(A(\omega),K(\omega))$-marginal of the joint PDF $f_{X,Y,Z}(x,y,z)$. This yields

$$f_{\mathrm{L}}(n,t)=\int_{\mathbb{R}^2}f_{N_{\mathrm{L}}^{0},A,K}\left(\frac{nk}{n+\mathrm{e}^{a(t-t_0)}(k-n)},a,k\right)\frac{\mathrm{e}^{a(t+t_0)}\,k^2}{(n\,\mathrm{e}^{at_0}+\mathrm{e}^{at}(k-n))^2}\,\mathrm{d}a\,\mathrm{d}k. \tag{11}$$

2.3 Gompertz Growth Model

The solution of the random Gompertz IVP (9) is

$$N_{\mathrm{G}}(t,\omega)=\exp\left(\frac{C(\omega)+\mathrm{e}^{-B(\omega)(t-t_0)}(B(\omega)\ln(N_{\mathrm{G}}^{0}(\omega))-C(\omega))}{B(\omega)}\right),\quad t\geq t_0.$$

In the following, $f_{N_{\mathrm{G}}^{0},B,C}(n^0,b,c)$ denotes the joint PDF of random vector of input parameters $(N_{\mathrm{G}}^{0}(\omega),B(\omega),C(\omega))$, which is considered known. Let us fix

$t \geq t_0$, then we apply the RVT method, stated in Theorem 1, with the following choice

$$\mathbf{U}(\omega) = (N_{\mathrm{G}}^0(\omega), B(\omega), C(\omega))^\top, \quad \mathbf{V}(\omega) = (X(\omega), Y(\omega), Z(\omega))^\top,$$

$$\mathbf{g} : \mathbb{R}^3 \to \mathbb{R}^3, \quad g(n_{\mathrm{G}}^0, b, c) = (g_1(n_{\mathrm{G}}^0, b, c), g_2(n_{\mathrm{G}}^0, b, c), g_3(n_{\mathrm{G}}^0, b, c))^\top = (x, y, z)^\top,$$

where

$$x = \exp\left(\frac{c + \mathrm{e}^{-b(t-t_0)}(b \ln(n_{\mathrm{G}}^0) - c)}{b}\right), \quad y = b, \quad z = c.$$

It is straightforward to check that the inverse mapping of $\mathbf{g}$ is defined by $\mathbf{h} : \mathbb{R}^3 \to \mathbb{R}^3$, being

$$\begin{aligned} n_{\mathrm{G}}^0 &= h_1(x, y, z) = \exp\left(\frac{z + \mathrm{e}^{y(t-t_0)}(y \ln(x) - z)}{y}\right), \\ b &= h_2(x, y, z) = y, \\ c &= h_3(x, y, z) = z. \end{aligned}$$

Then, the joint PDF of the random vector $\mathbf{V}(\omega) = (X(\omega), Y(\omega), Z(\omega))$ is

$$f_{X,Y,Z}(x, y, z) = f_{N_{\mathrm{G}}^0,B,C}\left(h_1(x, y, z), y, z\right) \frac{\mathrm{e}^{y(t-t_0)}\, h_1(x, y, z)}{x}.$$

Notice that in this last expression we have omitted the absolute value, $|x|$, in the Jacobian since $x > 0$. Marginalizing with respect to the random vector $(B(\omega), C(\omega))$ and taking $t \geq t_0$ arbitrary the 1-PDF of the solution $N_{\mathrm{G}}(t, \omega)$ is given by

$$f_G(n, t) = \int_{\mathbb{R}^2} f_{N_{\mathrm{G}}^0,B,C}\left(h_1(n, b, c), b, c\right) \frac{\mathrm{e}^{b(t-t_0)}\, h_1(n, b, c)}{n} \mathrm{d}b\, \mathrm{d}c. \tag{12}$$

3 An Application to Modelling Real Data

In Sect. 2 expressions for the 1-PDFs of the three solution to the random IVPs (7)–(9) have been presented. Now, we will show how this key probabilistic information can be utilized in practice. First, we will assign reliable parametric PDF for each model parameter and then we will determine the parameters of each PDF by imposing that the solution stochastic process of the corresponding random IVP adjusts the real data using a minimization criterion. Specifically, we will minimize the mean square error between real data and the expectation of the solution which can be calculated via the expression $\mathbb{E}\left[N(t, \omega)\right]$ given (4) because we know the 1-PDF of the solution. Furthermore, confidence intervals will be also constructed.

Table 1 collects real data about the Spirulina sp. biomass production over, approximately, 330 h, in different time instants, [15]. Notice that the sample size is $N = 21$.

Table 1. Data of Spirulina sp. biomass production in a particular medium, n_i, in different time instants t_i, $1 \leq i \leq 21$. Source: [15].

t_i	0	5.627	17.629	23.569	71.841	89.484	114.115
n_i	0.0005	0.0003	0.15649	0.30068	0.45631	0.53661	0.57986
t_i	120.18	137.459	143.586	162.456	168.212	191.96	210.613
n_i	0.59229	0.59652	0.61243	0.66041	0.73996	0.8046	0.75662
t_i	234.076	240.119	263.92	282.544	306.132	311.829	330.582
n_i	0.8487	0.86061	0.7959	0.85233	0.90827	0.928	0.88017

Now, we apply the three growth models (7)–(9), with $t_0 = 0$, to describe the evolution over the time of the Spirulina sp. In order to apply formulas for the 1-PDFs (10)–(12), we have to choose specific probability distributions for the random vectors $(N_{\rm E}^0(\omega), R(\omega))$, $(N_{\rm L}^0(\omega), A(\omega), K(\omega))$ and $(N_{\rm G}^0(\omega), C(\omega), B(\omega))$. In the following, for the sake of simplicity in computations, independence among the random input parameters is considered. As the data collected in Table 1 lies between 0 and 1, the initial conditions $N_{\rm E}^0(\omega)$, $A_{\rm L}^0(\omega)$ and $N_{\rm G}^0(\omega)$ will also lie in the same interval. Then, we have made the decision of assuming that they will have beta distributions with certain positive parameters. For the growth rates, $R(\omega)$, $A(\omega)$ and $B(\omega)$, as they have to be positive, we choose truncated Gaussian distributions on the interval $T = (0, +\infty)$. As the parameter $K(\omega)$ represents the maximum number of individuals, so we choose a Uniform distribution on a given positive interval. Finally, the random input $C(\omega)$ is assumed to follow an exponential distribution truncated on the interval $S = (0, 0.0001]$ with a large enough parameter $\lambda_C > 0$. This choice has been motivated by the deterministic counterpart where the parameter c must be positive, thus we chose a distribution with positive domain. As we will see later, a deterministic model fitting provides a small value for the deterministic input $c > 0$. Therefore, in the random scenario this behaviour is retained, therefore we must truncate the exponential distribution to avoid computational problems. Here, we sum up the probability distributions assumed for the random input parameters:

- $N_{\rm E}^0(\omega) \sim {\rm Be}(a_{\rm E}, b_{\rm E})$, $N_{\rm L}^0(\omega) \sim {\rm Be}(a_{\rm L}, b_{\rm L})$ and $N_{\rm G}^0(\omega) \sim {\rm Be}(a_{\rm G}, b_{\rm G})$, where $a_{\rm E}, b_{\rm E}, a_{\rm L}, b_{\rm L}, a_{\rm G}, b_{\rm G} > 0$.
- $R(\omega) \sim {\rm N}_T(r_1, r_2)$, where $T = (0, +\infty)$ and $r_1, r_2 > 0$.
- $A(\omega) \sim {\rm N}_T(a_1, a_2)$, where $T = (0, +\infty)$ and $a_1, a_2 > 0$.
- $K(\omega) \sim {\rm U}([k_1, k_2])$, where $0 < k_1 < k_2$.
- $B(\omega) \sim {\rm N}_T(b_1, b_2)$, where $T = (0, +\infty)$ and $b_1, b_2 > 0$.
- $C(\omega) \sim {\rm Exp}_S(\lambda_C)$, where $S = (0, 0.0001]$ and $\lambda_C > 0$.

At this point, it is important to emphasize that the generality of the results obtained in Sect. 2 permits to choose other probability distributions for the input parameters, whenever they match the biological interpretation. We must stress that the optimal choice of these distributions is a difficult and critical step in modelling and in this paper we have made a decision which agrees with biological requirements such as positivity, boundedness, etc., although it might not be the optimal choice. Anyway, the spirit of the example is to illustrate how to apply our approach using reasonable distributions.

In a first stage, we are going to compare the results provided by the three deterministic models (1)–(3) when fitting the data tabulated in Table 1. This fitting is done by determining the values of model parameters that minimize the mean square error between the deterministic solution evaluated at the time instants t_i and the data n_i, $1 \le i \le 21$. This leads to the following optimization programs

$$\textbf{Exponential:} \quad \min_{0<n_{\mathrm{E}}^0<1, r>0} \sum_{i=1}^{21} \left(n_i - n_{\mathrm{E}}(t_i; n_{\mathrm{E}}^0, r)\right)^2 . \tag{13}$$

$$\textbf{Logistic:} \quad \min_{0<n_{\mathrm{L}}^0<1, a, k>0} \sum_{i=1}^{21} \left(n_i - n_{\mathrm{L}}(t_i; n_{\mathrm{L}}^0, a, k)\right)^2 . \tag{14}$$

$$\textbf{Gompertz:} \quad \min_{0<n_{\mathrm{G}}^0<1, b, c>0} \sum_{i=1}^{21} \left(n_i - n_{\mathrm{G}}(t_i; n_{\mathrm{G}}^0, b, c\right)^2 . \tag{15}$$

Solving these optimization programs, we obtain

$$n_{\mathrm{E}}^0 = 0.325602, \quad r = 0.0035502, \quad n_{\mathrm{L}}^0 = 0.127122, \quad a = 0.0205657,$$
$$k = 0.874279, \quad n_{\mathrm{G}}^0 = 0.124631, \quad b = 0.010733, \quad c = 4.7956 \cdot 10^{-12}.$$

The errors for the Exponential, Logistic and Gompertz deterministic models are 0.408203, 0.085812 and 0.083189, respectively. Notice that, Logistic and Gompertz models show a similar error since c is approximately zero. The square of the coefficient of determination in each case is 0.865757, 0.973318 and 0.975144. These results match with those established in [15], where authors indicate that the best results are obtained by the Gompertz model, whose goodness of fit coefficient is approximately $R = 0.98$. Therefore, Gompertz IVP (3) provides the best goodness of fit, although the Logistic IVP (2) gives also a good fitting. In Fig. 1 we show graphically the results.

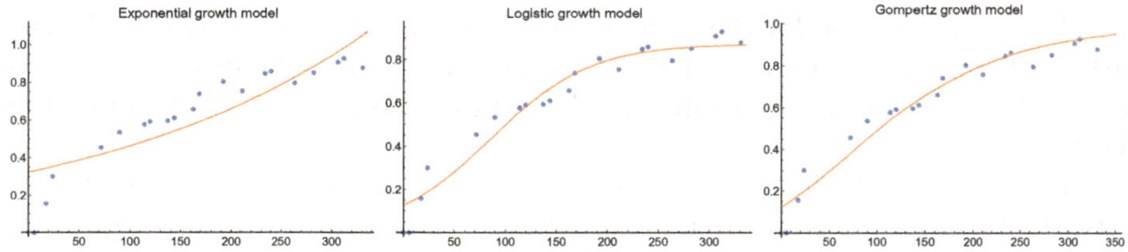

Fig. 1. Deterministic fitting: Exponential growth model (left), Logistic growth model (center) and Gompertz growth model (right).

Now, we present the result for the stochastic optimization approach proposed in this paper. Therefore, we will determine the parameters $a_{\mathrm{E}}, b_{\mathrm{E}}, a_{\mathrm{L}}, b_{\mathrm{L}}, a_{\mathrm{G}}, b_{\mathrm{G}}, r_1,$

$r_2, a_1, a_2, k_1, k_2, b_1, b_2$ and c associated to the previously chosen probability distributions. We minimize the mean square error between the data, n_i, and the expectation of the solution SP evaluated at each time instant t_i, $1 \leq i \leq 21$. It leads to the following optimization programs

$$\textbf{Exponential:} \quad \min_{a_{\rm E}, b_{\rm E}, r_1, r_2 > 0} \sum_{i=1}^{21} \left(n_i - \mathbb{E}[N(t_i, \omega; a_{\rm E}, b_{\rm E}, r_1, r_2)]\right)^2. \tag{16}$$

$$\textbf{Logistic:} \quad \min_{a_{\rm L}, b_{\rm L}, a_1, a_2 > 0, 0 < k_1 < k_2} \sum_{i=1}^{21} \left(n_i - \mathbb{E}[N(t_i, \omega; a_{\rm L}, b_{\rm L}, a_1, a_2, k_1, k_2)]\right)^2. \tag{17}$$

$$\textbf{Gompertz:} \quad \min_{a_{\rm G}, b_{\rm G}, b_1, b_2, \lambda_C > 0} \sum_{i=1}^{21} \left(n_i - \mathbb{E}[N(t_i, \omega; a_{\rm G}, b_{\rm G}, b_1, b_2, \lambda_C)]\right)^2. \tag{18}$$

According to (4) the above expectations can be computed, from the 1-PDF given in expressions (10)–(12), as

$$\mathbb{E}[N(t_i, \omega; a_{\rm E}, b_{\rm E}, r_1, r_2)] = \int_{-\infty}^{\infty} n f_{\rm E}(n, t_i) \mathrm{d}n,$$

$$\mathbb{E}[N(t_i, \omega; a_{\rm L}, b_{\rm L}, a_1, a_2, k_1, k_2)] = \int_{-\infty}^{\infty} n f_{\rm L}(n, t_i) \mathrm{d}n,$$

$$\mathbb{E}[N(t_i, \omega; a_{\rm G}, b_{\rm G}, b_1, b_2, \lambda_C)] = \int_{-\infty}^{\infty} n f_{\rm G}(n, t_i) \mathrm{d}n.$$

We have solve the optimization programs (16)–(18) by Mathematica® software. We have obtained the following results

- ▷ $a_{\rm E} = 37.044116$, $b_{\rm E} = 76.637711$, $r_1 = 0.003510$, $r_2 = 0.000516$.
- ▷ $a_{\rm L} = 15.991561$, $b_{\rm L} = 112.930464$, $a_1 = 0.021166$, $a_2 = 0.002337$, $k_1 = 0.848804$, $k_2 = 0.904245$
- ▷ $a_{\rm G} = 10.235329$, $b_{\rm G} = 72.049236, b_1 = 0.010910$, $b_2 = 0.001089, \lambda_C = 2.068077 \cdot 10^8$.

The errors obtained for the Exponential, Logistic and Gompertz models have been 0.411880, 0.084059 and 0.082454, respectively. These figures are in agreement with the deterministic results, so the randomized Gompertz model provides the best fitting.

Figure 2 shows the 1-PDFs given by (10)–(12) using the optimal values for the parameters of the probability distributions assumed for the input random variables in each model. In the three models we observe that the quantity of biomass increases. In fact, while in the Logistic and Gompertz models the dispersion reduces as time increases, in the Exponential model the variability augments. This behaviour is better observed in Fig. 3 where the mean and the variance have been plotted.

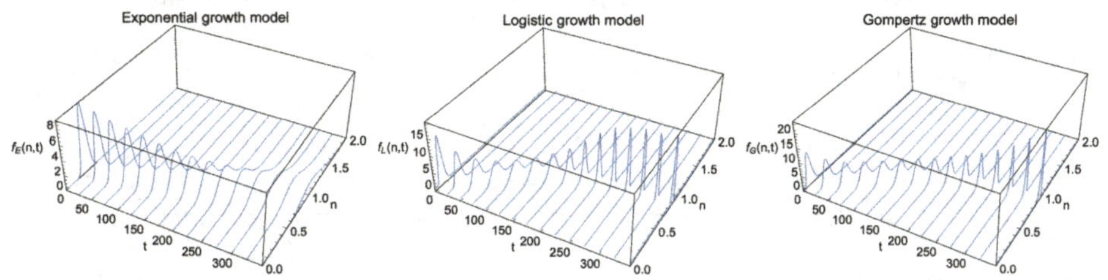

Fig. 2. Plot of the 1-PDFs for different values in the time interval [0, 350]: Exponential growth model (left), Logistic growth model (center) and Gompertz growth model (right).

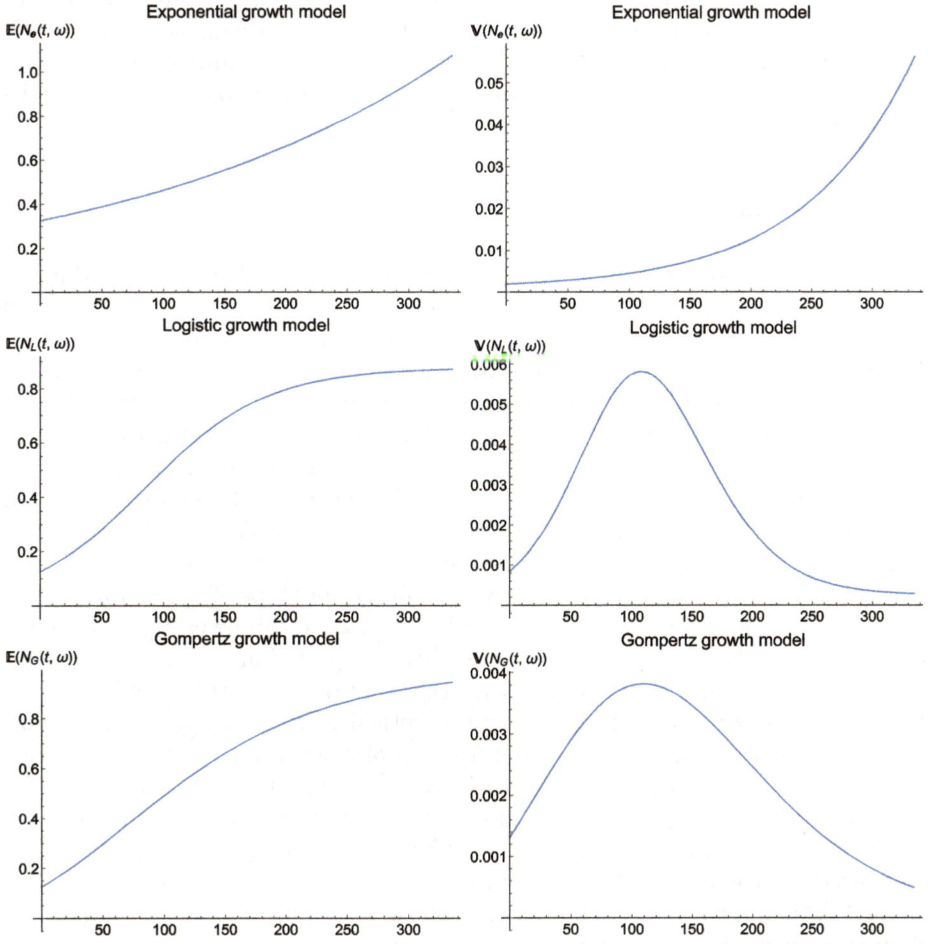

Fig. 3. Plot of the expectation (left panel) and the variance (right panel) in the time interval [0, 335] for the three growth models: Exponential (top), Logistic (center) and Gompertz (bottom).

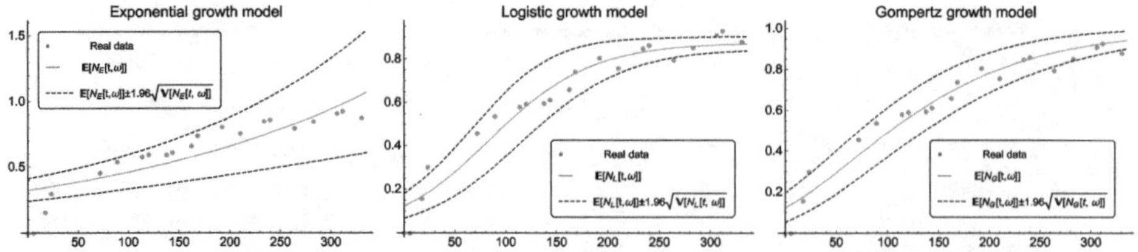

Fig. 4. Expectation (solid line) and confidence intervals (dotted lines) in the time interval [0, 335] for the three growth models: Exponential model (left), Logistic model (center) and Gompertz growth model (right).

To validate graphically the model, we have built confidence intervals according to the general rule mean plus/minus 1.96 standard deviation. The results are shown in Fig. 4. From this plot, we can observe that almost all the data lies within the confidence intervals, hence the model is capable of capturing the variability of the real data.

4 Conclusion

In this paper we have studied, from a probabilistic point of view, three fundamental models in Biology, the Exponential, the Logistic and the Gompertz growth models. We have given a full probabilistic description of the solution stochastic process for each model. The proposed methodology permits to compute the first probability density function of the solution under very general hypotheses. The involved parameters have been considered dependent continuous random variables having arbitrary probability density functions. We have applied the variable transformation technique to determine an explicit expression for the first probability density function of the solution. From this crucial deterministic function, the solution stochastic process is completely characterized in each time instant. In addition, from it, any one-dimensional statistical moment can be computed. In particular, the mean and the variance can be easily determined as well as confidence intervals. Additionally, the probability density functions permits to calculate the probability that the solution lies within any interval of interest. To show the usefulness of our theoretical findings, we have shown how they can be applied to describe the dynamics of *Spirulina* sp. biomass production in a particular medium using real data retrieved from the literature. To do that, we have assigned reliable parametric distributions to the randomized model coefficients and initial conditions. Afterwards, then we have estimated the parameters of the distributions by minimizing the mean square error between the real data and the expectation of the solution stochastic process evaluated in different time instants. These expectations have been calculated because the previous computation of the first probability density function of the solution. Once the probability distributions of the model parameters (coefficients and initial condition) have been obtained, we have completely determined and plotted the first probability density function, the mean, the variance and confidence intervals of the solution. We have show that the results are adjust the real data and capture their uncertainties. As it happens when the deterministic approach is applied, randomized Gompertz model seems to fit better than the other two randomized models. Finally, we want to point out that the proposed methodology can be extended to other important models in the biological framework.

Acknowledgments. This work has been supported by the Spanish Ministerio de Economía, Industria y Competitividad (MINECO), the Agencia Estatal de Investigación (AEI) and Fondo Europeo de Desarrollo Regional (FEDER UE) grant MTM2017–89664–P.

References

1. Malthus, T.R.: An Essay of the Principle of Population. J. Johnson, London (1798)
2. Verhulst, P.F.: Notice sur la loi que la population pursuit dans son accroissement. Corresp. Math. Phys. **10**, 113 (1838)
3. Gompertz, B.: On the nature of the function expressive of the law of human mortality, and on a new mode of determining the value of life contingencies. Philos. Trans. R. Soc. **123**, 513 (1825)
4. Cooper, R.G.: Growth in the ostrich (Struthio camelus var. domesticus). Anim. Sci. **76**, 1–4 (2005). https://doi.org/10.1111/j.1740-0929.2005.00230.x
5. Kozusko, F., Bajzer, Z.: Combining Gompertzian growth and cell population dynamics. Math. Biosci. **185**(2), 153–167 (2003). https://doi.org/10.1016/s0025-5564(03)00094-4
6. Neeff, T., Dos Santos, J.R.: A growth model for secondary forest in Central Amazonia. For. Ecol. Manag. **216**, 270–282 (2005). https://doi.org/10.1016/j.foreco.2005.05.039
7. Pirjol, D., Jafarpour, D., Iyer-Biswas, S.: Phenomenology of stochastic exponential growth. Phys. Rev. E **95**, 062406 (2017). https://doi.org/10.1103/PhysRevE.95.062406
8. Kink, P.: Some analysis of a stochastic logistic growth model. Stoch. Anal. Appl. **36**, 240–256 (2018). https://doi.org/10.1080/07362994.2017.1393343
9. Liu, M., Wang, K.: On a stochastic logistic equation with impulsive perturbations. Comput. Math. Appl. **63**, 871–886 (2012). https://doi.org/10.1016/j.camwa.2011.11.003
10. Albano, G., Giorno, V., Román-Román, P., Romero, D., Torres-Ruiz, F.: Inferring the effect of therapy on tumors showing stochastic Gompertzian growth. J. Theor. Biol. **276**, 67–77 (2011). https://doi.org/10.1016/j.jtbi.2011.01.040
11. Gutiérrez-Jáimez, R., Román, P., Romero, D., Serrano, J.J., Torres, F.: A new Gompertz diffusion process with application to random growth. Math. Biosci. **208**, 147–165 (2007). https://doi.org/10.1016/j.mbs.2006.09.020
12. Casella, G., Berger, R.: Statistical Inference. Brooks/Cole, New York (2002)
13. Kadry, S.: On the generalization of the probabilistic transformation method. Appl. Math. Comput. **190**(15), 1284–1289 (2007). https://doi.org/10.1016/j.amc.2011.12.088

14. Soong, T.T.: Random Differential Equations in Science and Engineering. Academic Press, New York (1973)
15. Vasquez-Villalobos, V., Vergaray, D., Suarez, S., Valladares, J., Zamora, A., Gaspar, K., Escurra, X.: Proportional influence of sea water en bicarbonate on the production of Spirulina sp. biomass with light-emitting diode lighting. Scientia Agropecuaria **5**, 199–209 (2014). https://doi.org/10.17268/sci.agropecu.2014.04.04

Uncertainty Propagation in Wind Turbine Blade Loads

Wilson J. Veloz(✉), Hongbo Zhang, Hao Bai, Younes Aoues, and Didier Lemosse

INSA de Rouen, Mechanical Laboratory of Normandy, Normandy University, 685 Avenue de l'Université, 76800 Saint-Étienne-du-Rouvray, France
wilson.veloz_parra@insa-rouen.fr

Abstract. Minimizing the cost and enhancing the lifespan of wind turbines entails the optimization of the material distribution of wind turbine components (blades, tower, etc.) without compromising their structural safety. Wind turbines are often design using the IEC 61400-1 standard to provide an appropriate level of protection against damage from all hazards during the planned lifetime. Typically, aero-elastic simulations codes are used to determine loads and displacements time history in the wind turbine. To predict the fatigue damage limit of the wind turbine blade, it is important to quantify and model all relevant uncertainties but it requires a considered amount of simulation time, and a surrogate model can substitute this simulation, to decrease this time consuming part of the problem. In this study, Monte Carlo simulation and FAST code are used to simulate different wind conditions. Here, 10-min of effective simulations generate a time history for all forces and moments acting in 10 selected gages of the blade. Subsequently, we quantify the uncertainty of their maximum value using a Gaussian process (Kriging) and Deep Neural Network (DNN), fitting this maximum output values with their correspondent input values. For Kriging and DNN a good fitting was found for almost all output variables.

Keywords: Uncertainty propagation · Wind turbine blade · Gaussian process · Deep Neural Network

1 Introduction

The wind turbine has evolved over the last 35 years. They are more cost-effective and more reliable. It cannot be concluded that the evolutionary period is over, because the trending growth in power (kW) continues in process, as the 10 MW wind turbine developed by DTU [2] in 2013 and more recently 20 MW offshore wind turbine [21]. There is pressure for designers to minimize the cost and enhance the lifespan of wind turbines developing new materials, optimizing their design and improving engineer methods for the analysis. Typically, numerical tools are used to facilitate the design of wind turbine blades, usually loads calculation is carried out with a beam Finite Element Model (FEM),

J. E. S. De Cursi (Ed.): Uncertainties 2020, LNME, pp. 337–346, 2021.
https://doi.org/10.1007/978-3-030-53669-5_24

taking into account aero-elastic behavior, turbine control commands and also the hydrodynamic behavior for off-shore turbines. Some examples of these codes are the Fatigue, Aerodynamics, Structures, and Turbulence (FAST) [10] from National Renewable Energy Laboratory and Horizontal Axis Wind turbine simulation Code 2nd generation (HAWC2) [12]. These multi-physics and multi-body aero-servo-hydro-elastic beam finite element codes can run a great number of the design situations described by certification bodies [6], taking into account all different extreme loads acting in the structure, generating the loading history used to basic or detailed analysis. In the context of the creation of a surrogate model to predict the fatigue damage limit of the wind turbine blade, it is important to quantify and model all relevant uncertainties.

In this study, Monte Carlo simulation and FAST code are used to simulate different wind conditions, first, a steady simulation of the 10 MW WT for 10 min is performed only taking to account the uncertainty of the wind speed, second an unsteady simulation for the same machine is executed, but in this case is added the uncertainty in wind speed standard deviation, shear exponent and yaw angle. These outputs a time history for all forces and moments acting along the blade in 10 selected locations (gages). Subsequently, after 300 s of simulation is extracted the maximum value for each simulation generating a data distribution. Two different methods are used to quantify the uncertainty, two machine learning techniques are used to predict these outputs, a Gaussian process [16] and a Deep Neural Network (DNN) [18].

In steady and unsteady cases, for machine learning techniques it is possible to predict these outputs with a relative error of 0.01 in the best cases. In these cases, for some outputs, DNN gives a better prediction than the trained Gaussian process. All cases and results are described in detail below.

2 Simulation Description

In this section is explained the process of each simulation, 2 different scenarios are analyzed using the same tools to verify their applicability in this kind of problem.

2.1 Wind Turbine Description

The wind turbine 10 MW DTU RWT [2] is the reference machine for this study, represented in the finite element software *Code_Aster* [5] as a shell model. Some changes were implemented to reduce the complexity of composite material distribution for future analysis implementations. The composite materials QQ1 and P2B extracted from the composite material database SNL/MSU/DOE [13] and Balsa wood are distributed constantly in 3 defined regions of the blade as shown in Fig. 1. Subsequently, the BEam Cross-section Analysis Software (BECAS) [3] determines cross-section stiffness properties that will be used for the aero-elastic simulations.

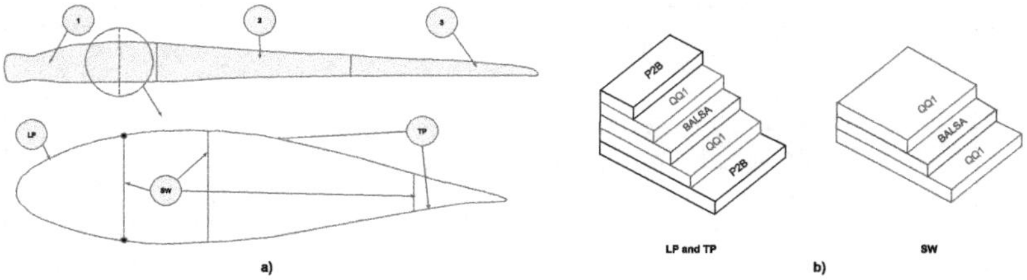

Fig. 1. Composite material distribution a) by region and b) stacking order composite laminates per section.

2.2 Case 1: Steady Aero-Elastic Simulation

Wind Conditions: Measurements of wind speed during the last three years where extracted from open-data Engie Renewable project at la haute borne, Grand Est, France [17]. This data is collected for a 2 MW wind turbine at a hub height of 80 m. The wind turbine used in this study has a hub height of 119 m and cut-in and cut-out wind speeds of [4, 25] m/s respectively. In order to use this measurements, wind speed is translated to DTU hub height using a power law [23] and Hellmann's exponent coefficient $q = 0.27$ for a condition of unstable air above human inhabited areas [11]:

$$\bar{u}(z) = \bar{u}(z_{ref}) \left(\frac{z}{z_{ref}} \right)^q \tag{1}$$

Then, a Kolmogorov-Smirnov test [15] is applied to determine a goodness of fit for the wind speed measurements. Statistics SciPy library [8] is used to perform this test overall defined continuous probability distribution function (PDF), giving us the best fit a loglogistic PDF or Fisk.

Aero-Elastic Steady Simulation: The environmental conditions described before are inputs to FAST [10] and a 10 min steady study with a time step $\Delta t = 0.015$ s is launched for $n = 1025$ wind speed generated using Monte-Carlo sampling from his PDF. The wind turbine controller as described by Hansen [7] was implemented as a DLL to interact with the FAST model. Based on proportional-integral control with additional filters, the controller uses the collective blade pitch angle and electromagnetic generator torque to control the wind turbine.

Subsequently, a time history response of the forces and moments acting on 10 nodes of the blade are extracted as outputs from FAST model resulting in a total of 60 outputs variables described as F_i^{Beam} and M_i^{Beam}, where $i = x, y, z$ for each node. After 300 simulated seconds, the maximum value of each variable is extracted for each output (as shown in Fig. 2) and this is the output data that will be used to calculate their PDF or create a surrogate model.

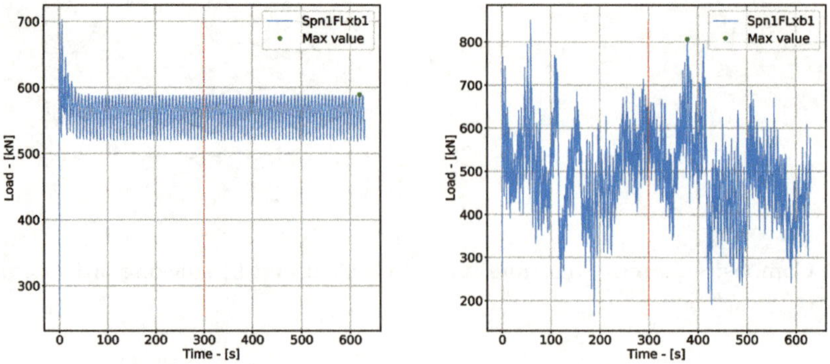

a) b)

Fig. 2. Load $Spn1FLxb1$ maximum value extraction for a) steady simulation and b) unsteady simulation.

2.3 Case 2: Unsteady Aero-Elastic Simulation

Wind and Turbulent Conditions: The dependency between wind speed (WS) and the standard deviation of the wind speed (σ_1) is defined in the Normal Turbulence Model described in the IEC 61400-1 [6]. In our study case is used a reference ambient turbulence intensity of a site Class 1A: $TI_{ref} = 0.16$. This dependency is given by the local statistical moments of σ_1 as: $\mathbf{E}(\sigma_1|WS) = TI_{ref}(0.75WS + 3.8)$ and $\mathbf{V}(\sigma_1|WS) = (1.4TI_{ref})^2$.

The parameters of the σ_1 density probability distribution are given in equations as a function of WS.

$$\sigma_{\sigma_1} = \left[ln\left(\frac{\mathbf{V}(\sigma_1|WS)}{\mathbf{E}^2(\sigma_1|WS)} + 1\right)\right]^{1/2} \tag{2}$$

$$\mu_{\sigma_1} = ln[\mathbf{E}(\sigma_1|WS)] - \frac{\sigma_{\sigma_1}^2}{2} \tag{3}$$

The correlation between WS and mean shear exponent α is based on the simplified joint distribution defined by Dimitrov et al. [4].

$$\mu_\alpha = 0.088(ln(WS) - 1) \tag{4}$$

$$\sigma_\alpha = 1/(WS) \tag{5}$$

For the yaw miss-align angle ϕ an independent normal distribution was used, with parameters as shown in Table 1.

Table 1. Wind turbine surrogate model inputs parameters.

Input parameter	PDF	PDF Parameter
10-min mean hub height wind speed	Loglogistic	$\mathbf{E}(WS) = 7.28\,\mathrm{m/s}$
Std. of the wind speed during the 10-min simulation	Lognormal	$\mu_{\sigma 1}(WS)$, $\sigma_{\sigma 1}(WS)$
10-min mean shear exponent	Normal	$\mu_{\alpha}(WS)$, $\sigma_{\alpha}(WS)$
10-min mean yaw miss-align	Normal	$\mu_{\phi} = 0°$, $\sigma_{\phi} = 5°$

Rosenblatt Transformation: To build the surrogate model with multiple correlated inputs, it is required to transform the correlated input space (X_i) into an uncorrelated space (u_i). The Rosenblatt transformation [19] gives a solution to this kind of problems, but the condition to use this transformation, is that is obligatory to know the joint distribution $F_{X_1,X_2,...,X_n}(x_1, x_2, ..., x_n)$. The transformation is written:

$$x_i \rightarrow u_i = T_i(x_j)$$

with:

$$\begin{aligned} u_1 &= \Phi^{-1}(F_{X_1}(x1)) \\ u_2 &= \Phi^{-1}(F_{X_2}(x2|x1)) \\ &\vdots \\ u_i &= \Phi^{-}1(F_{X_i}(xi|x1, x2, ..., x_{i-1})) \end{aligned} \tag{6}$$

In these expressions, $F_{X_i}(xi|x1, x2, ..., x_{i-1})$ is the conditional cumulative distribution of X_i conditioned by $X_1 = x1$, $X_2, x2$, ..., $X_{i-1} = x_{i-1}$.

In this study were generated the same number of inputs variables as uncorrelated unitary uniform variables and with the inverse transformation were calculated the values of the dependent variables (σ_1 and α).

Aero-Elastic Turbulent Simulation: In this second case, a turbulent simulation is made using a full-field flow that contains a proper spatio-temporal turbulent velocity field generated using TurbSim [9], afterward is used as a FAST input to simulate the inflow turbulent environment that incorporates many of the important fluid dynamic features known to adversely affect turbine aeroelastic response and loading. As the first case, all forces and moments on the blade are extracted, but only for $n = 715$ simulations.

2.4 Modeling Methods

For each case, all outputs are fitted using 2 different techniques: using a Gaussian process (Kriging) to predict the output data given an input value and a deep neural network (DNN) for regression. All two methods are described below.

Gaussian Process: Is a stochastic algorithm that assumes that the model output $y(x)$ is a realization of a Gaussian process that interpolates a wide range of complex functions. A Kriging surrogate model is described by the following equation [20]:

$$y^K(x) = \beta^T f(x) + \sigma^2 Z(x, \omega) \quad (7)$$

The first term in Eq. 7, $\beta^T f(x)$, is the mean value of the Gaussian process (i.e., its trend); it consists of P arbitrary functions $f_j; j = 1, ..., P$ and the corresponding coefficients β_j; $j = 1, ..., P$. The second term in Eq. 7 consists of the (constant) variance of the Gaussian process σ^2 and a zero-mean, unit-variance, stationary Gaussian process $Z(x, w)$. The underlying probability space is represented by ω and is defined in terms of a correlation function R (i.e. correlation family and its hyper-parameters θ. The correlation function $R = R(x, x'; \theta)$, in turn, describes the correlation between two sample points in the output space, that depend on x, x' the hyper-parameters θ. To accomplish this method was used the Matlab toolbox UQLab [14]. The surrogate model has an ordinary trend β_0 and a correlation family Matérn $5 - 2$. For more information about these parameter see reference.

Deep Neural Network (DNN): is an artificial neural network (ANN) with multiple layers between the input and output layers that can model complex functions [18]. DNN is composed of one (pass-through) input layer, one or more layers of linear threshold unit (LTU) called hidden layers and one final layer of LTUs called the output layer. Every layer except the output layer includes a bias neuron and is fully connected to the next layer. The LTU computes a weighted sum of its inputs ($z = w_1x_1 + w_2x_2 + ... + w_nx_n = w^T \cdot x$), that for each training instance it makes a prediction output, if that produced a wrong prediction, it reinforces the connection weights from the inputs that would have contributed to the correct prediction.

For train the DNN was used Tensorflow [1], and is a sequential fully connected neural network with 4 hidden layer, with (500, 500, 400, 400) neurons per layer respectively, each layer have as *leaky_Relu* activation function.

3 Results

After extracting the maximum loads and moments for 10 span locations in the blade at X, Y and Z axis two methods are tested to fit these outputs: Kriging, and DNN.

3.1 Kriging and DNN for both Cases: Steady and Unsteady

In order to compare the prediction given by both methods, the relative error $e = abs(y - \hat{y})/y$ where, y is the validation output and $\hat{y}$ is the prediction,

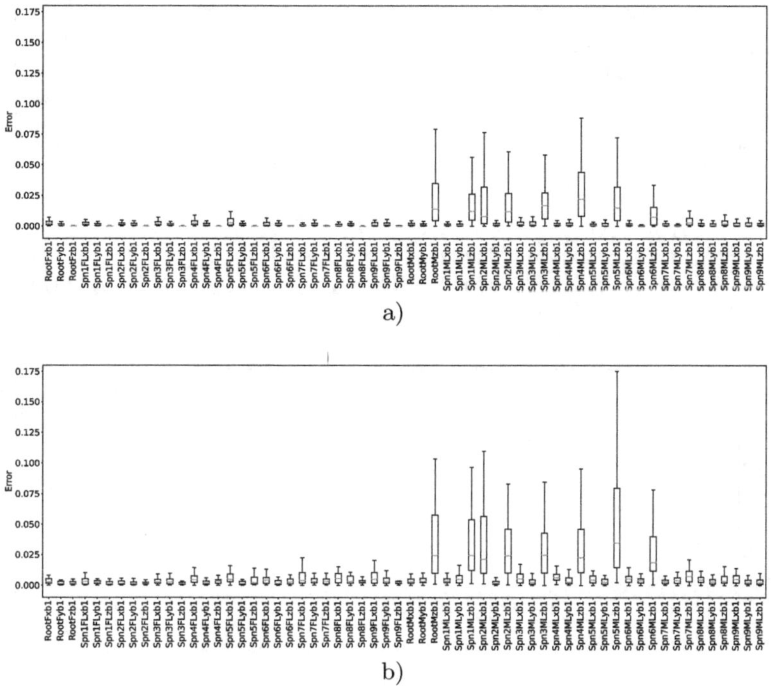

Fig. 3. Relative error for each output load for a steady study using a) Kriging and b) DNN.

is calculated for each output as shown in Fig. 3 for steady case and Fig. 4 for unsteady case.

In steady case, a Kriging has been trained for each output and as can be seen in Fig. 3a, their relative errors are contained in a range of $[0, 0.01]$ relative error for the forces in all three directions but some moments in Z direction present relative error between $[0, 0.1]$, giving an excellent prediction for each output by considering the wind speed. For DNN, given the capacity to manage multiple outputs and inputs only 1 network is trained to predict all the outputs given the wind speed. The relative error is founded in a range of $[0, 0.175]$, but not for all outputs, in the cases of moments in Z direction at different span location the prediction is not well fitted compared to Kriging, but for the other outputs is in a range of $[0, 0.025]$, that represent a good prediction given by the DNN.

In the unsteady case, Another Kriging model per output was trained using 4 inputs (see Table 1), the relative error in this case is between $[0, 0.9]$, but some cases shows a good prediction having errors between $[0, 0.2]$. Using DNN the relative error is less compared to Kriging, the relative error is between $[0, 0.45]$, not all output has this kind of error, similar to the steady case, moments in Z direction has a bad prediction, the outputs of the other loads have an error between $[0, 0.2]$. As adding more inputs variables can be seen as DNN performs better than Kriging in some cases.

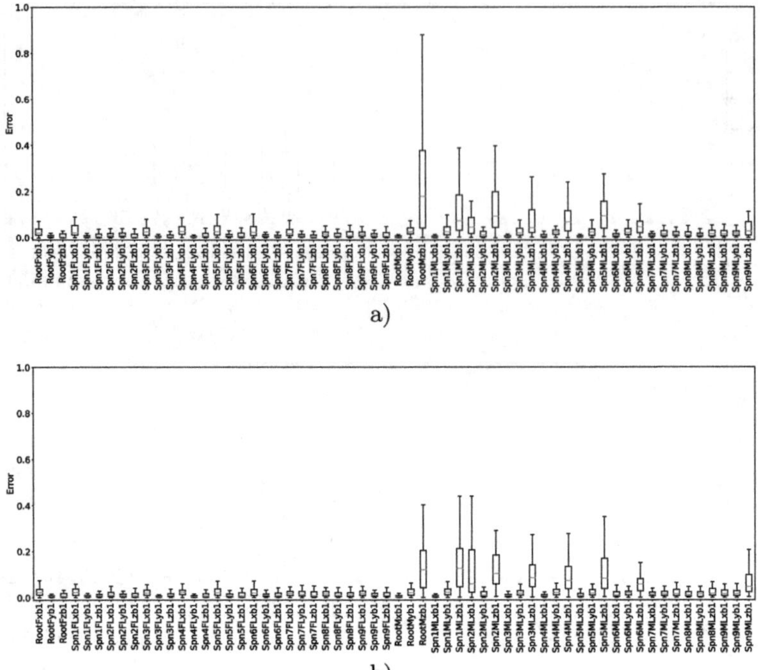

a)

b)

Fig. 4. Relative error for each output load for an unsteady study using a) Kriging and b) DNN.

4 Conclusion

A study of uncertainty propagation was made for the maximum simulated loads acting in a wind turbine blade, for two cases: steady and unsteady simulations using aero-elastic FAST code to generate time history load. Their maximum value was extracted and three different methods were applied to quantify their uncertainty.

First, a Gaussian process was used to predict the output load values, for the steady study it gave a good response for all output variables but in the unsteady case the relative error increased in some outputs. Finally, a DNN was implemented to predict the output load values, compared to the Gaussian process, in the steady study it gave a similar relative error prediction, but for others outputs the response has a greater error, and for unsteady case, DNN gave a better prediction for some outputs compared to those given by Kriging. To improve this results, in the unsteady case is needed to run more simulations at one wind speed changing it's seed number to better represent the extreme value distribution of the random process, and select a better method to sample the input space as done by [22].

DNN was trained in less time compared to Kriging, and it can handle all outputs in one neural network, different than the Gaussian process that has to be trained individually for each output. After these results are verified the applicability of machine learning methods to engineers problems. Adding more data and optimizing the hyperparameters of the DNN, will decrease the relative error for each output and give a better prediction.

As future works, it is planned to use this output loads to apply them in a blade shell model, and use a DNN to predict the stress level given by all outputs.

References

1. Abadi, M., Agarwal, A., Barham, P., Brevdo, E., Chen, Z., Citro, C., Corrado, G.S., Davis, A., Dean, J., Devin, M., Ghemawat, S., Goodfellow, I., Harp, A., Irving, G., Isard, M., Jia, Y., Jozefowicz, R., Kaiser, L., Kudlur, M., Levenberg, J., Mané, D., Monga, R., Moore, S., Murray, D., Olah, C., Schuster, M., Shlens, J., Steiner, B., Sutskever, I., Talwar, K., Tucker, P., Vanhoucke, V., Vasudevan, V., Viégas, F., Vinyals, O., Warden, P., Wattenberg, M., Wicke, M., Yu, Y., Zheng, X.: TensorFlow: large-scale machine learning on heterogeneous systems (2015). http://tensorflow.org/
2. Bak, C., Zahle, F., Bitsche, R., Kim, T., Yde, A., Henriksen, L.C., Hansen, M.H., Blasques, J.P.A.A., Gaunaa, M., Natarajan, A.: The DTU 10-MW reference wind turbine. In: Danish Wind Power Research 2013 (2013)
3. Blasques, J.P.A.A.: User's manual for BECAS: a cross section analysis tool for anisotropic and inhomogeneous beam sections of arbitrary geometry (2012)
4. Dimitrov, N., Natarajan, A., Kelly, M.: Model of wind shear conditional on turbulence and its impact on wind turbine loads. Wind Energy **18**(11), 1917–1931 (2015)
5. de France, E.: Finite element *code_aster*, analysis of structures and thermomechanics for studies and research, year = 1989–2017. Open source on www.code-aster.org
6. Guideline, G., Lloyd, G.: Guideline for the Certification of Wind Turbines. Germanischer Lloyd Wind Energie Gmb H, Hamburg (2010)
7. Hansen, M.H., Henriksen, L.C.: Basic DTU wind energy controller (2013)
8. Jones, E., Oliphant, T., Peterson, P., et al.: SciPy: open source scientific tools for Python (2001). http://www.scipy.org/
9. Jonkman, B.J., Buhl Jr, M.L.: Turbsim user's guide. Tech. rep., National Renewable Energy Lab(NREL), Golden, CO (United States) (2006)
10. Jonkman, J.M., Buhl Jr, M.L.: Fast user's guide-updated August 2005. Tech. rep., National Renewable Energy Lab (NREL), Golden, CO (United States) (2005)
11. Kaltschmitt, M., Streicher, W., Wiese, A.: Renewable Energy: Technology, Economics and Environment. Springer, Cham (2007)
12. Larsen, T.J., Hansen, A.M.: How 2 HAWC2, the user's manual (2007)
13. Mandell, J., Samborsky, D.: SNL/MSU/DOE composite material fatigue database mechanical properties of composite materials for wind turbine blades version 25.0. Montana State University, Bozeman (2016)
14. Marelli, S., Sudret, B.: UQLAB: a framework for uncertainty quantification in Matlab. In: Vulnerability, Uncertainty, and Risk: Quantification, Mitigation, and Management, pp. 2554–2563 (2014)
15. Massey Jr., F.J.: The Kolmogorov-Smirnov test for goodness of fit. J. Am. stat. Assoc. **46**(253), 68–78 (1951)
16. Rasmussen, C.E.: Gaussian processes in machine learning. In: Summer School on Machine Learning, pp. 63–71. Springer (2003)

17. Renewables, E.: La haute borne data (2017–2020). https://opendata-renewables.engie.com/explore/index. Accessed 22 Oct 2019
18. Rosenblatt, F.: The perceptron, a perceiving and recognizing automaton project para. Cornell Aeronautical Laboratory (1957)
19. Rosenblatt, M.: Remarks on a multivariate transformation. Ann. Math. Stat. **23**(3), 470–472 (1952)
20. Santner, T.J., Williams, B.J., Notz, W., Williams, B.J.: The design and analysis of computer experiments, vol. 1. Springer (2003)
21. Sartori, L., Bellini, F., Croce, A., Bottasso, C.: Preliminary design and optimization of a 20mw reference wind turbine. J. Phys.: Conf. Ser. **1037**, 042003 (2018)
22. Slot, R.M., Sørensen, J.D., Sudret, B., Svenningsen, L., Thøgersen, M.L.: Surrogate model uncertainty in wind turbine reliability assessment. Renew. Energy **151**, 1150–1162 (2019)
23. Touma, J.S.: Dependence of the wind profile power law on stability for various locations. J. Air Pollut. Control Assoc. **27**(9), 863–866 (1977)

Influence of Temperature Randomness on Vibration and Buckling of Slender Beams

Everton Spuldaro, Luiz Fabiano Damy, and Domingos A. Rade(✉)

Department of Mechanical Engineering, Aeronautical Institute of Technology, São José dos Campos, SP, Brazil
rade@ita.br

Abstract. Previous studies have demonstrated the influence of thermal stresses on the static and dynamic behavior of structures. In most cases of practical interest, temperature variations are governed by complex combinations of heat transfer mechanisms. As a result, the temperature values at different points of a structure can be considered as random variables. The present paper addresses the stochastic modeling of the influence of space-dependent temperature variations on the natural frequencies of beams. For this purpose, based on the hypotheses of the classical Euler-Bernoulli beam theory, a finite element model is constructed for the bending vibrations of beams, accounting for thermal influences. A particular scenario is considered in which the beam is subjected to random linearly-varying temperature fields, parameterized by two random variables. A probabilistic model is derived, which provides the PDF of the thermally-induced axial force from the PDFs of the random variables. Numerical simulations are performed for a clamped aluminum beam. Sampling-based statistics for the thermal axial load and the first six natural frequencies of the beam are presented. In addition, since thermal stresses can induce buckling, the probability of failure by this mechanism is also computed. Results enable to conclude that temperature uncertainty can be significant upon the vibration and buckling behavior of beams, which justifies its consideration in structural analyses.

Keywords: Thermal stresses · Stress-stiffening · Uncertainty quantifcation · Thermoelasticity

1 Introduction

Many structural systems are subjected to temperature variations, which can result either from environmental conditions or thermal effects induced by their operation. It is worth mentioning the case of viscoelastic materials, the mechanical properties of which (stiffness and damping) dependent strongly on temperature [10], which can vary due to either modifications of the ambient or self-heating resulting from energy dissipation within the material [3]. Less obvious is

J. E. S. De Cursi (Ed.): Uncertainties 2020, LNME, pp. 347–360, 2021.
https://doi.org/10.1007/978-3-030-53669-5_25

the influence of thermally-induced stresses, which are comprehensively covered in [1,6,7].

In ref. [9], Mead addressed the natural vibration modes and frequencies of a fully free rectangular thin plate subjected to in-plane stresses generated by prescribed non-uniform temperature distributions. Based on the Rayleigh-Ritz approach, and the use of an Airy's stress function, a model was derived for the problem. The results of numerical simulations enabled qualitative and quantitative understanding of the influence of thermal stresses the mode shapes and natural frequencies of the plate.

More recent, Cui and Hu [5] investigated buckling and natural vibration characteristics of a uniformly heated rectangular thin plate with two adjacent in-plane frictional sliding edges. Besides thermal stresses, temperature dependence of material properties and friction coefficients were taken into consideration. The study showed that it is possible to adjust the in-plane forces and the clearances in the edges to improve the mechanical performance of the plates.

The overwhelming majority of previous investigations, including those cited above, assume deterministic temperature distributions over the structures and, as result, deterministic thermal stress distributions. Although this can be an acceptable assumption in certain cases, it might not be applicable to most of the situations of practical interest. In fact, very often structures are subjected to rather complex combinations of heat transfer mechanisms with the surrounding media. In such cases, temperature can not be controlled and it becomes appropriate to consider temperature variations, and thermal stresses, as a result, as random quantities.

In this context, the present paper suggests an strategy for uncertainty quantification regarding the influence of random temperature fields on the natural frequencies of straight axially constrained beams, modeled according to the Euler-Bernoulli theory. In the remainder, the underlying theory is presented, followed by a case study in which an uniform aluminum beam is subjected to a random linearly-varying temperature field. For this case, the pertaining formulation and results of numerical simulations are presented and discussed.

2 Theory

In this section, the formulation enabling to assess the vibration and buckling behavior of beams under temperature variations is developed. This is performed in two phases: first, the expressions for internal axial loads associated to thermal stresses are derived. Then, the equations of motion accounting for the presence of axial loads are obtained, from which natural frequencies and buckling loads can be computed.

2.1 Internal Forces Induced by Thermal Effects

In Fig. 1 one considers a clamped-clamped beam of length L, which is assumed to be subjected to a temperature field $T(x)$. The beam is assumed to be thin, meaning that temperature is uniform over each of its cross sections. In addition, steady-state thermal conditions are assumed. Under these hypotheses, the heat transfer problem is considered as unidimensional, being governed by the well-known Fourier's law [8]:

$$q(x) = -k(x)A(x)\frac{\partial T(x)}{\partial x}, \tag{1}$$

where $A(x)$ is the cross-section area $k(x)$ is the thermal conduction coefficient.

Equation (1) is subjected to Neumann- and/or Dirichlet-type boundary conditions, expressing different thermal conditions imposed at the extremities of the beam.

Aiming at considering more general situations, it is assumed that the geometrical and material properties of the beam can vary along its axis, i.e., cross section area: $A = A(x)$; second moment of area with respect to the centroidal axis about which bending occurs: $I = I(x)$; Young's modulus: $E = E(x)$; mass density: $\rho = \rho(x)$; and coefficient of linear thermal dilatation: $\alpha = \alpha(x)$.

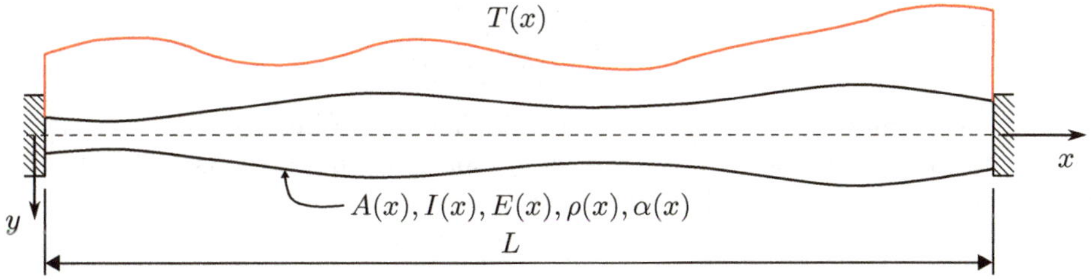

Fig. 1. Illustration of a non-uniform clamped-clamped beam subjected to space-varying temperature distribution.

Denoting by $T_0(x)$ the initial temperature field under which the beam is free of thermal stresses, when temperature evolves to $T(x)$, the free thermal elongation in the longitudinal direction is given by

$$\Delta L = \int_0^L \alpha(x)\tau(x)dx, \tag{2}$$

where

$$\tau(x) = T(x) - T_0(x), \tag{3}$$

and the free thermal strain is given by

$$\epsilon_T = \frac{1}{L}\int_0^L \alpha(x)\tau(x)dx, \tag{4}$$

When an internal axial force and a temperature change are applied simultaneously to the beam, the normal strain is given by

$$\epsilon = \epsilon_s + \frac{1}{L}\int_0^L \alpha(x)\tau(x)dx, \tag{5}$$

where

$$\epsilon_s = \frac{N_T}{L}\int_0^L \frac{1}{E(x)A(x)}dx, \tag{6}$$

As a result, the total axial strain of the beam writes:

$$\epsilon = \frac{N_T}{L}\int_0^L \frac{1}{E(x)A(x)}dx + \frac{1}{L}\int_0^L \alpha(x)\tau(x)dx. \tag{7}$$

Given that the beam is assumed to be axially constrained at both ends, the total strain must be null. Hence, from Eq. (7), the internal axial force applied to the beam is found as:

$$N_T = -\frac{\displaystyle\int_0^L \alpha(x)\tau(x)dx}{\displaystyle\int_0^L \frac{1}{E(x)A(x)}dx}. \tag{8}$$

It is worth mentioning that, in the particular case of beams whose geometric and material properties do not vary along the longitudinal axis, subjected to space-varying temperature field, one has:

$$N_T = -\frac{EA\alpha}{L}\int_0^L \tau(x)dx. \tag{9}$$

or

$$N_T = -\frac{EA\alpha}{L}\bar{\tau}. \tag{10}$$

where

$$\bar{\tau} = \int_0^L \tau(x)dx \tag{11}$$

is the mean value of the temperature difference field $T(x) - T_0(x)$ along the beam.

It can seen that if the temperature distribution is such that $\bar{\tau} = 0$, no internal force will be induced by thermal effects, i.e., $N_T = 0$.

2.2 Lateral Vibrations of Beams Under Thermally-Induced Axial Loads

It can be demonstrated, from basic dynamic principles, that the lateral vibrations of beams subjected to an axial force N_T, according to the Euler-Bernoulli beam theory, are governed by the following fourth-order partial differential equation [4]:

$$E(x)I(x)\frac{\partial^4 w(x,t)}{\partial x^4} + N_T\frac{\partial^2 w(x,t)}{\partial x^2} + \rho(x)A(x)\frac{\partial^2 w(x,t)}{\partial t^2} = q(x,t). \tag{12}$$

where $w(x,t)$ and $q(x,t)$ denote the transverse displacement field and distributed load, respectively.

Here, this equation is treated by using the finite element (FE) method. For this purpose, the beam is assumed to be discretized with n elements. As illustrated in Fig. 2, each element has two nodes, indicated by superscripts L and R and two degrees-of-freedom (DOFs) per node, namely, transverse displacements $u_i^L(t)$ and $u_i^R(t)$, and cross-section rotations $\theta_i^L(t)$ and $\theta_i^R(t)$, which are regrouped in the vector of nodal DOFs at element level:

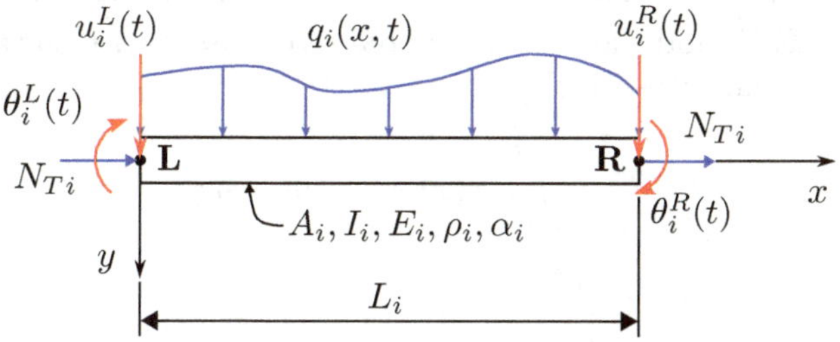

Fig. 2. Illustration of an individual Euler-Bernoulli beam element, accounting for axial load.

$$\mathbf{u}_i(t) = [u_i^L(t) \;\; \theta_i^L(t) \;\; u_i^R(t) \;\; \theta_i^R(t)]^T. \tag{13}$$

The interpolation of the transverse displacement within the i-th element is made by using third-order polynomials, as follows:

$$w_i(x,t) = \boldsymbol{\Psi}(x)\mathbf{u}_i(t), \tag{14}$$

where $\boldsymbol{\Psi}(x)$ is the matrix of shape functions, given as:

$$\boldsymbol{\Psi}(x) = [\psi_1(x) \;\; \psi_2(x) \;\; \psi_3(x) \;\; \psi_4(x)]^T \tag{15}$$

where $\psi_1(x) = 1 - 3(x/L_i)^2 + 2(x/L_i)^3$, $\psi_2(x) = x_i - 2L_i(x/L_i)^2 + L_i(x/L_i)^3$, $\psi_3(x) = 3(x/L_i)^2 - 2(x/L_i)^3$ and $\psi_4(x) = -L_i(x/L_i)^2 + L_i(x/L_i)^3$.

The equations of motion will be derived from Lagrange's Equations [4]. For this purpose, the kinetic and strain energies (this later accounting for the presence of axial loads) are written, respectively, as:

$$T(t) = \frac{1}{2}\int_0^{L_i} \rho(x)A(x)\left(\frac{\partial w}{\partial t}\right)^2 dx. \tag{16}$$

$$U(t)=\int_0^{L_i}\left\{\frac{1}{2}E(x)I(x)\left(\frac{\partial^2 w}{\partial x^2}\right)^2-q(x,t)w(x,t)+\frac{N_i}{2}\left(\frac{\partial w}{\partial x}\right)^2\right\}dx \qquad (17)$$

It should be mentioned that, in the absence of axial loads other than those induced by thermal effects, in Eq. (17) all the elements are subjected to $N_i = N_T$, this later defined in Eq. (8).

Associating Eqs. (14), (16) and (17), one gets:

$$T(t)=\frac{1}{2}\dot{\mathbf{u}}_i^T(t)\mathbf{M}_i\dot{\mathbf{u}}_i, \qquad (18)$$

$$U(t)=\frac{1}{2}\mathbf{u}_i^T(t)\left(\mathbf{K}_i^{(E)}+\mathbf{K}_i^{(G)}\right)\mathbf{u}_i(t)-\mathbf{u}_i^T(t)\mathbf{Q}_i(t), \qquad (19)$$

where the mass matrix $\mathbf{M}_i$, the elastic stiffness matrix $\mathbf{K}_i^{(E)}$, the geometric stiffness matrix $\mathbf{K}_i^{(G)}$ and the vector of nodal forces associated to the distributed load $\mathbf{Q}_i(t)$ are given, respectively, as:

$$\mathbf{M}_i=\int_0^{L_i}\rho(x)\boldsymbol{\Psi}^T(x)\boldsymbol{\Psi}(x)dx, \qquad (20)$$

$$\mathbf{K}_i^{(E)}=\int_0^{L_i}E(x)I(x)\boldsymbol{\Psi}''^T(x)\boldsymbol{\Psi}''(x)dx, \qquad (21)$$

$$\mathbf{K}_i^{(G)}=N_i\int_0^{L_i}\boldsymbol{\Psi}'^T(x)\boldsymbol{\Psi}'(x)dx, \qquad (22)$$

$$\mathbf{Q}_i(t)=\int_0^{L_i}q(x,t)\boldsymbol{\Psi}'^T(x)dx. \qquad (23)$$

Using Lagrange's Equations, accounting for Eqs. (18) and (19), the equations of motion at element level are obtained under the form:

$$\mathbf{M}_i\ddot{\mathbf{u}}_\mathbf{i}(t)+\left(\mathbf{K}_i^{(E)}+\mathbf{K}_i^{(G)}\right)\mathbf{u}_\mathbf{i}(t)=\mathbf{Q}_i(t). \qquad (24)$$

It is worth pointing-out that the influence of axial force provoked by thermal effects is conveyed by the geometric stiffness matrix $\mathbf{K}_i^{(G)}$, defined by Eq. (22).

After assembling of the equations of motion at element level, accounting for element connectivity, the corresponding equations of motion at global level are expressed as follows:

$$\mathbf{M}\ddot{\mathbf{u}}(t)+\left(\mathbf{K}^{(E)}+\mathbf{K}^{(G)}\right)\mathbf{u}(t)=\mathbf{Q}(t). \qquad (25)$$

From the matrices appearing in Eq. (25), different types of static and dynamic analyses can be performed. Of interest here are the eigenproblems to be solved for the evaluation of the vibration natural frequencies and mode shapes, and the buckling loads of beams subjected to thermally-induced axial forces, expressed, respectively, as:

$$\left(\mathbf{K}^{(E)} + \mathbf{K}^{(G)} - \omega^2 \mathbf{M}\right) \mathbf{U} = \mathbf{0}, \tag{26}$$

$$\left(\mathbf{K}^{(E)} - \lambda \mathbf{K}^{(G)}\right) \mathbf{V} = \mathbf{0}. \tag{27}$$

It can be shown that the critical buckling condition is also characterized by one of the natural frequencies, solution of Eq. (26), becoming null. This fact is the basis of the dynamic criterion for assessing stability analysis [2].

3 Case Study: Uniform Beam Subjected to Random Linearly-Varying Temperature Field

In this section a case study involving stochastic linearly-varying temperature field is presented. For this particular condition, the formulation developed above is adequately adapted.

One considers a uniform beam (i.e. all the physical and geometrical properties do not vary along the axis) subjected to a linearly varying temperature field, given as:

$$\tau(x) = \tau_L(\theta)(1 - x/L) + \tau_R(\theta)(x/L), \tag{28}$$

where $\tau_L(\theta)$ and $\tau_R(\theta)$ are, respectively, the random variables describing the temperature at the left- and right-hand side extremities of the beam. It should be mentioned that this type of distribution is found when the beam is subjected to heat sources (including convection heat exchange) or prescribed temperature values at one or both of its extremities. In this case, Eq. (8) becomes:

$$N_T(\theta) = \frac{EA\alpha}{2} \left[\tau_L(\theta) + \tau_R(\theta)\right]. \tag{29}$$

From basic probability properties, it can be shown that the mean and variance of the axial force are given as:

$$\mu_{N_T} = \frac{EA\alpha}{2} \left(\mu_{\tau_L} + \mu_{\tau_R}\right). \tag{30}$$

$$\sigma^2_{N_T} = \left(\frac{EA\alpha}{2}\right)^2 \left(\sigma^2_{\tau_L} + \sigma^2_{\tau_R} + 2\sigma_{\tau_L \tau_R}\right). \tag{31}$$

A even more particular case is found when the uniform beam is subjected to a random space-invariant temperature field, which is obtained from (28) by making $\tau_L(\theta) = \tau_R(\theta) = \tau(\theta)$. In such a case, one has the axial force described by a single random variable, as follows:

$$N_T(\theta) = EA\alpha T(\theta). \tag{32}$$

$$\mu_{N_T} = EA\alpha\mu_\tau. \tag{33}$$

$$\sigma^2_{N_T} = \left(\frac{EA\alpha}{2}\right)^2 \sigma^2_\tau \tag{34}$$

4 Numerical Simulations

In this section, numerical simulations are presented, exploring the formulation previously developed for the case of interest.

One considers here a uniform aluminum beam, illustrated in Fig. 3 the characteristics of which are given in Table 1, subjected to a random, linearly-varying temperature field. The structural model of the beam, constructed according to the formulation presented in Sect.2.2, comprises 12 elements and 13 nodes (two of those coinciding with the clamped extremities of the beam).

The temperature at the extremities of the beam are modeled as independent uniformly distributed random variables, $\tau_L \sim \mathcal{U}(5, 14)\,°C$ and $\tau_R \sim \mathcal{U}(10, 15)\,°C$.

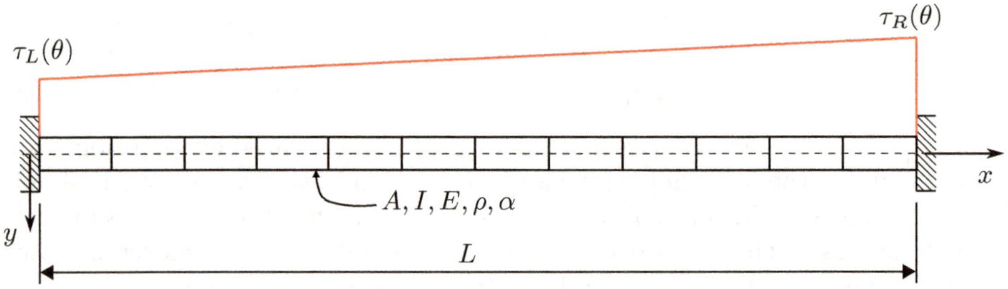

Fig. 3. Illustration of the uniform beam subjected to linearly distributed random temperature field.

Table 1. Characteristics of the beam subjected to a random linearly-varying temperature field.

Properties	Symbol	Value	Unit
Length	L	1	m
Area	A	5.0×10^{-4}	m^2
Second moment of area	I	4.17×10^{-9}	m^4
Young's modulus	E	6.9	GPa
Volumetric mass density	ρ	2710	kg/m^3
Coefficient of linear thermal expansion	α	2.4×10^{-5}	1/°C

Figure 4 shows the variations of the first six natural frequencies of the beam for increasing values of the sum $\tau_L + \tau_R$, which defines the axial load, according to Eq. (29). It can be seen that all the natural frequencies, except the first one, decay almost linearly with the increase of $\tau_L + \tau_R$. Table 2 shows the values of the natural frequencies of the beam under two particular conditions, namely: when the temperature is uniform with $\tau_L = \tau_L = 0\,^\circ$C (in this case, one has $\tau(x) \equiv 0\,^\circ$C, meaning that no temperature effects exist); and when the temperatures at the extremities take their mean values, i.e., $\tau_L = 9.5\,^\circ$C and $\tau_L = 12.5\,^\circ$C. It can be seen that the values of the natural frequencies vary considerably from one condition to the other.

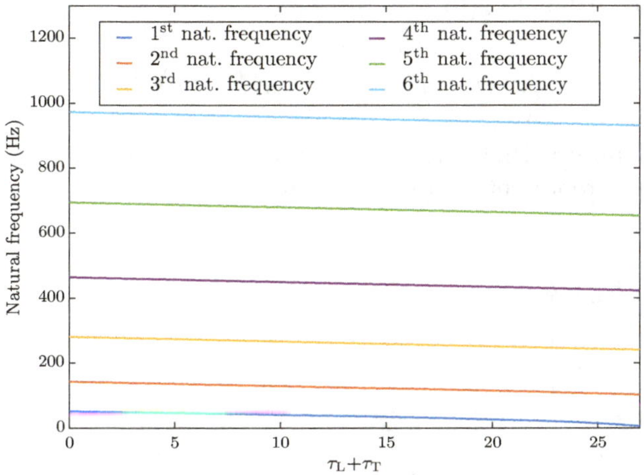

Fig. 4. Variation of the first six natural frequencies as functions of deterministic values of $\tau_L + \tau_R$

Table 2. Values of the natural frequencies of the uniform beam under two deterministic temperature distributions.

Mode	$\tau_L = \tau_L = 0$	$\tau_L = 9.5\,^\circ$C, $\tau_L = 12.5\,^\circ$C
1st	51.96	23.43
2nd	143.26	112.23
3rd	280.94	249.02
4th	464.79	432.08
5th	695.39	662.13
6th	973.72	940.09

It is also interesting to mention that, for a clamped-clamped beam, the critical buckling load, according to the Euler's linear buckling theory is [12]

$$N_{crit} = \frac{4\pi^2 EI}{L^2}. \tag{35}$$

Hence, accounting for Eq. (29), the beam undergoes buckling if

$$\tau_L(\theta) + \tau_R(\theta) \geq \frac{8\pi^2 I}{AL\alpha} = 27.44\,°\text{C}. \tag{36}$$

Introducing

$$Z(\theta) = \tau_L(\theta) + \tau_R(\theta), \tag{37}$$

according to Papoulis and Pillai [11], the PDF of this random variable is given in terms of the PDFs of $\tau_L(\theta)$ and $\tau_R(\theta)$ as:

$$f_Z(z) = \int_0^z f_{\tau_L}(z - \tau_R) f_{\tau_R}(\tau_R) d\tau_R. \tag{38}$$

From the uniform PDFs assigned to τ_L and τ_R, the PDF $f_Z(z)$ takes the trapezoidal form represented graphically in Fig. 5.

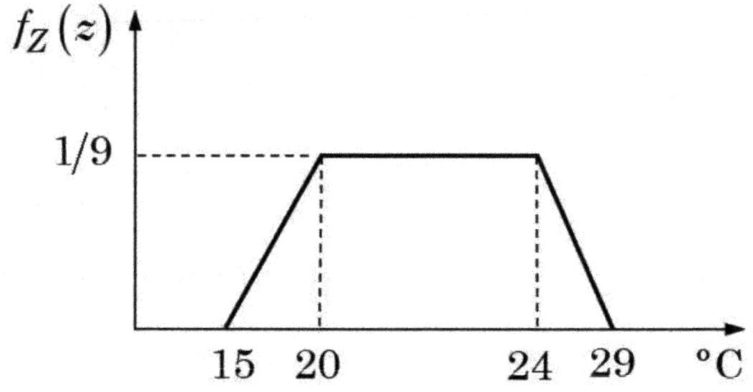

Fig. 5. PDF of the random variable $Z = \tau_L + \tau_R$.

Given the buckling condition expressed by Eq. (36), the probability of thermal buckling is computed as:

$$P_{buckl.} = 1 - \int_{-\infty}^{27.44} f_Z(z) dz, \tag{39}$$

which can be easily computed by evaluating the area under the curve shown in Fig. 5, corresponding to the integral indicated, which gives:

$$P_{buck.} = 2.70\%$$

At this point is worth mentioning that, although the statistics of the thermal axial load could be obtained analytically, the same does not apply to the statistics of the natural frequencies. In fact, the relation between the thermal load and the natural frequency is established by the eigenproblem expressed by Eq. (26). Hence, Monte Carlo simulation was performed by generating $N_{samp.} = 100,000$ samples of the random variables $\tau_L(\theta)$ and $\tau_R(\theta)$ and, for each pair of samples, computing the axial load as per Eq. (9). Then, for each value of this load, the FE model developed according to the formulation presented in Subsect. 2.2 was used to compute corresponding set of the first six natural frequencies. The results are presented in Figs. 6 and 7, respectively, in terms of histograms (normalized as PDFs) of the axial force and natural frequencies of interest. It should be explained that negative values of the axial force indicate compressive forces, according to the usual convention in structural mechanics.

In addition, Tables. 3 and 4 provides the mean values and standard deviations of the natural frequencies, estimated from the Monte Carlo samples.

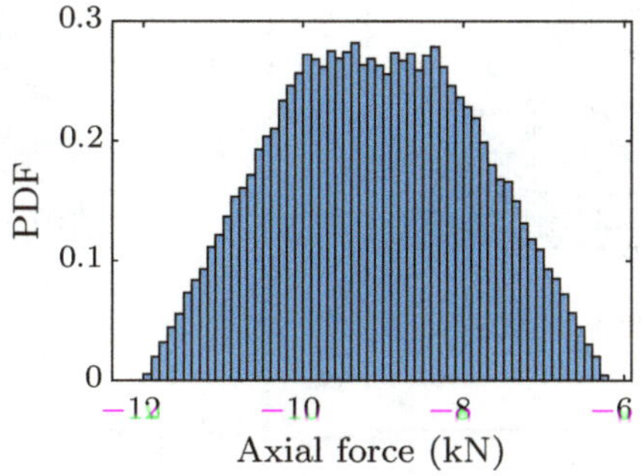

Fig. 6. PDF of the thermal axial force.

For some of the temperature samples, the corresponding thermal axial load is large enough to lead the beam to buckle. Numerically, this condition is identified by the presence of null or complex natural frequencies. Hence, the sampled values of the natural frequencies have been used to estimate the probability of buckling failure, as follows:

$$\tilde{P}_{buck.} = \frac{N_{buck.}}{N_{samp.}} \tag{40}$$

where $N_{buck.}$ is the number of Monte Carlo samples that resulted in null or complex natural frequencies.

For the present case, it was found that

$$\tilde{P}_{buck.} = 2.685\%$$

which is quite close to the theoretical value predicted by Eq. (39).

The results above show that, as the natural frequencies, except for the first one, depend linearly on $\tau_L + \tau_R$, as indicated in Fig. 4, their PDFs have the same

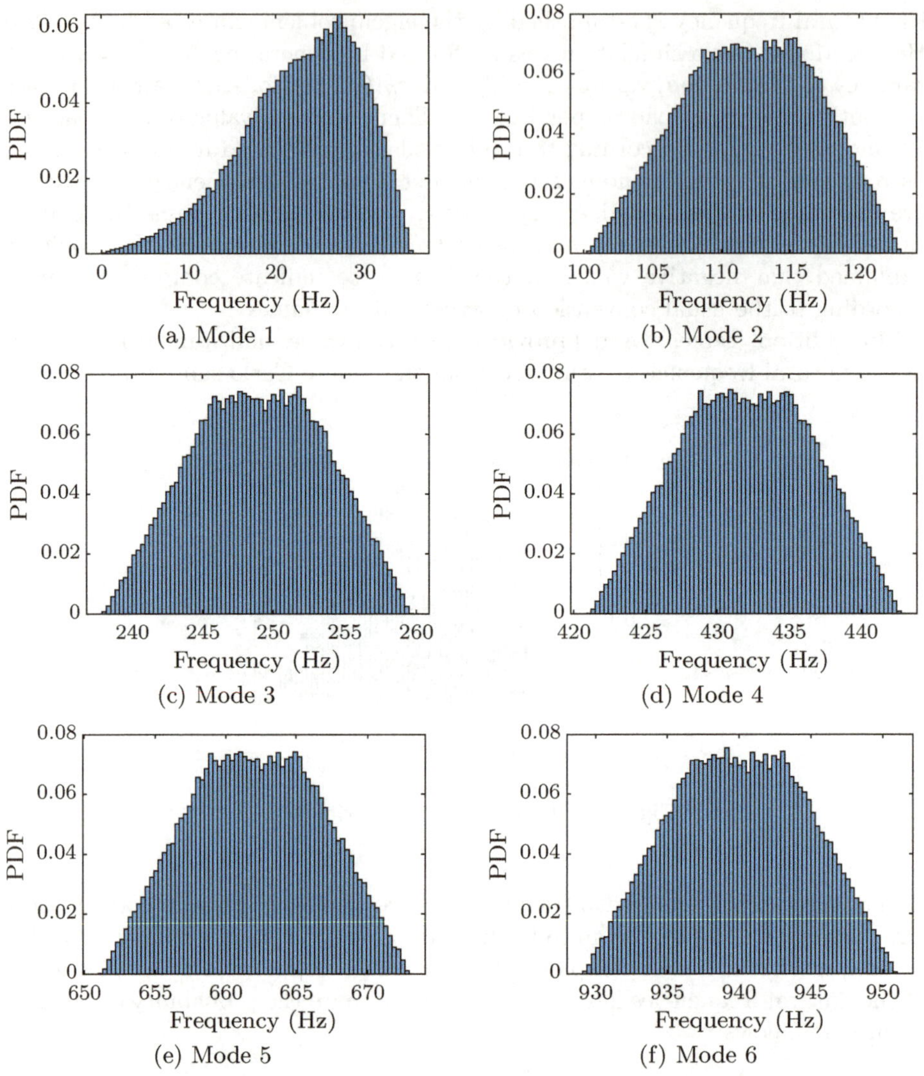

(a) Mode 1 (b) Mode 2 (c) Mode 3 (d) Mode 4 (e) Mode 5 (f) Mode 6

Fig. 7. PDFs of the first six natural frequencies of the uniform beam subjected to linearly distributed random temperature field (N = 100,000 samples).

Table 3. Mean value and standard deviation of the axial force.

Axial force (N_T)	
Mean (kN)	−9.107
Standard deviation (kN)	1.227

Table 4. Mean values and standard deviations of the first six natural frequencies.

Mode	Mean frequency (Hz)	Standard deviation (Hz)	Coeff. of variation (%)
1st	22.79	6.68	22.3
2nd	112.12	4.81	4.29
3rd	248.97	4.58	1.84
4th	432.06	4.57	1.06
5th	662.12	4.59	0.69
6th	940.08	4.61	0.49

form as that of the random variable $Z(\theta) = \tau_L(\theta) + \tau_R(\theta)$, which is depicted in Fig. 5. Differently, for the first natural frequency, the similarity of the PDFs is not observed. In addition, the coefficients of variation indicate that the sensitivity of the natural frequencies with respect to temperature randomness decrease with the order of the vibration mode.

5 Conclusions

In the present paper an investigation was carried-out on the influence of random temperature variations on the vibration and buckling characteristics of slender beams. The most important conclusion to be drawn is that temperature randomness can induce significant stochastic variations of those characteristics and, as a result, should be accounted for in situations in which uncontrolled temperature distributions are likely to occur. From the design standpoint, to comply with robustness and reliability requirements, the consideration of random temperature variations can be of utmost importance. Modeling and analyses similar to the ones presented here can be performed to other types of structural elements. In this sense, the authors have been investigating the influence of thermal effects on the dynamic behavior of more complex structures such as metallic and composite plates.

Acknowledgments. E. Spuldaro express his gratitude to FAPESP for his M.Sc. scholarship (grant #2016/16480-0). L.F. Damy is grateful to the Brazilian Army for supporting his research work. D.A. Rade gratefully acknowledges FAPESP (grant #2015/20363-6) and CNPq (grant #402238/2013-3 and #310633/2013-3) for the support to his research activities.

References

1. Boley, B.A., Weiner, J.H.: Theory of Thermal Stresses. Dover Publications, USA (1960)
2. Bolotin, V.: Dynamic Stability of Structures. Springer, Cham (1995)

3. de Cazenove, J., Rade, D., de Lima, A., Araújo, C.: A numerical and experimental investigation on self-heating effects in viscoelastic dampers. Mech. Syst. Signal Process. **27**, 433–445 (2012). https://doi.org/10.1016/j.ymssp.2011.05.004
4. Craig Jr., R.R., Kurdila, A.J.: Fundamentals of Structural Dynamics, 2nd edn. Wiley, Hoboken (2006)
5. Cui, D., Hu, H.: Thermal buckling and natural vibration of a rectangular thin plate with in-plane stick-slip-stop boundaries. J. Vib. Control **22**(7), 1950–1966 (2014). https://doi.org/10.1177/1077546314546394
6. Eslami, M.R., Hetnarski, R.B., Ignaczak, J., Noda, N., Sumi, N., Tanigawa, Y.: Theory of Elasticity and Thermal Stresses, vol. 197. Springer, Cham (2013)
7. Gatewood, B.E.: Thermal Stresses - with Applications to Airplanes, Missiles, Turbines, and Nuclear Reactors, 1st edn. McGraw-Hill, New York (1957)
8. Incropera, F.P., Lavine, A.S., Bergman, T.L., DeWitt, D.P.: Principles of Heat and Mass Transfer. Wiley, Hoboken (2013)
9. Mead, D.: Vibration and buckling of flat free–free plates under non-uniform in-plane thermal stresses. J. Sound Vib. **260**(1), 141–165 (2003). https://doi.org/10.1016/s0022-460x(02)00919-7
10. Nashif, A.D., Jones, D.I.G., Henderson, J.P.: Vibration Damping. Wiley, Hoboken (1985)
11. Papoulis, A., Pillai, S.U.: Probability, Random Variables and Stochastic Processes, 4th edn. McGraw Hill, New York (2002)
12. Timoshenko, S.P., Gere, J.M.: Theory of Elastic Stability. Courier Corporation, North Chelmsford (2009)

Investigating the Influence of Mechanical Property Variability on Dispersion Diagrams Using Bayesian Inference

Luiz Henrique Marra Silva Ribeiro[1(✉)], Vinícius Fonseca Dal Poggetto[1(✉)], Danilo Beli[2(✉)], Adriano T. Fabro[3(✉)], and José Roberto F. Arruda[1(✉)]

[1] School of Mechanical Engineering, University of Campinas, Campinas, SP 13083-860, Brazil
luiz.marra@outlook.com, vdalpoggetto@gmail.com, arruda@fem.unicamp.br
[2] São Carlos School of Engineering, University of São Paulo, São Carlos, SP 13566-590, Brazil
beli.danilo@gmail.com
[3] Department of Mechanical Engineering, University of Brasilia, Brasilia, DF 70910-900, Brazil
fabro@unb.br

Abstract. Phononic crystals are periodic structures that exhibit frequency ranges where elastic waves cannot propagate, also known as Bragg scattering band gaps. This study aims at simulating dynamic response of a periodic plane frame structure using Bayesian statistics to understand the bandgap behavior with respect to the variability in the Young's modulus and mass density. The spectral element method is used to compute Frequency Response Functions (FRF). The spatial distribution of simulated FRFs at each frequency is used as a virtual measurement, from which the wavenumbers are obtained via a Prony type method. A Markov chain Monte Carlo (MCMC) algorithm which considers the difference between the simulated and virtually observed wavenumbers is used to simulate the posterior distribution of the mechanical properties. Stochastic dispersion diagrams are simulated via Monte Carlo from these posterior distributions. The MCMC algorithm, by using the additional information related to the wavenumber, presented a more precise wavenumber dispersion estimation. These investigations are relevant due to the expected variability of mechanical properties of periodic frame structures made by additive manufacturing.

Keywords: Band gap · Bayesian inference · Informative prior · Uncertainty quantification · Plane frame

1 Introduction

Even though the studies about the dynamics of periodic structures in structural engineering can be traced back to the 1970's, only in recent decades researchers

J. E. S. De Cursi (Ed.): Uncertainties 2020, LNME, pp. 361–373, 2021.
https://doi.org/10.1007/978-3-030-53669-5_26

have started to analyse periodic structures using dispersion diagrams and wave bandgap concepts [15]. More recently, cases where the material property variability, e.g. caused by additive manufacturing, substantially influences the dynamic response and have to be considered were reported [9].

Uncertainties related to the model response can be quantified via mathematical or statistical approaches [1,16]. Stochastic analysis has been used to estimate and infer model parameters, and to predict their influence in the model response [3,4,9,13,21]. Moreover, in order to represent variability, a correct selection of the probability distribution and a correct sample size are necessary [14]. The Bayesian approach yields less biased estimates and improves precision and accuracy [19].

In this work we develop spectral element models for arbitrarily shaped two-dimensional frame elements and show how to compute FRFs with them. Then a Prony type method is used to estimate wavenumbers from a set of spatially distributed FRFs that were numerically simulated. Given this type of input data, a statistical framework is presented to investigate the effect of material property variability in dispersion diagrams using a Bayesian approach via Markov chain Monte Carlo (MCMC).

2 Methodology

2.1 Deterministic Model

Governing Equations of Motion: A planar frame structure consists of frame elements, which support longitudinal and transversal loads. It is possible to obtain the equations of motion of the elementary rod and Euler-Bernoulli beam given by [7]:

$$\frac{\partial}{\partial x}\left(EA\frac{\partial u_x}{\partial x}\right) - \rho A\frac{\partial^2 u_x}{\partial t^2} = -q_x(x,t) \text{ and } \frac{\partial^2}{\partial x^2}\left(EI\frac{\partial^2 u_y}{\partial x^2}\right) + \rho A\frac{\partial^2 u_y}{\partial t^2} = q_y(x,t), \tag{1}$$

where the mechanical properties E, ρ, and G are, respectively, the material Young's modulus, mass density, and shear modulus. The geometric properties A, and I are, respectively, the element cross-sectional area, and second moment of area. The quantities u_x, and u_y are, respectively, the longitudinal, and transversal displacements of the in-plane frame element. For a straight element the longitudinal an bending behavior are uncoupled. They are coupled by the connection of frame elements at arbitrary angles.

Spectral Element Method: From Eqs. (1), applying the time domain Fourier transform, so that the equations become ordinary differential equations in x, knowing the expressions for the element forces and/or moments, and after some manipulations, one can find the spectral element dynamic stiffness matrix and the wave number for each frequency ω_j. Further details about this procedure for the rod and beam elements can be found in Doyle (1992) [7].

The dynamic stiffness matrix, obtained via the spectral element for each frequency, relates dynamic forces and displacements in the frequency domain (stationary responses) as follows [3]:

$$\begin{Bmatrix} \boldsymbol{F}(0) \\ \boldsymbol{F}(L) \end{Bmatrix} = \begin{bmatrix} \boldsymbol{D}_{11} & \boldsymbol{D}_{12} \\ \boldsymbol{D}_{21} & \boldsymbol{D}_{22} \end{bmatrix} \begin{Bmatrix} \boldsymbol{u}(0) \\ \boldsymbol{u}(L) \end{Bmatrix}, \tag{2}$$

where **F**, **D**, and **u** are, respectively, the force, dynamic stiffness matrix, and displacement on the frequency domain.

Frequency Response Function: The forced dynamic response is commonly obtained in the form of Frequency Response Functions (FRF) [13] that can be written as [12]:

$$H(x, \omega_j) = U(x, \omega_j)/F(x_f; \omega_j), \tag{3}$$

where $F(x_f, \omega_j)$ is the force applied at point x_f, $U(x, \omega_j)$ is the displacement that can the obtained at any x, and $H(x, \omega_j)$ is the FRF. Equations (2) for each structural element can be assembled as a global dynamic stiffness matrix using the direct stiffness method [7]. The assembled linear system of equations for each ω_j can be solved and the forced response obtained. The FRFs can be used to identify the structural model parameters using stochastic methods [13,21].

Prony's Method: Considering that there are q FRF observations along x of both longitudinal or transverse displacements in the frame, Prony's method can be applied to estimate the "observed" wavenumber.

The analytical solution for each frequency ω_j can be written as [7]:

$$W(x, \omega_j) = A(\omega_j)e^{-ikx} + B(\omega_j)e^{ikx} + \cancelto{\simeq 0}{C(\omega_j)} e^{-kx} + \cancelto{\simeq 0}{D(\omega_j)} e^{kx}, \tag{4}$$

where k is the wavenumber for frequency ω_j. In Eq. (4) the first and second terms represent the propagating waves, and the third and fourth terms are related to the evanescent waves, important only at discontinuities such as ends. Therefore, in the far field, far from the ends or discontinuities, it can be considered that $C(\omega_j) \simeq D(\omega_j) \simeq 0$ [2]. For $q+1$ FRFs measured at equally spaced distances (Δx) (that we make equal to the periodic cell length), $x_1 = 0, \ldots, x_{q+1} = q\Delta x$ [8], with $\lambda_1 = -k$, $\lambda_2 = k$, and $V_i = e^{i\lambda_i \Delta x}$ it is possible to write, for $q+1 = 3$, and after multiplying each line $n+1$ by an arbitrary parameter η_n:

$$\begin{aligned} \eta_0 W(x_0, \omega_j) &= \eta_0 A(\omega_j) + \eta_0 B(\omega_j) \\ \eta_1 W(x_1, \omega_j) &= \eta_1 A(\omega_j) V_1^1 + \eta_1 B(\omega_j) V_2^1. \\ \eta_2 W(x_2, \omega_j) &= \eta_2 A(\omega_j) V_1^2 + \eta_2 B(\omega_j) V_2^2 \end{aligned} \tag{5}$$

Now, making $A(\omega_j) = A_1$ and $B(\omega_j) = A_2$, and adding all these equations:

$$\sum_{n=0}^{2} \eta_n W(x_n, \omega_j) = \sum_{j=1}^{2} \left[A_j \sum_{n=0}^{2} \left(\eta_n V_j^n \right) \right], \tag{6}$$

where the parameters η_n are the coefficients of the polynomial $\sum_{n=0}^{2}(\eta_n V_j^n) = 0$. Hence, from Eq. (6), it is possible to obtain the following relations:

$$\sum_{n=0}^{2} \eta_n V_j^n = 0, \tag{7}$$

$$\sum_{n=0}^{2} \eta_n W(x_n, \omega_j) = 0. \tag{8}$$

Without loss of generality, parameter η_2 can be considered equal to one. Then, Eq. (8) can be written as [10]

$$\{W(x_0, \omega_j)\ W(x_1, \omega_j)\} \begin{Bmatrix} \eta_0 \\ \eta_1 \end{Bmatrix} = -W(x_2, \omega_j). \tag{9}$$

When $q + 1 > 3$, the same procedure is used to assemble this matrix equation:

$$\begin{bmatrix} W(x_0, \omega_j) & W(x_1, \omega_j) \\ W(x_1, \omega_j) & W(x_2, \omega_j) \\ \vdots & \vdots \\ W(x_q - 1, \omega_j) & W(x_q, \omega_j) \end{bmatrix} \begin{Bmatrix} \eta_0 \\ \eta_1 \end{Bmatrix} = - \begin{Bmatrix} W(x_2, \omega_j) \\ W(x_3, \omega_j) \\ \vdots \\ W(x_{q+1}, \omega_j) \end{Bmatrix}, \tag{10}$$

or

$$[\boldsymbol{W}]\{\boldsymbol{\eta}\} = -\{\tilde{\boldsymbol{W}}\}. \tag{11}$$

The Moore–Penrose generalized inverse can be used to determine $\{\boldsymbol{\eta}\}$ [6]:

$$\boldsymbol{W}^{\dagger} = (\boldsymbol{W}^{*}\boldsymbol{W})^{-1}\boldsymbol{W}^{*}, \tag{12}$$

and premultiplying both sides of Eq. (11) by Eq. (12), $\{\boldsymbol{\eta}\}$ is obtained in a least-squares sense:

$$\{\boldsymbol{\eta}\} = -[\boldsymbol{W}]^{\dagger}\{\tilde{\boldsymbol{W}}\}. \tag{13}$$

After η_0 and η_1 are computed, they can be used in Eq. (7), where the roots V_j^n may be found with the use of the companion matrix, and the wavenumber can be found using the relationship $V_i = e^{-ik\Delta x}$.

2.2 Statistical Analysis

Statistical analysis is used to check the robustness of a bang gap due manufacturing variability [3]. Bayesian statistics, analogously to the Bayes' theorem, has its basis on the posterior distribution that can be obtained using [20]:

$$\pi(\boldsymbol{\theta} \mid \boldsymbol{x}) = \frac{\pi(\boldsymbol{x} \mid \boldsymbol{\theta})\pi(\boldsymbol{\theta})}{\pi(\boldsymbol{x})}, \tag{14}$$

where $\pi(\boldsymbol{\theta} \mid \boldsymbol{x})$ is the posterior distribution, which carries all information about the parameters, $\pi(\boldsymbol{x} \mid \boldsymbol{\theta})$ is the likelihood function, which carries the information about the observations, $\pi(\boldsymbol{\theta})$ is the prior distribution, where all the information available prior to the observations should be added, and $\pi(\boldsymbol{x})$ is the normalizing constant. In Eq. (14), we assume that the likelihood function and the prior distribution are independent [20].

Prior Distribution: The prior distributions for the stochastic parameters (E and ρ) was built according to the prior information values presented in the literature for the material used (in this work, nylon). For each parameter, a Gamma distribution was fitted considering the minimum value found in literature as the 2.5% quantile and the maximum value its 97.5% quantile.

Likelihood Function: Given that the random mechanical properties E and ρ cannot be negative, a Gamma probability density function (pdf) was assumed, expressed as

$$f(x;\alpha,\beta) = \frac{\beta^\alpha}{\Gamma(\alpha)} x^{\alpha-1} e^{-\beta x} I_{[0,\infty)}, \tag{15}$$

where Γ is the gamma function, and $I_{[0,\infty)}$ is the indicator function. The Gamma is a versatile distribution with shape ($\alpha > 0$) and scale parameter ($\beta > 0$) [20].

Assuming that all the observations come from random variables that are independent and identically distributed (IID), the likelihood function is given as their pdf product:

$$\pi(\boldsymbol{x} \mid \alpha,\beta) = \prod_{i=1}^{N} f(x;\alpha,\beta), \tag{16}$$

where N is the sample size. The natural logarithm of $\pi(\boldsymbol{x} \mid \alpha,\beta)$, can be used in the estimation and inference as well, with the advantage of being easier in terms of mathematical manipulations [9]. The log-likelihood in this case is given by:

$$l(\boldsymbol{x} \mid \alpha,\beta) = N\alpha ln(\beta) - N ln(\Gamma(\alpha)) + (\alpha - 1)\sum_{i=1}^{N} ln(x_i) - \beta \sum_{i=1}^{N} x_i. \tag{17}$$

FRF and Wavenumber Simulation: Considering that E and ρ follow a Gamma distribution, and their values found in the literature were used to estimate, via maximum likelihood, $\alpha_{L,E}$, $\beta_{L,E}$, $\alpha_{L,\rho}$, and $\beta_{L,\rho}$. Using these estimations, N values of E and ρ were simulated as observed values of these distributions. The mean and variance of these samples can be calculated as:

$$\bar{X} = \frac{\sum_{i=1}^{m} x_i}{m}, \text{ and } \quad s^2 = \frac{\sum_{i-1}^{m} (x_i - \bar{X})^2}{m-1}, \tag{18}$$

the mean and variance of these Gamma distributions can also be calculated, and with $E(X)$ and $Var(X)$, it is possible to estimate the parameter:

$$E(X) = \frac{\alpha}{\beta}, \quad Var(X) = \frac{\alpha}{\beta^2} \quad \rightarrow \quad \beta = \frac{E(X)}{Var(X)}, \quad \alpha = E(X)\beta. \tag{19}$$

With the N simulated values of E and ρ, N FRFs can be evaluated and, with Prony's method, N wavenumber values can be estimated from these FRFs.

Posterior Distribution: In order to simulate the posterior distribution from any prior distribution, the MCMC algorithm can be used [20]. This method uses a Monte Carlo simulation combined with a Markov chain random walk in the parameter space to simulate the posterior distribution. Hence, it needs a large number of simulations to converge, and several methods exist to verify its convergence [18]. This methodology generates vectors (chains) that represent the parameter posterior distribution.

Markov Chain Monte Carlo: In the present study, the following algorithm was used to obtain the posterior distribution of E and ρ through the model that generates the wavenumber.

1) while E and ρ vectors are smaller than their defined size:
2) initial values for E (E_0) and ρ (ρ_0) are determined from a range of values that the simulated material usually presents;
3) new values for E (E_1) and ρ (ρ_1) are, respectively, obtained, via random sampling of positive values of these uniform distributions:

$$U(\mathrm{E}_0 - 0.30, \mathrm{E}_0 + 0.30), \tag{20}$$

$$U(\rho_0 - 0.30, \rho_0 + 0.30). \tag{21}$$

   In practice, a sample is taken from these distributions and, if the sample value is negative, it is discarded and a new sample is taken until a positive value is obtained;
4) using Eq. (19), the sample variance as $Var(X)$, and E_i and ρ_i with $i = 0, 1$ as $E(X)$, α and β are given to use in $l(\boldsymbol{x} \mid \alpha, \beta)$, and the following metric (R_i) is calculated for E_0 and ρ_0, and for E_1 and ρ_1:

$$R_i = \left(\frac{1}{\mathrm{MAE}}\right) \times [\pi(\boldsymbol{x} \mid E) \times \pi(E)] \times [\pi(\boldsymbol{x} \mid \rho) \times \pi(\rho)], \quad i = 0, 1. \tag{22}$$

   where MAE is the mean absolute error for each frequency of the generated longitudinal and flexural wavenumbers for the N observed values;
5) R_{10} is calculated:

$$R_{10} = \frac{R_1}{R_0}; \tag{23}$$

6) one value (u) is sampled from the distribution $u \sim U(0; 1)$;
7) the decision of whether discarding or including E_1 and ρ_1 into E and ρ vectors, which represent their distributions, is taken:
   i) include E_1 and ρ_1 in their distribution vectors, and make $\mathrm{E}_0 = \mathrm{E}_1$ and $\rho_0 = \rho_1$ if $R_{10} > 1$, or if $R_{10} > u$;
   ii) reject E_1 and ρ_1 if $R_{10} < u$ and restart from step (ii) with the original values of E_0 and ρ_0.

It is suggested to cut out the first part of the chain ("burn in"), because an initial value distant from the mean influences negatively the parameter representation. In order to make sure that the Markov chain converges, the observations must be independent; however, being a Markov chain, it is dependent of the immediately previous position. A way to overcome this issue is to select the elements that will represent the posterior distribution making periodic jumps in the Markov chain ("thin") [20].

The so-called "burn in" used here was of 500 elements, the so-called "thin" was 3, and the final chain had 5.000 elements, resulting in a raw chain of 15.500 elements.

MCMC Convergence Criteria: There are three methods that present good results for testing the MCMC convergence, and they are already implemented in R package *coda* [17]: Geweke's criterion (|ZG|), where values more extreme than |1.96| indicate lack of convergence, Heidelberger and Welch test (HW-p), which verifies stationarity, where p-values below 0.05 can be interpreted as lack of stationarity, and Raftery and Lewis (1992) factor (RL), where values below 5 indicates no sign of dependence.

Bayesian Estimate and Inference: The estimation and inference on the model parameters consider the posterior distribution in a similar way as for frequentist statistics [20]. Once the posterior distribution is obtained, it can be used to simulate and predict the stochastic model response, through the Monte Carlo method.

One estimator for the posterior distribution is its mean [18]. It is can be defined numerically with the vector that represents the posterior distribution.

The Bayesian γ% credible interval is similar to the frequentist γ% confidence interval, but its interpretation is more intuitive: the probability that the interval contains the "true" parameter. The highest posterior density (HPD) interval is the credible interval (lower HPD_L and upper HPD_U) that contains γ% of the probability with the smallest interval [20].

The Bayes factor (BF) is a tool that makes it possible to test, from the marginal posterior and prior distribution of a parameter θ, the null and alternative hypotheses [11]:

$$H_0 : \theta \in \theta_0 \text{ vs } H_1 : \theta \in \theta_1, \tag{24}$$

where $\theta_0 \cup \theta_1 = \Theta$, with Θ being the parameter space, $\theta_0 \cap \theta_1 = 0$. Given that the Markov chain represents the distribution of θ, it is possible to calculate $p(\theta \in \theta_0)$ through proportion, and, consequently, $p(\theta \in \theta_1) = 1 - p(\theta \in \theta_0)$. The Bayes factor is a ratio of the prior, $o(1,0)$, and posterior, $o(1,0 \mid \boldsymbol{x})$, distributions [20]:

$$FB_{1,0} = \frac{o(1,0 \mid \boldsymbol{x})}{o(1,0)}, \tag{25}$$

where,

$$o(1,0 \mid \boldsymbol{x}) = \frac{\pi(\theta_1 \mid \boldsymbol{x})}{\pi(\theta_0 \mid \boldsymbol{x})}, \text{ and } \quad o(1,0) = \frac{\pi(\theta_1)}{\pi(\theta_0)}. \tag{26}$$

The quantitative BF values can be transformed into qualitative evidence, creating what is called Jeffrey's scale of evidence (Table 1).

Table 1. Jeffrey's scale of evidence.

BF value	Evidence against H_0
1–3	Weak
3–10	Moderate
10–30	Substantial
30–100	Strong
>100	Decisive

3 Results

The considered plane frame element is presented in Fig. 1. In this simple case the longitudinal and flexural behaviors are uncoupled. Combining such frame elements into a planar frame couples these behaviors. Here only this simple case is presented for clarity.

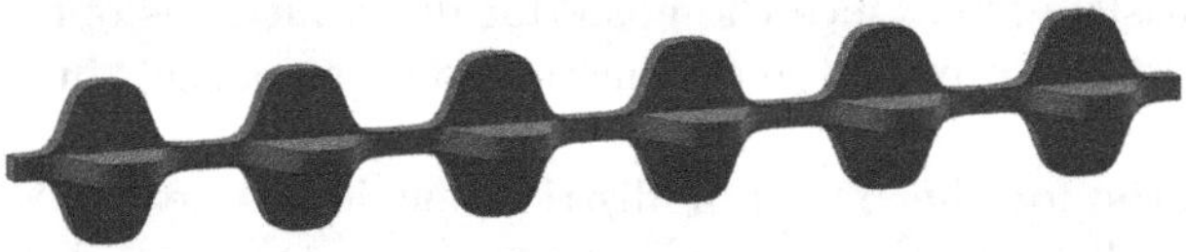

Fig. 1. Plane frame element used in the present study.

Prior Distribution: In the literature, the values of E for the used material (nylon) were 3.5 GPa and 3.8 GPa [22], and 0.72 GPa and 0.86 GPa [9]. For ρ, they were $1.38\,\mathrm{kg/m^3}$ and $1.68\,\mathrm{kg/m^3}$ [5], and $0.70\,\mathrm{kg/m^3}$ and $1.00\,\mathrm{kg/m^3}$ [9]. Then, using the procedure stated in the methodology section, these two prior distributions were identified as:

$$\pi(E) \sim Gamma(6.0269, 3.0806), \qquad \pi(\rho) \sim Gamma(20.54, 18.05), \tag{27}$$

$X \sim Gamma(\alpha, \beta)$ means X follows a Gamma pdf with parameters α and β.

Likelihood Function and Simulated Observation: Using the four values obtained in the literature, Gamma pdfs were fitted for the parameters, as their maximum likelihood estimation. Their parameters $\alpha_{P,L}$ and $\beta_{P,L}$ were:

$$\alpha_{E,L} = 0.0009375, \quad \beta_{E,L} = 0.000625, \quad \alpha_{\rho,L} = 0.000234, \text{ and } \quad \beta_{\rho,L} = 0.000195. \tag{28}$$

Table 2. Simulated values of E and ρ and some necessary statistics.

Property	Simulated values as observation								$\Sigma_{i=1}^{N}(ln(X_i))$	$\Sigma_{i=1}^{N} X_i$
E	1.5981	1.5563	1.4843	1.3478	1.5275	1.4939	1.4812	1.5129	3.2365	12.002
ρ	1.2071	1.1838	1.1423	1.2027	1.2646	1.2055	1.1990	1.2058	1.4648	9.6108

These distributions were used to simulate 8 values, here considered as the observations, $X_1, ..., X_8$, for each parameter E and ρ presented in Table 2. These values were used to obtain the log-likelihood (Eq. (17)).

For each value of E and ρ, longitudinal and flexural FRFs were computed and the "observed" wavenumber was estimated via Prony's method.

Posterior Distribution: Using the software OpenBUGS [23], the posterior distributions were simulated. These posteriors are summarized in Table 3.

Table 3. Estimation, inference and convergence for E and ρ, using OpenBUGS.

Parameter	Estimation and credible interval			Convergence tests		
	mean	$HPD_{95\%,L}$	$HPD_{95\%,U}$	|ZG|	HW-p	RL
E	1.9758	0.1351	5.1538	0.8526	0.2170	1.0000
ρ	0.9176	0.1028	4.9338	0.1045	0.8216	1.0000

Table 4 summarizes the posterior distributions obtained using the MCMC algorithm. Comparing Tables 4 and 3, it can be noticed that the additional information considerably reduced the posterior distribution amplitudes. Figure 2 presents the parameter posterior distributions obtained via the MCMC.

Table 4. Estimation, inference and convergence for E and ρ, for the used MCMC algorithm.

Parameter	Estimation and credible interval			Convergence tests		
	mean	$HPD_{95\%,L}$	$HPD_{95\%,U}$	|ZG|	HW-p	RL
E	1.7022	0.9609	2.4255	0.0278	3.1500	0.1590
ρ	1.2551	0.9099	1.5917	0.1776	1.1200	0.8260

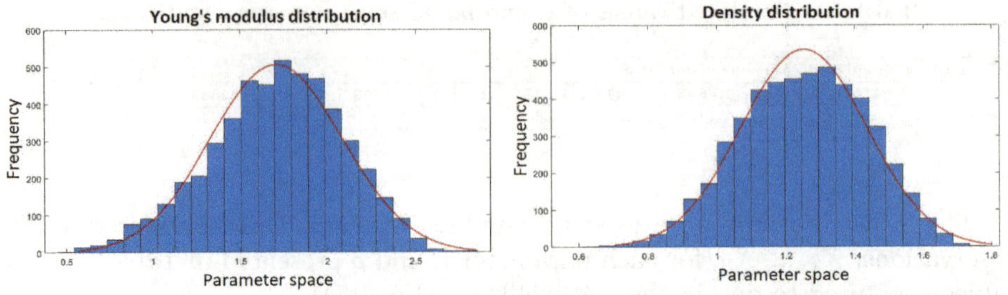

(a) Posterior density and histogram for the Young's modulus.

(b) Posterior density and histogram for the mass density.

Fig. 2. Parameters posterior distribution with the used MCMC.

Figure 3 represents the deterministic wavenumber values (real and imaginary parts) for the proposed plane frame using the mean value for E and ρ obtained via the used MCMC method. In the stochastic analysis, only the imaginary part of the wavenumber values will be used to identify bandgaps.

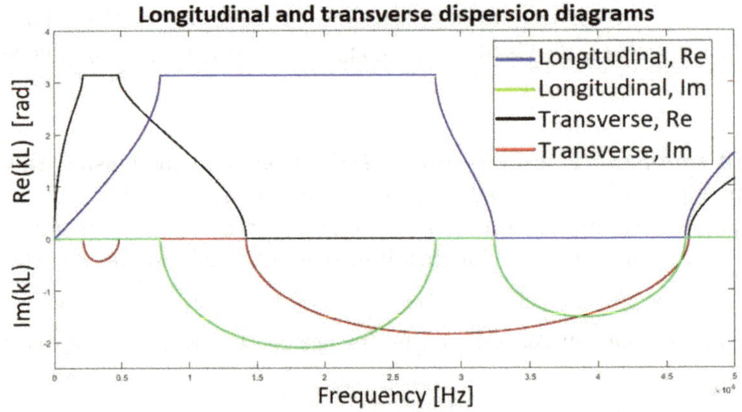

Fig. 3. Deterministic wavenumber values for longitudinal and transverse waves using mean values obtained via the used MCMC algorithm.

Stochastic Response: The convergence of the Monte Carlo method was verified graphically. Using the Monte Carlo method and the distribution presented in Fig. 2, the BFs were used to create intervals against the null hypothesis. Here, the null hypothesis means that the "true" maximum absolute imaginary part of the wavenumber is greater than the tested value. In practice, in Fig. 4, there is a weak evidence that there is a bandgap (7.5–50 kHz), i.e., there is no strong evidence of the bandgap. In Fig. 5a, it is possible to verify that, for longitudinal waves, there are, actually, two bandgaps (8–23 kHz and 33–38 kHz). Figure 5b

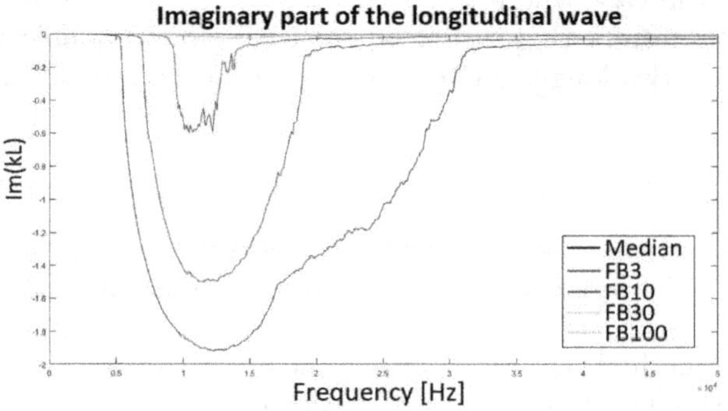

Fig. 4. BF used to infer about the imaginary part of the longitudinal wavenumber.

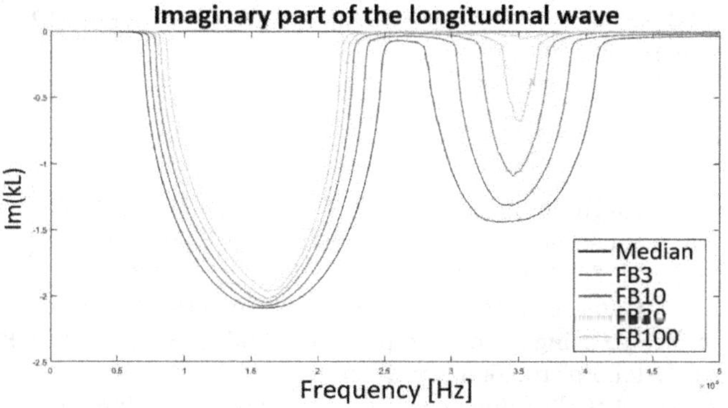

(a) BF used to infer about the imaginary part of the longitudinal wavenumber.

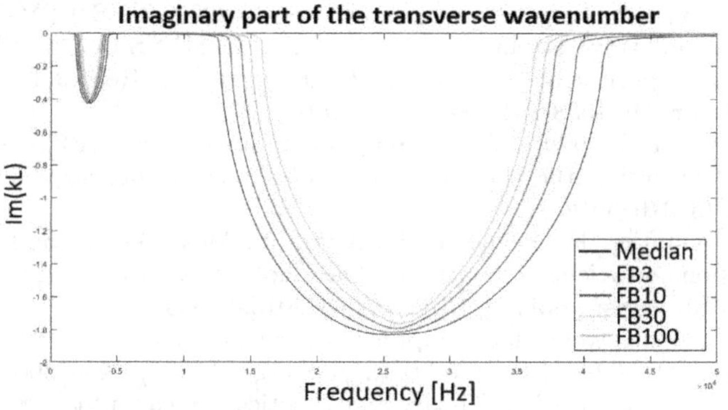

(b) BF used to infer about the imaginary part of the flexural wavenumber.

Fig. 5. Stochastic wavenumber simulated from the posterior distributions obtained via the used MCMC method.

corresponds to the case where using the posterior with the used MCMC also presented better results than using OpenBUGS. It is possible to verify, with decisive evidence, two longitudinal bandgaps (2–4 kHz and 15.5–38 kHz).

4 Conclusions

It was verified that there are two complete and robust bang gaps in a periodic frame element, which will occur in almost all waveguides manufactured with this design.

The MCMC method that incorporated the use of the estimated wavenumbers presented a better behavior, reducing uncertainties when compared to the one using only the prior and the likelihood function.

Acknowledgments. The authors gratefully acknowledge the financial support of the São Paulo Research Foundation (FAPESP) through process number 2019/00315-8 and process number 2018/15894-0, and the Brazilian National Council of Research CNPq (Grant Agreement ID: 420304/2018-5).

References

1. Alkmim, M.H., Fabro, A.T., Morais, M.V.G.: Response variability with random uncertainty in a tuned liquid column damper. Revista Interdisciplinar de Pesquisa em Engenharia. **2**(16), 1–11 (2020). https://doi.org/10.26512/ripe.v2i16.21611
2. Arruda, J.R.F., Campos, J.P., Piva, J.I.: Experimental determination of flexural power flow in beams using a modified prony method. J. Sound Vib. **197**(3), 309–328 (1996). https://doi.org/10.1006/jsvi.1996.0534
3. Beli, D., Arruda, J.R.F.: Influence of additive manufacturing variability in elastic band gaps of beams with periodically distributed resonators. In: Proceedings of the 3rd International Symposium on Uncertainty Quantification and Stochastic Modeling, vol. 1, p. 10 (2016). https://doi.org/10.20906/CPS/USM-2016-0019
4. Beli, D., et al.: Wave attenuation and trapping in 3D printed cantilever-in-mass metamaterials with spatially correlated variability. Sci. Rep. **9**(1), 1–11 (2019). https://doi.org/10.1038/s41598-019-41999-0
5. Castilho, M., et al.: Fabrication of computationally designed scaffolds by low temperature 3D printing. Biofabrication **5**(3), 035012 (2013). https://doi.org/10.1088/1758-5082/5/3/035012
6. Denis, V., Gautier, F., Pelat, A., Poittevin, J.: Measurement and modelling of the reflection coefficient of an acoustic black hole termination. J. Sound Vib. **349**, 67–79 (2015). https://doi.org/10.1016/j.jsv.2015.03.043
7. Doyle, J.F.: A spectral element approach to wave motion in layered solids. J. Vib. Acoust. **114**(4), 569–577 (1992). https://doi.org/10.1115/1.2930300
8. Ewins, D.J.: Modal Testing: Theory and Practice. Research Studies Press, Letchworth (1984)
9. Fabro, A.T., et al.: Uncertainty analysis of band gaps for beams with periodically distributed resonators produced by additive manufacturing. In: Proceedings of the International Conference on Noise and Vibration Engineering (2016)
10. Grosh, K., Williams, E.G.: Complex wave-number decomposition of structural vibrations. J. Acoust. Soc. Am. **93**(2), 836–848 (1993)

11. Jeffreys, H.: Theory of Probability. Oxford University Press, Oxford (1961)
12. Lee, U., Shin, J.: A frequency response function-based structural damage identification method. Comput. Struct. **80**(2), 117–132 (2002). https://doi.org/10.1016/S0045-7949(01)00170-5
13. Machado, M.R., et al.: Estimation of beam material random field properties via sensitivity-based model updating using experimental frequency response functions. MSSP **102**, 180–197 (2018). https://doi.org/10.1016/j.ymssp.2017.08.039
14. Mood, A.M.: Introduction to the Theory of Statistics. McGraw-Hill, New York (1950)
15. Nobrega, E.D., et al.: Vibration band gaps for elastic metamaterial rods using wave finite element method. MSSP **79**, 192–202 (2016). https://doi.org/10.1016/j.ymssp.2016.02.059
16. Nunes, R.F., Klimke, A., Arruda, J.R.F.: On estimating frequency response function envelopes using the spectral element method and fuzzy sets. J. Sound Vib. **291**(3), 986–1003 (2006). https://doi.org/10.1016/j.jsv.2005.07.024
17. Plummer, M., et al.: coda: output analysis and diagnostics for MCMC. Cran (2019). https://cran.r-project.org/web/packages/coda/index.html. Accessed 15 oct 2019
18. Ribeiro, L.H.M.S., et al.: Bayesian modelling of the effects of nitrogen doses on the morphological characteristics of braquiaria grass. Agro@mbiente **12**(4), 245–257 (2018). https://doi.org/10.18227/1982-8470ragro.v12i4.5166
19. Ribeiro, L.H.M.S., et al.: Modelling of ISO 9001 certifications for the American countries: a Bayesian approach. Total Qual. Manag. **30**, 1–26 (2019). https://doi.org/10.1080/14783363.2019.1696672
20. Robert, C.: The Bayesian Choice: From Decision-theoretic Foundations to Computational Implementation. Springer, New York (2007)
21. Souza, M.R., et al.: A Bayesian approach for wavenumber identification of metamaterial beams possessing variability. MSSP **135**, 106437 (2020). https://doi.org/10.1016/j.ymssp.2019.106437
22. Stafford, C.M., et al.: A buckling-based metrology for measuring the elastic moduli of polymeric thin films. Nat. Mater. **3**, 545 (2004). https://doi.org/10.1038/nmat1175
23. Thomas, A., O'Hara, R.B.: OpenBUGS (2004). http://www.openbugs.net/w/Downloads

A Computational Procedure to Capture the Data Uncertainty in a Model Calibration: The Case of the Estimation of the Effectiveness of the Influenza Vaccine

David Martínez-Rodríguez, Ana Navarro-Quiles, Raul San-Julián-Garcés, and Rafael-J Villanueva(✉)

Instituto Universitario de Matemática Multidisciplinar,
Universitat Politécnica de Valéncia, Edificio 8G, piso 2, accesos A y C,
Camino de vera s/n, 46022 Valencia, Spain
{damarro3,annaqui,rausanga,rjvillan}@upv.es,
https://www.imm.upv.es

Abstract. In this paper we propose a technique to estimate the effectiveness of the influenza vaccine. The effectiveness of the vaccine is estimated every year, when the influenza season has finished, analyzing samples of a high number of patients in the emergency departments of the hospitals, and this makes it very expensive.

To do so, our proposal consists of a difference equations model that simulates the transmission dynamics of the influenza where the vaccine effectiveness is included as a parameter to be determined. The proposed technique will calibrate the model parameters taking into account the uncertainty of the data provided from reported cases of influenza. The calibration will return an estimation of the vaccine effectiveness with a 95% confidence interval.

Keywords: Influenza dynamics model · Computational uncertainty quantification · Vaccine effectiveness

1 Introduction

Influenza is an acute disease of the respiratory system caused by the infection of the influenza virus. This disease not only affects to the respiratory system but also includes general symptoms such as headache, fever and weakness, affecting the circulatory and muscular systems.

Influenza virus belong to the Orthomyxoviridae family, and it can be divided into three different genders (A, B and C). The widest and serious infections are caused by A viruses. B viruses are less serious than the A viruses and C viruses symptoms are similar to a common cold [8].

J. E. S. De Cursi (Ed.): Uncertainties 2020, LNME, pp. 374–382, 2021.
https://doi.org/10.1007/978-3-030-53669-5_27

The World Health Organization (WHO) strongly recommends to the population to be vaccinated every year against the influenza, in order to prevent the possibility of infection and to reduce the number of infected people [15].

Vaccination is the main resource to face this disease. The effectiveness of the vaccine is about 50%–80% if there is relationship between the inoculated viruses of the vaccine and the virus strain which circulates in the outbreak. If there is not relationship, the effectiveness might be around 5%–40%. However, the vaccination coverage rates are not as high as desirable, especially in older age groups, the most affected by the influenza and its consequences [5,14].

Every year there are reports with estimations of the influenza vaccine effectiveness in USA and Europe [4,7]. Until now, several techniques have been proposed to estimate the effectiveness of the influenza vaccine *a posteriori* [3]. However, these studies are very expensive and their reliability is currently under discussion because of possible biases of the studies [3,10,13].

The aim of this paper is to propose a technique based on a difference equations system which models the transmission dynamics of the influenza where the vaccine effectiveness is a model parameter to be determined.

Our data do not come from ramdomized controlled trials or observational studies as usual [3], but also from reported cases of influenza [12]. Then, the model is calibrated taking into account the data uncertainty and consequently, the vaccine effectiveness is estimated providing a mean value with a 95% confidence interval.

The paper is organized as follows. Section 2 is devoted to the introduction of the data and their uncertainty. The model is designed and built in the Sect. 3. The description of the procedure to calibrate the model taking into account the data uncertainty is done in Sect. 4. In the Sect. 5 the results of the calibration are presented. The paper finishes with the Sect. 6 where some conclusions are given.

2 Data

The data available represent the weekly reported cases of influenza in the Spanish region of Valencia [1] over a period of 33 weeks between the week 40 (October 2016) ($t = 1$) to week 20 (May 2017) ($t = 33$) [12]. Assuming that an individual is infected of influenza or not, every value of reported cases may be distributed following a binomial whose mean is this weekly datum value. Then, we are able to find the 95% confidence intervals (CI95%) by sampling the binomial distributions and calculating the percentiles 2.5 and 97.5 of these binomial distributions. We denote by $m_1^d, m_2^d, \ldots, m_{33}^d$ the data (mean) of reported cases of influenza and $p_1^d, p_2^d, \ldots, p_{33}^d$ and $P_1^d, P_2^d, \ldots, P_{33}^d$ the percentiles 2.5 and 97.5 in each time instant t, respectively. The data with their CI95% can be shown in the Fig. 1[1].

Note that the data provided consists of reported cases, not total number of cases. This adds difficulty to the problem, because (i) the number of total

[1] The study [12] gives the reported cases per 100 000 people. We have multiplied these values by 10.

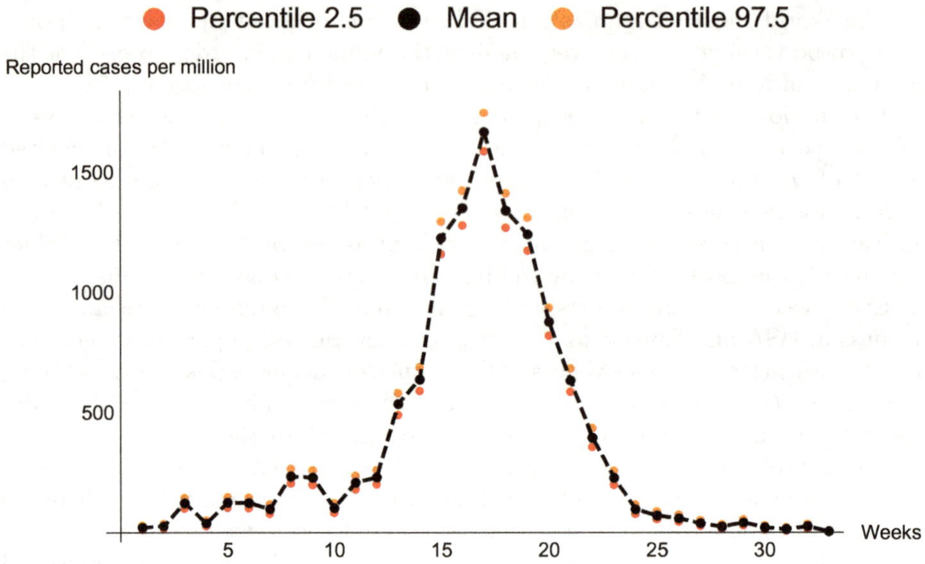

Fig. 1. Weekly reported cases of the influenza in the season 2016–2017 in the Spanish region of Valencia [12] per million people. Black points are the data retrieved from [12] and red and orange points determine the percentiles 2.5 and 97.5 of the binomial distributions describing the data uncertainty.

cases, reported and non reported, are unknown; (ii) the non-reported individuals infected/infectious may also infect other individuals; (iii) the proportion or scale of reported cases respect the total infected individuals is unknown and it will be an additional parameter in the model.

3 Model Building

First, we are going to describe the underlying demographic model. We consider the total population $n_T = 1\ 000\ 000$. Also, taking into account that we are going to study the influenza during 33 weeks, we assume constant population without deaths nor births.

Respect to the influenza, an individual can be: susceptible, when the individual is healthy and has not been previously infected during the current season; infected/infectious, when the individual gets sick by contact with other infected individuals; recovered, when the individual gets healthy again after the infection and acquires total immunity, at least, for the rest of the season. Also, susceptible individuals may be vaccinated. Those who are vaccinated and the vaccine is effective, get total protection against the infection. Thus, taking time t in weeks, we introduce the following subpopulations:

- $S(t)$ denotes the number of healthy individuals in the week t;
- $V(t)$ denotes the number of people vaccinated and the vaccine has been effective obtaining a total protection, in the week t;
- $I(t)$ denotes the number of infected people in the week t;
- $R(t)$ denotes the number of recovered people that will not be infected again, in the week t.

Thus, from the demographic model, $S(t) + V(t) + I(t) + R(t) = n_T = 1\ 000\ 000$, for all $t = 1, 2, \ldots, 33$. We assume the hypothesis of population homogeneous mixing [2,11], that is, any infectious individual can contact and infect any susceptible individual.

A susceptible individual may get infected by contact with an infected individual. Under the hypothesis of homogeneous mixing this transmission can be modeled by the term

$$\beta_t S(t) \frac{I(t)}{n_T},$$

where β_t is the transmission term in the week t.

Also, a susceptible individual may get vaccinated and the vaccine may be effective, becoming protected against the infection. The probability to be vaccinated in the week t, $\zeta(t)$, is usually called coverage, and the effectiveness τ measures how well the influenza vaccine protects against the illness. Thus, the protection by vaccination can be modeled by the term $\zeta(t) \times \tau \times S(t)$.

Furthermore, an infected individual recovers from the disease, cannot be infected again during the season and moves to the recovered stated after a while. This can be modeled by the term $\gamma I(t)$, where $1/\gamma$ is the average time to get recovered [2].

Then, the following nonlinear system of difference equations given by (1) describes the transmission dynamics of the influenza disease over the time.

$$\begin{aligned} S(t+1) &= S(t) - \beta_t S(t) \tfrac{I(t)}{n_T} - \zeta(t) \times \tau \times S(t), \\ V(t+1) &= V(t) + \zeta(t) \times \tau \times S(t), \\ I(t+1) &= I(t) + \beta_t S(t) \tfrac{I(t)}{n_T} - \gamma I(t), \\ R(t+1) &= R(t) + \gamma I(t). \end{aligned} \tag{1}$$

The influenza dynamics can be described graphically as shown in Fig. 2. Some of the model parameters have known values. The average time to recover from the influenza is $1/\gamma = 1$ week [9]. Also, the vaccination campaign lasts 14 weeks, from the week 43 of 2016 ($t = 4$) to the week 4 of 2017 ($t = 17$), and the global vaccine coverage is 43.55% [12]. Therefore, if we assume an equitable distribution of the vaccinated people per week, we have,

$$\zeta(t) = \begin{cases} \frac{0.4355}{14} & \text{if } 4 \leq t \leq 17, \\ 0 & \text{otherwise.} \end{cases}$$

4 Model Calibration with Uncertainty

The unknown model parameters are the transmission ones $\beta_1, \ldots, \beta_{33}$ and the effectiveness τ. Given a set of model parameter values

$$\Gamma = (\beta_1, \ldots, \beta_{33}, \tau),$$

we substitute them into the model, taking as initial condition a sampled value of the binomial distribution of the data initial condition $t = 1$, and run it. We, then, obtain the model output for the subpopulations $S(t)$, $V(t)$, $I(t)$ and $R(t)$ for $t = 1, 2, \ldots, 33$. As we are especially interested on the infected $I(t)$ subpopulation in order to compare to the data, for a given set of parameters $\Gamma = (\beta_1, \ldots, \beta_{33}, \tau)$ we denote

$$M(\Gamma) = (I_1^{\Gamma}, I_2^{\Gamma}, \ldots, I_{33}^{\Gamma}),$$

only the number of infected population every week returned by the model.

The goal of this section is to find sets of model parameter values $\Gamma_1, \ldots, \Gamma_N$ whose infected outputs $M(\Gamma_1), \ldots, M(\Gamma_N)$ capture the data uncertainty given by the CI95% appearing in the Fig. 1. To do so, we propose two main steps: (i) we look for model outputs that lie inside or are close to the data CI95%; (ii) among all of the models outputs generated, we select those which mean and CI95% be as close as possible to the data mean and CI95%.

First, we need to define the distance from a point to an interval as follows: let $J = [a, b]$ a real interval and $x \in \mathbb{R}$, the distance from x to J is

$$d(x, J) = d(x, [a, b]) = \begin{cases} 0 & \text{if } x \in J, \\ \min\{|x - a|, |x - b|\} & \text{if } x \notin J. \end{cases}$$

To address (i), we should search model outputs in a targeted way and we propose the computational optimization algorithm Random Particle Swarm Optimization (rPSO) [6]. We consider as the fitness function

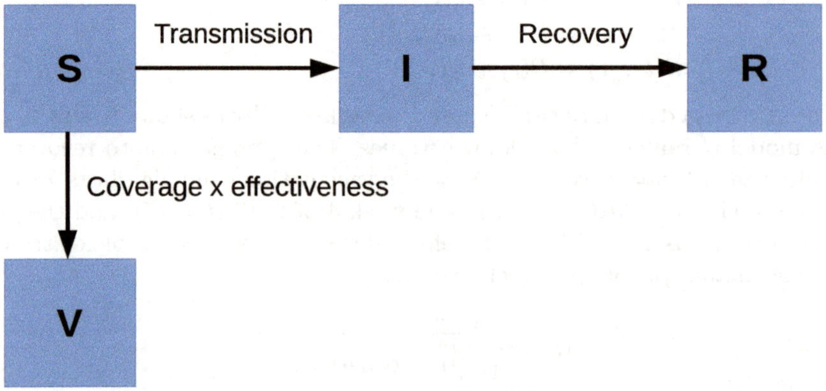

Fig. 2. Flow diagram of the influenza transmission dynamics.

$$F(\Gamma) = \sum_{t=1}^{33} d(I_t^{\Gamma}, [s \times p_t^d, s \times P_t^d]),$$

where $M(\Gamma) = (I_1^{\Gamma}, I_2^{\Gamma}, \dots, I_{33}^{\Gamma})$ and the scale s, that is, the value that multiply the reported cases to convert them into infected cases. s is also calibrated. Therefore, a set of model parameter values Γ should include the scale s.

In order to minimize this fitness function, we use rPSO algorithm to perform several calibrations. We store the results and the initial number of infected people of all the evaluations. Remember that the initial condition is sampled from the binomial distribution of the datum $t = 1$. The aim of performing several calibrations with rPSO is that we need to guarantee that we have model outputs (infected) around the data CI95%.

Once we have enough model outputs (infected), we are going to select the appropriate model outputs, associated to their corresponding model parameter values, in such a way their mean and CI95% be as close as possible to the data CI95%. To perform this selection, we propose an algorithm inspired in PSO and named PSO_S that uses groups of model outputs (infected) and select some of them such that the mean and the CI95% be as close as possible to the mean and CI95% of the data. The algorithm is as follows:

Input: S, the number of model outputs to be selected; $m_1^d, \dots, m_{33}^d$, $p_1^d, \dots, p_{33}^d$ and $P_1^d, \dots, P_{33}^d$ the data mean and the percentiles 2.5 and 97.5 in each time instant t (as described in Sect. 2, Fig. 1).

Step 1: *(particles initialization)* Initialize $\Theta_1, \dots, \Theta_n$ sets of S model outputs (particles) randomly chosen from the model outputs (infected) obtained in the rPSO calibrations.

Step 2: *(fitness)* For each particle Θ_i, $i = 1, \dots, n$, calculate, for every week $t = 1, \dots, 33$, the mean and the percentiles 2.5 and 97.5 of the S model outputs in Θ_i, obtaining the vectors of mean $m_1, \dots, m_{33}$, percentiles 2.5 $p_1, \dots, p_{33}$ and percentiles 97.5 $P_1, \dots, P_{33}$. Then, the fitness (error) function is given by

$$\sum_{t=1}^{33} (m_t - m_t^d)^2 + (p_t - p_t^d)^2 + (P_t - P_t^d)^2.$$

This fitness function measures the proximity of the model outputs of Θ_i to data, considering the uncertainty.

Step 3: *(updating the local best)* As in the regular PSO, for $i = 1, \dots, S$, if the fitness of Θ_i is below the local best particle Θ_i^{local} fitness, Θ_i^{local} is updated.

Step 4: *(updating the global best)* As in the regular PSO, if there is a particle Θ_i which fitness is below the global best particle Θ^{global} fitness, Θ^{global} is updated.

Step 5: *(updating the particles)* For each particle Θ_i, $i = 1, \dots, n$, update the particle choosing randomly S model outputs from the union of

- the current particle Θ_i,

- its local best particle Θ_i^{local},
- the best global particle Θ^{global} and
- n models outputs chosen randomly from all the evaluations.

Step 6: If the maximum number of iterations is reached, the algorithm stops and returns the current global. Otherwise, go to Step 2.

The inclusion of the n models outputs chosen randomly in Step 5, introduces variety and space exploration in the algorithm.

Once the best global set of model outputs Θ^{global} has been returned by the algorithm, the model parameter values associated to these model outputs will be the ones which outputs best capture the data uncertainty. Then, taking these model parameter values, for every model parameter we can calculate their mean and their CI95%. These will be considered as estimations (with CI95%) of the vaccine effectiveness and the transmission rate of the disease every week.

5 Results

We perform 10 calibrations of the model as indicated in the previous section using rPSO. In total, we perform 50 000 evaluations of the model and calculate their fitnesses.

Then, we apply the selection algorithm PSO_S to select the $n = 100$ model outputs among the 50 000 that best capture the data uncertainty. Taking into account these 100 model outputs (infected), in Fig. 3 we can see the comparison between

- the mean and the CI95% of the reported infected data multiplied by the scale, during the season 2016–2017, and
- the mean and the CI95% of the model output (infected) of the calibrated model.

It can be observed that the calibration confidence interval of the real data is included in the confidence interval of the model outputs. It is clear that the tendency in the dynamics of both is quite similar.

Regarding the 100 model parameter values corresponding to the model outputs selected by PSO_S, we obtain that the scale, that is, the factor that transform the number of the reported cases into number of real cases, is 1.27 CI95% $[1.20, 1.38]$ and the influenza vaccine effectiveness is 17.6% CI95% $[7.50\%, 26.33\%]$. As we can see, the effectiveness is low.

6 Conclusion

Here we propose a novel technique based on a difference equations system model to determine the effectiveness of the influenza vaccine in the population the Spanish region of Valencia during the influenza season 2016–2017. This technique takes into account the data uncertainty and the fact that the data correspond to reported cases instead of real number of cases.

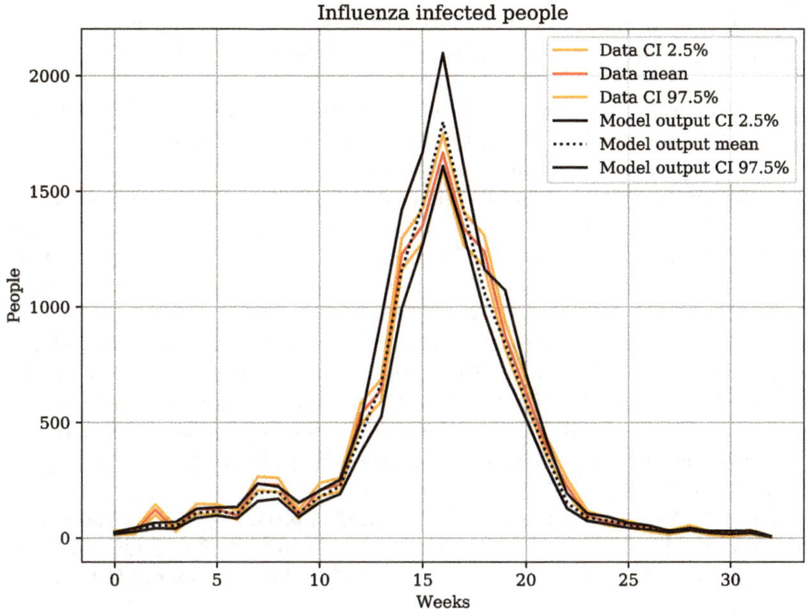

Fig. 3. Reported data of the infected multiplied by the scale compared with the model output (infected) of the 100 calibrated and selected by the PSO_S.

If we compare our results 17.6% CI95% [7.50%, 26.33%] with the report of the I-MOVE multicentre case control studies at primary care and hospital levels in Europe [7] 25.7% CI95% [1.5%, 43.9%], our results are not far, taking into account that we use reported cases and they collect data from hospitals in the whole Europe, that is, the methods are very different. Nevertheless, we must say that the uncertainty provided by our procedure is more reduced. Anyway, both methods confirm the suboptimal performance of inactivated influenza vaccine against influenza in the season 2016–2017. Also, our technique is cheaper.

This technique might help the public health policymakers to establish a proper vaccination calendar, and even make cost-effectiveness studies of the vaccination policies implemented by the public authorities.

Acknowledgments. – This paper has been supported by the Spanish Ministerio de Economia y Competitividad grant MTM2017-89664-P.

– This paper has been supported by the European Union through the Operational Program of the [European Regional Development Fund (ERDF)/European Social Fund (ESF)] of the Valencian Community 2014–2020. Files: GJIDI/2018/A/010 and GJIDI/2018/A/009.

References

1. Autonomous Community of Valencia, Spain. https://en.wikipedia.org/wiki/Valencian_Community/

2. Brauer, F., Castillo-Chavez, C.: Mathematical Models in Population Biology and Epidemiology. Springer, New York (2012). https://doi.org/10.1007/978-1-4614-1686-9
3. Center for Disease Control and Prevention (CDC): How Flu Vaccine Effectiveness and Efficacy are Measured. Questions & Answers. https://www.cdc.gov/flu/vaccines-work/effectivenessqa.htm. Accessed 22 Jan 2020
4. Center for Disease Control and Prevention (CDC): Seasonal influenza vaccine effectiveness (2016–2017). https://www.cdc.gov/flu/vaccines-work/2016-2017.html
5. Clar, C., Oseni, Z., Flowers, N., Keshtkar-Jahromi, M., Rees, K.: Influenza vaccines for preventing cardiovascular disease. Cochrane Database Syst. Rev. (2015). https://doi.org/10.1002/14651858.cd005050.pub3
6. Khemka, N., Jacob, C.: Exploratory toolkit for evolutionary and swarm-based optimization. Math. J. **11**(3), 376–391 (2010). https://doi.org/10.3888/tmj.11.3-5
7. Kissling, E., Rondy, M.: Early 2016/17 vaccine effectiveness estimates against influenza A(H3N2): I-MOVE multicentre case control studies at primary care and hospital levels in Europe. Eurosurveillance **22**(7), 30464 (2017). https://doi.org/10.2807/1560-7917.es.2017.22.7.30464
8. Krammer, F., Smith, G.J.D., Fouchier, R.A.M., Peiris, M., Kedzierska, K., Doherty, P.C., Palese, P., Shaw, M.L., Treanor, J., Webster, R.G., García-Sastre, A.: Influenza. Nat. Rev. Dis. Primers **4**, 1 (2018). https://doi.org/10.1038/s41572-018-0002-y
9. Longo, D.L., Fauci, A.S., Kasper, D.L., Hauser, S., Jameson, J.L., Loscalzo, J., et al.: Harrison's Principles of Internal Medicine, vol. 2, pp. 1209–1214. McGraw-Hill, New York (2015)
10. McLean, H.Q., Thompson, M.G., Sundaram, M.E., Meece, J.K., McClure, D.L., Friedrich, T.C., Belongia, E.A.: Impact of repeated vaccination on vaccine effectiveness against influenza A(H3N2) and B during 8 seasons. Clin. Infect. Dis. **59**(10), 1375–1385 (2014). https://doi.org/10.1093/cid/ciu680
11. Murray, J.D. (ed.): Mathematical Biology. Springer, New York (2004). https://doi.org/10.1007/b98868
12. Portero, A., Alguacil, A.M., Sanchis, A., Pastor, E., López, A., Miralles, M.T., Sierra, M.M., Alberich, C., Roig, F.J., Lluch, J.A., Paredes, J., Vanaclocha, H.: Prevención y vigilancia de la gripe en la Comunitat Valenciana. Temporada 2016–2017 (Prevention and surveillance of the influenza in the Community of Valencia. Season 2016–2017. Generalitat Valenciana. Conselleria de Sanitat Universal i Salut Pública (2017). http://publicaciones.san.gva.es/publicaciones/documentos/IS-150.pdf
13. Sullivan, S.G., Kelly, H.: Stratified estimates of influenza vaccine effectiveness by prior vaccination: caution required. Clin. Infect. Dis. **57**(3), 474–476 (2013). https://doi.org/10.1093/cid/cit255
14. Warren-Gash, C., Smeeth, L., Hayward, A.C.: Influenza as a trigger for acute myocardial infarction or death from cardiovascular disease: a systematic review. Lancet Infect. Dis. **9**(10), 601–610 (2009). https://doi.org/10.1016/s1473-3099(09)70233-6
15. World Health Organization (WHO): Recommended composition of influenza virus vaccines for use in the 2019–2020 northern hemisphere influenza season. https://www.who.int/influenza/vaccines/virus/recommendations/2019_20_north/en/

Optimization

Uncertainty Quantification of Pareto Fronts

Mohamed Bassi[1(✉)], Emmanuel Pagnacco[1(✉)], Roberta Lima[1,2], and Eduardo Souza de Cursi[1(✉)]

[1] LMN, INSA Rouen Normandie, Normandie Université, 76801 Saint-Etienne du Rouvray, France
bassi.mohamed@gmail.com, {Emmanuel.Pagnacco,souza}@insa-rouen.fr
[2] Department of Mechanical Engineering, PUC - Rio, Rio de Janeiro RJ, Brazil
robertalima@puc-rio.br

Abstract. Uncertainty quantification of Pareto fronts introduces new challenges connected to probabilities in infinite dimensional spaces. Indeed, Pareto fronts are, in general, manifolds belonging to infinite dimensional spaces: for instance, a curve in bi-objective optimization or a surface in three objective optimization. This article examines the methods for the determination of means, standard deviations and confidence intervals of Pareto fronts. We show that a punctual mean is not adapted and that the use of chaos expansions may lead to difficulties. Then we propose an approach based on a variational characterization of the mean and we show that it is effective to generate statistics of Pareto fronts. Finally, we examine the use of expansions to generate large samples and evaluate probabilities connected to Pareto fronts.

Keywords: Uncertainty quantification · Multiobjective optimization · Pareto fronts · Chaos expansions

1 Introduction

In the framework of multiobjective optimization, the solutions are generally represented by Pareto fronts giving the possible tradeoffs. If uncertainty has to be considered, Pareto fronts become random variables and we are interested in the analysis of their uncertainty and robustness. For instance, we may look for, on the one hand, their probability distribution s and, on the other hand, confidence intervals and statistics such as their mean or the variance.

Such an analysis face some difficulties connected to the operational use of probabilities in infinite dimensional objects. For instance, the construction of confidence interval needs the definition of quantiles in convenient functional space containing all the possible Pareto fronts - which is an infinite dimensional vector space (recall that a confidence interval is a region having a given probability, so that confidence intervals are closely connected to quantiles). We underline that the situation is analogous when considering orbits and limit cycles

J. E. S. De Cursi (Ed.): Uncertainties 2020, LNME, pp. 385–395, 2021.
https://doi.org/10.1007/978-3-030-53669-5_28

of uncertain dynamical systems: we are interested in the distributions of probabilities and the associated statistics for manifolds - both the situations lead to a main difficulty concerning the definition of probability distributions and statistics in infinite dimensional spaces having functions as elements.

For instance, let us consider a biobjective problem involving the objective functions (f_1, f_2): in such a situation, the Pareto front Π is generally a curve in $\mathbb{R}^2$, which is generally described by one of two following manners:

i) implicit description: Π is the set of solutions of the equation $\psi(\mathbf{x}) = 0$, i. e., $\Pi = \{\mathbf{x} \in S \subset \mathbb{R}^2 : \psi(\mathbf{x}) = 0\}$;
ii) parametric description: the points of Π are generated by a vector function associating an interval $I \subset \mathbb{R}$ to a set of points in $\mathbb{R}^2$ $(I \ni t \to \mathbf{x}(t) \in \mathbb{R}^2)$.

We are interested in the situation where the curve depends upon a random variable $\mathbf{Z} \in \mathcal{Z} \subset \mathbb{R}^m$: denoting by $\mathbf{z}$ a variate from $\mathbf{Z}$, we have

i) implicit description: the equation becomes $\psi(\mathbf{x}, \mathbf{z}) = 0,\ \mathbf{x} \in S(\mathbf{z}) \subset \mathbb{R}^2$, so that $\Pi(z) = \{\mathbf{x} \in S(\mathbf{z}) \subset \mathbb{R}^2 : \psi(\mathbf{x}, \mathbf{z}) = 0\}$;
ii) parametric description: the map reads as $\mathbf{x}(t, \mathbf{z}) : I(\mathbf{z}) \to \mathbb{R}^2$. In the sequel, we denote $\mathbf{x}_t$ the function $\mathbf{z} \to \mathbf{x}(t, \mathbf{z})$ and $\mathbf{x_z}$ the function $t \to \mathbf{x}(t, \mathbf{z})$.

To produce statistics of $\mathbf{x_z}$, a first idea consists in applying the following procedures:

i) implicit description: take the conditional mean of x_1 for a given x_2: the "mean" curve is given by the equation $\overline{\psi}(x, y) = x - E(x_1 \mid x_2 = y) = 0$. This is equivalent to take the mean of x_1 for a fixed x_2. Conversely, we may consider the mean of x_2 for a fixed x_1: the "mean" curve is $\overline{\psi}(x, y) = y - E(x_2 \mid x_1 = x) = 0$. If S is independent from Z, this approach is equivalent to consider the mean equation $E(\varphi(\mathbf{x}, \mathbf{z})) = 0,\ \mathbf{x} \in S$. When the equation reads as $\psi(\mathbf{x}, \mathbf{z}) = x_2 - \phi(x_1, \mathbf{z}) = 0$, this approach corresponds to taking $y = \int \phi(x, \mathbf{z}) P(\mathbf{z} \in d\mathbf{z})$. Analogously, if $x_1 = \phi(x_2, \mathbf{z})$, it corresponds to taking the mean of x_1 for a fixed x_2: $x = \int \phi(y, \mathbf{z}) P(\mathbf{z} \in d\mathbf{z})$.
ii) parametric description: consider $\mathbf{x}(t, \mathbf{z}) \in \mathbb{R}^2$ as being a stochastic process $\{\mathbf{x}_t : t \in I\}$ indexed by t and look for the distribution of $\mathbf{x}_t$ for a given t, what furnishes a mean $E(\mathbf{x}_t)$ for each t.

Unfortunately, both these approaches lead to unsatisfactory results. Let us illustrate this difficulty by considering the situation where $f_1(u, Z) = u^2$, $f_2(u, Z) = (u - Z)^2$, where Z is uniformly distributed on $(0, 1)$. In this simple situation, the Pareto's front may be determined analitically and represented by both the ways introduced above, but no one of the representations is unique, since the family of curves may be represented by using different parameterizations or different implicit equations. Let z be a variate from Z: to determine the Pareto's front, we may consider convex combinations $f_\theta(u, z) = (1 - \theta) f_1(u, z) + \theta f_2(u, z)$ and determine the value u^* corresponding to the minimum of f_θ for $0 \le \theta \le 1$ - we have $u^* = z\theta$, so that $f_1(u^*, z) = \theta^2 z^2$, $f_2(u^*, z) = (1 - \theta)^2 z^2$. To apply the implicit approach, we may set $x_1 = \theta^2 z^2$ and $x_2 = (1-\theta)^2 z^2$, so that $\theta z = \sqrt{x_1}$ and $z - \theta z = \sqrt{x_2}$. Thus, the Pareto's front is given by $\psi(\mathbf{x}, z) = \sqrt{x_1} + \sqrt{x_2} - z = 0$. We may write the front under the form $x_2 = \phi(x_1, \mathbf{z})$ with $\phi(x, z) = x - 2z\sqrt{x} + z^2$. Then, this approach furnishes as mean Pareto's front the equation $y = x - \sqrt{x} + 1/3$, which is not a representative of the family of curves.

When considering the parametric approach, the first difficulty is the choice of the parameterization. For instance, we may consider a first parameterization $\mathbf{x}(t, z) = \left(t^2 z^2, (1-t)^2 z^2\right)$, $t \in (0, 1)$: in this case, $E(\mathbf{x}_t) = \left(t^2, (1-t)^2\right)/3$, $t \in (0, 1)$ - this result corresponds to the equation $y = x - 2\sqrt{x} + 1$, which corresponds to the curve of the family generated by $z = 1/\sqrt{3}$. But we may also consider a second parameterization $\mathbf{x}(t, z) = \left(t, t - 2z\sqrt{t} + z^2\right)$, $t \in (0, 1)$: in this case, $E(\mathbf{x}_t) = \left(t, t - \sqrt{t} + 1/3\right)$, which corresponds to a mean Pareto's front having as equation $y = x - \sqrt{x} + 1/3$ - the same result as in the implicit approach. A third parameterization $\mathbf{x}(t, z) = \left(t - 2z\sqrt{t} + z^2, t\right)$, $t \in (0, 1)$ will furnish a result different from the preceding. Thus, on the one hand, the result is not independent from the parameterization and, on the other hand, it may be not a representative of the family. Figure 1 illustrates these considerations.

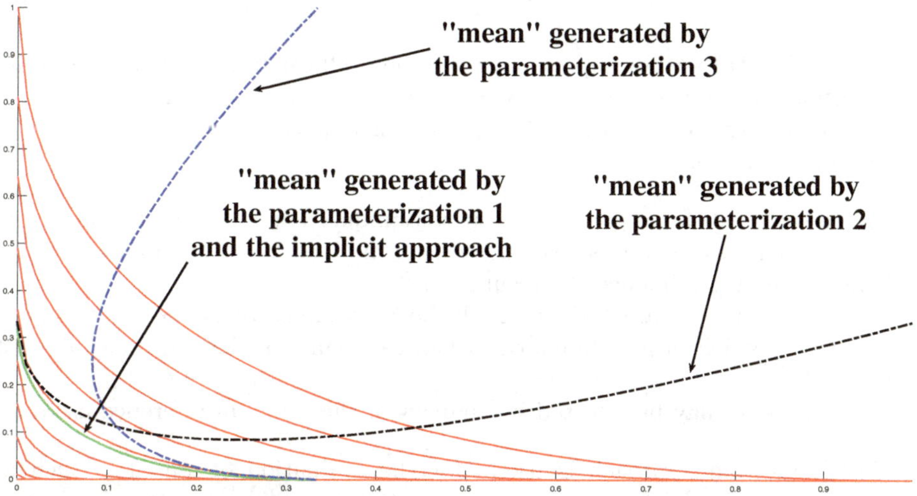

Fig. 1. Inadequacy of the punctual approaches: the results may depend on the parameterization and do not represent the family of curves.

Indeed, these approaches are not adequate in general situations. To get adequate results, we may consider two alternative approaches *using the parametric approach*:

i) The Hilbertian approach, which brings the analysis to probabilities on spaces formed of sequences of real numbers, such as, ℓ^2 [8,9]. It produces means, envelopes and suggests a method for infinite dimensional optimization [10,11]. For this approach, hard mathematical questions remain unsolved: independence from the basis, choice of the parameterizations.

ii) A second approach consists in the use of the variational characterization of the mean and of the median [4].

In the sequel, these approaches are introduced. For the sake of simplicity, we consider curves, but the arguments exposed extend to manifolds.

2 The Hilbertian Approach

Most of the arguments in this section were developed in [2]. Assume that $\mathbf{x}$ is an element from a Hilbert space V of inner product $(\bullet, \bullet)$. Consider a basis or total family $\Phi = \{\varphi_i\}_{i \in \mathbb{N}^*}$. Then,

$$\mathbf{x} = \sum_{i \in \mathbb{N}^*} \mathbf{x}_i \varphi_i \approx P_n \mathbf{x} = \sum_{1 \leq i \leq n} \mathbf{x}_i \varphi_i \tag{1}$$

We have

$$\lim_{n \to \infty} P_n \mathbf{x} = \mathbf{x}. \tag{2}$$

As usual in Hilbertian approximations, the coefficients $\mathbf{X} = (\mathbf{x}_1, \ldots, \mathbf{x}_n)$ are the solutions of a linear system $\mathbf{AX} = \mathbf{B}$, with $A_{ij} = (\varphi_i, \varphi_j)$, $B_i = (\mathbf{x}, \varphi_i)$. If the family is orthonormal, then $\mathcal{A}$ is the identity matrix and $\mathbf{x}_i = (\mathbf{x}, \varphi_i)$. The main difficulties are:

1. the choice of Φ: for a given V, it is not unique. For instance, in dimension one, we may use wavelets, trigonometric functions or polynomial functions. The choice may influence the results.
2. the choice of the parameterization: all the curves must be defined on the same interval I, which brings difficulties when each curve is defined on a different interval.
3. the expansion may be affected by changes in the frame of reference used.

The Hilbertian approach is useful for the evaluation of statistics such as means and covariances, but not for the generation of confidence intervals - it generates envelopes containing predetermined quantile, but not confidence intervals (see, for instance, [2]).

2.1 Classical UQ Approach

The classical UQ approach uses $V = L^2(\Omega, \mu) = \left\{\mathbf{U} : E\left(|\mathbf{U}|^2\right) < \infty\right\}$, where Ω is the universe and μ is a probability measure on Ω. The inner product is $(\mathbf{U}, \mathbf{V}) = E(\mathbf{U}.\mathbf{V})$. It assumes that $\mathbf{Z} \in V$ and, for any t, $\mathbf{x}_t \in V$. Considering

$\Phi = \{\varphi_i(\mathbf{z})\}_{i\in\mathbb{N}^*}$, the random vector $\mathbf{x}_t$ has an expansion given by Eq. (1) for each t. Thus, we have (see [1,3,9]):

$$\mathbf{x}(t,\mathbf{z}) = \sum_{i\in\mathbb{N}^*} \mathbf{x}_i(t)\varphi_i(\mathbf{z});\ P_n\mathbf{x}(t,\mathbf{z}) = \sum_{1\leq\ i\ \leq n} \mathbf{x}_i(t)\varphi_i(\mathbf{z}). \tag{3}$$

In this case, $\mathbf{X}(t) = (\mathbf{x}_1(t), \ldots, \mathbf{x}_n(t))$ and the linear system reads as $\mathcal{A}\mathbf{X}(t) = \mathcal{B}(t)$, where $\mathcal{A}_{ij} = (\varphi_i,\ \varphi_j)$ and $\mathcal{B}_i(t) = (\mathbf{x}(t), \varphi_i)$. We estimate the mean as

$$E(\mathbf{x}(t,\mathbf{Z})) = \sum_{i\in\mathbb{N}^*} \mathbf{x}_i(t)E(\varphi_i(\mathbf{Z})) \approx \sum_{1\leq\ i\ \leq n} \mathbf{x}_i(t)E(\varphi_i(\mathbf{Z})) = E(P_n\mathbf{x}(t,\mathbf{Z})). \tag{4}$$

Other statistics may be defined into an analogous way. It is expected that $E(P_n\mathbf{x}(t,\mathbf{z})) \longrightarrow E(\mathbf{x}(t,\mathbf{z}))$ for $n \to \infty$. Envelopes may be generated by solving the optimization problem

$$\text{Find } \mathbf{y}^* = (y_1^*, \ldots, y_n^*) = \arg\max \left\{ \left\| \sum_{1\leq\ i\ \leq n} \mathbf{x}_i(t)y_i \right\| : y_i \in C_i \right\}. \tag{5}$$

where, for each i, C_i is a confidence interval for $\varphi_i(\mathbf{Z})$. Notice that the problem formulated in Eq. 5 admits multiple solutions, which must be all determined - for each t - to generate a correct envelope - this is a severe limitation of this approach to generate envelopes when n increases. Nevertheless, it is useful to generate means and covariances.

2.2 Inverted UQ Approach

[5] proposed a different approach, using an expansion on the variable t and coefficients depending on $\mathbf{z}$, what is equivalent to consider $V = L^2(0,T)$, with the inner product $(\mathbf{u},\mathbf{v}) = \int_0^T \mathbf{u}(t).\mathbf{v}(t)dt$. Then, $\Phi = \{\varphi_i(t)\}_{i\in\mathbb{N}^*}$ and

$$\mathbf{x}(t,\mathbf{z}) = \sum_{i\in\mathbb{N}^*} \mathbf{x}_i(\mathbf{z})\varphi_i(t). \tag{6}$$

In this case, we evaluate the mean as

$$E(\mathbf{x}(t,\mathbf{Z})) = \sum_{i\in\mathbb{N}^*} E(\mathbf{x}_i(\mathbf{Z}))\varphi_i(t) \approx \sum_{1\leq\ i\ \leq n} E(\mathbf{x}_i(\mathbf{Z}))\varphi_i(t) = E(P_n\mathbf{x}(t,\mathbf{z})). \tag{7}$$

Here, confidence intervals may be determined for the coefficients $\mathbf{x}_i$, so that we may find envelopes by an analogous way:

$$\text{Find } \mathbf{y}^* = (\mathbf{y}_1^*, \ldots, \mathbf{y}_n^*) = \arg\max \left\{ \left\| \sum_{1\leq\ i\ \leq n} \mathbf{y}_i\varphi_i(t) \right\| : \mathbf{y}_i \in C_i \right\}. \tag{8}$$

Here, C_i is a confidence interval for $\mathbf{x}_i(\mathbf{Z})$. Analogously to the preceding one, this approach has a severe limitation when the dimensions increase, but remains useful to generate means and covariances.

2.3 Mixed Approach

In the mixed approach, we consider expansions on the couple $(t, \mathbf{z})$: in this case, $V = L^2(0,T) \times L^2(\Omega, \mu)$ and

$$\mathbf{x}(t, \mathbf{z}) = \sum_{i \in \mathbb{N}^*} \mathbf{x}_i \varphi_i(t, z). \tag{9}$$

The basis Φ may be generated by a tensor product between a basis of $L^2(0,T)$ and a basis of $L^2(\Omega, \mu)$.

3 The Variational Approach

The difficulties found in the Hilbertian approach are connected with the parameterization. In [4], a variational approach generates solutions that are independent from the parameterization: the mean corresponds to the curve of the family minimizing the distance to the rest of the family. Thus, for a random variable A defined on a discrete universe $\Omega = \{\omega_1, ..., \omega_k\}$, the mean satisfies

$$E(A) = \arg\min \left\{ \frac{1}{k} \sum_{i=1}^{k} (A(\omega_i) - y)^2 : y \in \mathbb{R} \right\}. \tag{10}$$

This equality suggests the determination of the mean by the solution of an optimization problem: the mean of A minimizes the distance to all the possible values of A. In terms of curves, we may determine the curve that minimizes the distance to the whole family of curves:

$$E(\mathbf{x}) = \arg\min \left\{ \text{dist}(\mathbf{x}, \mathbf{y}) : \mathbf{y} : \mathrm{I} \to \mathbb{R}^2 \right\}. \tag{11}$$

It often occurs that the mean *is not* an element of the family $\mathcal{F} = \{\mathbf{x}(t, \mathbf{z}) : \mathbf{z} \in \mathcal{Z}\}$. If we are interested in determining a member of the family, we may consider

$$\text{med}(\mathbf{x}) = \arg\min \{\text{dist}(\mathbf{x}, \mathbf{y}) : \mathbf{y} \in \mathcal{F}\}. \tag{12}$$

In this case, the membership to $\mathcal{F}$ is a constraint in the optimization procedure: we will determine an element of $\mathcal{F}$ occupying a central position - analogously to a *median*. Even if, in general, this problem may not admit any solution, approximate solutions may be determined by using samples from $\mathbf{Z}$.

The distance is, in principle, arbitrary. For instance, we may use the distance dn generated by the inner product defined on V:

$$\text{dn}(\mathbf{x}, \mathbf{y}) = E(\|\mathbf{x}_\mathbf{z} - \mathbf{y}\|). \tag{13}$$

In this case, the difficulties previously state arise, such as the dependence on the parameterization. In order to get independence of the parameterization, we must consider set distances, such as the Hausdorff distance dh, which is defined as

$$\mathrm{dh}\left(\mathbf{x}, \mathbf{y}\right) = E\left(\mathrm{d}\left(\mathbf{x_z}, \mathbf{y}\right)\right). \tag{14}$$

Here,

$$\mathrm{d}\left(\mathbf{x_z}, \mathbf{y}\right) = \max\left\{\sup\left\{\delta\left(\mathbf{x_z}, \mathbf{y}(s)\right) : s \in I\right\}, \sup\left\{\delta\left(\mathbf{x_z}(t), \mathbf{y}\right) : t \in I\right\}\right\}, \tag{15}$$

where

$$\delta\left(\mathbf{x_z}, \mathbf{y}(s)\right) = \inf\left\{\|\mathbf{x}(t, \mathbf{z}) - \mathbf{y}(s)\| : t \in I\right\} \tag{16}$$

and

$$\delta\left(\mathbf{x_z}(t), \mathbf{y}\right) = \inf\left\{\|\mathbf{x}(t, \mathbf{z}) - \mathbf{y}(s)\| : s \in I\right\}. \tag{17}$$

In this approach, we may generate confidence intervals: once $E\left(\mathbf{x}\right)$ or med $\left(\mathbf{x}\right)$ is determined, we may look for a region including this object and containing a given proportion $1 - \alpha$ of the family.

4 Using Samples

In general, only samples from the objects are available in practice. In this case, we must evaluate the means, medians envelopes and confidence intervals by using a finite number of objects: assuming that a sample $\mathcal{S} = \{\mathbf{x_{z_i}} : 1 \le i \le n_s\}$ of n_s curves is available, we may estimate the punctual mean as

$$E\left(\mathbf{x}\right) \approx \bar{\mathbf{x}}_p(t) = \frac{1}{n_s} \sum_{i=1}^{n_s} \mathbf{x_{z_i}}(t).$$

To estimate the Hilbertian mean, we may consider the inverted UQ expansions of each element of the sample and take the means of the coefficients:

$$\mathbf{x_{z_i}} = \sum_{k \in \mathbb{N}^*} \mathbf{x}_{i,k} \varphi_k(t); \; P_n \mathbf{x_{z_i}} = \sum_{1 \le \; k \; \le n} \mathbf{x}_{i,k} \varphi_k(t).$$

Then

$$E\left(\mathbf{x}\right) \approx \bar{\mathbf{x}}_h(t) = \sum_{k \in \mathbb{N}^*} \bar{\mathbf{x}}_k \varphi_k(t) \approx \sum_{1 \le \; k \; \le n} \bar{\mathbf{x}}_k \varphi_k(t), \; \bar{\mathbf{x}}_k = \frac{1}{n_s} \sum_{i=1}^{n_s} \mathbf{x}_{i,k}.$$

To estimate the variational mean, we use the median of the sample:

$$\mathrm{med}\left(\mathbf{x}\right) \approx \mathrm{med}_{\mathcal{S}}\left(\mathbf{x}\right) = \arg\min \left\{ \frac{1}{\mathrm{n_s}} \sum_{\mathrm{i}=1}^{\mathrm{n_s}} \mathrm{d}\left(\mathbf{x_{z_i}}, \mathbf{y}\right) : \mathbf{y} \in \mathcal{S} \right\}$$

The envelope and the confidence interval are determined by using the sample: we evaluate the distances $dn\left(E\left(\mathbf{x}\right), \mathbf{x_{z_i}}\right)$ or $dh\left(\mathrm{med}\left(\mathbf{x}\right), \mathbf{x_{z_i}}\right)$ for $1 \le i \le n_s$. The confidence interval of level $1 - \alpha$ is defined by the $(1 - \alpha)n_s$ curves of $\mathcal{S}$ corresponding to the smaller $(1 - \alpha)n_s$ distances. The envelope is analogously determined: we exclude the largest distances to keep the smaller ones.

4.1 Transformation of Small Sets of Data in Large Samples

One of the interesting features of the Hilbertian approach is the possibility of generating large samples from a few data. When a small set of data (i.e., n_s is small), we may use the classical UQ approach to generate a representation of the family $\mathbf{z}$ and, then, use the representation to generate new curves, corresponding to other values of $\mathbf{z}$. In the UQ framework, a procedure exists for the situation where the bounds of $\mathbf{z}$ are unknown or, eventually, $\mathbf{z}$ itself is unknown - the reader may found it in [9]. Large samples may be useful when we look for the probability distribution of random variables connected to the family of curves. They are also useful to estimate probabilities of events connected to $\mathbf{z}$. They may also be used to improve the evaluations of the means, medians and confidence intervals.

5 An Example

Let us consider the design of a 5-bar truss structure (Fig. 2) for two simultaneous objectives: the minimum mass w and minimum of the maximum displacement u [7].

We assume that the structure remains in the domain of linear elasticity, subjected only to axial forces. The geometric and material parameters are: length $\ell = 9.3144$ m, cross-section $a = 0.01419352$ m^2, load $p = 448.2$ kN, Young's modulus $e = 68.95$ GPa, density $\rho = 2,768$ Kg/m^3, yield stress $\overline{s} = 172.4$ MPa.

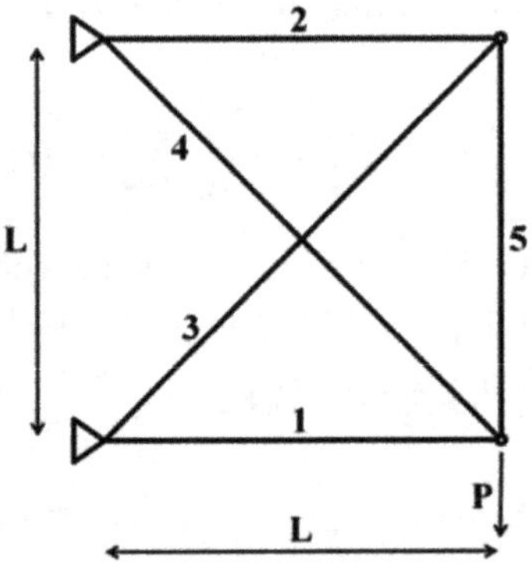

Fig. 2. An example involving a 5 bar truss structure

Denoting $\mathbf{x} \in \mathbb{R}^5$ the vector of the topological and sizing optimization parameters, such that $0 \leq x_i \leq 1$ for $i \in \{1, 2, \dots, n\}$ where $n = 5$ is the number of elements, we must solve the following problem

$$\underset{\mathbf{x}\in\mathbb{R}^5}{\text{Minimize}} \begin{cases} f_1(\mathbf{x}) = w = \sum_{i=1}^{5} \rho a \ell_i x_i \\ f_2(\mathbf{x}) = u = \max\left(\mathbf{u}^* = \text{argmin}\left(\frac{1}{2}\mathbf{u}^t \mathbf{k}(\mathbf{x})\,\mathbf{u} - \mathbf{u}^t \mathbf{f}\right)\right) \end{cases}$$

$$\text{Such that}: \; s_i \leq \overline{s}, \; 1 \leq i \leq n$$

where **k** is the stiffness matrix and **f** the vector of loads of the finite element (FE) model and s_i is the stress of the ith bar.

The multiobjective optimization problem is solved by the variational method presented in [6,10] with a polynomial of degree 6: we obtain the Pareto front in Fig. 3.

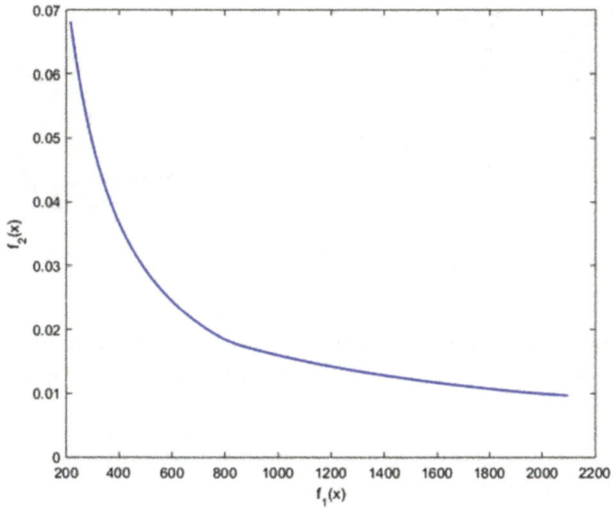

Fig. 3. Pareto's front of the 5 bar truss structure

Let us introduce uncertainties:

1. the force p becomes uncertain (denoted P) following a normal distribution with 10% for the coefficient of variation.
2. the Young's modulus e becomes uncertain too (denoted E) following a truncated normal distribution defined on [60.68, 77.22] GPa with 3% for the coefficient of variation.

ns = 200 problems are generated and the result we obtained is on Fig. 4:

Results are "as expected": the mean is in the middle of the curves set, while hypersurfaces beyond the 90%-quantile (in blue) are located at the exterior of the curves

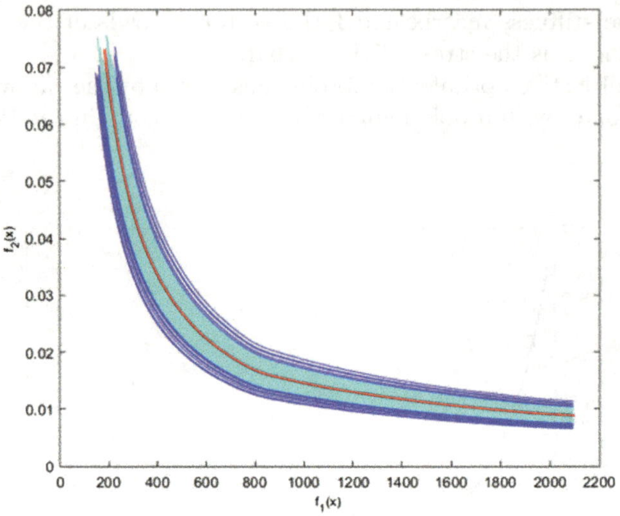

Fig. 4. Pareto fronts of the 5 bar truss structure with uncertainties for ns = 200 sample size: The mean Pareto front is in red and the Pareto fronts beyond the 90%-quantile are in blue

6 Concluding Remarks

The analysis of the uncertainty of curves and surfaces involves probabilities in infinite dimensional spaces, introduces operational and conceptual difficulties: for instance, if $\mathbf{x}$ is a family of curves depending upon a random vector $\mathbf{Z}$, giving a signification to $E(\mathbf{x})$ requires the definition of probability on an infinite dimensional space. To be effective to calculate, two basic approaches may be considered: on the one hand, the use of Hilbert basis and, on the other hand, approaches based on the variational characterization of the mean. The first approach is effective to calculate and furnishes a method for the generation of large samples: if $\mathbf{x}$ is represented by an expansion in a Hilbert basis, we may generate a sample from the random variable $\mathbf{Z}$ and use the representation to generate a sample from $\mathbf{x}$. However, it produces results that may depend on the parameterization. The Hilbertian approach is useful to evaluate covariances. The second approach is independent from the parameterization and is also effective to calculate. Among its advantages, it produces the median and a confidence interval - both these quantities may be difficulty to obtain by the Hilbertian approach. It may be used to determine confidence intervals for Pareto fronts in multiobjective optimization.

References

1. Bassi, M.: Quantification d'incertitudes et objets en dimension infinie. Ph.D. thesis, Normandie University, INSA Rouen Normandie (2019)
2. Bassi, M., Pagnacco, E., de Cursi, E.S.: Uncertainty quantification and statistics of curves and surfaces, pp. 17–37. Springer, Cham (2020)

3. Bassi, M., Souza de Cursi, J.E., Ellaia, R.: Generalized Fourier series for representing random variables and application for quantifying uncertainties in optimization. In: 3rd International Symposium on Uncertainty Quantification and Stochastic Modeling (2016)
4. Bassi, M., Cursi, E.S.D., Pagnacco, E., Ellaia, R.: Statistics of the pareto front in multi-objective optimization under uncertainties. Lat. Am. J. Solids Struct. **15**(11), 1015–1036 (2018)
5. Rémi, C., de Cursi, E.S.: Statistics of uncertain dynamical systems. In: Topping, B.H.V., Adam, J.M., Pallares, F.J., Bru, R., Romero, M.L. (eds.) Proceedings of the Tenth International Conference on Computational Structures Technology, pp. 541–561. Civil-Comp Press, Stirlingshire (2010)
6. de Cursi, E.S.: Variational Methods for Engineers with Matlab, 1st edn. Wiley, London (2015)
7. Ellaia, R., Habbal, A., Pagnacco, E.: A new accelerated multi-objective particle swarm algorithm. Applications to truss topology optimization. In: 10th World Congress on Structural and Multidisciplinary Optimization, Orlando, FL, United States (May 2013)
8. de Cursi, E.S.: Representation of solutions in variational calculus. In: Tarocco, E., de Souza Neto, E.A., Novotny, A.A. (eds.) Variational Formulations in Mechanics: Theory and Applications. International Center for Numerical Methods in Engineering (CIMNE), Barcelona, Espagne, pp. 87–106 (2007)
9. de Cursi, E.S., Sampaio, R.: Uncertainty Quantification and Stochastic Modeling with Matlab, 1st edn. Elsevier Science Publishers B. V., Amsterdam (2015)
10. Zidani, H.: Representation de solutions en optimisation continue, multiobjectif et applications. Ph.D. thesis, INSA Rouen (2013)
11. Hafidi, Z., de Cursi, E.S., Ellaia, R.: Numerical approximation of the solution in infinite dimensional global optimization using a representation formula. J. Glob. Optim. **65**(2), 261–281 (2016)

Robust Optimization for Multiple Response Using Stochastic Model

Shaodi Dong[1(✉)], Xiaosong Yang[2], Zhao Tang[1(✉)], and Jianjun Zhang[2]

[1] State Key Laboratory of Traction Power, Southwest Jiaotong University, Chengdu 610031, China
1216823479@qq.com, tangzhao@swjtu.edu.cn
[2] National Centre for Computer Animation, Bournemouth University, Poole BH12 5BB, UK

Abstract. Due to a lot of uncertainties in the robust optimization process, especially in multiple response problems, many random factors can cost doubt on results. The aim of this paper is to propose a robust optimization method for multiple response considering the random factors in the robust optimization design to solve the aforementioned problem. In this paper, we research the multi-response robustness optimization of the anti-rolling torsion bar using a stochastic model. First, the quality loss function of the anti-rolling torsion bar is determined as the optimization object, and the diameters of the anti-rolling torsion bar are determined as the design variables. Second, the multi-response robust optimization model, considering random factors (such as the loads), is established by using the stochastic model. Finally, the Monte Carlo sampling method combined with a non-dominated sorting genetic algorithm II (NSGA II) is adopted to solve this robust optimization problem, and then the robust optimization solution is obtained. The research results indicate that the anti-rolling torsion bar weight decreases, and the stiffness and fatigue strength increase. Furthermore, the quality performance of the anti-rolling torsion bar gets better, and the anti-disturbance ability of the anti-rolling torsion bar gets stronger.

Keywords: Robust optimization · Multiple response · Stochastic model · Anti-rolling torsion bar

1 Introduction

The anti-rolling torsion bar devices installed between the body and the bogie of the railway vehicle is an important part of the second series suspension device, which plays an important role in the safety of the vehicle. To improve the quality characteristics, robust optimization design is used to achieve the torsions' robustness through adjusting the control design variables and their tolerances, as that the quality characteristics insensitive to design variables [1]. Consequently, this work will have very important theoretical and application value for improving the robustness and safety of train and vehicle structures.

Until now, the researchers have made many good kinds of research about robust optimization [2–4]. Pishvaee et al. realized robust optimization considering random

J. E. S. De Cursi (Ed.): Uncertainties 2020, LNME, pp. 396–405, 2021.
https://doi.org/10.1007/978-3-030-53669-5_29

uncertainty in actual engineering [2]. Then they obtained a robust and reliable optimization model, and achieved the robust optimization for multiple responses [3, 4]. However, most all of these methods have high computational complexity, poor convergence, and almost nothing is involved in the torsion bar device.

In this paper, we propose a stochastic multi-response robust optimization method of the anti-rolling torsion bar devices. First, the stochastic model is established fully considering the randomness. Then the Monte Carlo method combined with the non-dominated sorting genetic algorithm II (NSGA II) is applied to solve the model. Finally, the results indicate that the computational complexity is reduced, and the convergence is closer to the Pareto optimal layer. Accordingly, we could obtain a robust and reliable optimization solution of the anti-rolling torsion bar devices.

2 Stochastic Modelling

To evaluate the robustness of product quality characteristics, a stochastic model is established according to the stochastic modeling principles [5], and then the optimal solution can be obtained by adjusting the design variables and controlling its tolerance (the allowed maximum deviation). Generally, there are mainly three product quality design criteria in the robust design [6].

The first product quality design criterion is the probability of unsatisfactory products, as expressed in Eq. 1.

$$P = \{\bigcap_{j=1}^{q} (y_{0j} + \Delta_{yj}^{-} > y_j \cup y_j + \Delta_{yj}^{-})\} \to \min \tag{1}$$

Where y_{oj} is the design quality characteristic target value; Δ_{yj}^{-}, Δ_{yj}^{+} is the allowable tolerance of the quality characteristic; y_j is the output quality characteristic value; q is the number of satisfying quality characteristics; and $P = \{\cdot\} \in (0, 1)$.

The second product quality design criterion is the sensitivity index or expected loss function, expressed as follows:

$$SI = \prod_{j=1}^{q} \frac{2\left|\mu_{yj} - y_{oj}\right|}{\left|\Delta_{yj}^{-}\right| + \left|\Delta_{yj}^{+}\right|} \to \min \tag{2}$$

$$LQ = \sum_{j=1}^{q} w_j E\left\{\left(y_j - y_{oj}\right)^2\right\} = \sum_{j=1}^{q} w_j\left[\left(\mu_{yj} - y_{oj}\right)^2 + \sigma_{yj}^2\right] \to \min \tag{3}$$

Where w_j is the weight coefficient corresponding to each quality characteristic; μ_{yj} is the statistical mean value of the quality characteristics; σ_{yj}^2 is the statistical variance of the quality characteristics.

The third product quality design criterion is the constraint feasibility criterion, that is, the random constraints on other aspects of the product (raw materials, conditions of use, etc.) should meet the requirements when random variables become worse, as expressed in Eq. 4.

$$E\{g_j(X, Z)\} + \Delta g_j \le 0 \quad j = 1, 2, \ldots, q \tag{4}$$

Where $g_j(X, Z)$ is the constraint of the response j; Δg_j is the random constraint of the response j. As $g(X, Z)$ is a non-normal distribution or a constraint random correlation, the calculation formula can be expressed in Eq. 5.

$$p\{\bigcap_{j=1}^{q} g_j(X, Z) \leq 0\} \geq \alpha_0 \tag{5}$$

Where α_0 is a predetermined satisfaction probability value.

So the stochastic robust optimization model is built according to the product quality design guidelines and the boundary constraints.

$$\min_X SI(X) \text{ or } \min_X LQ(X)$$
$$s.t. \begin{cases} p\{\bigcap_{j=1}^{q} g_j(X, Z) \leq 0\} \geq \alpha_0 \\ X^L + \Delta_X \leq X \leq X^U - \Delta_X \end{cases} \tag{6}$$

Where X is design variables; X^U, X^L is the upper limits and the lower limits of design variables, respectively; Δ_X is the manufacturing error of design variables.

3 Multi-response Robust Optimization

The multi-response optimization can make multiple quality characteristics insensitive to interference factors. So the robustness of multiple quality characteristics could be achieved by rationally selecting the design variables and controlling their tolerances.

To achieve the respective robustness of multiple quality characteristics, the multi-response robust optimization is proposed. For the consideration of the influencing factors randomness, a stochastic robust optimization model with multiple quality characteristics is built, and then the robustness of multiple quality characteristics is achieved by reducing the quality loss of each quality characteristic.

3.1 The Anti-rolling Torsion Bar Model

The connecting rod of the vehicle body is applied a load P_1 through the upper rubber joint, then torque T2 is transmitted by the connecting rod and the torsion bar. And the torsion bar also subjected to the load P2 and the bending moment M. Its main working principle is that the load P1, according to the good torsional deformation and rebound characteristics of the torsion bar shaft, is adjusted to meet the requirements of anti-rolling, while not affecting the vibration characteristics (such as heave, yaw, shaking, telescoping and nodding) of the vehicle [7]. The force analysis of the torsion bar shaft is shown in Fig. 1.

The specific parameters of the anti-rolling torsion bar device in a bogie of the vehicle [8] are shown in Table 1.

Table 1. The parameters of anti-rolling torsion bar

Serial number	Parameters	Symbol	Unit	Value
1	Elastic modulus	E	Mpa	2.06 × 105
2	Tensile limit	σ_b	Mpa	1274
3	Bow to extremes	σ_s	Mpa	1127
4	Fatigue limit	σ_{-1}	Mpa	648.27
5	Allowable shear stress	$[\tau]$	Mpa	70
6	Nominal length of torsion bar	2a + b	mm	2370
7	Effective length of torsion bar	B	mm	2182
8	Effective length of torsion arm	R	mm	215

Optimization Design Variables. The controllable factors (also known as design variables) and uncontrollable factors (also known as noise factors) [9] are determined in the optimization design of the anti-rolling torsion bar. Table 2 shown the value range of anti-rolling torsion bar design variables, which were determined according to the actual situation and experience.

Table 2. The value range of design variables

Design variables	d_1/mm	d_2/mm	d_3/mm
Upper limit	40	40	120
Lower limit	60	60	140

The cross-sectional diameters of the torsion bar shaft's working area, the transition area, the connecting are determined as design variables d_1, d_2, d_3, respectively. And the shear modulus G and the applied load P are determined as the noise factors.

At the same time, it is assumed that the design parameters are random variables, which obey the normal distribution and are independent of each other. Among the controllable factors: d_1 (mm) $\sim N(49.0, 3.0)$, d_2 (mm) $\sim N(50.5, 3.0)$, d_3 (mm) $\sim N(130, 3.0)$; The uncontrollable shear modulus G (Mpa) $\sim N(7.85E4,\ 70.0)$. The load P(N) $\sim N(1.13E4, 50.0)$ and requires $\alpha_0 = 99\%$.

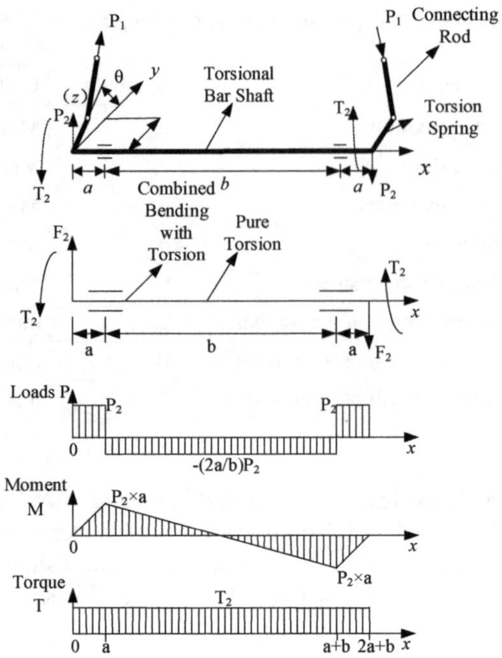

Fig. 1. Analysis of force applied to a torsion bar

Optimization Objective Function. To achieve robust optimization, we try to reduce the fluctuation and the error between the actual value and the target value of the anti-rolling torsion bars' volume (or weight), stiffness and fatigue strength. And then the loss function of the anti-rolling torsion bars' volume (or weight), stiffness and fatigue strength can be used as the robust optimization target in the stochastic model.

Robust optimization goals for volume $L(v)$: $L(v) = w_1\bar{v}^2 + w_2 s_v^2 \to \min$

Where $v = \frac{\pi d_1^2}{4} l_1 + \frac{\pi d_2^2}{4} l_2 + \frac{\pi d_3^2}{4} l_3$; $\bar{v}$, s_v is the statistical mean and statistical variance of the volume, respectively; and w_1, w_2 is weight coefficients, respectively.

Robust optimization goal for stiffness $L(k)$: $L(k) = w_3(\frac{1}{k})^2 + w_4 \frac{3S_k^2}{(\bar{k})^4} \to \min$

Where $\frac{1}{k} = \frac{32 l_1}{\pi d_1^4 G} + \frac{32 l_2}{\pi d_2^4 G} + \frac{32 l_3}{\pi d_3^4 G}$; $\bar{k}$, s_k is the statistical mean and statistical variance of the stiffness, respectively; and w_3, w_4 is weight coefficients, respectively.

Robust optimization goals for fatigue strength $L(N)$:

$$L(\sigma_{-1}) = w_5\left(\frac{1}{\overline{\sigma_{-1d}}}\right)^2 + w_6 \frac{3S_{\sigma_{-1d}}^2}{(\overline{\sigma_{-1d}})^4} \to \min$$

Where $\sigma_{-1d} = \sigma\text{max} = \frac{4p}{\pi d_1^2}$, $\overline{\sigma_{-1d}}$, $s_{\sigma_{-1d}}$ is the statistical mean and statistical variance of the fatigue strength, respectively; and w_5, w_6 is weight coefficients, respectively.

Optimization Constraints. The constraint feasibility criterion, considering the fatigue strength randomness, is applied to constrain. And its formula is expressed as in Eq. 7.

$$P\left\{g_1(X, \Delta X, Z) = \frac{4p}{\pi d^2} - \left[\sigma_{-1}\right] \leq 0\right\} \geq \alpha \tag{7}$$

The random constraints of stress conditions can be determined according to the third strength theory. And its formula as expressed in Eq. 8.

$$E\left\{g_1(X, \Delta X, Z) = \frac{1}{W_Z}\sqrt{M^2 + T^2} - [\sigma_b]\right\} \leq 0 \tag{8}$$

Then the boundary constraint is also determined, as shown in Eq. 9 and Eq. 10.

$$X = (x_1, x_2, x_3) = (d_1, d_2, d_3) \in (\Omega, J, P) \subset R^{n+k} \tag{9}$$

$$Z = (\mathrm{z}_1, \mathrm{z}_2) = (\mathrm{G,\ P}) \in (\Omega_1, J_1) \subset R^{n+k} \tag{10}$$

4 Results of Analyses

The Monte Carlo sampling statistical method combined with the NSGA II [10] is applied to perform statistical analysis and the multi-response optimization. The specific optimization process is shown in Fig. 2.

(1) Monte Carlo method [11]: Assuming the uncertainty factors of the system as random variables, the probability distribution characteristics of the system response could be estimated by random sampling of random variables (Mean/standard deviation, etc.) while we know the probability distribution. And this method can improve the efficiency of parallel computing.

(2) NSGA-II method [10]: Sudden mutation operation mechanism could play the global search role and improve the global search ability of genetic algorithm through using crossover. The distribution in the target space is uniform, which means that it has convergence and robustness.

The specific algorithm parameters can be set as follows: the normal distribution parameters of Monte Carlo samples are 100, the initial population size in the NSGA-II algorithm is 36, the genetic generation is 20, the cross-initial rate is 0.9, and the mutation distribution index is 10. Then 361 times were need to calculated, if the weight of each objective function is equal, and it is $w_i = 0.5$.

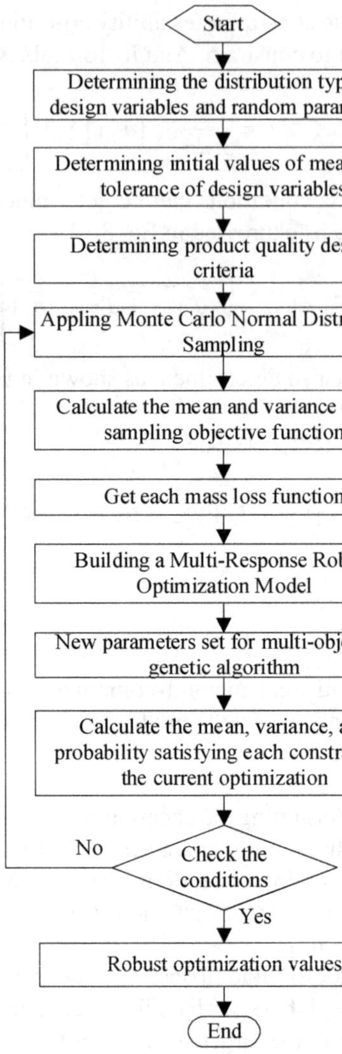

Fig. 2. The flowchart of multi-response robust optimization using the stochastic model

Eventually, the probability distribution characteristics (mean, standard deviation, etc.) of design variables and noise variables could be obtained (see Fig. 3).

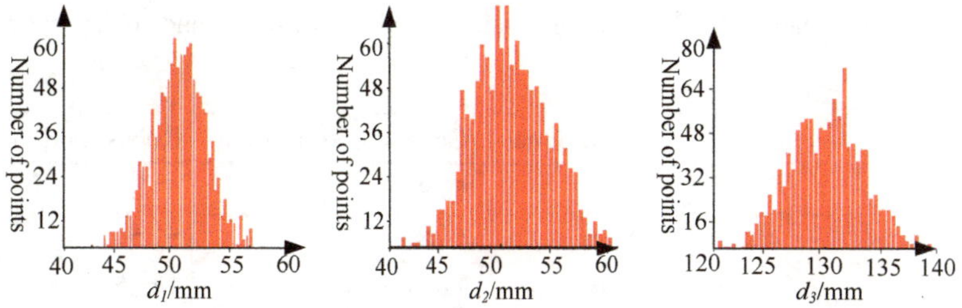

(a) Random distribution map of d_1 (b) Random distribution map of d_2 (c) Random distribution map of d_3

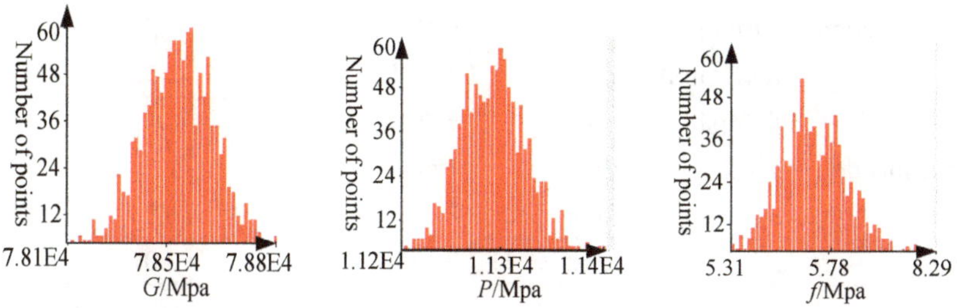

(d) Random distribution map of G (e) Random distribution map of P (*f*) Random distribution

Fig. 3. Random distribution map of the Monte Carlo method

Among them, the NSGA-II algorithm can be used to quickly find the Pareto front, which mainly calculates the crowding distance by ranking the crowding distance. Then the superiority and the individual weakness is evaluated by evaluating the Pareto superiority, thereby the Pareto optimal solution set could be obtained [12].

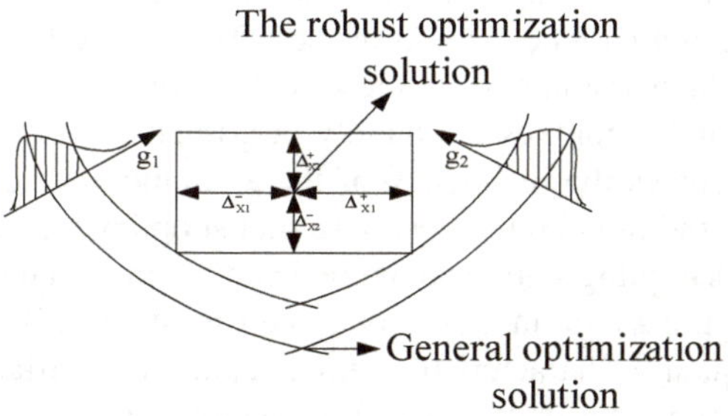

Fig. 4. Robust optimization solution of the stochastic model

If the design variable internal is $[X^L + \Delta_X X^U - \Delta_X]$ and noise factor Z is a random variable that obeys the normal distribution, the robust optimization solution of the stochastic model is as shown in Fig. 4. And the final robust optimization Pareto solution set is obtained, see in Table 3, and then each optimization result is compared.

Table 3. The optimization results of traditional, general optimization and robust optimization design

Variables		Traditional	General optimization	Robust optimization
d_1/mm		49.0	50.9	51.5
d_2/mm		50.5	51.8	52.5
d_3/mm		130	129.2	128.5
G/(Mpa)		7.84E4	7.84E4	7.84E4
P/(N)		11300	11300	11300
Objective function value	V/(mm^3)	6.6E6	6.4E6	6.5E6
	K/(N/mm)	2.0E7	3.0E7	2.5E7
	f/(Mpa)	5.94	5.99	6.13
Objective function variance	S_V	1.68	1.68	1.5
	S_K	8.46	8.46	6.8
	S_F	0.025	0.025	0.01

The results show that the optimization objective value of the torsion bar has been improved, and the statistical variation of the quality characteristics is reduced. We demonstrate that the performance fluctuation caused by random factors is more stable and reliable, and then the robustness of the anti-rolling torsion bars effectively ensured.

5 Conclusion

A multi-response robust optimization method based on a stochastic model considering the randomness of design parameters is proposed. This method not only considers the randomness of the influencing factors, but also realizes the robustness of multiple quality characteristics, which improves the quality performance of the anti-rolling torsion bar device, and overcomes the deficiency of the assumption of design parameters as determined values in traditional multi-response robust optimization.

The Monte Carlo sampling statistical method combined with the NSGA II is applied to solve the robust optimization model to obtain a robust optimization solution, which could improve the quality characteristic value, reduce the variance fluctuation, and improve the anti-jamming ability of anti-rolling torsion bar.

The robust optimization case of the anti-rolling torsion bar device illustrates that the method is feasible, and can be extended to robust optimization with random multiple responses in the project.

References

1. Ben-Tal, F., Aharon, S., El Ghaoui, L.T., et al.: Robust Optimization. Princeton University Press, Princeton (2009)
2. Pishvaee, M.S.F., Rabbani, M.S., Torabi, S.A.T.: A robust optimization approach to closed-loop supply chain network design under uncertainty. Appl. Math. Model. **35**(2), 637–649 (2011)
3. Chen, S.F., Chen, W.S., Lee, S.T.: Level set based robust shape and topology optimization under random field uncertainties. Struct. Multidiscip. Optim. **41**(4), 507–524 (2010)
4. Changizi, N.F., Jalalpour, M.S.: Robust topology optimization of frame structures under geometric or material properties uncertainties. Struct. Multidiscip. Optim. **56**(4), 791–807 (2017)
5. Yu Chian-Son, F., Li Han-Li, S.: A robust optimization model for stochastic logistic problems. Int. J. Prod. Econ. **64**(1–3), 385–397 (2000)
6. Chen, L.F.: Robust Design, pp. 237–263. China Machine Press, Beijing (2000). (in Chinese)
7. Ruo, X., Zhou, F.: Application of the anti-rolling torsion bar in urban railway cars. Roll. Stock **12** (2001)
8. Sun, M.F., Luan, T.S., Liang, L.T.: RBF neural network compensation-based adaptive control for lift-feedback system of ship fin stabilizers to improve anti-rolling effect. Ocean Eng. **163**, 307–321 (2018)
9. Thomas, V.F., Pondard, J.S., Bengio, E.T., et al.: Independently controllable factors. arXiv preprint arXiv:1708 (2017)
10. Pires, D.F.F., Antunes, C.H.S., Martins, A.G.T.: NSGA-II with local search for a multi-objective reactive power compensation problem. Int. J. Electr. Power Energy Syst. **43**(1), 313–324 (2012)
11. Sobol, I.M.F.: A Primer for the Monte Carlo Method. CRC Press, Boca Raton (2018)
12. Detchusananard, T.F., Sharma, S.S., Maréchal, F.T., et al.: Generation and selection of Pareto-optimal solution for the sorption enhanced steam biomass gasification system with solid oxide fuel cell. Energy Convers. Manag. **196**, 1420–1432 (2019)

Robust Design of Stochastic Dynamic Systems Based on Fatigue Damage

Ulisses Lima Rosa(✉), Lauren Karoline S. Gonçalves, and Antonio M. G. de Lima

School of Mechanical Engineering, Federal University of Uberlândia, Campus Santa Mônica, Uberlândia, MG 38400-902, Brazil
ulisses.ufu@hotmail.com

Abstract. The design of engineering structures goes through several phases before arriving at a final concept or prototype. Comfort, safety and reliability are desirable characteristics to be achieved, but when it comes to obtaining lighter structures, this can be a problem as to the effects of mechanical vibration, such as fatigue in metals. Material properties related to fatigue are only obtained experimentally using standardized specimens and controlled environments. Results obtained by these tests present a statistical character and carry uncertainties and measurement errors that may significantly affect the final fatigue failure condition. In this context, the inclusion of uncertainties in a computational model becomes essential. This work presents a methodology for fatigue failure prediction by applying the Sines' fatigue criterion and allowing fatigue analysis to be performed numerically during the design phase. Uncertainties are included in the model using the stochastic finite elements method, with random fields discretized by the so-called Karhunen-Loève expansion method. As stochastic analysis demands multiple function evaluations, the computational cost involved becomes high. Due to this, the application of a model condensation procedure is necessary. After presenting the theory, a robust multiobjective optimization procedure is performed to enhance fatigue life of a thin plate subjected to cyclic loads, which is directly in conflict with reducing its mass. This procedure seeks for not only one point in the search space, but a whole group of solutions where all are treated as optimal and are less susceptible to parameter fluctuations. Numerical results are presented in terms of FRF, stress responses in the frequency domain and Sines fatigue index for each finite element composing the plate.

Keywords: Uncertainties · Stochastic finite elements method · Fatigue · Robust optimization · Model condensation

1 Introduction

The conception of engineering projects classically goes through some stages where the structure is analyzed for its failure under different conditions, whether static, regarding concepts of resistance of materials or dynamic, regarding concepts of classical mechanics. Both of these approaches aim to characterize and quantify structural failure when

J. E. S. De Cursi (Ed.): Uncertainties 2020, LNME, pp. 406–423, 2021.
https://doi.org/10.1007/978-3-030-53669-5_30

its resistance limits are exceeded under loadings. Designing a system in this way is usually not a complex process since resistance limits and modulus are available for most materials [1].

Despite this classical approach, practical engineering applications usually require machines and equipment to operate under conditions where they are constantly subjected to dynamic disturbances and cyclic loads, which can lead to undesirable vibration levels and noise. Fatigue cracks usually appear at critical areas such as geometry changes, load application points and boundary conditions [2]. Besides that, they nucleate and propagate under stresses much lower than the static limits, leading to definitive fractures or total collapse of the structure [3].

It is not usual to characterize fatigue by its failure mode, which involves microstructural defects, persistent slip bands and dislocations propagation. The practical way of evaluating multiaxial fatigue consists in applying methods based on experimental data, that are already consolidated in literature [2, 4]. Among these criteria, Sines' criterion is the most adapted for the case presented herein. It presents accuracy and has a simpler formulation than others [4], involving two material properties related to fatigue, which are obtained experimentally using standardized specimens and controlled environments. These procedures may carry and propagate uncertainties through the model.

Computational models for engineering analysis must present fidelity to real applications. One way to obtain more accuracy is to include effects caused by the presence of aleatory variables and uncertainties such as those inherent to material properties, geometry, loading and environmental condition. These uncertainties affect the system's responses and sensibility as well as propagate through the calculations involved in the computational model [5].

In this paper, uncertainties are taken into account in the model using the stochastic finite elements method, with random fields discretized by the so-called Karhunen-Loève expansion method [6], which performs a parametric inclusion of random variables through the modification of mass and stiffness matrices [7]. Aleatory effects inserted via this method are not only present in the random variables but also affect the integration process of elementary matrices, allowing its propagation throughout the entire structural domain [8].

The development of such a methodology for fatigue life prediction allows the durability analysis to be performed before the manufacture, enabling major modifications in the project phase. The design of a dynamic system sometimes runs into an optimization problem where two or more conflicting characteristics are present. As an example, thinner and lighter materials normally present lower inherent damping rates, aggravating vibration problems.

Performing robust optimization envisaging increasing fatigue life involves a large number of function evaluations and consequently high computational costs. Due to this reason, a model condensation technique, initially proposed to viscoelastically damped structures [9] was modified and extended to mechanical systems subjected to uncertainties. This procedure ensured that the numerical simulations were carried out in a timely manner.

The optimization problem associated with the analysis of a stochastic structure whose objective is fatigue life analysis is treated as a multiobjective problem, since the desired global solution, be it a maximum or a minimum, is not obtained for a single point in the search space but for a set of optimal solutions. To solve problems of this type, what is recommended in the literature is to use heuristic methods or the so-called metaheuristic methods [10]. Finally, it is up to the designer to establish a final compromise and adopt a solution among the set that best fits the project [11].

2 Stochastic Finite Element Formulation

There are two main approaches for performing fatigue analysis: in time or frequency domain. Both are feasible, but the first requires large stress data historic [1] and its calculation involves convolution integrals for the numeric procedure, generation high computational cost and storage capacity. This becomes even more critical in the presence of random loads. As an alternative, the use of stress responses in the frequency domain via Power Spectral Density (PSD) [2] is recommended. The following subsections are devoted to present the formulation applied in the development of the stochastic finite element model, the calculation of stress PSDs and estimation of structural fatigue life via Sines' criterion [12], as well as the model condensation procedure developed to enable calculations to be performed in a timely manner.

2.1 Stress Response in the Frequency Domain

The model of a thin rectangular plate is summarized herein, based on the original hypothesis proposed by Kirchhoff [13]. Under these conditions, the plate is considered as in a plane stress state, making the transverse displacement coordinate-independent along thickness. Therefore, the normal strain in this direction is zero as well as both shear strains $\left(\varepsilon_z = 0,\ \gamma_{xz} = 0,\ \gamma_{yz} = 0\right)$. The adopted finite element is rectangular and composed by four nodes and five degrees-of-freedom per node: two in-plane displacements in directions X and Y (u_0, v_0), one transverse displacement in direction Z (w_0) and two cross-section rotations $\left(\theta_x, \theta_y\right)$, as depicted in Fig. 1.

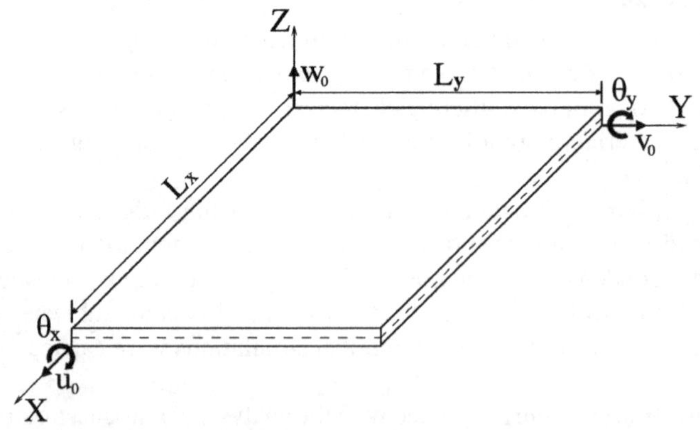

Fig. 1. Four-nodes rectangular finite elements with five degrees-of-freedom per node

Let **M** and **K** be the global system mass and stiffness matrices, by definition square and symmetric with dimension equivalent to the total number of degrees-of-freedom. They are classically calculated by the integration of shape functions $\mathbf{N}(x, y)$ for each element with elementary volume $V_{e,}$ based on energy variational calculus. After, they are globally assembled by means of node connectivity [14].

Adding an inherent structural damping matrix $\mathbf{C} = \beta\mathbf{K}$ proportional to stiffness with a coefficient β, the system's movement equation in time domain can be written as follows:

$$\mathbf{M}\ddot{\mathbf{u}}(t) + \mathbf{C}\dot{\mathbf{u}}(t) + \mathbf{K}\mathbf{u}(t) = \mathbf{f}(t) \tag{1}$$

where $\mathbf{u}(t)$ is the displacement vector and $\mathbf{f}(t)$ is the external forces vector.

An external force, in the form of a harmonic excitation $\mathbf{f}(t) = \mathbf{F}e^{j\omega t}$ also generates a harmonic response $\mathbf{u}(t) = \mathbf{U}e^{j\omega t}$ and it is possible to calculate the displacement frequency response function (FRF) by Eq. (2).

$$\mathbf{G}(\omega) = \left|\frac{\mathbf{U}}{\mathbf{F}}\right| = \frac{1}{\left[\mathbf{K} + j\omega\mathbf{C} - \omega^2\mathbf{M}\right]} \tag{2}$$

Spectral density functions can be defined via correlation functions with the assumption that autocorrelation and cross-correlation and exists [15]. The formulation of the displacement power spectral density matrix $\phi_u(\omega)$ is shown in the sequence:

$$\phi_u(\omega) = \int_{-\infty}^{+\infty} R_u(\tau)e^{-j\omega t}d\tau \tag{3}$$

$$R_u(\tau) = \int_{-\infty}^{+\infty}\int_{-\infty}^{+\infty} g(\lambda)R_f(\tau + \lambda - \zeta)g^T(\zeta)d\lambda d\zeta \tag{4}$$

where R_u is the autocorrelation function of the displacement field, submitted to a load $f(t)$ that, in turn, has an autocorrelation function R_f and $g(t) = \int_{-\infty}^{+\infty} \mathbf{G}(\omega)e^{j\omega t}d\omega$ is the system's impulse response.Using Fourier transform definitions and spectral density properties [2], Eq. (3) can be rewritten in terms of displacement PSD $\phi_u(\omega)$, load PSD $\phi_f(\omega)$ and FRF $\mathbf{G}(\omega)$.

$$\phi_u(\omega) = \mathbf{G}(\omega)\left(\int_{-\infty}^{+\infty} R_f(\tau)e^{-j\omega t}d\tau\right)\mathbf{G}^H(\omega) = \mathbf{G}(\omega)\phi_f(\omega)\mathbf{G}^H(\omega) \tag{5}$$

The stress power spectral density $\phi_s(\omega)$ is finally calculated by applying Hooke's law to Eq. (5), as shown in Eq. (6).

$$\phi_s(\omega) = \mathbf{HBG}(\omega)\phi_f(\omega)\mathbf{G}^H(\omega)\mathbf{B}^\mathbf{T}\mathbf{H}^\mathbf{T} \tag{6}$$

where **H** is the elasticity tensor and **B** is the derivative terms matrix.

2.2 Karhunen-Loève Expansion Method

The so-called Karhunen-Loève (KL) expansion (or decomposition) is a continuous representation for random fields $X(x, y, \theta)$, for the plate's bi-dimensional case. It is composed by a set of continuous parameters (x, y), which represents the structure's physical geometry and variables θ, composing the space of aleatory events. The method itself is a field discretization method that belongs to the called *Series Expansion Methods* [6], and seeks to approximate the random field $X(x, y, \theta)$ by a Fourier-type series $\hat{X}(x, y, \theta)$ truncated at n terms:

$$X(x, y, \theta) \approx \hat{X}(x, y, \theta) = \mu(x, y) + \sum_{r=1}^{n} \sqrt{\lambda_r} f_r(x, y)\xi_r(\theta) \tag{7}$$

where $\mu(x, y)$ is the expected value for the physical field, $\{\xi_r(\theta),\ r = 1, \ldots, n\}$ is a set of random variables and $\left(\sqrt{\lambda_r}, f_r(x, y)\right)$ are the eigen solutions of a problem associated to the covariance function. Ghanem and Spanos [6] present the complete methodology for solving this problem and obtain the eigenvalues λ_r and the eigenfunctions $f_r(x, y)$. It was also previously applied by the authors [7, 8].

After obtaining all the parameters presented previously in Eq. (7), it is possible to perform the integration of the new elementary mass and stiffness matrices $\mathbf{M}^{(e)}(\theta)$ and $\mathbf{K}^{(e)}(\theta)$, now composed of random variables θ. The assemblage procedure is shown in Eqs. (8) and (9) and the integration of the part inside the sum is presented in Eqs. (10) and (11).

$$\mathbf{M}^{(e)}(\theta) = \mathbf{M}^{(e)} + \sum_{r=1}^{n} \Delta\mathbf{M}_r^{(e)}\xi_r(\theta) \tag{8}$$

$$\mathbf{K}^{(e)}(\theta) = \mathbf{K}^{(e)} + \sum_{r=1}^{n} \Delta\mathbf{K}_r^{(e)}\xi_r(\theta) \tag{9}$$

$$\Delta\mathbf{M}_r^{(e)} = \sqrt{\lambda_r} \int_{\Omega_x}\int_{\Omega_y} f_r(x, y)\rho\mathbf{N}(x, y)^T\mathbf{N}(x, y)d\Omega_y d\Omega_x \tag{10}$$

$$\Delta\mathbf{K}_r^{(e)} = \sqrt{\lambda_r} \int_{\Omega_x}\int_{\Omega_y} f_r(x, y)\mathbf{N}'(x, y)^T\mathbf{H}\mathbf{N}'(x, y)d\Omega_y d\Omega_x \tag{11}$$

where ρ is the mass density and Ω_x, Ω_y are the geometric domains.

After the integration, the system matrices are assembled by means of the system connectivity, similarly to the deterministic approach. By using the notation $\mathbf{M}(\theta)$ and $\mathbf{K}(\theta)$ for stochastic mass and stiffness matrices, Eqs. (1) and (6) can be rewritten to indicate the presence of uncertainties:

$$\mathbf{G}(\omega, \theta) = \frac{1}{\left[\mathbf{K}(\theta) + j\omega\mathbf{C}(\theta) - \omega^2\mathbf{M}(\theta)\right]} \tag{12}$$

$$\phi_s(\omega, \theta) = \mathbf{HBG}(\omega, \theta)\phi_f(\omega, \theta)\mathbf{G}^H(\omega, \theta)\mathbf{B}^\mathbf{T}\mathbf{H}^\mathbf{T} \tag{13}$$

2.3 Stochastic Condensation Procedure

Although the study carried out in this work comprises a didactic structure that is not as complex as real-world engineering models, the necessary calculations carry a high computational burden. The major difficulty in these cases is the calculation time spent in matrices inversions and operations as the one shown in Eqs. (13) and (14). However, it is possible to obtain responses in a faster way by applying model condensation procedures.

The aim of these methods is to apply component-mode synthesis [16] in order to set a basis that represents the dynamic behavior of the nominal model. When applied to mass and stiffness matrices, this basis reduces the effective number of degrees of freedom, decreasing the computational burden and storage memory required.

Assuming that the exact system responses provided by the full model can be approximated by projections on a reduced vector basis in an alternative subspace, as follows:

$$\mathbf{U}_r = \mathbf{T}\mathbf{U} \tag{14}$$

where the matrix $\mathbf{T} \in C^{NxNR}$ is a reduction basis, that, when applied to the vector $\mathbf{U}$ generates a reduced response $\mathbf{U}_r \in C^{NR}$ with $NR << N$, being NR the number of component-modes retained in the basis and N the total numbers of structural degrees of freedom.

This basis, applied in system's mass and stiffness matrices generates a new system with similar dynamic behavior:

$$\mathbf{M}_R = \mathbf{T}^T\mathbf{M}\mathbf{T} \tag{15}$$

$$\mathbf{K}_R = \mathbf{T}^T\mathbf{K}\mathbf{T} \tag{16}$$

where $\mathbf{M}_R$ and $\mathbf{K}_R$ are the so-called reduced mass and stiffness matrices.

Different techniques depending on the structure are adopted to obtain reduction basis. For linear structures with well-known physical characteristics, it is appropriate to use Ritz modal basis [17], composed by the nominal system's first vibration mode shapes (φ_0). It is usual to include all mode shapes that are inside a wider range at least 1.5 times the maximum desired frequency. Classically, $\varphi_0 = [\varphi_1, \varphi_2, \ldots, \varphi_{NR}]$ is obtained through the solution of the following eigenproblem associated with the nominal system:

$$[\mathbf{K} - \lambda_i\mathbf{M}]\varphi_i = \mathbf{0} \tag{17}$$

The standard Ritz basis $\mathbf{T}_0$ is assembled including static residues $\bar{\mathbf{U}} = \mathbf{K}^{-1}\mathbf{F}$ of a unitary load as shown in Eq. (18). The whole process is summarized in Fig. 2.

$$\mathbf{T}_0 = \left[\varphi_0\ \bar{\mathbf{U}}\right] \tag{18}$$

With this approach, matrices have their dimensions reduced and computational gain is achieved when performing operations such as matrix inversions as presented in Eq. (2). On the other hand, solving stochastic problems implies multiple function evaluations. Avoiding the solution of the eigenproblem presented in Eq. (17) for each one of the samples is a reasonable way to decrease computational efforts even more.

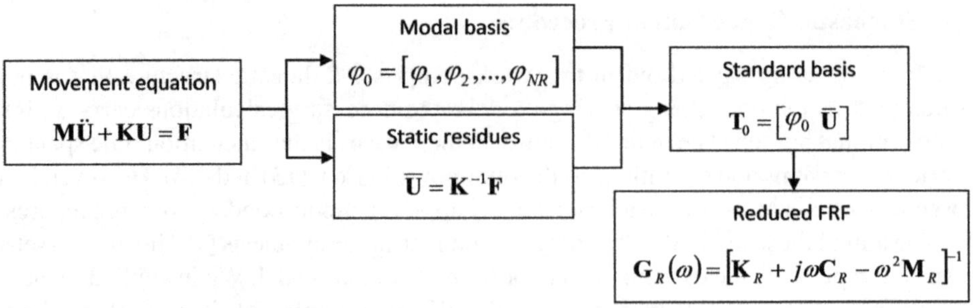

Fig. 2. Flowchart of Ritz basis assembling process

De Lima et al. [9] proposed a condensation strategy to be applied in viscoelastically damped structures. It is updated every time the system undergoes structural modifications. This procedure was extended to stochastic systems, considering parameter fluctuations (stochasticity) as structural modifications.

Consider Eqs. (8) and (9) rewritten in the global form:

$$\mathbf{M}(\theta) = \bar{\mathbf{M}} + \Delta\mathbf{M} \tag{19}$$

$$\mathbf{K}(\theta) = \bar{\mathbf{K}} + \Delta\mathbf{K} \tag{20}$$

where $\bar{\mathbf{M}}, \ \bar{\mathbf{K}}$ refer to nominal system (with mean properties) and variations $\Delta\mathbf{M}, \ \Delta\mathbf{K}$ denote system modifications.

Introducing Eqs. (19) and (20) in Eq. (1) without structural damping, it possible to write Eq. (21).

$$\left(\bar{\mathbf{M}} + \Delta\mathbf{M}\right)\ddot{\mathbf{u}}(t) + \left(\bar{\mathbf{K}} + \Delta\mathbf{K}\right)\mathbf{u}(t) = \mathbf{f}(t) \tag{21}$$

It is possible to rewrite Eq. (21), retaining only the parameters related to the nominal system on the left side:

$$\bar{\mathbf{M}}\ddot{\mathbf{u}}(t) + \bar{\mathbf{K}}\mathbf{u}(t) = \mathbf{f}(t) - \mathbf{F}_{\Delta\mathbf{M}} - \mathbf{F}_{\Delta\mathbf{K}} \tag{22}$$

where $\mathbf{F}_{\Delta\mathbf{M}} = \Delta\mathbf{M}\ddot{\mathbf{u}}(t)$ and $\mathbf{F}_{\Delta\mathbf{K}} = \Delta\mathbf{K}\mathbf{u}(t)$ are vectors of no-exact forces associated to the modifications.

Since this dynamic behavior and system responses are not known, these vectors cannot be computed exactly. The robust condensation procedure consists first in generating a pre-evaluation of no-exact forces to represent the subspace containing them [18]. These vectors are then introduced in the standard basis to represent the dynamic behavior of the modified model. Finally, these two steps are repeated for each sample subjected to parameter fluctuations.

Being a system subjected to structural modifications on a parameter p_i, it is possible to formulate a vector of pre-evaluation forces:

$$\mathbf{F}_{\Delta p_i} = \left[\ \mathbf{F}^{\mathbf{M}}_{\Delta p_i}\ \mathbf{F}^{\mathbf{K}}_{\Delta p_i}\ \right] \tag{23}$$

where $\mathbf{F}^{\mathbf{M}}_{\Delta p_i} = \Delta\mathbf{M}\varphi_0\Lambda_0$, $\mathbf{F}^{\mathbf{K}}_{\Delta p_i} = \Delta\mathbf{K}\varphi_0$ and $\Lambda_0 = diag(\lambda_1, \lambda_2, \ldots, \lambda_{NR})$ ate the eigenvalues associated to the problem presented in Eq. (17).

Equation (24) summarizes the calculation of static residues due to force modifications caused by each parameter p_i.

$$\mathbf{R}_{\Delta p_i} = \bar{\mathbf{K}}^{-1}\mathbf{F}_{\Delta p_i} \tag{24}$$

Finally, the basis is obtained by introducing residues calculated in Eq. (24) into the standard Ritz basis, generating the *stochastic condensation basis* $\mathbf{T}_S$:

$$\mathbf{T}_S = \left[\varphi_0\ \bar{\mathbf{U}}\ \mathbf{R}_\Delta\right] \tag{25}$$

where $\mathbf{R}_\Delta = \left[\mathbf{R}_{\Delta p_1}, \ \mathbf{R}_{\Delta p_2}, \ldots, \ \mathbf{R}_{\Delta p_{np}}\right]$ and *np* is the number of random variables.

Although reduced mass and stiffness matrices calculated by this procedure present larger dimensions than those calculated by the classic Ritz method, computational time reduction is expected owing to only one solution of the eigenproblem of the nominal system being necessary. For each function evaluation, the basis is reupdated based on the iterative process shown in Fig. 3.

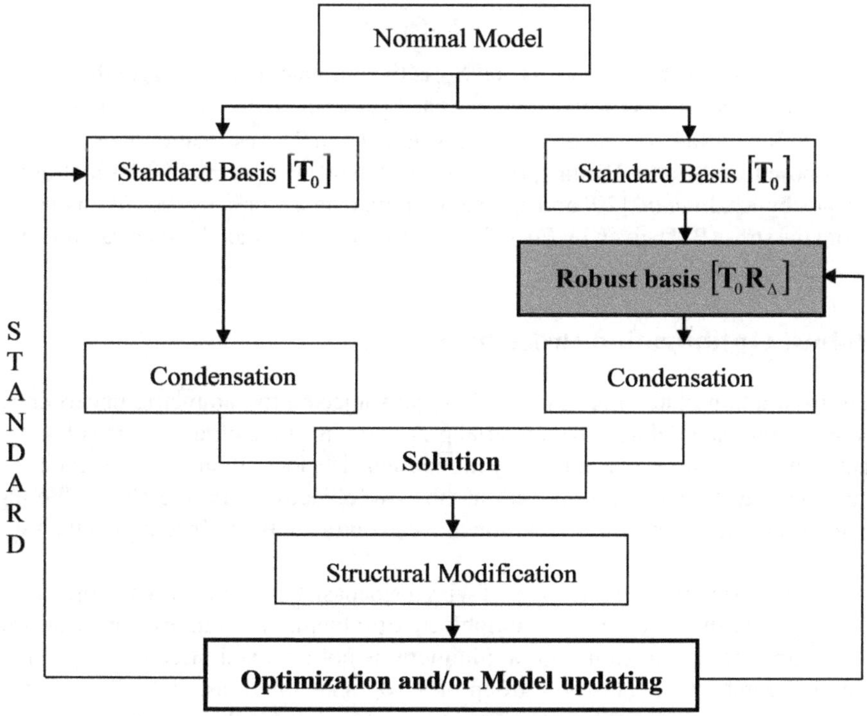

Fig. 3. Block-diagram of the condensation robust procedure (adapted from de Lima et al. [17])

3 Sines Criterion for Fatigue Analysis

The estimation of uniaxial fatigue life can be evaluated directly from S-N diagrams [1]. On the other hand, other criteria have to be applied if the structure is subjected to multiaxial loadings, which represents the greatest part of real situations.

Weber [4] studied some of these methods, comparing them for accuracy [2]. The conclusion was that the criterion proposed by Sines [12] is highly suitable for this kind of analysis. Despite being the second best among nine in terms of general accuracy, its formulation is simpler, concerning only two material properties related to multiaxial fatigue: alternate torsional $\left(t_{\_1}\right)$ and alternate traction $\left(f_{\_1}\right)$ fatigue resistances.

This criterion consists in the calculation of a coefficient D_{Sines}, function of the square root of the second invariant of the stress deviatoric tensor $\left(\sqrt{J_{2,a}}\right)$. Sines' index must be lower than 1 to indicate that no crack initiation will occur. One may note that, when performing analysis in the presence of random loads, the procedure implies a calculation of a mathematical hope rather than direct estimation, as show in the following equation:

$$E[D_{Sines}] = \frac{E\left[\sqrt{J_{2,a}}\right] + \alpha E\left[p_{hid}(t)\right]}{t_{\_1}} \tag{26}$$

$$\alpha = \frac{3t_{\_1}\left(S_{ut} + f_{\_1}\right)}{S_{ut}f_{\_1}} \tag{27}$$

where $p_{hid}(t)$ is the hydrostatic stress and S_{ut} is the ultimate material strength.

When subjected to a zero-mean excitation, $E\left[p_{hid}(t)\right] = 0$, and the procedure for obtaining $E[D_{Sines}]$ falls on the calculation of $E\left[\sqrt{J_{2,a}}\right]$. The estimation method, originally proposed by Li e de Freitas [19], was later expanded for multiaxial and out of phase loads by Khalij *et al.* [20] and applied in more recent works by the authors [8, 21, 22]. From the stress PSD given by Eq. (13), it is possible to extract the spectral moments involved in this calculation [15].

4 Robust Optimization Procedure

Starting from a thin plate under random loading subjected to parametric uncertainties with known nominal thickness and material properties, the final objective is to obtain an optimal value that ensures non-failure by fatigue and also does not increase too much the system's mass. It is also easily noticed that the two objectives are directly conflicting. This means that there is no single solution that can optimize both objectives at the same time.

The optimization problem associated with a stochastic structure whose objective is fatigue analysis is, then, treated as a multiobjective problem. This implies that the desired global solution, be it a maximum or a minimum, is not obtained directly for a unique point in the search space, but for a group of all-optimal solutions, called Pareto front [23]. To solve this type of problem, heuristic or metaheuristic methods are recommended [10, 11]. A multiobjective problem is formulated as follows:

$$\begin{cases} \min\limits_{x} F(x) = (f_1(x), f_2(x), \ldots, f_n(x)) \\ \quad g_j(x) \leq 0 \quad j = 1, \ldots, m \\ \quad x_L \leq x \leq x_U \quad x \in C \end{cases} \tag{28}$$

where $F(x)$ is the objective functions vector to be optimized, $n \geq 2$ is the number of objective functions, $x = (x_1, x_2, \ldots, x_k)$ is the project variables vector, $C \subset \mathbb{R}^k$ is the project space associated to the restrictions defined by $g_j(x)$ and the lower (x_L) and upper (x_U) boundaries.

A stochastic problem such as the one addressed herein, usually presents high sensibility. Due to this, coupled with possible discontinuities, a slight parameter fluctuation may affect abruptly the regions close to a global optimum [24]. Thus, the chosen optimization routine must have the lowest sensibility as possible. In other words, robustness is required.

The designer must also be aware that the optimum to be chosen may not always be the multiobjective global optimum. The curve depicted in Fig. 4 illustrate two different optimal solutions. Solution A is the global optimum, obtained by a deterministic optimization procedure and solution B is the called robust optimum [25]. When the function $f(x)$ is subjected to a slight perturbation denoted here as Δx_1, the curve behavior around point A presents a greater variation than around point B.

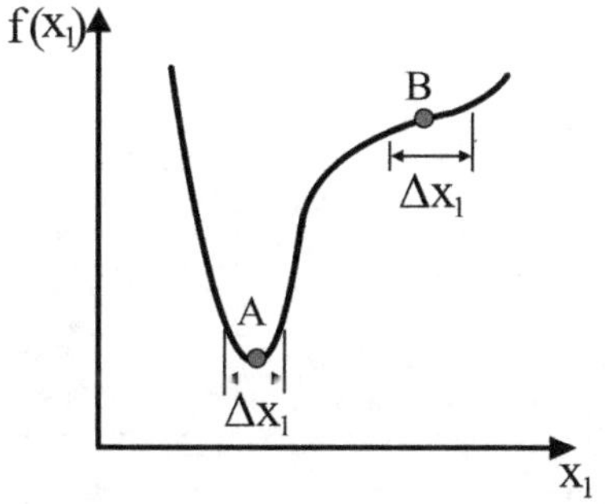

Fig. 4. Representation of optimal solutions (adapted from Lee and Park [25])

Applying a robust optimization procedure generates results that are less sensitive to small design variations when project and cost functions are subjected to parametric uncertainties. Thus, it is possible to obtain a solution that, despite not being the global optimum, presents higher stability when concerning random effects.

The robustness is achieved by adding vulnerability functions $f_i^v(x)$ to the objective functions to be optimized. Robust optimal solutions are those that simultaneously optimize the initial cost functions and at the same time maximize its robustness (or minimize its vulnerability). For each objective $f_i(x)$, the corresponding vulnerability function is calculated by the inverse of dispersion, as follows:

$$f_i^v(x) = \left(\sigma_{f_i} / \mu_{f_i}\right)^{-1} \tag{29}$$

where σ_{f_i} and μ_{f_i} are the standard deviation and mean of $f_i(x)$.

A robust multiobjective problem is presented in the following general form:

$$\begin{cases} \min_x F(x) = \left(f_1(x), f_1^v(x), f_2(x), f_2^v(x), \ldots, f_n(x), f_n^v(x)\right) \\ g_j(x) \leq 0 \quad j = 1, \ldots, m \\ x_L \leq x \leq x_U \quad x \in C \end{cases} \tag{30}$$

In the present work, the multiobjective optimization problem is solved by using the so-called NSGA (Non-Dominated Sorting Genetic Algorithm) in a MatLab® toolbox [26].

After obtaining the set of solutions (Pareto front), the designer must choose the solution that best fits the project, weighting the pros and cons of each point, achieving a compromise between the objectives.

5 Numerical Applications

All the numerical simulations presented herein were performed over a thin rectangular aluminum plate with mechanical properties given in Tab. (1). Its dimensions are 654 mm × 527 mm, boundary condition is double-clamped along y axis, loading is applied and displacement is measured in the central node. The chosen loading is a random gaussian noise with zero mean and variance $85 \times 10^3 N^2/Hz$. The system was discretized in an 8 × 8 mesh of rectangular elements composed by four nodes with five degrees of freedom per node, as shows in Fig. 5 (Table 1).

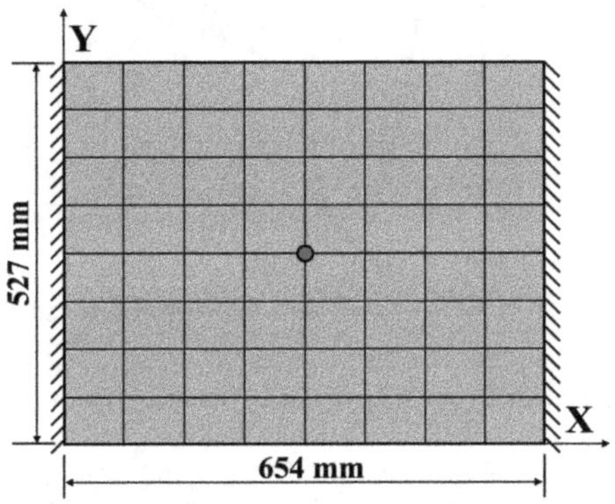

Fig. 5. General layout and meshing of the structure

Table 1. Mechanical properties of the system

Thickness	Young's Module (E)	Mass Density (ρ)	Poisson Ratio (ν)	Ultimate Strength Limit (S_{ut})	Alternate Torsional Limit[(a)] ($t_{\_1}$)	Alternate Traction Limit[(a)] ($f_{\_1}$)
[mm]	[GPa]	[kg/m^3]		[MPa]	[MPa]	[MPa]
1.7	70	2700	0.33	343	92	132

[(a)]obtained for 2.0×10^6 cycles

The first analysis concerns the proposed condensation basis. The aim here is to compare FRF envelopes obtained by the full system, reduced system by Ritz basis and by stochastic basis. For this first analysis, $n_S = 500$ samples were generated by using Latin Hypercube Sampling (LHS) [27], with an uncertainty level of 5% over a three-sigma deviation in thickness.

As it can be seen in Fig. 6, the three curves are stacked, showing an excellent approximation when analyzed by the naked eye. Figure 7 proves the efficiency of both condensation basis, showing a maximum error of 0.016% for the Ritz basis and 0.025% for the stochastic basis. Both errors are minimal, not affecting the model in a significant way.

The standard Ritz basis presented a computational gain of 70.4% compared to the unreduced system while the stochastic basis provided a gain of 96.4%. These results are very important for the following procedure of fatigue life evaluation due to the necessity of multiple function evaluations. This becomes even more critical when performing robust optimization procedures. It is unfeasible to perform this analysis with the uncondensed model.

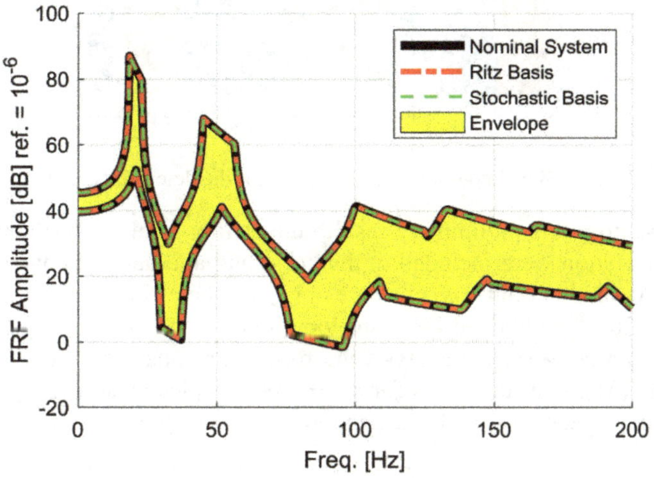

Fig. 6. Displacement FRF envelope for condensation basis comparison

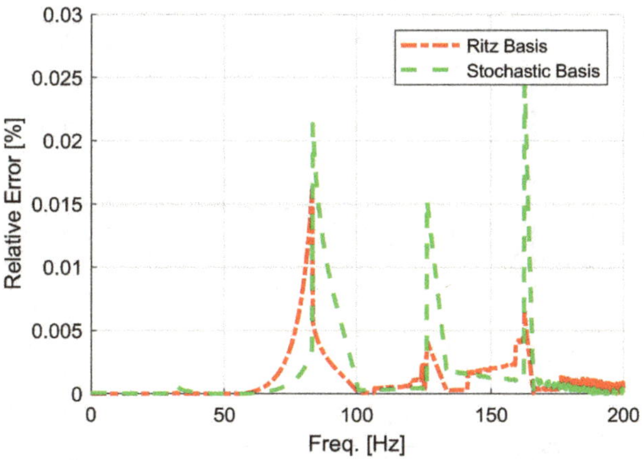

Fig. 7. Relative errors of both condensation basis with respect to the nominal system

Starting from a deterministic approach, that is, the plate with mean properties not subjected to random effects in physical properties, the obtained maximum value for Sines' index is 0.7130, not indicating fatigue failure before 2 million cycles. Figure 8 shows Sines' coefficient distribution for this case.

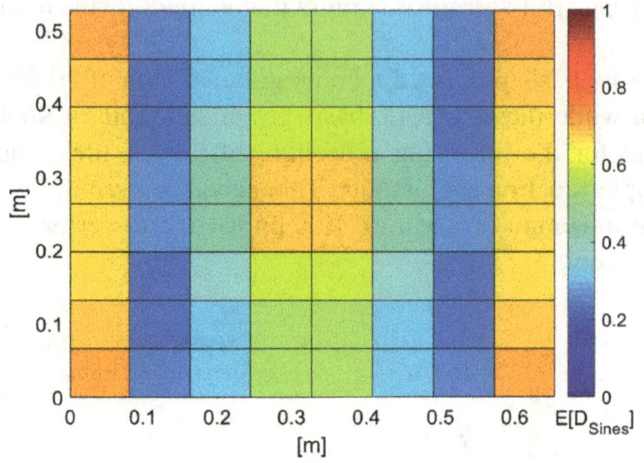

Fig. 8. Sines' coefficient distribution for the deterministic case

After presenting this preliminary result, uncertainty levels of 5% and 10% over a three-sigma deviation were included in thickness and alternate torsional fatigue limit, respectively. As fatigue limit has influence only over the estimation of Sines' index, as shown in Eq. (26), the obtained FRF envelope is similar to the one depicted in Fig. 6. Figure 9 shows a convergence analysis via root-mean-square deviation (RMSD) calculated by Eq. (31), over the FRFs for $n_S = 500$ samples generated by LHS [27]. A satisfactory convergence is noted at around $n_S = 300$ samples.

$$RMSD(n_S) = \sqrt{\frac{1}{n_S}\sum_{i=1}^{n_S}\left|G_i(\omega,\theta) - \bar{G}(\omega,\theta)\right|^2} \tag{31}$$

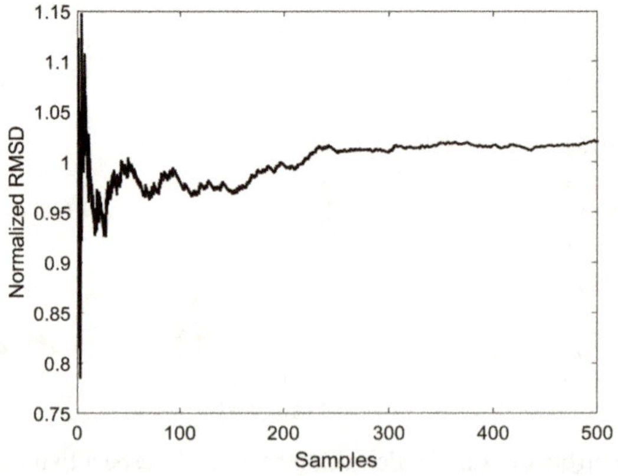

Fig. 9. Convergence analysis via normalized RMSD

Stress PSDs envelopes for the critical element and Sines' index maximum, mean and minimum distributions are shown in Figs. 10a and 10b. As it can be noted, in this case the maximum value for Sines' coefficient is 1.1964. This indicates fatigue failure in the border and central elements before 2 million cycles.

The result carried herein shows the change of the subjective analysis of non-failure in the deterministic approach to a failure condition when uncertainties were included in the model. This highlights the importance of considering random effects when performing numerical analysis, even more so in structures that can suffer catastrophic failures such as those subjected to dynamic loads, that may be affected by the fatigue process.

After presenting the preliminary results for the thin plate in question, the optimization procedure can be carried out. As the structure presented a fatigue failure condition when

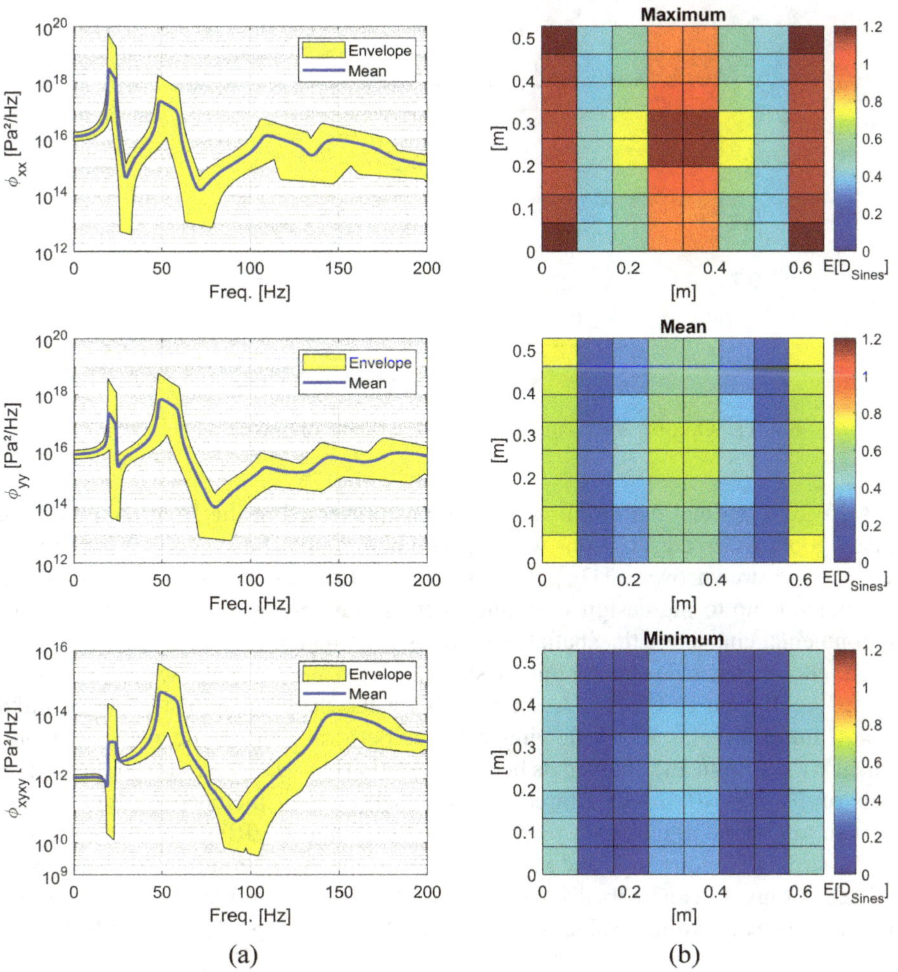

Fig. 10. Preliminary results, thickness 1.7 mm. (a) Stress PSD envelopes for the critical plate element, (b) Sines' index distribution

subjected to uncertainties, the main objective is to enhance its fatigue life. In order to guarantee this, Sines' maximum coefficient must not surpass 1.000 for the most critical element, which, as shown before, is one of the border elements. The conflicting objective is to minimize the system's mass (consequently its thickness) as well as vulnerability functions. With the objectives defined, after the optimization procedure, it is possible to visualize the trade-off curve in two dimensions, as shown in Fig. 11.

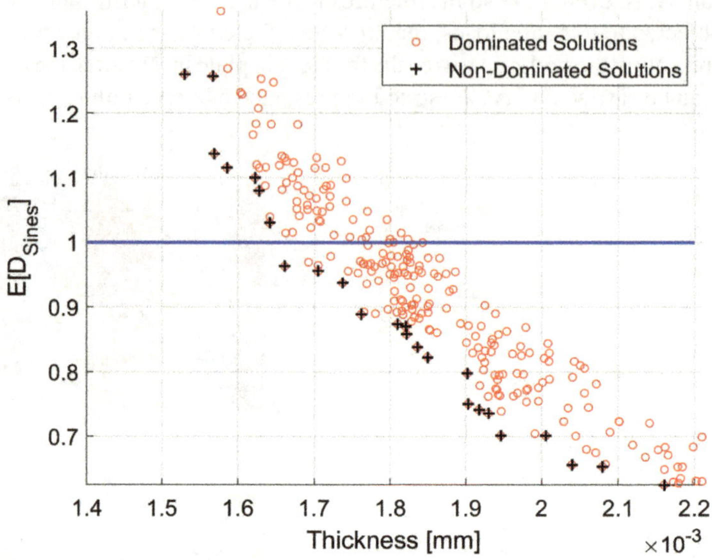

Fig. 11. Optimal solutions obtained in the robust optimization procedure

All Pareto optimal non-dominated solutions presented in the front are considered equally good. As no constrictions were defined before, the chosen point must be below the blue line drawn over $E[D_{Sines}] = 1.000$, limiting optimal values to be chosen. The choice is up to the designer as other factors must be taken into account, such as environmental characteristic, spatial limitations and financial cost, among others.

The chosen optimal value for thickness is 2.0 mm. The deterministic analysis showed a Sines' coefficient of 0.5177. As before, this value does not indicate fatigue failure before 2 million cycles. The following step is to perform a stochastic analysis, in order to disturb the system and analyze its behavior. Similarly to the previous test, uncertainty levels of 5% and 10% over a three-sigma deviation were included in thickness and alternate torsional fatigue limit, respectively. The obtained FRF envelope is shown in Fig. 12, stress PSDs envelope and Sines' maximum, mean and minimum distribution are depicted in Figs. 13a and 13b. As one may notice, even when subjected to uncertainties, the Sines' index maximum value is 0.8619, not characterizing a failure condition.

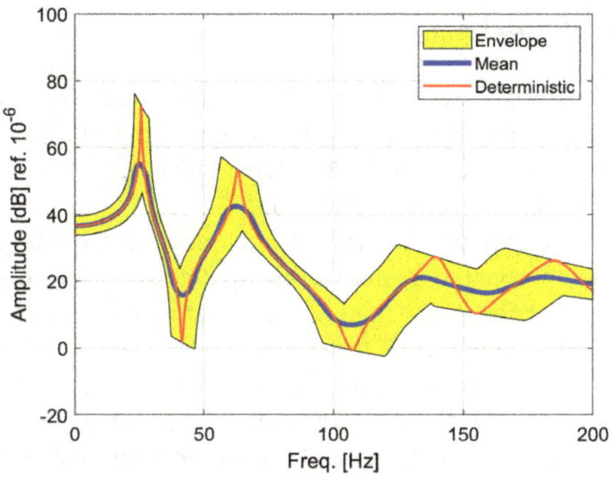

Fig. 12. Displacement FRF envelope

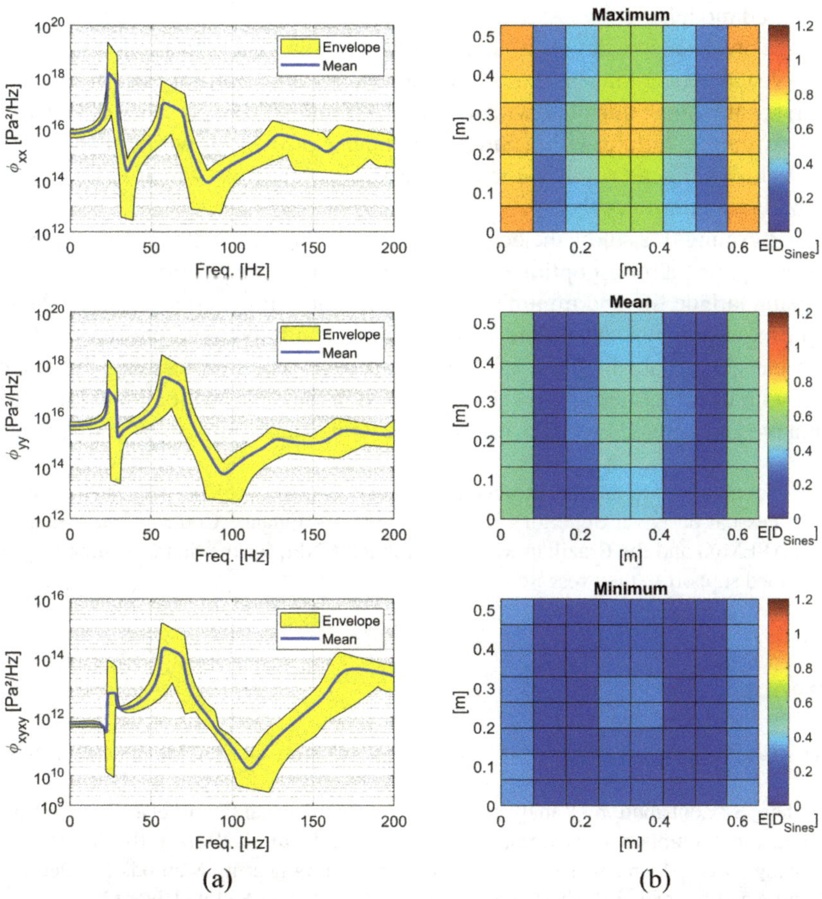

Fig. 13. Results for the optimized plate, thickness 2.0 mm. (a) Stress PSD envelopes for the critical plate element, (b) Sines' index distribution

6 Concluding Remarks

The aim of the methodology presented herein is to allow durability analysis to be performed in the project phase, giving a wider range of possibilities and tools for the designer to choose when dealing with structures subjected to dynamic loadings and random vibrations that may lead to fatigue failure.

The inclusion of uncertainties in structural geometry and loadings provides a more accurate and realistic approach than the classic applications. As fatigue strength limits are obtained strictly experimentally, they carry uncertainties and errors that can affect the fatigue failure analysis and impair the decision-making process at the design phase.

As this kind of analysis presents high computational costs, coupled with a huge number of function evaluations, a condensation procedure is necessary. A stochastic basis was proposed and evaluated, generating a computational gain of 96.4% when compared to the nominal system, while the classic Ritz basis provides a gain of 70.4%. These procedures become essential as it is unfeasible to perform this analysis with the uncondensed model.

Fatigue breakage is normally a catastrophic event that occurs in a sudden way without previous visual notice. This highlights the importance of performing stochastic analysis, as the first result for the thin plate with thickness 1.7 mm with no presence of uncertainties showed no fatigue failure with a Sines' coefficient of 0.7130, while in the stochastic approach this value can reach the maximum value of 1.1964, which indicated fatigue failure before 2 million cycles.

By performing this study, the designer is able to choose a better value for the plate's thickness applying a robust optimization procedure. The one performed herein aimed at maximizing fatigue life and minimize thickness, avoiding sensitive points in the search domain and changes in the failure or non-failure condition. The chosen thickness of 2.0 mm caused a mass increase of 17.6%. but even when performing the stochastic analysis with uncertainties in thickness itself and in the alternate torsional fatigue limit, it did not present failure before 2 million cycles with a maximum Sines' index of 0.8619.

Acknowledgements. The authors express their acknowledgments to Coordenação de Aperfeiçoamento de Pessoal de Nível Superior – Brasil (CAPES) – Finance Code 001, to Minas Gerais Agency FAPEMIG and the Brazilian Research Council CNPq for the financial support provided and continued support to their research activities.

References

1. Budynas, R., Nisbett, J.: Shigley's Mechanical Engineering Design. McGraw-Hill, New York (2011)
2. Lambert, S.: Contribution à l'analyse de l'endommagement par fatigue et au dimensionnement de structures soumises à des vibrations aléatoires. Ph.D. thesis, Rouen, INSA (2007)
3. Dowling, N.E.: Mechanical Behavior of Materials: Engineering Methods for Deformation, Fracture and Fatigue, 3rd edn. Pearson Prentice Hall, Upper Saddle River (2007)
4. Weber, B.: Fatigue multiaxiale des structures industrielles sous chargement quelconque. Ph.D. thesis, INSA de Lion, Lion (1999)

5. Stefanou, G.: The stochastic finite element method: past, present and future. Comput. Methods Appl. Mech. Eng. **198**(9–12), 1031–1051 (2009)
6. Ghanem, R.G., Spanos, P.D.: Stochastic finite elements: a spectral approach. Courier Corporation, North Chelmsford (2003)
7. de Lima, A.M.G., Rade, D.A., Bouhaddi, N.: Stochastic modeling of surface viscoelastic treatments combined with model condensation procedures. Shock Vib. **17**(4–5), 429–444 (2010)
8. Rosa, U.L., de Lima, A.M.G.: Fatigue analysis of stochastic systems subjected to cyclic loading in the frequency domain. In: 3rd International Symposium on Uncertainty Quantification and Stochastic Modeling. Maresias, Brazil (2016)
9. de Lima, A.M.G., da Silva, A.R., Rade, D.A., Bouhaddi, N.: Component mode synthesis combining robust enriched Ritz approach for viscoelastically damped structures. Eng. Struct. **32**, 1479–1488 (2010)
10. Lobato, F.S.: Otimização multi-objetivo para o projeto de sistemas de engenharia, Ph.D. thesis, UFU, Uberlândia (2008)
11. Rao, S.S.: Engineering Optimization: Theory and Practice. Wiley, New York (2009)
12. Sines, G.: Behavior of metals under complex static and alternating stresses. Metal Fatigue **1**, 145–169 (1959)
13. Zienkiewicz, O.C., Taylor, R.L.: The Finite Element Method for Solid and Structural Mechanics. Elsevier, Amsterdam (2005)
14. Moaveni, S.: Finite element analysis theory and application with Ansys®. Upper Saddle River, New Jersey, Vol. 7458 (1999)
15. Bendat, J., Piersol, A.: Random Data. Wiley Series in Probability and Statistics. Wiley, Hoboken (2010)
16. Craig Jr., R.R., Kurdila, A.J.: Fundamentals of Stuctural Dynamics. Wiley, Hoboken (2006)
17. Gèrges, Y.: Méthodes de reduction de modèles en vibroacoustique non-linéaire. PhD thesis, Université de Franche-Comté, Besançon (2013)
18. Masson, G., Ait, Brik B., Cogan, S., Bouhaddi, N.: Component mode synthesis (CMS) based on an enriched Ritz approach for efficient structural optimization. J. Sound Vib. **296**, 845–860 (2006)
19. Li, B., de Freitas, M.: A procedure for fast evaluation of high-cycle fatigue under multiaxial random loading. J. Mech. Des. **124**, 558–563 (2002)
20. Khalij, L., Pagnacco, E., Lambert, S.: A measure of the equivalent shear stress amplitude from a prismatic hull in the principal coordinate system. Int. J. Fatigue **32**(12), 1977–1984 (2010)
21. Rosa, U.L.: Análise de fadiga de sistemas estocásticos submetidos a carregamentos aleatórios multiaxiais, M.Sc. dissertation, UFU, Uberlândia (2016)
22. Gonçalves, L.K.S., Rosa, U.L., de Lima, A.M.G.: Fatigue damage investigation and optimization of a viscoelastically damped system with uncertainties. J. Brazilian Soc. Mech. Sci. Eng. **41**(9), 382 (2019
23. Eschenauer, H., Koski, J., Osyczka, A.: Multicriteria Design Optimization: Procedures and Applications. Springer, Cham (2012)
24. Moreira, F.R.: Otimização robusta multiobjetivo para o projeto de sistemas em engenharia, Ph.D. thesis, UFU, Uberlândia (2015)
25. Lee, K.H., Park, G.J.: Robust optimization considering tolerances of design variables. Comput. Struct. **79**(1), 77–86 (2001)
26. Lobato, F., Borges, R. and Steffen Jr., V.: Bio-inspired Optimization Methods: Modeling, Design, Inverse Problem and Robust Optimization of a Representative Mechanical System. Lambert Academic Publishing (2015)
27. Florian, A.: An efficient sampling scheme: updates Latin Hypercube sampling. Probab. Eng. Mech. **7**(2), 123–130 (1992)

Stochastic Gradient Descent for Risk Optimization

André Gustavo Carlon[1], André Jacomel Torii[2], Rafael Holdorf Lopez[3](✉), and José Eduardo Souza de Cursi[4]

[1] Computer, Electrical and Mathematical Sciences and Engineering Division (CEMSE), King Abdullah University of Science and Technology (KAUST), Thuwal 23955-6900, Saudi Arabia
agcarlon@gmail.com

[2] Latin American Institute for Technology, Infrastructure and Territory (ILATIT), Federal University for Latin American Integration (UNILA), Av. Tancredo Neves 6731, Foz do Iguaçu, Brazil

[3] Center for Optimization and Reliability in Engineering (CORE), Department of Civil Engineering, Federal University of Santa Catarina, UFSC, Rua João Pio Duarte, s/n, Florianopolis, Brazil
rafae.holdorf@ufsc.br

[4] Institut National des Sciences Appliquées (INSA) de Rouen, 76801 Saint Etienne du Rouvray Cedex, France

Abstract. This paper presents an approach for the use of stochastic gradient descent methods for the solution of risk optimization problems. The first challenge is to avoid the high-cost evaluation of the failure probability and its gradient at each iteration of the optimization process. We propose here that it is accomplished by employing a stochastic gradient descent algorithm for the minimization of the Chernoff bound of the limit state function associated with the probabilistic constraint. The employed stochastic gradient descent algorithm, the Adam algorithm, is a robust method used in machine learning training. A numerical example is presented to illustrate the advantages and potential drawbacks of the proposed approach.

Keywords: Risk optimization · Stochastic gradient descent · Chernoff bound · Stochastic optimization

1 Introduction

It is often the case in engineering that uncertainties need to be considered when designing a project. It becomes even more critical when optimization is applied for project design, leading to optimization under uncertainty problems. In the optimization under uncertainty literature, two design philosophies are of particular interest for the civil/mechanical engineering community: Reliability Based Design Optimization (RBDO) and Risk Optimization (RO) [1,2]. In RBDO, reliability constraints are imposed to the design. On the other hand, in RO,

J. E. S. De Cursi (Ed.): Uncertainties 2020, LNME, pp. 424–435, 2021.
https://doi.org/10.1007/978-3-030-53669-5_31

expected costs of failure are explicitly incorporated in the objective function, and consequently, the reliability level is implicitly defined during the optimization procedure.

In this paper, we focus on the solution of RO problems, whose most critical step is the evaluation of the probabilities of failure (or reliability analysis) of the engineering systems under analysis. Usually, the methods used in reliability analysis rely on approximations of the limit-state equations or in Monte Carlo (MC) sampling. It well known that the former introduces bias in the optimization problem and the later leads high computational cost. Since a large number of reliability analyses are required for design optimization, if MC sampling is used, the optimization procedure can become inviable.

Sampling based schemes have been historically employed for the solution of RO problems [3–5], which normally require a huge computational cost. Recently, approximate models were developed for this purpose. For example, Torii, Lopez, and Miguel [6] successfully proposed a gradient-based polynomial chaos expansion as a surrogate model for the state-limit functions and their gradients in the solution of RO problems. However, these approaches may become inviable for large scale problems, due to, for instance, the curse of dimensionality inherent of surrogate models. On the other hand, considering approximate models for RO, another approach by Torii et al. [7], consists on casting the RO problem as a sequence of RBDO problems and employing a performance measure approach to solve them. Although computationally efficient, this approach is based on First Order Reliability Methods, thus, introducing bias in the solution if the limit state function is not linear or the random variables are not independent and Gaussian.

For the general case of RO, where linearity of the state-limit functions can not be assumed and the number of random variables is large, there are not many alternatives to using MC sampling for the computation of failure probabilities. In this context, we propose in this paper an initial study for the use of stochastic gradient descent methods [8] for the solution of RO problems. Stochastic gradient methods are iterative methods used to optimize the expected value of functions by using as little as one singleton gradient sample per iteration. In engineering problems, stochastic gradient methods were recently applied for the solution of optimal design of experimental problems [9] and the robust design of passive control devices [10], reaching promising results. In this paper, we propose the construction of the RO problem in a manner that it avoids the high-cost evaluation of the failure probability and its gradient at each iteration of the optimization process. Indeed, here we approximate the probability of failure by its Chernoff bound, thus, being able to use stochastic gradient methods. For the solution of the resulting problem, we employ the Adam algorithm [11], which is a robust stochastic descent algorithm used in machine learning training. A numerical example is then presented to illustrate the potential advantages and drawbacks of the use of the Chernoff bound and stochastic gradient for the solution of RO problems.

2 Risk Optimization (RO)

RO can be presented in the general form as the search of the design parameters $\boldsymbol{d}$ in the search space $\mathcal{D}$ that minimize the objective function f, i.e.,

$$\text{Find } \boldsymbol{d}^* = \underset{\boldsymbol{d}\in\mathcal{D}}{\arg\min}\, f(\boldsymbol{d}), \tag{1}$$

where

$$f(\boldsymbol{d}) \stackrel{\text{def}}{=} C_0(\boldsymbol{d}) + \sum_{i=1}^{m} C_i(\boldsymbol{d}) P_{f,i}(\boldsymbol{d}), \tag{2}$$

for constants $\{C_i\}_{i=1}^m$ and probabilities of failure $\{P_{f,i}\}_{i=1}^m$. Each constant C_i and probability of failure $P_{f,i}$ are related to a mode of failure given by a state limit function G_i, hence, minimizing f implies in minimizing the weighted sum of the probabilities of failure added by C_0. To account for uncertainties, we define a random vector $\boldsymbol{X} \in \mathcal{X}$ such that the probability of failure is

$$P_{f,i}(\boldsymbol{d}) \stackrel{\text{def}}{=} P(G_i(\boldsymbol{d}, \boldsymbol{X}) > 0) \tag{3}$$

$$= \int_{G_i(\boldsymbol{d},\boldsymbol{X})\geq 0} \mu_{\boldsymbol{X}}(\boldsymbol{d}, \boldsymbol{X}) d\boldsymbol{X}, \tag{4}$$

where $\mu_{\boldsymbol{X}}$ is the probability density of $\boldsymbol{X}$. For a limit state function $G_i \stackrel{\text{def}}{=} S_i - R_i$, as used in [12],

$$P_{f,i}(\boldsymbol{d}) = P\left(S_i(\boldsymbol{d}, \boldsymbol{X}) > R_i(\boldsymbol{d}, \boldsymbol{X})\right) \tag{5}$$

$$= P\left(\frac{S_i(\boldsymbol{d}, \boldsymbol{X})}{R_i(\boldsymbol{d}, \boldsymbol{X})} > 1\right). \tag{6}$$

Evaluating the probability of failure can be expensive, possibly requiring expensive methods like MC simulation, or approximations of the limit state functions.

3 Chernoff Bound

The Chernoff bound is a direct application of the Markov inequality over the exponential of a random variable [13]. Assuming Y is a random variable, Chernoff's bound can be written as

$$P(Y \geq a) \leq \frac{\mathbb{E}(e^{tY})}{e^{ta}}, \; t > 0. \tag{7}$$

We can use (7) to bound the probability of failure. Thus, from (6), we can write

$$P_{f,i}(\boldsymbol{d}) \leq \hat{P}_{f,i}(\boldsymbol{d}) \stackrel{\text{def}}{=} \min_{t_i>0} \frac{\mathbb{E}\left[e^{t_i \frac{S_i(\boldsymbol{d},\boldsymbol{X})}{R_i(\boldsymbol{d},\boldsymbol{X})}}\right]}{e^{t_i}}, \; t_i > 0. \tag{8}$$

In order to avoid direct evaluation of the probability of failure, we employ the bound $\hat{P}_{f,i}$ during the optimization procedure. Then, instead of directly minimizing f as presented in (2), we minimize $\hat{f}$ given by

$$\hat{f}(\boldsymbol{d}) \stackrel{\text{def}}{=} C_0(\boldsymbol{d}) + \sum_{i=1}^{m} C_i(\boldsymbol{d})\hat{P}_{f,i}(\boldsymbol{d}). \tag{9}$$

The main advantage of minimizing (9) instead of (2) is that the first does not require the evaluation of the probability of failure and its sensitivities, which are known to be cumbersome to evaluate from the computational point of view.

4 Stochastic Gradient Descent (SGD)

The SGD is a method for minimizing the expected value of functions while avoiding MC sampling. SGD is a direct application of Robbins and Monro [8] stochastic approximation in convex stochastic optimization problems.

To use SGD, one necessary condition is that an unbiased estimator of the gradient of the function to be minimized is available. To minimize $\hat{f}$ in (9), we need to define a function F such that

$$\mathbb{E}[F(\boldsymbol{d}, \boldsymbol{X})] = \hat{f}(\boldsymbol{d}). \tag{10}$$

Thus, we define F as

$$F(\boldsymbol{d}, \boldsymbol{X}) \stackrel{\text{def}}{=} C_0(\boldsymbol{d}) + \sum_{i=1}^{m} C_i(\boldsymbol{d}) \frac{e^{t_i \frac{S_i(\boldsymbol{d},\boldsymbol{X})}{R_i(\boldsymbol{d},\boldsymbol{X})}}}{e^{t_i}}. \tag{11}$$

Note that, from (10),

$$\begin{aligned}\nabla_{\boldsymbol{d}} \hat{f}(\boldsymbol{d}) &= \nabla_{\boldsymbol{d}} \mathbb{E}[F(\boldsymbol{d}, \boldsymbol{X})] &(12)\\ &= \mathbb{E}[\nabla_{\boldsymbol{d}} F(\boldsymbol{d}, \boldsymbol{X}) + F(\boldsymbol{d}, \boldsymbol{X}) \nabla_{\boldsymbol{d}} \log(\mu_{\boldsymbol{X}}(\boldsymbol{d}, \boldsymbol{X}))] &(13)\end{aligned}$$

(cf. Appendix B in [10]), hence, by defining the unbiased gradient estimator

$$\nabla_{\boldsymbol{d}} \mathcal{F}(\boldsymbol{d}, \boldsymbol{X}) \stackrel{\text{def}}{=} \nabla_{\boldsymbol{d}} F(\boldsymbol{d}, \boldsymbol{X}) + F(\boldsymbol{d}, \boldsymbol{X}) \nabla_{\boldsymbol{d}} \log(\mu_{\boldsymbol{X}}(\boldsymbol{d}, \boldsymbol{X}))], \tag{14}$$

it can be seen that $\nabla_{\boldsymbol{d}} \mathcal{F}$ is an unbiased gradient estimator of $\hat{f}$. In what follows, we use $\nabla_{\boldsymbol{d}} \mathcal{F}$ as a search direction for SGD. For the minimization of the expected value of a function $F : \mathcal{D} \times \mathcal{X} \mapsto \mathbb{R}$ such that $\mathbb{E}[F(\boldsymbol{d}, \boldsymbol{X})] = \hat{f}(\boldsymbol{d})$, the SGD update is

$$\begin{aligned}&\text{Sample } \boldsymbol{X} \sim \mu_{\boldsymbol{X}} &(15)\\ &\boldsymbol{d}_{k+1} = \boldsymbol{d}_k - \alpha_k \nabla_{\boldsymbol{d}} \mathcal{F}(\boldsymbol{d}, \boldsymbol{X}), &(16)\end{aligned}$$

where α_k is a sequence of decreasing step-sizes [10].

Developing the gradient of F in (11),

$$\nabla_{\boldsymbol{d}} F(\boldsymbol{d}, \boldsymbol{X}) = \nabla_{\boldsymbol{d}} C_0(\boldsymbol{d}) + \nabla_{\boldsymbol{d}} \sum_{i=1}^{m} C_i(\boldsymbol{d}) \frac{e^{t_i \frac{S_i(\boldsymbol{d},\boldsymbol{X})}{R_i(\boldsymbol{d},\boldsymbol{X})}}}{e^{t_i}} \tag{17}$$

$$= \nabla_{\boldsymbol{d}} C_0(\boldsymbol{d}) + \sum_{i=1}^{m} \left(\nabla_{\boldsymbol{d}} C_i(\boldsymbol{d}) \frac{e^{t_i \frac{S_i(\boldsymbol{d},\boldsymbol{X})}{R_i(\boldsymbol{d},\boldsymbol{X})}}}{e^{t_i}} + C_i(\boldsymbol{d}) \frac{\nabla_{\boldsymbol{d}} e^{t_i \frac{S_i(\boldsymbol{d},\boldsymbol{X})}{R_i(\boldsymbol{d},\boldsymbol{X})}}}{e^{t_i}} \right). \tag{18}$$

Derivatives of the constants C_i depend on the problem formulation and should not be expensive to calculate. The gradient inside the summation on the right hand side of (18) is obtained from S_i, R_i, and their gradients as

$$\nabla_{\boldsymbol{d}} e^{t_i \frac{S_i(\boldsymbol{d},\boldsymbol{X})}{R_i(\boldsymbol{d},\boldsymbol{X})}} = t_i e^{t_i \frac{S_i(\boldsymbol{d},\boldsymbol{X})}{R_i(\boldsymbol{d},\boldsymbol{X})}} \frac{\nabla_{\boldsymbol{d}} S_i(\boldsymbol{d}, \boldsymbol{X}) R_i(\boldsymbol{d}, \boldsymbol{X}) - S_i(\boldsymbol{d}, \boldsymbol{X}) \nabla_{\boldsymbol{d}} R_i(\boldsymbol{d}, \boldsymbol{X})}{R_i(\boldsymbol{d}, \boldsymbol{X})^2}. \tag{19}$$

Hence, to use SGD to minimize the Chernoff bound one uses only S_i, R_i, and its derivatives, without requiring any MC sampling. This results in an optimization procedure with a small cost per iteration when compared to a classical approach, e.g., using any classical optimization method with MC with sample average approximation (SAA) to evaluate the probability of failure.

4.1 Adam Algorithm

In the numerical section, we use an algorithm of the family of SGD named Adam [11]. Adam uses the following update rule

$$\text{Sample } \boldsymbol{X} \sim \mu_{\boldsymbol{X}} \tag{20}$$

$$\boldsymbol{m}^{(k+1)} = \beta_1 \boldsymbol{m}^{(k)} + (1 - \beta_1) \nabla_{\boldsymbol{d}} \mathcal{F}(\boldsymbol{d}^{(k)}, \boldsymbol{X}) \tag{21}$$

$$\boldsymbol{v}^{(k+1)} = \beta_2 \boldsymbol{v}^{(k)} + (1 - \beta_2) (\nabla_{\boldsymbol{d}} \mathcal{F}(\boldsymbol{d}^{(k)}, \boldsymbol{X}))^2 \tag{22}$$

$$\hat{\boldsymbol{m}} = \frac{\boldsymbol{m}^{(k+1)}}{1 - \beta_1^{k+1}} \tag{23}$$

$$\hat{\boldsymbol{v}} = \frac{\boldsymbol{v}^{(k+1)}}{1 - \beta_2^{k+1}} \tag{24}$$

$$\boldsymbol{d}^{(k+1)} = \boldsymbol{d}^{(k)} - \alpha_k \frac{\hat{\boldsymbol{m}}}{\sqrt{\hat{\boldsymbol{v}}} + \epsilon}, \tag{25}$$

where $0 < \beta_1 < \beta_2 < 1$ and ϵ are constants to be defined. Adam is more robust than SGD, being efficient in dealing with noisy gradients, the reason why we chose to use it in the numerical section.

4.2 Setting the Parameter t_i of Chernoff's Bound

Note that tightness of Chernoff's bound depends on the parameter $t > 0$. In order to get an optimal Chernoff's bound, the parameter that minimizes the bound should be employed. Thus, for each limit state, the parameter t_i at iteration (k) is obtained from the minimization problem

$$\text{Find } t_i^{(k)} = \underset{t_i>0}{\arg\min} \frac{\mathbb{E}\left[e^{t_i \frac{S_i(d,X)}{R_i(d,X)}}\right]}{e^{t_i}}. \tag{26}$$

To solve the optimization sub-problem from (26), we independently sample t uniformly from a fixed interval between positive values many times and verify the value that minimizes the bound. Tuning t requires sampling $\boldsymbol{X}$ and evaluating S and R, what can be expensive. Thus, we update t if the distance of $\boldsymbol{d}^{(k)}$ to $\boldsymbol{d}^{(\overline{k})}$, being $\overline{k}$ the iteration when t was last tuned, exceeds a decaying value,

$$\text{Find } t^{(k)} \text{ if } \left\| \boldsymbol{d}^{(k)} - \boldsymbol{d}^{(\overline{k})} \right\| \geq \frac{\epsilon_t}{\sqrt{k}}, \tag{27}$$

with ϵ_t being a parameter to be set. Note that the decay in the right hand side of (27) is the same used for the step-size in stochastic optimization.

The parameter $t_i^{(k)}$ as defined above is expected to oscillate due to sampling errors. To mitigate the noise in t, we use a running average $\overline{t}$ instead of t, defined as

$$\overline{t}_i^{(k)} \stackrel{\text{def}}{=} \frac{2}{k} \sum_{j=\frac{k}{2}}^{k} t_i^{(j)}. \tag{28}$$

5 Numerical Example

5.1 Cantilever Beam

To test the proposed methodology, we solve one numerical example, a cantilever beam subject to a point load P at its end. This benchmark problem has been presented by Torii et al. [7]. We consider a rectangular cross-section whose dimensions, width b and height h, are the design parameters, i.e., $\boldsymbol{d} = (b, h)$. The failure modes considered are the elastic torsional buckling and the plastic yielding of the material with, respectively, state limit functions

$$G_1 = \frac{S_1}{R_1} = \frac{M_p}{PL} \tag{29}$$

$$G_2 = \frac{S_2}{R_2} = \frac{P_{\text{cr}}}{P}. \tag{30}$$

According to [14], the plastic moment M_p is given by

$$M_p = Z f_y, \tag{31}$$

where

$$Z = \frac{bh^2}{4} \tag{32}$$

is the plastic modulus and f_y is the yielding strength of the material. The critical torsional buckling load P_{cr} is given by [15] as

$$P_{\text{cr}} = 4.013\frac{\sqrt{EI_yC}}{L^2} \tag{33}$$

for beam length L, Young modulus E, cross-sectional inertia over the vertical axis

$$I_y = \frac{b^3h}{12}, \tag{34}$$

torsional constant of a thin cross section

$$C = \frac{1}{3}b^2hG, \tag{35}$$

and shear modulus G.

The parameters f_y, G, L, and P are modeled as truncated Gaussian-distributed random variables with means and standard deviations as presented in Table 1.

Table 1. Mean and standard deviation for the random parameters of the cantilever beam example.

Parameter	Mean	Std
f_y (MPa)	35.0	3.5
G (MPa)	7700.0	385.0
L (mm)	200.0	10.0
P (kN)	8.0	1.6

Box constraints are imposed on b and h,

$$0.1 \leq b, h \leq 30.0. \tag{36}$$

A Poisson constant of 0.364 is used to calculate E from G. The objective function to be minimized in this example is

$$f(\boldsymbol{d}) = bh\left(1 + 2\sum_{i=1}^{2} P_{f,i}\right) + 200\sum_{i=1}^{2} P_{f,i}. \tag{37}$$

5.2 Setup

We approximate (37) as

$$\hat{f}(\boldsymbol{d}) = bh\left(1 + 2\sum_{i=1}^{2} \hat{P}_{f,i}\right) + 200\sum_{i=1}^{2} \hat{P}_{f,i}. \tag{38}$$

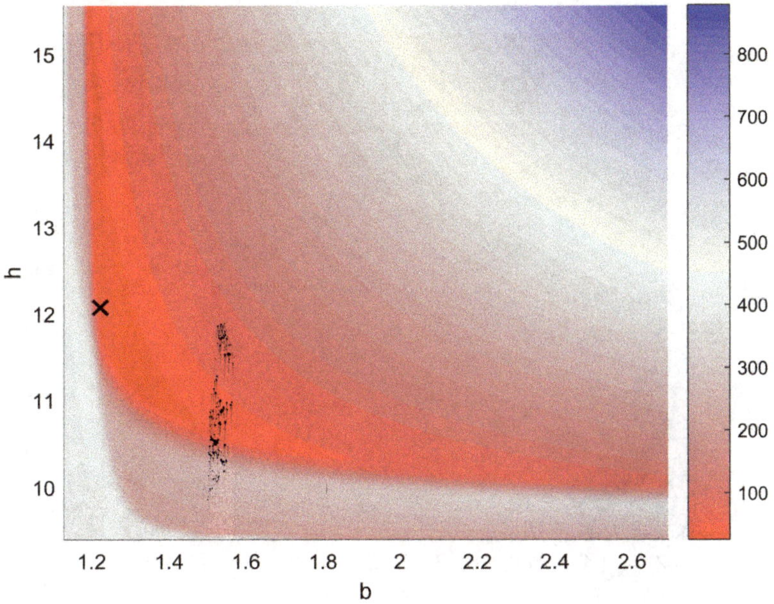

(a) f

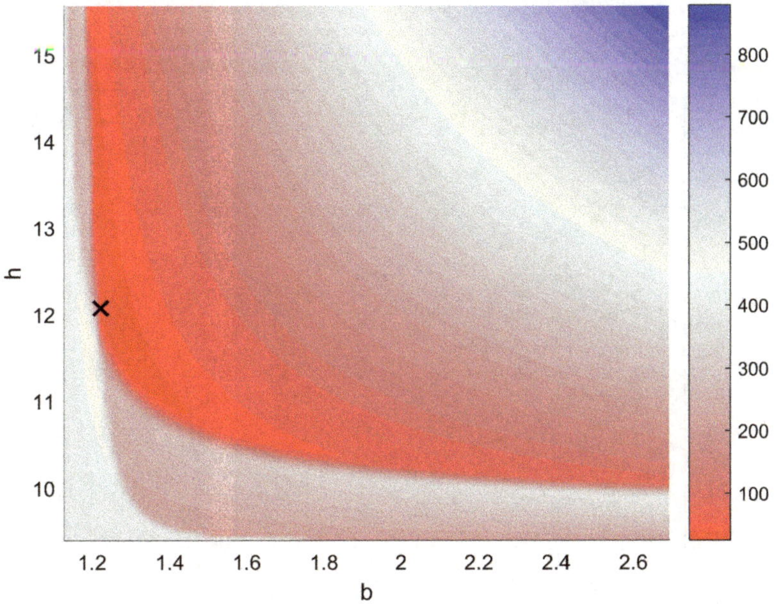

(b) $\hat{f}$

Fig. 1. Contour of f and $\hat{f}$ with respect to b and h for the cantilever beam in Sect. 5.1. The optimum in each case is marked with a black $\times$.

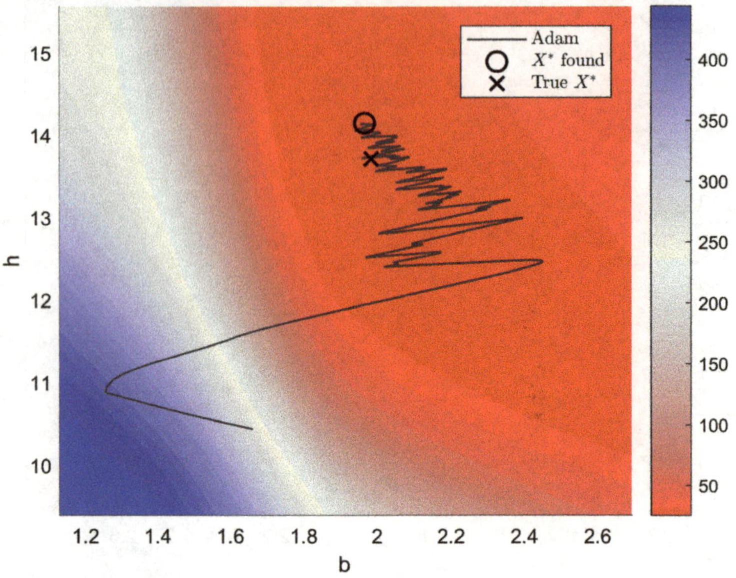

Fig. 2. Optimization path presented over the contour of f. The true optimum is marked with a black $\times$ while the solution found is denoted with a black circle.

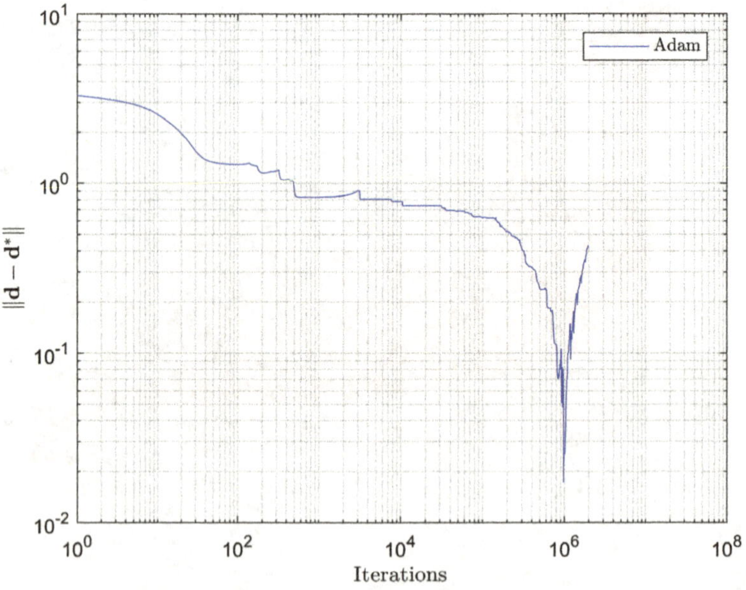

Fig. 3. Distance to the optimum per iteration.

From S_i and R_i, as defined in (29) and (30), and their gradients, we evaluate, at each iteration, the search direction $\nabla_d \mathcal{F}$ using (14), (18) and (19).

We run the optimization using Adam until 10^7 model gradients are evaluated. The initial step-size used in this example is $\alpha_0 = 0.2$ and the tolerance for the distance between tunings is set as $\epsilon_t = 2.0$. For the tuning of t, we use a Monte Carlo sample of size 10^3. We set Adam's parameters as $\beta_1 = 0.9$, $\beta_2 = 0.999$, and $\epsilon = 10^{-12}$. We start optimization from the same location as [7], at $\boldsymbol{d}_0 = (1.6608, 10.4370)$.

5.3 Results

Figure 1 presents the contours of f and $\hat{f}$ for the cantilever example, with the minimum value marked with an "×". It can be seen that the objective function using the Chernoff bound in Fig. 1b resembles the original objective function, presented in Fig. 1a. Moreover, the optimum of $\hat{f}$ is close to the one of f, with a small difference due to the Chernoff bound gap.

In Fig. 2, we present the optimization path for Adam over the contour of f. It can be seen that Adam converges to a biased optimum close to the real optimum. For this example, the difference between the objective function in the true optimum and the biased one found is small, as can be observed in Table 2.

Table 2. Results for Example 1.

	b^*	h^*	$P(g_1 \geq 1)$	$P(g_2 \geq 1)$	$f(b^*, h^*)$
True	1.982	13.715	1.99×10^{-4}	1.05×10^{-4}	27.262
Found	1.964	14.145	7.24×10^{-5}	8.44×10^{-5}	27.827

The distance of $\boldsymbol{d}^{(k)}$ to the optimum $\boldsymbol{d}^*$ versus iteration is presented in Fig. 3. It can be observed that the optimization gets close to the real optimum and then proceeds to converge to the biased one, as can be seen in Fig. 2.

For the 10^7 gradients used in 2×10^6 iterations, 1.7×10^6 extra gradients were needed to tune the Chernoff parameter, less than 15% of the total optimization cost. Still, if Monte Carlo sampling was to be used to evaluate the probabilities of failure, we would not be able to use gradient methods, requiring less efficient methods for optimization. A simplex method with a Monte Carlo sample of size 10^4 would require 3×10^4 evaluations of the model per iteration, what means that it would only be able to perform 390 iterations for the same computational cost. Moreover, convergence of the simplex method to local optimum is only guaranteed if SAA is used, what includes a bias in the optimum found.

6 Conclusion

This paper proposed an initial study of the use of the Chernoff bound on the probability of failure to perform risk optimization. Since we are able to derive gradients of the Chernoff bound with respect to the design variables, stochastic gradient methods can be used to perform optimization. How distant is the approximated optimum from the true one depends on how tight is the Chernoff bound; in other words, how well is the tuning parameter set. To tighten the Chernoff bound, we proposed a heuristic procedure to control the value of the tuning parameter during optimization.

Testing the methodology on a numerical example, a cantilever beam, we observed that the stochastic gradient method converged to a solution close to the true one. Moreover, it converged with as little as 5 model evaluations per iteration, which is much less than using Sample Average Approximation together with classical deterministic optimization methods. The drawback of our approach lies in setting the Chernoff parameter; however, the heuristic that we proposed needed less than 15% of the optimization cost. In the example studied, the advantages of using Adam with the gradient of the Chernoff bound far surpassed the extra cost of tuning the parameter.

For future research, we suggest testing different methods for tuning the constant t; a discussion on the calibration of t will be presented in a future work to be published by the authors along with complexity analyses. Other suggestions are the use of different optimization algorithms like the Nesterov accelerated gradient descent and the use of importance sampling to reduce the variance in gradient estimates and thus improve convergence, e.g., with information from the first-order reliability method. Finally, the performance of the proposed method in high-dimensional problems must be studied.

References

1. Lopez, R.H., Beck, A.T.: RBDO methods based on form: a review. J. Braz. Soc. Mech. Sci. Eng. **34**(4), 506–514 (2012). https://doi.org/10.1590/S1678-58782012000400012
2. Beck, A.T., Gomes, W.J.: A comparison of deterministic, reliability-based and risk-based structural optimization under uncertainty. Probab. Eng. Mech. **28**, 18–29 (2012)
3. Gomes, W.J.S., Beck, A.T.: Global structural optimization considering expected consequences of failure and using ann surrogates. Comput. Struct. **126**, 56–68 (2013). https://doi.org/10.1016/j.compstruc.2012.10.013
4. Beck, A.T., Gomes, W.J.S., Lopez, R.H., Miguel, L.F.F.: A comparison between robust and risk-based optimization under uncertainty. Struct. Multidisc. Optim. **52**(3), 479–492 (2015). https://doi.org/10.1007/s00158-015-1253-9
5. Gomes, W.J.S., Beck, A.T.: The design space root finding method for efficient risk optimization by simulation. Probab. Eng. Mech. **44**, 99–110 (2016). https://doi.org/10.1016/j.probengmech.2015.09.019
6. Torii, A., Lopez, R., Miguel, L.: A gradient-based polynomial chaos approach for risk and reliability-based design optimization. J. Braz. Soc. Mech. Sci. Eng. **39**(7), 2905–2915 (2017)

7. Torii, A.J., Lopez, R.H., Beck, A.T., Miguel, L.F.F.: A performance measure approach for risk optimization. Struct. Multidisc. Optim. **60**(3), 927–947 (2019)
8. Robbins, H., Monro, S.: A stochastic approximation method. Ann. Math. Stat. **22**, 400–407 (1951)
9. Carlon, A.G., Dia, B.M., Espath, L.F., Lopez, R.H., Tempone, R.: Nesterov-aided stochastic gradient methods using Laplace approximation for Bayesian design optimization. arXiv preprint arXiv:1807.00653
10. Carlon, A.G., Lopez, R.H., da Espath Rosa, L.F., Miguel, L.F.F., Beck, A.T.: A stochastic gradient approach for the reliability maximization of passively controlled structures. Eng. Struct. **186**, 1–12 (2019)
11. Kingma, D.P., Ba, J., Adam: a method for stochastic optimization. arXiv preprint arXiv:1412.6980
12. Melchers, R.E., Beck, A.T.: Structural Reliability Analysis and Prediction. Wiley, Hoboken (2018)
13. Lehman, E., Leighton, T., Meyer, A.R.: Mathematics for computer science, Technical report, 2006. Lecture notes (2010)
14. Gere, J.M.: Mechanics of Materials, 6th edn. Cengage Learning, Boston (2004)
15. Timoshenko, S.P., Gere, J.M.: Theory of Elastic Stability, Courier Corporation. Dover Publication, New York (2009)

Time-Variant Reliability-Based Optimization with Double-Loop Kriging Surrogates

Hongbo Zhang(✉), Younes Aoues, Didier Lemosse, Hao Bai, and Eduardo Souza De Cursi

INSA Rouen Normandie, LMN, Normandie University, Rouen, France
hongbo.zhang@insa-rouen.fr

Abstract. Time-variant reliability (TvR) analysis allows capturing the time-dependence of the probability of failure and the uncertainty of the deterioration process. The assessment of TR of existing structures subjected to degradation is an important task for taking decisions on inspection, maintenance and repair actions. The Time-variant Reliability-Based Design Optimization (TvRBDO) approach aims at searching the optimal design that minimizes the structural cost and to ensure a target reliability level during the operational life. However, for engineering problems, the TvRBDO problems may become computationally prohibitive when complex simulation models are involved (ie. Finite element method). This work proposes a surrogate-assisted double-loop approach for TvRBDO, where the outer-loop optimizes the objective function and the inner loop calculates the time-dependent reliability constraints. The time-dependent reliability of the inner loop is calculated by Monte Carlo simulations at discreted time intervals. To reduce the number of function evaluations, an inner Kriging is used to predict the response of limit state functions. The so-called single-loop Kriging surrogate method (SILK) developed by Hu and Mahadevan is adopted to calculate the time-variant reliability. The reliability results of the inner-loop are then used to train an outer-loop Kriging, and Expected Feasible Function (EFF) is used to improve its accuracy. After the outer-loop Kriging is trained, TvRBDO is conducted based on it, and the program stops when the difference between two optimum is less then allowance error. Two examples are used to validate this method.

Keywords: Time-variant reliability-based optimization · Kriging surrogate · Monte Carlo simulation · SILK method

1 Introduction

Reliability design optimization (RBDO) has been widely used in engineering applications [2]. However in many engineering cases, due to the degradation, stochastic loads, and system deterioration with time, time-variant Reliability

J. E. S. De Cursi (Ed.): Uncertainties 2020, LNME, pp. 436–446, 2021.
https://doi.org/10.1007/978-3-030-53669-5_32

(TvRBDO) must be considered [11]. Time-variant reliability (TvR) is the probability that a product performs under its intended functionality within its serving time. The general form of TvRBDO can be formulated as:

$$\begin{aligned} &\min C(\boldsymbol{d}, \boldsymbol{\theta}_{\boldsymbol{X}_d}) \\ &st. \begin{cases} \text{Prob}\left\{G_i\left(\mathbf{d}, \mathbf{X}_\mathrm{d}, \mathbf{X}_\mathrm{c}, \mathbf{Y}(t), t\right) < 0\right\} \leq p_i^* \\ H_j\left(\mathbf{d}, \boldsymbol{\theta}_{\mathrm{x_d}}\right) \leq 0 \end{cases} \\ &i = 1, 2, ...N_p;\; j = 1, 2, ...N_d \end{aligned} \tag{1}$$

C is the cost function that can be a combination of deterministic variables $\mathbf{d}$, and probability distribution parameters (statistical moments, etc.) $\boldsymbol{\theta}$ of random variables $\boldsymbol{X}$. G_i is the ith performance function that can be consisted of the design variables $\boldsymbol{X}_d$, the random variables $\boldsymbol{X}_c$, the stochastic process $F(t)$ and time t. The cost function C is minimized under N_p probability constraints of $Prob(G_i) < p_i^t$, with p_i^* being the allowable probability of failure and N_d deterministic constraints H_j.

For the evaluation of TvR constraints, existing methods can be classified into two categories: out-crossing methods and extreme value methods [5,13].

The out-crossing is defined as the events of performance function passing through a failure threshold. The out-crossing methods are based on Rice's formula [10] that the distribution of out-crossing is assumed to follow Poisson distribution. Under this assumption, the probability of failure can be bounded by the sum of the initial failure plus the integral of time as:

$$P_f(0, T) \leq P_f(0, 0) + \int_0^t v(t)dt \tag{2}$$

where $v(t)$ is the out-crossing rate defined as:

$$v(t) = \lim_{\Delta t \to 0} \frac{\Pr(G(t) > 0 \bigcap G(t + \Delta t) \leq 0)}{\Delta t} \tag{3}$$

To calculate the out-crossing rate, the most popular method is the PHI2 method [1]. The PHI2 method calculates the outer-crossing rate using two component parallel system at successive time instants and estimates the bivariate normal integral with FORM method. Later the PHI2 method is improved to integrate the analytical gradient solution of the bivariate normal integral as PHI2+ method [12]. Compared with PHI2 method, the PHI2+ method is less sensitive to the increments of time step. Though the PHI2 and PHI2+ methods are computationally efficient, their accuracy highly rely on the presumption that the out-crossings are independent. When the reliability is low, there may exist dependence of out-crossings, these two methods will include large errors.

The extreme value based method try to evaluate the performance functions at the maximum values of each trajectory of the random variables through time. In most cases, there is no analytical solution for the extreme value of the performance functions, so it will become prohibitive for engineering problems. In this case, surrogate models are used, for example Zhen Hu has proposed a mixed-EGO method uses a Kriging surrogate to approximate the extreme value. The Kriging is first updated with AK-MCS method [4] and then MCS simulations are used to calculate the time-variant reliability [6]. Mixed EGO uses double-loop Kriging with inner-loop represents the performance functions and outer-loop represents the extreme value of the inner-loop. The double-loop Kriging of EGO method will amply the error of the inner-loop and it's computational expensive to identify new training points. To alleviate this problem, a single-loop Kriging surrogate method (SILK) is proposed [8], In SILK, the time interval is discretized into small time steps at each random design variables and uses a learning function U to decide whether to update the inner Kriging along this track of the random variable. This paper is based on the SILK method, double-loop Kriging surrogates are used for analysis of TvRBDO, that the reliability of inner-loop is calculated with SILK. Results of inner-loop are used to train the outer-loop, and Expected Feasible Function (EFF) is used to update the outer-loop Kriging.

So far, not many papers have focused on TvRBDO, due to the high computational cost for the evaluation of TvR constraints and lack of accurate analytical TvR methods. Zequn Wang and Pingfeng Wang have proposed a nested extreme response surface method (NERS) which uses a nested EGO approach to conduct TvRBDO. The TvR of NERS is calculated with EGO method from the inner-loop [14]. Zhen Hu and Xiaoping Du (2016) have expanded the time-independent RBDO method SORA to solve TvR problems [7]. Lara Hawchar & Charbel-Pierre EI Soueidy have based SILK and proposed a global Kriging surrogate method that utilizes the Polak-He algorithm [9] to accelerate convergence. For NRES, EGO method itself is a double-loop method, using TvR for TvRBDO may amplify the final error. Though surrogates are used, the evaluation of TvR constraints is still very time-consuming, and the TvRBDO involves multiple iterations of searching in the design space. This will become prohibitive for high reliability problems.

So this work is based on the SILK method and adds another Kriging as the outer-loop to solve TvRBDO problems. In this method, The inner-loop calculates the TvR using SILK, some improvements are made to SILK to consider multiple-constraints; the outer-loop is another Kriging surrogate trained by the results of inner-loop, after the outer-loop is trained, Expected Feasible Function (EFF) is then used to update this outer-Kriging. The time-variant reliability of all sample points of outer-loop are evaluated with the same inner Kriging with SILK method. And after the outer-Kriging is built, the computational cost for TvRBDO is minimal.

2 Brief Review of Time-Variant Reliability with SILK

In this Section, firstly, the Kriging surrogate and SILK is briefly introduced. Then SILK method is modified to consider multiple constraints.

2.1 Kriging Introduction

Kriging (KRG) is a surrogate model based on regression using observed data sets. The Kriging model can be described as:

$$\widehat{G}(\boldsymbol{X}) = \sum_{i=1}^{k} \beta_i h_i(\boldsymbol{X}) + Z(\boldsymbol{X}) \tag{4}$$

where $\widehat{G}(\boldsymbol{X})$ is a Gaussian distribution with mean $\mu\widehat{G}(\boldsymbol{X})$, and standard deviation $\sigma\hat{G}(\mathbf{X})$. It's a linear combination of base function $h(\mathbf{x})$ and their coefficients βi, plus stochastic process $Z(X)$ of zero mean and standard deviation σ_Z.

2.2 Modified SILK to Consider Multiple Constraints

The probability of failure in Eq. 1 can be calculated with Monte Carlo Simulation (MCS):

$$p_f(t_0, t_e) = \Pr\{G(\mathbf{X}, t) > 0, \exists t \in [t_0, t_e]\} = \sum_{i=1}^{N} I_t\left(\mathbf{x}^{(i)}\right)/N \tag{5}$$

where N is the number of samples, I_t is the indicator for failure for each random $\boldsymbol{x}$ during the time interval of $[t_0,\ t_e]$:

$$I_t\left(\mathbf{x}^{(i)}\right) = \begin{cases} 1, & \text{if } G\left(\mathbf{x}^{(i)}, t\right) > 0,\ \exists t \in [t_0, t_e],\ \forall i = 1, 2, \cdots, N \\ 0, & \text{otherwise} \end{cases} \tag{6}$$

For each random variable, the time is discretized into N_t small time steps, Eq. 6 becomes:

$$I_t\left(\mathbf{x}^{(i)}\right) = \begin{cases} 1, & \text{if any } I\left(G\left(\mathbf{x}^{(i)}, t^{(j)}\right)\right) = 1,\ \forall j = 1, 2, \cdots, N_t \\ 0, & \text{otherwise} \end{cases} \tag{7}$$

The fundamental idea of SILK is build a surrogate $\hat{G}(\mathbf{X}, t)$ to evaluate Eq. 5 to Eq. 7. To make sure the surrogate is accurate for the classification, a U criterion is used:

$$U^{(k)}\left(\mathbf{x}^{(i)}, t^{(j)}\right) = \frac{\left|\mu\hat{G}^{(k)}\left(\mathbf{x}^{(i)}, t^{(j)}\right)\right|}{\sigma\hat{G}^{(k)}\left(\mathbf{x}^{(i)}, t^{(j)}\right)} \tag{8}$$

here k is the number of constraints. The probability of making a mistake for the sign of $G\left(\mathbf{x}^{(i)}, t^{(j)}\right)$ is given by:

$$P_{\text{error}} = \Phi\left(-U\left(\mathbf{x}^{(i)}, t^{(j)}\right)\right) \tag{9}$$

in the paper, the limit for U is set to be bigger than 2.5, which means the probability of making a error is smaller than 0.62%.

Equation 7 can then be transformed rewritten as:

$$I_t\left(\mathbf{x}^{(i)}\right)=\begin{cases}1, & \text{if } \hat{G}\left(\mathbf{x}^{(i)},\, t^{(j)}\right)>0 \text{ and } U\left(\mathbf{x}^{(i)},\, t^{(j)}\right)\geq 2.5,\ \exists j=1,2,\cdots,N_t\\ 0, & \text{if } \hat{G}\left(\mathbf{x}^{(i)},\, t^{(j)}\right)\leq 0 \text{ and } U\left(\mathbf{x}^{(i)},\, t^{(j)}\right)\geq 2.5,\ \forall j=1,2,\cdots,N_t\end{cases} \tag{10}$$

If a failure is identified in the track of $\boldsymbol{x}^{(i)}$, $U^{(k)}(\boldsymbol{x}^{(i)},:)$, $k=1,....,N_p$ will be replaced with a value u_e bigger than 2.5, because no more updating is needed in this track of $\boldsymbol{x}^{(i)}$:

$$U_{\min}\left(\mathbf{x}^{(i)}\right)=\begin{cases}u_e\begin{cases}\text{if } \hat{G}^{(k)}\left(\mathbf{x}^{(i)}t^{(j)}\right)>0\\ \text{and } U^{(k)}\left(\mathbf{x}^{(i)},\, t^{(j)}\right)\geq 2.5,\ \exists j=1,\, 2,\cdots,N_t\end{cases}\\ \min\limits_{j=1,2,\cdots,N_t}\left\{U^{(k)}\left(\mathbf{x}^{(i)},\, t^{(j)}\right)\right\}, \text{ otherwise}\end{cases} \tag{11}$$

When the surrogate fails to find a accurate point that $\hat{G}(\mathbf{X},t)>0$, or not accurate enough to prove $\hat{G}(\mathbf{X},t)<0$ for all time steps. Points are needed to update the Kriging. The updating points are decided by searching all the random $x^{(i)}$ by minimizing $U(x^{(i)},\, t^{(j)})$ for each time step t:

$$[i_{\min}, j_{\min}]=\underset{i=1,2,\cdots,N}{\arg\min}\left\{\underset{j=1,2,\cdots,N_t}{\arg\min}\left\{U^{(k)}\left(\mathbf{x}^{(i)},t^{(j)}\right)\right\}\right\} \tag{12}$$

To avoid the clustering of new points and current training points, a correlation criterion is added before the searching. If the correlation between the points $\left[x^{(i)},\, t^{(j)}\right]$ and existing training points are bigger than a limit value, for example 0.95, $U\left(\mathrm{x}^{(i)},\, t^{(j)}\right)$ will be replaced with a big value, say $U\left(\mathrm{x}^{(i)},\, t^{(j)}\right)=10$. The correlation is calculated with:

$$\rho_c=\max\left\{(R(\theta,\, [x^{(i)},\, t^{(j)}],\, [\mathbf{x}^{train},\, \mathbf{t}^{train}]))\right\} \tag{13}$$

here R is the correlation function of Kriging, θ is the parameters of current Kriging. $[\mathbf{x}^{train},\, \mathbf{t}^{train}]$ is the existing training points.

After identify $i_{\min}$ and $j_{\min}$, a new sample $\left[x^{(i_{\min})},\, t^{j_{\min}}\right]$ will be added into the current training set. The stop criterion for the updating is that the maximum prediction error is lower than a limit value, 5% for example. The maximum error is calculated as:

$$\varepsilon_r^{\max}=\max_{N_{f2}^*\in[0,N_2]}\left\{\frac{\left|N_{f2}-N_{f2}^*\right|}{N_{f1}+N_{f2}^*}\times 100\%\right\} \tag{14}$$

Here N_{f1} is the number of samples for all random x^i, $i=1,\cdots,N$ and time steps $t^{(j)}$, $j=1,\cdots N_t$ that satisfy $U(\mathbf{x},t^{(j)})\geq 2.5$, which are considered to be accurate; $N_{f2}=N*Nt-N_{f1}$ is the number of the other part of samples.

After the Kriging is built, MCS is used to calculate the reliability.

$$I_t\left(\mathbf{x}^{(i)}\right)=\begin{cases}1,\begin{cases}\text{if } \hat{G}^{(k)}\left(\mathbf{x}^{(i)}t^{(j)}\right)>0\\ \text{and } U^{(k)}\left(\mathbf{x}^{(i)},\,t^{(j)}\right)\geq 2.5,\ \exists j=1,2,\cdots,N_t\end{cases}\\ 0,\begin{cases}\text{if } \hat{G}^{(k)}\left(\mathbf{x}^{(i)},\,t^{(j)}\right)\leq 0\\ \text{and } U^{(k)}\left(\mathbf{x}^{(i)},\,t^{(j)}\right)\geq 2.5,\ \forall j=1,2,\cdots,N_t\end{cases}\end{cases} \tag{15}$$

Coefficient of variation is used as the stopping criteria of the SILK method, that $Cov_{pf} < 0.02$:

$$\mathrm{Cov}_{\mathrm{pf}}=\sqrt{\left(1-p_f\left(t_0,\,t_e\right)\right)/\left(p_f\left(t_0,\,t_e\right)\right)/N_{\mathrm{MCS}}} \tag{16}$$

3 TRBDO Using Double-Loop Kriging

This section introduces the process of TRBDO. This method consist of building two separate Kriging surrogates, the inner-loop Kriging is used to calculate the TR with SILK method, the outer-loop Kriging is used to conduct TRBDO.

3.1 Initial Sampling from Augmented Reliability Space for Inner Loop

In this work, the lower and upper bound for augmented PDF is used [3], which is given as:

$$\begin{cases} q_{X_i}^{-}=\min\limits_{\theta\in D\theta}\theta F_{X_i}^{-1}\left(\Phi\left(-\beta_{0,i}\right)|\theta\right)\\ q_{X_i}^{+}=\max\limits_{\theta\in D\theta}F_{X_i}^{-1}\left(\Phi\left(+\beta_{0,i}\right)|\theta\right)\end{cases} \tag{17}$$

where $q_{X_i}^{-}$ and $q_{X_i}^{+}$ are the lower and upper bound respectively, where $\boldsymbol{H}^{-1}\boldsymbol{X}_i$ is the quantile function of the margin of $\boldsymbol{X}_i$, Φ is the CDF of normal distribution, β is the reliability level. After the bound is calculated, the training samples are drawn uniformly using Latin Hypercube Sampling (LHS), to train the initial inner Kriging.

3.2 Updating the Kriging for Outer-Loop Optimization

To build the outer-loop Kriging, first LHS sampling is used to train the outer Kriging, the failure rate is calculated using SILK in the inner loop:

$$p_f=\mathrm{SILK}(\mathrm{d},\,\mathbf{X}_{\mathrm{d}},\,\mathbf{X}_{\mathrm{c}},\,\mathbf{Y}(t)) \tag{18}$$

After the outer-loop Kriging is built, expected feasible function (EFF) is used to update the Kriging. The EFF function is given as:

$$EF[\widehat{G}(\boldsymbol{u})] = \left(\mu\hat{G} - \bar{z}\right)\left[2\Phi\left(\frac{\bar{z} - \mu\hat{G}}{\sigma\hat{G}}\right) - \Phi\left(\frac{z^{-} - \mu\hat{G}}{\sigma\hat{G}}\right) - \Phi\left(\frac{z^{+} - \mu\hat{G}}{\sigma\hat{G}}\right)\right.$$
$$- \sigma\hat{G}\left[2\phi\left(\frac{\bar{z} - \mu\hat{G}}{\sigma\hat{G}}\right) - \phi\left(\frac{z^{-} - \mu\hat{G}}{\sigma\hat{G}}\right) - \phi\left(\frac{z^{+} - \mu\hat{G}}{\sigma\hat{G}}\right)\right]$$
$$+ \varepsilon\left[\Phi\left(\frac{z^{+} - \mu\hat{G}}{\sigma\hat{G}}\right) - \Phi\left(\frac{z^{-} - \mu\hat{G}}{\sigma\hat{G}}\right)\right] \tag{19}$$

z is taken to be p^* in this case. The updating stops when the maximum EFF is smaller than 1e−6 or the maximum number of samples is reached. After the outer surrogate is built, it's then used to realize TvRBDO. At each iteration of TvRBDO of the outer-loop, the optimum point of this iteration is added into the outer-loop Kriging training set to update the outer-loop Kriging. The TvRBDO process stops when the change of optimum is smaller than ϵ.

4 Example and Results

In this section, two examples are used to validate the proposed method.

4.1 A Two-Variable Example

The first example is an numerical example that consists two random variables: $X_1 \sim N(\mu_1, 0.6)$ and $X_2 \sim N(\mu_2, 0.6)$, the design variables are the mean values of the two variables. The time interval for this problem is set to be $[0, 5]$. The objective function and constraints are summarized as follows:

$$\begin{aligned} &\min \mathrm{C}(\mu_1, \mu_2) = \mu_1 + \mu_2 \\ &st. \begin{cases} \mathrm{Prob}\{G_1(X_1, X_2, t) < 0\} \le p_i^*, & i = 1, 2, 3 \\ 0 \le \mu_j \le 10 & , j = 1, 2 \end{cases} \end{aligned} \tag{20}$$

The limit failure rate p^* is set to be 0.1 for all probability constraints. The performance functions are given as:

$$\begin{cases} G_1(X_1, X_2, t) = X_1^2 X_2 - 5X_1 t + (X_2 + 1)t^2 - 20 \\ G_2(X_1, X_2, t) = (X_1 + X_2 - 0.1t - 5)^2/30 + (X_1 - X_2 + 0.2t - 12)^2/120 - 1 \\ G_3(X_1, X_2, t) = 90/\left((X_1 + 0.05t)^2 + 8(X_2 + 0.1t) - \sin(t) + 5\right) - 1 \end{cases} \tag{21}$$

The three performance functions with different time instants are shown in Fig. 1.

The program starts by sampling 64 LHS samples $[\boldsymbol{x}_1^{\mathrm{outer}}, \boldsymbol{x}_2^{\mathrm{outer}}]$ for outer-loop and 36 LHS samples $[\boldsymbol{x}_1^{\mathrm{Inner}}, \boldsymbol{x}_2^{\mathrm{Inner}}, \boldsymbol{t}]$ for inner-loop from the augmented PDF. The failure rates $\boldsymbol{f}$ of the samples of outer-Kriging are evaluated with SILK

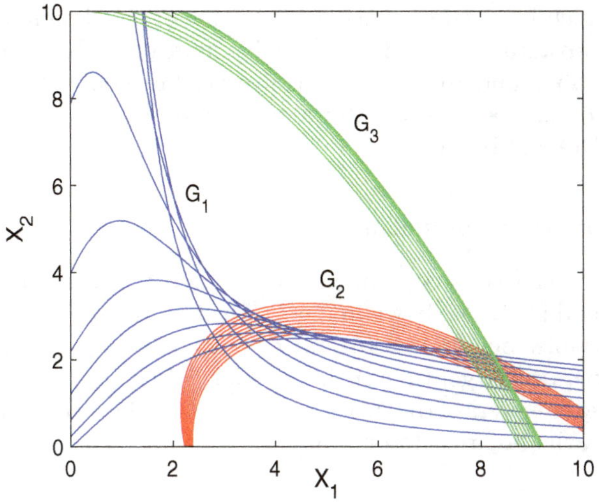

Fig. 1. Performance functions with different time instants

method from the inner-loop Kriging, then the failure rates and the outer-loop samples are used to train the outer-Kriging. Before the TvRBDO optimization, the outer-loop is first updated with points that maximize EFF function, the maximum number for updating outer-loop is set to be 50. After the EFF updating of the outer-loop, this outer-Kriging is used to conduct TvRBDO.

The optimum results from each iteration of TvRBDO of the outer-loop is also added into the outer-loop Kriging training sets. The TvRBDO optimization stops when the change between iterations smaller than 1e−6. The final results are shown in Table 1.

Table 1. Comparison of Results with NERS and TRDLK

	NERS	TRDLK
Initial design	[2.5000, 4.3555]	[0, 0]
Optimal design	[3.6390, 4.0352]	[3.6243 4.1553]
Optimal cost	7.6642	7.7797
$P_f^1(0, 5)$	0.1216	0.0983
$P_f^2(0, 5)$	0.0836	0.0580
$P_f^3(0, 5)$	0	0
N_{func}	336	137
N_{iter}	4	9

The Time-variant Reliability-based optimization using Double-Loop Kriging (TRDLK) proposed in this work converges with less function evaluations (137) than NERS (which is based on the extreme value of the double-loop EGO TvR method). This advantage comes from the advantage of SILK. TRDLK uses more function evaluations and iterations to converge, but compared with NERS, it should be noted that, after the outer-loop Kriging is built, the computational cost for TvRBDO will be minimal.

4.2 Two-Bar Frame Structure

Another example is a two-bar frame structure subject to stochastic force $F(t)$ taken from Hu and Du [7] as shown in Fig. 2. The two distances of O_1O_3 and O_1O_2 are random variables denoted as L_1 and L_2. Failure occurs when the maximum stress in either bars are larger than their material yield stress S_1 and S_2. The design variables are the diameters of the two bars D_1 and D_2. The distribution parameters of the random variables are listed in Table 2.

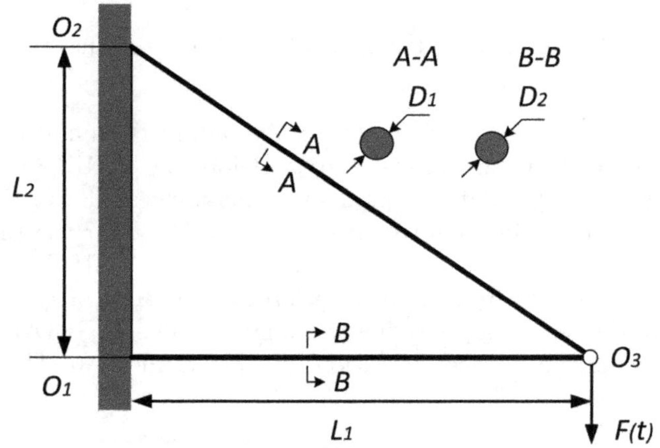

Fig. 2. A two-bar frame structure

Table 2. Random parameters of two-bar frame structure

Parameters	Distribution	Mean	Standard deviation
D_1	Normal	μ_{D_1}	1e−3
D_2	Normal	μ_{D_2}	1e−3
L_1	Normal	0.4 m	1e−3
L_2	Normal	0.3	1e−3
S_1	Log-normal	1.7e8Pa	1.7e7Pa
S_2	Log-normal	1.7e8Pa	1.7e7Pa

$F(t)$ is considered as a stationary Gaussian process with exponential square auto-correlation function given by:

$$\rho_F(t_i, t_j) = \exp\left(-\left(\frac{t_j - t_i}{0.1}\right)^2\right) \tag{22}$$

The mean value and standard deviation of the Gaussian process are stationary as $\mu_F = 2.2e6$ and $\sigma_F = 2.0e5$. The problem can be summarized as:

$$\begin{aligned} &\min C(\mu_{D_1}, \mu_{D_2}) = \pi \mu_{L_1} \mu_{D_1}^2/4 + \pi \sqrt{\mu_{L_1}^2 + \mu_{L_2}^2}\, \mu_{D_3}^2/4 \\ &st. \begin{cases} \text{Prob}\{G_i(\mathbf{X}_1, F(t), t) < 0\} \le p_i^*, & i = 1,2 \\ 0.07 \le \mu_{D_j} \le 0.25, & j = 1,2 \end{cases} \end{aligned} \tag{23}$$

The limit failure rate p_i^* for two performance functions are set to be 0.01 and 0.001 that are given as:

$$\begin{aligned} G_1(\mathbf{X}, t) &= \pi D_1^2 S_1/4 - \sqrt{L_1^2 + L_2^2} F(t)/L_2 \\ G_2(\mathbf{X}, t) &= \pi D_2^2 S_2/4 - L_1 F(t)/L_2 \end{aligned} \tag{24}$$

Similar to example 1, the TvRBDO is started by 64 LHS samples $[\boldsymbol{x}_d^{\text{outer}}, \boldsymbol{x}_c^{\text{outer}}]$ for outer-loop Kriging and 64 LHS samples $[\boldsymbol{x}_d^{\text{inner}}, \boldsymbol{x}_c^{\text{inner}}, \boldsymbol{F}_t^{\text{inner}}]$ for inner-loop. The stochastic process F_t is discretized over 10 intervals. The final results are shown in Table 3. As shown from the results, this method uses fewer function evaluations than NERS and the accuracy is similar.

Table 3. Comparison of Results with NERS and TROSK

	NERS	TRDLK	Analytical
Optimal design	[0.2102 0.1964]	[0.2017 0.1886]	[0.2027 0.1894]
Optimal cost	0.0290	0.027	0.0267
$P_f^1(0, 5)$	0.0094	0.0102	0.01
$P_f^2(0, 5)$	0.0009	0.0009	0.001
N_{func}	715	471	\

5 Conclusion

This paper proposed a double-loop Kriging surrogate method to conduct time-variant reliability optimization. This method consists of two separate Kriging surrogates, the inner-loop Kriging evaluates the time-variant reliability using SILK method, and the outer-loop Kriging is trained with the results of the inner-loop. After the out-loop Kriging is built, it is then updated with EFF function at allowable failure rate. Two examples are used to validate the effectiveness of the method. Compared to other existing methods, this method is simple, accurate and easy to implement. The computational cost mainly comes from the searching of points to update the inner-Kriging. After the outer-Kriging is built, the computational cost for optimization will be minimal. However, due to Kriging is less competent for high-linear problems and high-dimensional problems, this method may me less effective for such problems. Also the inner-loop Kriging error will be accumulated to outer-loop Kriging, it's recommended to use smaller a Coefficient of variation (Cov) value, this will inevitably increase the total computational cost.

References

1. Andrieu-Renaud, C., Sudret, B., Lemaire, M.: The PHI2 method: a way to compute time-variant reliability. Reliab. Eng. Syst. Saf. **84**(1), 75–86 (2004)
2. Aoues, Y., Chateauneuf, A.: Benchmark study of numerical methods for reliability-based design optimization. Struct. Multidiscip. Optim. **41**(2), 277–294 (2010)
3. Dubourg, V.: Adaptive surrogate models for reliability analysis and reliability-based design optimization. Ph.D. thesis (2011)
4. Echard, B., Gayton, N., Lemaire, M.: AK-MCS: an active learning reliability method combining kriging and Monte Carlo simulation. Struct. Saf. **33**(2), 145–154 (2011)
5. Gong, C., Frangopol, D.M.: An efficient time-dependent reliability method. Struct. Saf. **81**, 101864 (2019)
6. Hu, Z., Du, X.: Mixed efficient global optimization for time-dependent reliability analysis. J. Mech. Des. **137**(5) (2015)
7. Hu, Z., Du, X.: Reliability-based design optimization under stationary stochastic process loads. Eng. Optim. **48**(8), 1296–1312 (2016)
8. Hu, Z., Mahadevan, S.: A single-loop kriging surrogate modeling for time-dependent reliability analysis. J. Mech. Des. **138**(6) (2016)
9. Polak, E.: Optimization: Algorithms and Consistent Approximations, vol. 124. Springer, Cham (2012)
10. Rice, S.O.: Mathematical analysis of random noise. Bell Syst. Tech. J. **23**(3), 282–332 (1944)
11. Singh, A., Mourelatos, Z.P., Li, J.: Design for lifecycle cost using time-dependent reliability. J. Mech. Des. **132**(9), 091008 (2010)
12. Sudret, B.: Analytical derivation of the outcrossing rate in time-variant reliability problems. Struct. Infrastruct. Eng. **4**(5), 353–362 (2008)
13. Wang, Y., Zeng, S., Guo, J.: Time-dependent reliability-based design optimization utilizing nonintrusive polynomial chaos. J. Appl. Math. **2013** (2013)
14. Wang, Z., Wang, P.: A nested extreme response surface approach for time-dependent reliability-based design optimization. J. Mech. Des. **134**(12), 121007 (2012)

A Variational Approach for the Determination of Continuous Pareto Frontier for Multi-objective Problems

Hafid Zidani[1,2(✉)], Rachid Ellaia[1], and Edouardo Souza De Cursi[2]

[1] Laboratory of Study and Research in Applied Mathematics (LERMA), Mohammadia School of Engineers, Mohammed V University of Rabat, Ibn Sina Avenue, Agdal, BP 765, Rabat, Morocco
hafidzidani@yahoo.fr, ellaia@emi.ac.ma
[2] LOFIMS Laboratory, INSA-Rouen, Université Avenue, BP. 08, 76801 St Etienne du Rouvray Cedex, France
eduardo.souza@insa-rouen.fr

Abstract. In this paper, a novel approach is proposed to generate set of Pareto points to represent the optimal solutions along the Pareto frontier. This approach, which introduces a new definition of dominance, can be interpreted as a representation of the solution of the multi-objective optimization under the form of the solution of a problem in variational calculus. The method deals with both convex and non-convex problems. In order to validate the method, multi-objective numerical optimization problems are considered.

Keywords: Multiple objective programming · Multi-objective optimization · Variational calculation

1 Introduction

Multi-objective optimization problems are common in engineering design. In this field, we may be brought to some of the hardest optimization problems, which arise in many real-world applications. There exists a large fan of methods intending to solve this class of problems, most of which use an iterative process to generate a set of points approximating the Pareto's front either in a single simulation run (such as, for instance, Evolutionary Multi-objective Optimization methods [7,18,19]) or by transforming the multi-objective problems into a series of single optimization problems (such as, for instance, weighting sums [5], $\varepsilon-$constraint [10], Normal Boundary Intersection [6], Normal constraint [11],...). In all these approaches, we are interested in the approximation of an approximated Pareto's front by generating a finite set of Pareto nondominated points, giving an accurate representation of the Pareto's front, what generally supposes a large number of points. The quality of the results is evaluated by their ability to generate an evenly distributed curve [6]. Thus, a suitable strategy must be adopted to

J. E. S. De Cursi (Ed.): Uncertainties 2020, LNME, pp. 447–469, 2021.
https://doi.org/10.1007/978-3-030-53669-5_33

avoid prohibitive computational costs. In this paper, a novel approach, called Variational Approach for Multi-objective Optimization (VAMO), is presented to generate Pareto points to represent the Pareto's front with a good quality. The VAMO method may be interpreted as a representation of the solution of the multi-objective optimization under the form of the solution of a problem in variational calculus. The multi-objective optimization problem is transformed into a different one: the minimization of an integral involving the objective functions. The method deals with both convex and non-convex problems. To illustrate its efficiency and accuracy, we consider typical multi-objective test functions from the global optimization literature. VAMO is compared to some popular multi-objective optimization methods used in engineering optimization. Namely, we consider the following algorithms to carry out this comparison: Non-dominated Sorting Genetic Algorithm (NSGA II), Normalized Normal Constraint (NNC) and Normal Boundary Intersection (NBI).

2 Multi-objective Optimization

Multi-objective optimization(MOO) involves the simultaneous optimization of non comparable, eventually incommensurable, and often competing, multiple objectives. In the absence of any information about the decision maker's preference, the classical approach consists in look for the set of the non-dominated points, id est, the set of the points such that one of the objectifs cannot be improved without degradation of another one: such a set of points form the Pareto's optimal set or Pareto's efficient set - the concept is classical and will not be developed here (see, for instance,[3]) In general, a multi-objective problem consists of a vector-valued objective function to be minimized, and of some equality or inequality constraints, we consider in the sequel a MOO defined as follows: let

- $x = (x_1, ..., x_n) \in \mathbf{R}^n$ be a decision vector;
- $f = (f_1, \cdots, f_\ell)$ be a vector of objectives, with $f_i : \mathbf{R}^n \to \mathbf{R}, for 1 \leq i \leq \ell$;
- $g = (g_1, \cdots, g_m)$ be a vector of inequality constraints, with $g_i : \mathbf{R}^n \to \mathbf{R}, for 1 \leq i \leq m$;
- $h = (g_1, \cdots, g_p)$ be a vector of equality constraints, with $h_i : \mathbf{R}^n \to \mathbf{R}, for 1 \leq i \leq p$;
- a non-empty feasible set $C = \{x \in \mathbf{R}^n : g(x) \leq 0 \, and \, h(x) = 0\} \subset \mathbb{R}^n$ (as usual in the framework of MOO, vectorial inequalities are interpreted as inequalities for all the components).

Then, we look for

$$x = argmin\{f(y) : y \in C\} \tag{1}$$

Since there are multiple objectives, which may be conflicting, the problem formulated in Eq. (1) admits only solutions in the sense of Pareto optimality - concept firstly proposed by Edgeworth and Pareto [14], which is defined in the sequel: let us recall that the objective vector f takes his values on R^ℓ: we call the objective space the set of all the possible values of the objective vector:

$$OS(f) = \{y \in \mathbb{R}^\ell : y = f(x)\} \tag{2}$$

Pareto's optimality is based on the idea of dominance, which defines a partial ordering on OS(f) [7,12]:

Definition 1. A vector $u \in \mathbb{R}^\ell$ is said to dominate another vector $v \in \mathbb{R}^\ell$, denoted as $u \prec v$, if and only if: $u \leq v$ and there is $i \in \{1, \cdots, l\}$, such that $u_i \leq v_i$.

Definition 2. Let $x \in \mathbb{R}^n$. We sayy that x is a Pareto optimum associated to (1) if and only if, $x \in C$, and, there is not any feasible point $z \in C$ such that $f(z) \prec f(x)$. The set of all the Pareto optima is called the *Pareto's optimal set.* It is denoted $POS(f)$ in the sequel.

As suggested by this definition, Pareto's optima lead to the determination of the points $f(x)$ which are not dominated by any $f(z)$:

Definition 3. The Pareto's front is the set of the points:

$$PF(f) = \{y \in \mathbb{R}^n : y = f(x) : x \in OS(f)\} \subset OS(f) \subset \mathbb{R}^l. \tag{3}$$

We are interested in the situation where $PF(f)$ is a manifold having dimension $n - 1$. For instance, when considering bi objective optimization, it is a curve; when considering three-objective optimization, it is a surface. In general, the solution of an MOO problem involves the determination of $PF(f)$. We may find in the literature many methods intended to solve determine Pareto's front.

3 Variational Approach for Multi-objective Optimization (VAMO)

3.1 Introduction

In the field of single-objective optimization, the literature furnishes some results concerning the representation of the solutions (see, for instance, [1,4,9,13,16] and [2]). These results usually characterize the solutions of a single objective optimization problem by a formula to be evaluated or an equation to be solved. At the best of our knowledge, analogous results do not exist in multi-objective optimization, although the single-objective results may be applied in combination with aggregation methods. The Variational Approach for Multi-Objective Optimization (VAMO) method characterizes the Pareto's front as the solution of a variational problem and, by this way, may be interpreted as a representation formula - since it furnishes a way to obtain the Pareto's front by solving a problem of variational calculus. We present the approach in the framework of bi-objective optimization. Mathematically, the variational representation extends to any finite number of objectives by using the concept of dominance of a manifold, derived from the concept of dominance of the curves exposed in the sequel. In practice, the determination of a Pareto's front for a large number of objectives lead to the solution of a variational problem in higher dimension, so that numerical difficulties and the curse of dimensionality may arise. It may be interesting to notice that NBI contains some elements of the approach, by looking for points which tends to the be closer to the utopia one.

3.2 Dominance of Curves

As observed, we manipulate dominance of curves (or, in general, manifolds) instead of the dominance of points, so that we introduce the following definitions:

Definition 4. A curve Γ_1 is said to dominate another curve Γ_2, denoted as $\Gamma_1 \prec\prec \Gamma_2$, if and only if, $\forall U_i \in \Gamma_1,\ \forall V_i \in \Gamma_2,\ U_i \prec V_i$.

Definition 4. A curve Γ is a *Pareto optimal solution* of the problem (1), if and only if, $\forall \Sigma$, a curve in the feasible space, $\Gamma \prec\prec \Sigma$.

3.3 Basic Idea

The VAMO approach is based on the idea that a Pareto optimal point is characterized by the fact that displacements tending to transform it into a point closer to one of the axis in the Pareto space lead to a non-optimal point., so that the hypervolume limited by the manifold formed by the Pareto's front is minimal among the possible ones. For a bi-objective optimization problem, the Pareto's front becomes a curve and the area limited by the Pareto curve and the two axes, $f_1 = f(x_2^*)$ and $f_1 = f(x_1^*)$,), appears as minimal among all the possible curves formed by points of the Pareto space, joining the ideal points A and B (see Fig. 1)

3.4 Formulation

Let us consider the following bi-objective optimization problem:

$$\begin{cases} \min\limits_{x} \ f(x) = (f_1(x), f_2(x))^T \\ \text{subject to } x \in C \end{cases} \tag{4}$$

where f is the decision vector to be optimized, $x \in \mathbb{R}^n$ is the variables of the decision and C is the feasible set of decision vectors, associated to equality and inequality constraints and explicit bounds.

$$C = \{\ x \in \mathbb{R}^n \ / \ h(x) = 0, \ \ g(x) \leq 0 \ \text{ and } \ x^L \leq x \leq x^U\ \}$$

$S = f(C) = \{(f_1(x), f_2(x)), x \in C\}$ is the set of image of C (objectifs space) (see Fig. 1).

We consider a Hilbert space $H^1(a,b)$ ($(a,b) \in \mathbb{R}^2$), and a functional of the form:

$$J(v) = \int_a^b v(t)dt, \qquad v \in H^1(a,b).$$

Let be $E = \{(t, v(t)), t \in [a,b]\}$, where $a = f_1(x_1^*)$, $\quad b = f_1(x_2^*)$, (x_1^* and x_2^* are the optimal solutions for f_1 and f_2 respectively.)

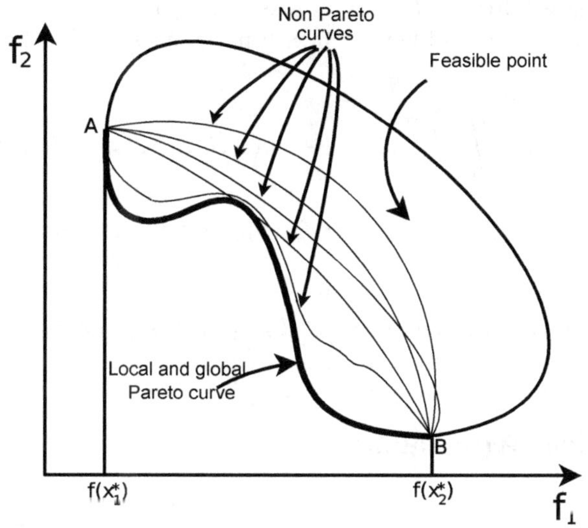

Fig. 1. VAMO method: Basic idea

The solution to this problem using the VAMO method can be formulated as a variational calculus problem:

$$\begin{cases} \min\limits_{v} \;\; J(v) = \min \;\; \displaystyle\int_a^b v(t)dt \\ \text{subject to} \\ v(a) = f_2(x_1^*), \;\; v(b) = f_2(x_2^*), \text{and } E \subset S. \end{cases} \tag{5}$$

Similarly, for a tri-objective optimization problem, the formulation of the problem can be written:

$$\begin{cases} \min\limits_{x} \;\; f(x) = (f_1(x), f_2(x), f_3(x))^T \\ \text{subject to } x \in C \end{cases} \tag{6}$$

The solution is the Pareto surface that passes through the three points $f(x_1^*)$, $f(x_2^*)$ and $f(x_3^*)$ and minimizes the volume between this surface and its projection on the plane (f_1, f_2). The new problem of the variational calculus to be solved is:

$$\begin{cases} \min\limits_{v} \quad J(v) = \min\limits_{v} \quad \int_{a_1}^{b_1}\int_{a_2}^{b_2} v(s,t)dsdt \\ \textit{subject to} \\ v(a_1,c_2) = f_3(x_1^*), v(c_1,a_2) = f_3(x_2^*), \text{ and } v(a_1,c_2) = f_3(x_3^*), \\ E \subset S \end{cases} \quad (7)$$

where $E = \{(t, s, v(t)), t \in [a_1, b_1]$ and $s \in [a_2, b_2]\}$, $S = f(C)$, $a_1 = f_1(x_1^*)$, $a_2 = f_2(x_2^*)$, $b_1 = f_1(x_3^*)$, $b_2 = f_2(x_3^*)$, $c_1 = f_1(x_2^*)$, $c_2 = f_2(x_1^*)$ (x_1^*, x_2^* and x_3^* are the optimal variables for f_1, f_2 and f_3 respectively).

For a multi-objective problem with m functions ($m \geq 2$), The VAMO formulation is :

$$\begin{cases} \min\limits_{v} \quad J(v) = \min\limits_{v} \quad \int_{a_1}^{b_1}\int_{a_2}^{b_2} \cdots \int_{a_{m-1}}^{b_{m-1}} v(t_1, t_2, \ldots, t_{m-1})dt_1, dt_2, \ldots, dt_{m-1} \\ \text{subject to} \\ v(f_1(x_i^*), f_2(x_i^*), \ldots, f_{m-1}(x_i^*)) = f_m(x_i^*), \text{ for } i = 1 \textit{ to } m, \text{ and } E \subset S \end{cases} \quad (8)$$

where $S = f(C)$, and
$E = \{(t_1, t_2, t_{m-1}, \ldots, t_{m-1}, v(t_1, t_2, \ldots, t_{m-1})), \ t_i \in [f_i(x_i^*), f_i(x_m^*)\}$,
x_i^* is the optimal solution for f_i, for $i = 1, \cdots, m$.

4 Theoretical Approach

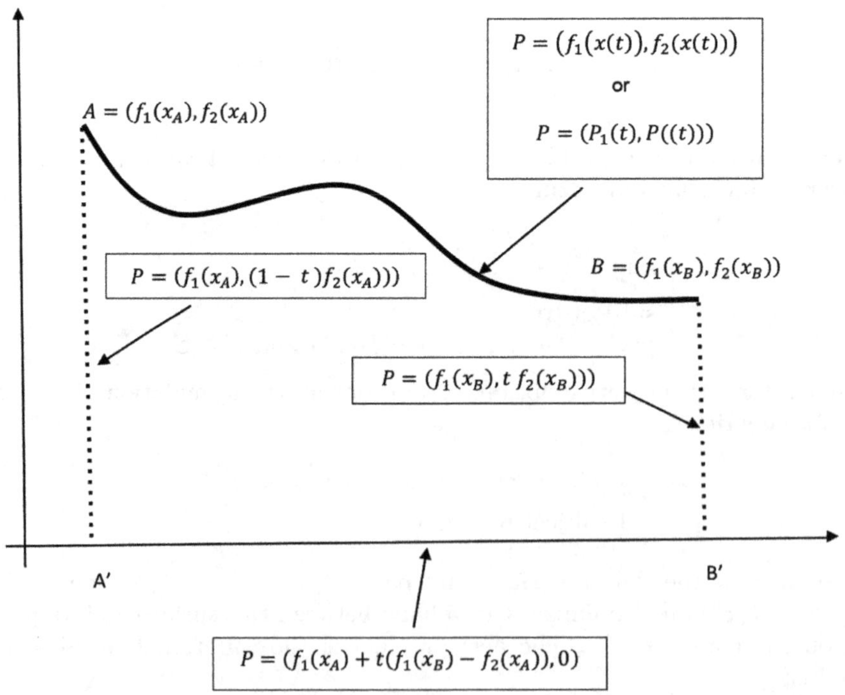

Fig. 2. VAMO: Illustration figure

We seek to minimize the area limited by the curve and the horizontal axis, provided that the variables belong to the feasible region. Let us consider a regular function (continuous at least) $\varphi : \mathbb{R}^2 \to \mathbb{R}$ and two disjoint areas

$$S = \{P \in \mathbb{R}^2 : \varphi(P) \geq 0\}; \text{ and } R = \{P \in \mathbb{R}^2 : \varphi(P) < 0\}.$$

The common border areas is: $\Sigma = \{P \in \mathbb{R}^2 : \varphi(P) = 0\}$.

Let us consider two distinct points $A \neq B$ such that $\varphi(A) = \varphi(B) = 0$ (therefore $\{A, B\} \subset \Sigma$). Let us consider the set of curves:
$\Im = \{P : (0,1) \to \mathbb{R}^2 : P$ is continuous, $P(0) = A, P(1) = B$ and $P(t) \in S$ for $t \in (0,1)\}$

Bearing in mind that, for $(X, Y) \in \mathbb{R}^2$, we have

$$X \leq Y \Leftrightarrow X_1 \leq Y_1 \text{ and } X_2 \leq Y_2,$$

$$X < Y \Leftrightarrow X \leq Y \text{ and } (X_1 < Y_1 \text{ or } X_2 < Y_2).$$

Let us consider a functional $J : \Im \to \mathbb{R}$ having the following property
$P, Q \in \Im; P \leq Q$ in $(0,1)$ and $P < Q$ in $I \subset (0,1)$ where $mes(I) > 0 \Rightarrow J(P) < j(Q)$. Therefore, we have

Theorem 1. Let $P^* = \text{argmin}\,\{J(P) : P \in \Im\}$, then $P^*(t) \in \Sigma$ for $t \in (0,1)$.

Proof. Since $P^* \in \Im$, we have $\varphi(P^*(t)) \geq 0$, in $(0,1)$. Assume, there exists a point $\bar{t} \in (0,1)$ such that $\varphi(P^*(\bar{t}) > 0$. We have $0 < t < 1$, because $P^*(0) = A \in \Sigma$ and $P^*(1) = B \in \Sigma$. By continuity, it exists $\delta > 0$ such that $\varphi(P^*(t)) > 0$ for $t \in (\bar{t} - \delta, \bar{t} + \delta)$. Let us consider the function

$$\theta(\alpha, t) = \varphi(P^*(t) - \alpha \begin{pmatrix} 0 \\ 1 \end{pmatrix} = \varphi(P_1^*(t), P_2^*(t) - \alpha).$$

We have

$$\theta(0, t) = \varphi(P^*(t)) > 0.$$

Let $\alpha(t) = min\{1, max\{\alpha > 0 : \theta(\alpha, t) > 0\}\}$. We have $\alpha(t) > 0$, for all $t \in (\bar{t} - \delta, \bar{t} + \delta)$; indeed, if $\alpha(t) = 0$, then there exists a sequence $\{\alpha_n\}_{n\in\mathbb{N}}$ such that $\alpha_n > 0, \forall n$ and $\alpha_n \to +\infty$, where $\theta(\alpha_n, t) \leq 0$, i.e.,

$$\varphi\left(P^*(t) - \alpha \begin{pmatrix} 0 \\ 1 \end{pmatrix}\right) \leq 0.$$

By using the limit for $n \to +\infty$, the continuity shows that

$$\varphi(P^*(t)) \leq 0,$$

so, we have $0 < \varphi(P^*(t)) = 0$, which is a contradiction. Let us consider a regular function $\psi : (t - \delta, t + \delta) \to (0,1)$ such that $\psi \geq 0, \psi(t - \delta) = \psi(t + \delta) = 0$, and

$\psi(s) = 1$, for all $s \in (a, b), \bar{t} - \delta < a < b < \bar{t} + \delta$. We extend ψ by zero outside $(\bar{t} - \delta, \bar{t} + \delta)$, and we set

$$P(t) = P^*(t) - \frac{1}{2}\psi(t)\alpha(t)\begin{pmatrix} 0 \\ 1 \end{pmatrix}.$$

On the other hand

$$P(t) \in S, \forall t \in (0, 1) \text{ (because } \psi(t)\alpha(t) \leq \frac{1}{2}\alpha(t) < \alpha(t) \; in \; (\bar{t} - \delta, \bar{t} + \delta)),$$

so, $P \in \Im$. On the other hand

$$P(t) \leq P^*(t), \forall t \in (0, 1); P(t) < P^*(t), \forall t \in (a, b).$$

Consequently,

$$J(P) < J(P^*).$$

But, $P^* = \text{argmin}\,\{J(P) : P \in \Im\}$, so that $J(P^*) \leq J(P) < J(P^*)$, which is absurd.

Corollary. Let $S \subset \mathbb{R}^2$ be a non null set such that $\exists m \in \mathbb{R} : X_1 \geq m$ and $X_2 \geq m, \forall X = (X_1, X_2) \in S$. We suppose that S is regular (i.e. a continuous application defined in S has a continuous extension to $\mathbb{R}^2$). Let

$$\Sigma = \{X \in S : \nexists Y \in S \text{ such that } Y < X\}.$$

We suppose that Σ is a curve $P : (0, 1) \to \mathbb{R}^2$ that verifies $P(0) = A, P(1) = B$. Then

$$P = \text{argmin}\,\{J(Q) : Q \in \Im\}.$$

Proof. Let $\phi(P) = \max\{P_1 - Y_1, P_2 - Y_2 : Y \in S\}$.

We have $\phi(P) \leq \max\{P_1 - m, P_2 - m\}$. Moreover, $P \in S, P \notin \Sigma \Rightarrow \exists Y \in S$, such that $Y < P \Rightarrow \varphi(P) > 0$,
$P \in \Sigma \Rightarrow \nexists Y \in S$, such that $Y < P \Rightarrow \varphi(P) = 0$.
φ is continuous; indeed, if $\| P - Q \| \leq \delta$, then $-\delta \leq P_1 - Q_1 \leq \delta$ and $-\delta \leq P_2 - Q_2 \leq \delta$, such that
$Q_1 - Y_1 - \delta \leq P_1 - Y_1 \leq Q_1 - Y_1 + \delta; Q_2 - Y_2 - \delta \leq P_2 - Y_2 \leq Q_2 - Y_2 + \delta$.

Thus, $\varphi(Q) - \delta \leq \varphi(P) \leq \varphi(Q) + \delta$. So, we have $\| P - Q \| \leq \delta$ and φ is continuous. We extend φ using a positive and regular function outside S : the result follows from the theorem.

These results can be applied to the functional J defining the area between the curve and the horizontal axis. For example, we can use the Stokes theorem : circulating a vector field $\overrightarrow{v}$ along a closed curve C is equal to the rotational flow $\overrightarrow{rot}(\overrightarrow{v})$ across the surface S limited by a curve, i.e.,

$$\iint_S \overrightarrow{rot}(\overrightarrow{v}).\overrightarrow{n}\,dS = \oint_C \overrightarrow{v}.d\overrightarrow{P}.$$

We should consider a vectorial field $\overrightarrow{v}$ such that $\overrightarrow{rot}(\overrightarrow{v}) = \overrightarrow{n} = \overrightarrow{e}_3$ for a curve in the plane$(0, \overrightarrow{e}_1, \overrightarrow{e}_2)$. In this case,

$$\iint_S 1dS = \oint_C \overrightarrow{v}.d\overrightarrow{P}.$$

For example

$$\forall\gamma \in \mathbb{R} : \overrightarrow{v} = (-\gamma x_2, (1-\gamma)x_1, 0) \Rightarrow \overrightarrow{rot}(\overrightarrow{v}) = (0, 0, 1),$$

Thus, possible choices for the fields $\overrightarrow{v}$ are :

$$\overrightarrow{v} = \frac{1}{2}(-x_2, x_1, 0); \overrightarrow{v} = (1, x_1, 0),$$

so, the theorem and corollary show that the solution is a curve $P^*(t) : t \in (0,1)$ such that

$$P^* = \text{argmin}\,\{\oint_C v.dP : P \in K\}$$

$$K = \{P : (0,1) \to \mathbb{R}^2 : P(0) = B, P(1) = A, P(t) \in S = Im(f) = f(\mathbb{R}^n)\}$$

$$x_A = \text{argmin}\,(f_1); x_B = \text{argmin}\,(f_2).$$

The second choice $\overrightarrow{v} = (x_2, 1, 0)$ corresponds to a curve issued from a function $P_2 = \eta(P_1)$. In this case, $P_1(t) = t$ and $P_2 = \eta(t) = \eta(P_1)$. The third choice corresponds to a curve issued from a function $P_1 = \eta(P_2)$. In this case, $P_2(t) = \eta(t) = \eta(P_2)$.

Example 1. Let $\overrightarrow{v} = \frac{1}{2}(x_2, 1, 0)$. On $(A, B) : v.dP = (P_1P_2^{'} - P_2P_1^{'})/2$ such that

$$\oint_C v.dP = cst + \frac{1}{2}\int_0^1 (P_1P_2^{'} - P_2P_1^{'})dt$$

Then, we can easily search $P^*(t) : t \in (0,1)$ such that

$$P^* = \text{argmin}\left\{\int_0^1 (P_2P_1^{'} - P_1P_2^{'})dt : P \in K\right\}$$

Another alternative is to put

$$P = (f_1(x(t)), f_2(x(t))).$$

In this case,

$$P_1^{'} = \nabla f_1(x)^t x^{'}, P_2^{'} = \nabla f_2(x)^t x^{'},$$

and

$$P_1P_2^{'} - P_2P_1^{'} = f_1(x)\nabla f_2(x)^t x^{'} - f_2(x)\nabla f_1(x)^t x^{'}$$

In this form, we look for a curve $x^*(t) : t \in (0,1)$, such that :

$$x^* = \text{argmin}\,\{\int_0^1 [x^{'}]^t T(x) : x \in C\}$$

$$T(x) = f_1(x)\nabla f_2(x) - f_2(x)\nabla f_1(x)$$
$$C = \{x : [0,1] \to \mathbb{R}^n : x(0) = x_B, x(1) = x_A\}$$

The Euler-Lagrange equation associated to this functional is

$$\int_0^1 ([\delta x^{'}]^t T(x) + [x^{'}]^t[[\delta T(x)]^t \delta x]) = 0, \forall x \quad \text{such that} \quad \delta x(0) = \delta x(1) = 0,$$

but

$$[\delta T(x)]^t \delta x = [(\delta x)^t \nabla f_1(x)]\nabla f_2(x) + f_1(x)[\nabla^2 f_2(x)\delta x] - [(\delta x)^t \nabla f_2(x)]\nabla f_1(x) + f_2(x)[\nabla^2 f_1(x)\delta x],$$

we notice that $\delta T(x)$ is a matrix. Rewriting

$$[x^{'}]^t[[\delta T(x)]^t \delta x] = [\delta x]^t \delta T(x) x^{'},$$

we have

$$(T(x))^{'} = \delta T(x) x^{'}; x(0) = x_B, x(1) = x_A,$$

i.e.,

$$\nabla T(x) x^{'} = \delta T(x) x^{'}; x(0) = x_B, x(1) = x_A.$$

Example 2. We take $\overrightarrow{v} = (-x_2, 0, 0)$. On $(A, B) : v.dP = -P_2 P_1^{'}$, such that

$$\oint_C v.dP = \text{constant} - \int_0^1 P_2 P_1^{'} dt,$$

thus, we can find $P^*, t \in (0,1)$ such that

$$P^* = \text{argmin} \left\{ \int_0^1 P_2 P_1^{'} dt : P \in K \right\}.$$

Like the previous example, We take $P = (f_1(x(t)), f_2(x(t)))$. In this case,

$$P_1^{'} = \nabla f_1(x)^t x^{'}, P_2^{'} = \nabla f_2(x)^t x^{'}$$

and

$$P_2 P_1^{'} = f_2(x) \nabla f_1(x)^t x^{'}.$$

In this case, we look for a curve $x^*(t) : t \in (0,1)$, such that

$$x^* = \text{argmin} \left\{ \int_0^1 [x^{'}]^t T(x) : x \in C \right\}$$

$$T(x) = -f_2(x)\nabla f_1(x)$$
$$C = \{x : [0,1] \to \mathbb{R}^n : x(0) = x_B, x(1) = x_A\}.$$

The Euler-Lagrange equation associated to this functional is

$$\int_0^1 ([\delta x^{'}]^t T(x) + [x^{'}]^t[[\delta T(x)]^t \delta x]) = 0, \forall x \text{ such that } \delta x(0) = \delta x(1) = 0,$$

but

$$[\delta T(x)]^t \delta x = -[(\delta x)^t \nabla f_2(x)]\nabla f_1(x) - f_2(x)[\nabla^2 f_1(x)\delta x].$$

Rewriting

$$[x^{'}]^t[[\delta T(x)]^t \delta x] = [\delta x]^t \delta T(x) x^{'},$$

we have

$$(T(x))^{'} = \delta T(x) x^{'}; x(0) = x_B, x(1) = x_A,$$

i.e.,

$$\nabla T(x) x^{'} = \delta T(x) x^{'}; x(0) = x_B, x(1) = x_A.$$

Example 3. We take $\overrightarrow{v} = (0, x_1, 0)$. On $(A, B) : v.dP = P_1 P_2^{'}$, such that

$$\oint_C v.dP = cst - \int_0^1 P_1 P_2^{'} dt,$$

thus, we can find $P^* : t \in (0, 1)$ such that

$$P^* = \text{argmin} \left\{ -\int_0^1 P_1 P_2^{'} dt : P \in K \right\},$$

where

$$P = (f_1(x(t)), f_2(x(t))).$$

We have

$$P_1^{'} = \nabla f_1(x)^t x^{'}, P_2^{'} = \nabla f_2(x)^t x^{'}$$

and

$$P_1 P_2^{'} = f_1(x) \nabla f_2(x)^t x^{'}.$$

Now, we look for a curve $x^*(t) : t \in (0, 1)$, such that

$$x^* = \text{argmin} \left\{ \int_0^1 [x^{'}]^t T(x) : x \in C \right\}$$

$$T(x) = f_1(x) \nabla f_2(x)$$

$$C = \{x : [0, 1] \to \mathbb{R}^n : x(0) = x_B, x(1) = x_A\}.$$

The Euler-Lagrange equation is

$$\int_0^1 ([\delta x^{'}]^t T(x) + [x^{'}]^t[[\delta T(x)]^t \delta x]) = 0, \forall x \text{ such that } \delta x(0) = \delta x(1) = 0,$$

or

$$[\delta T(x)]^t \delta x = [(\delta x)^t \nabla f_1(x)]\nabla f_2(x) + f_1(x)[\nabla^2 f_2(x)\delta x].$$

Rewriting

$$[x^{'}]^t[[\delta T(x)]^t \delta x] = [\delta x]^t \delta T(x) x^{'},$$

we have

$$(T(x))^{'} = \delta T(x) x^{'}; x(0) = x_B, x(1) = x_A,$$

i.e.,

$$\nabla T(x) x^{'} = \delta T(x) x^{'}; x(0) = x_B, x(1) = x_A.$$

5 Experimental Validation

To demonstrate the effectiveness and validity of this new method, two implementations have been proposed. The first one, based on a direct application of the approach, has the advantage of generating a sufficient number of points depending on the needs of the decision maker and the desired accuracy. This approach is valid only for convex problems. The second implementation consists in writing the decision variables as a polynomial function of a real and positive variable. It is valid for both convex and non convex multi-objective problems.

5.1 First Implementation

To illustrate the efficiency of the VAMO method, we will focus on the case of convex problems, through several test bi-objective problems. Solving the problem (5) can be done in several ways, either by constructing the Pareto curve point by point or by generating a set of points at one time. Algorithm 1 and 2 propose an iterative procedure based on the calculation of the surface by the trapezoidal method, and for each iteration we search the suitable point that can best represent the Pareto curve, taking into account the points identified earlier.

Algorithm 1. VAMO Algorithm

Require: Np : Maximum number of intended Pareto points, $Tolj$: Accepted Relative error for

Ensure: *sufficient Pareto Front*

$t = (f_1(x_1^*), f_1(x_2^*))$

$v = (f_2(x_1^*), f_2(x_2^*))$

$J = \frac{1}{2}.(t(2) - t(1)).(v(1) + v(2))$

$N = 0$

repeat

  $N = N + 1$

  Construct the function J(x) (see algorithm 2)

  Minimize $J(x)$ using a global algorithm, we obtain x_N^* et J_N^*

  $t \leftarrow [t, f_1(x_N^*)]$

  $v \leftarrow [v, f_2(x_N^*)]$

  $\delta_{Err} = |\frac{J_N^* - J}{J_N^*}|$

  $J = J_N^*$

until $\delta_{Err} \leq TolJ$ or $N = Np$

End

Algorithm 2. Function $J(x)$

Require: current t, v, and the variable x
Ensure: *Function of the surface approximation using the trapezoidal method*
$t^p \leftarrow [t, f_1(x)]$
$v^p \leftarrow [v, f_2(x)]$
Sort t^p and v^p
$J(x) = \frac{1}{2} \cdot \sum_{i=1}^{N+1} (t_{i+1}^p - t_i^p)(v_i^p + v_{i+1}^p)$
End

In this section, tree convex multi-objective problems were selected from the literature. The first problem is a single variable problem, it is used to illustrate the VAMO method and show the progress of the Pareto front construction. For the other test problem, a comparison with widely used methods in multi-objective optimization in the field of engineering will be considered. Namely: Normalized Normal Constraint (NNC), Normal Boundary Intersection (NBI), and Non-dominated Sorting Genetic Algorithm (NSGA II). The comparison is made by using these methods to generate the Pareto front for the second and the third problem.

Illustrative Example : Zhang Problem To illustrate the principle of this method, a bi-objective problem with a single variable is considered [15]. The problem formulation is:

$$\begin{cases} \min \quad F(\mathbf{x}) = (f_1(\mathbf{x}), f_2(\mathbf{x}))^T, \text{ where} \\ \qquad\qquad f_1(\mathbf{x}) = cosh(\mathbf{x}) \\ \qquad\qquad f_2(\mathbf{x}) = \mathbf{x}^2 - 12\mathbf{x} + 35 \\ \text{subject to} \qquad 0 \le \mathbf{x} \le 10. \end{cases} \tag{9}$$

Solving this problem, based on Algorithm 1, and using our algorithm RFNM (see the papers [16,17]), with a relative error of 10^{-4} leads to the Pareto front of Fig. 2. We observe that the points generated are rather focused on the curvature of the Pareto front, located between $t = 0.07$ and 0.40.

By increasing the number of points of the Pareto curve, we obtain a more accurate curve, with lower relative error. The Figs. 4, 5, 6 and 7 are obtained for a number of $1, 3, 5, 10, 20, 25, 30$ and 50 points. We notice that the quality of the Pareto curve is improved by increasing the number of points; an acceptable approximation of the Pareto curve is obtained as soon as the number of points is greater than 7; and a number of 30 points is enough to adequately represent the Pareto front (with a relative error of 0.52%). Figure 8 which represents the evolution of the relative error versus the required number of points, shows that the error decreases rapidly between 1 and 10 points; from 60 points, this error tends to zero. The relative error is $Err = \frac{J-J^*}{J^*} * 100$, where J^* is the exact value (for a large enough number N). The results for this first example, show that the VAMO method is effective for generating enough points to represent the Pareto front.

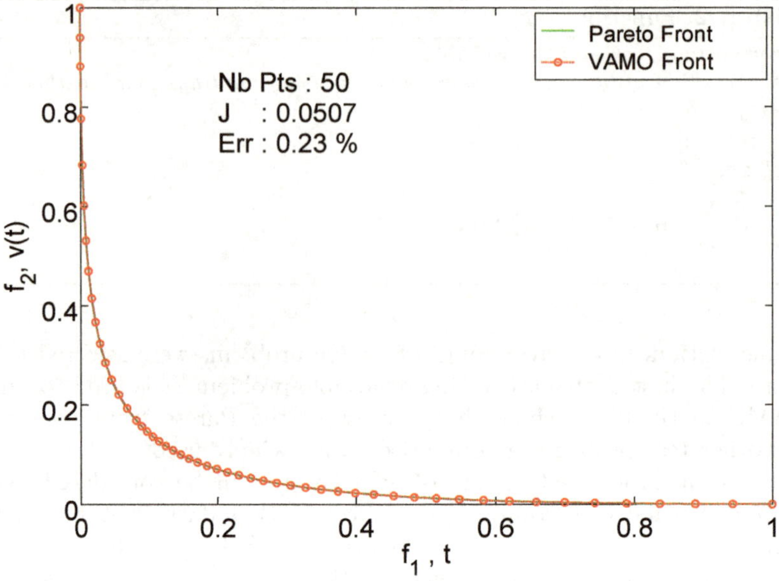

Fig. 3. Pareto curve for problem 1 (Zhang) using VAMO

Comparison with Other Multi-objective Methods. In this section, we propose a comparison of the VAMO method with methods which are widely used in multi-objective optimization in the field of engineering, namely the NNC, NBI and NSGA II methods.

* **Deb problem**

It is a bi-objective problem, with two variables [7,8]:

$$\begin{cases} \min \quad F(\mathbf{x}) = (f_1(\mathbf{x_1}, \mathbf{x_2}), f_2(\mathbf{x_1}, \mathbf{x_2}))^T, \qquad \text{with} \\ f_1(\mathbf{x_1}, \mathbf{x_2}) = \sin(\frac{\pi}{2}\mathbf{x_1}) \\ f_2(\mathbf{x_1}, \mathbf{x_2}) = \dfrac{(1 - e^{-\frac{(\mathbf{x_2}-0,1)^2}{0,0001}}) + (1 - 0,5e^{-\frac{(\mathbf{x_2}-0,8)^2}{0,8}})}{\arctan(100\mathbf{x_1})} \\ \text{subject to} \quad 0 \leq \mathbf{x_i} \leq 1, \ i = 1,2 \end{cases} \tag{10}$$

Parameters: Concerning NBI and NNC, the number of points in the Pareto front is set to 21 and our RFNM hybrid method is used to solve the generated subproblems. Regarding NM, we adopted the standard parameters recommended by the authors. The used stopping criteria are: maximum number of function evaluations, $MAXFUN = 6000$, maximum number of iterations, $maxiter = 6000$; tolerance on improving the function $Tolf = 10^{-4}$, and the tolerance on the relative variance of x is $tolx = 10^{-4}$.

Regarding the size of the sample for RFNM, it is set to a value less than 50.

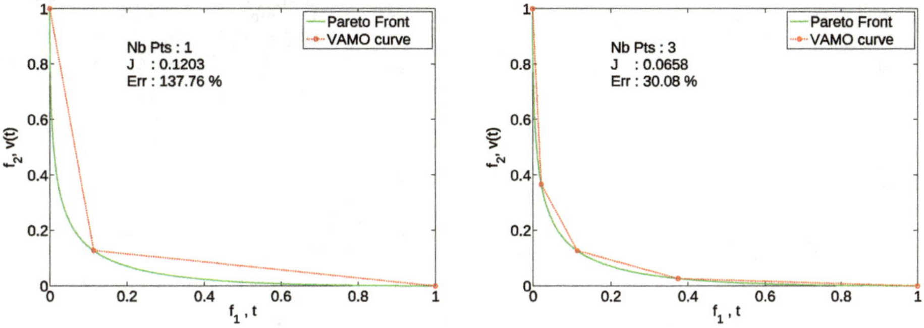

Fig. 4. Pareto front for 1 and 3 points

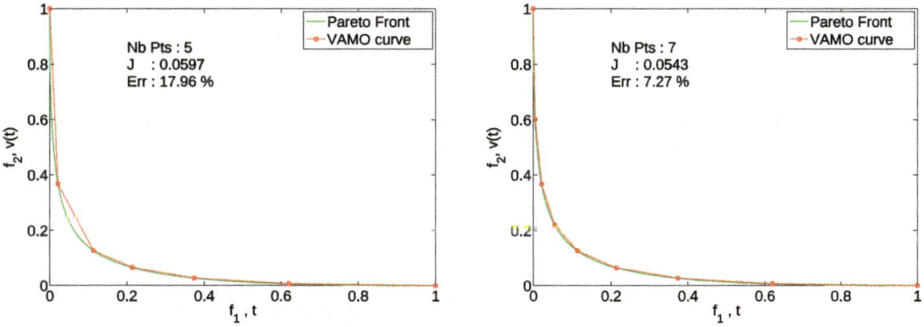

Fig. 5. Pareto front for 5 and 7 points

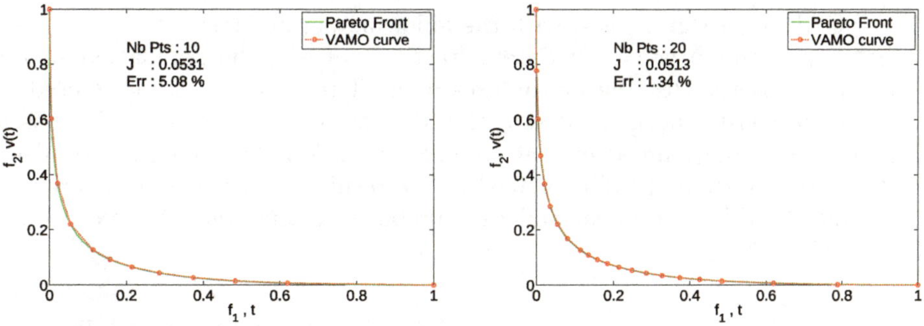

Fig. 6. Pareto front for 10 and 20 points

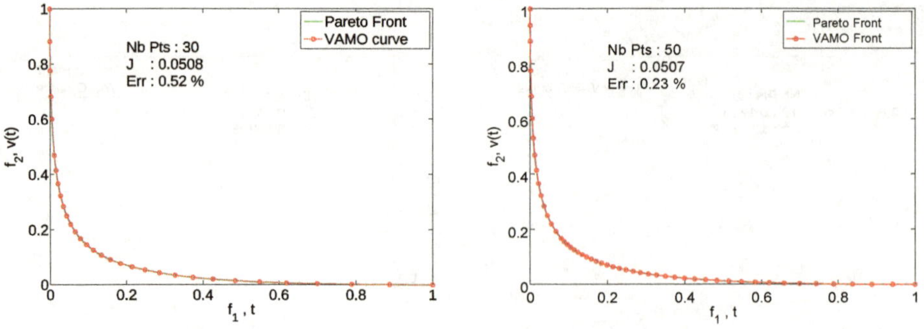

Fig. 7. Pareto front for 30 and 50 points

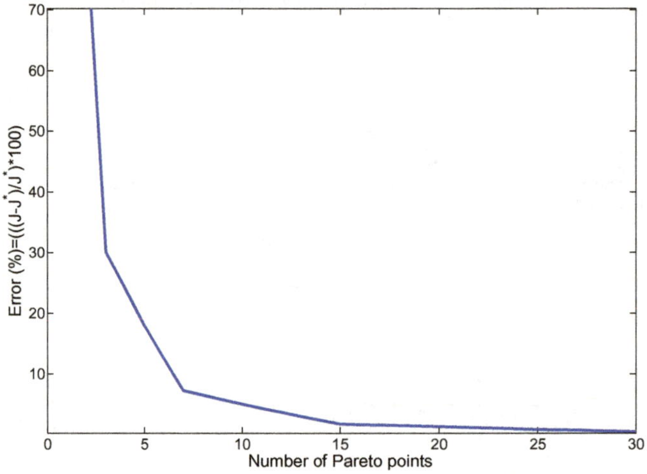

Fig. 8. Evolution of the relative error as a function of the number of the Pareto points, using VAMO

The NSGAII algorithm is used with the following parameters:
Probability of crossover = 0.8; Pareto fraction = 0.45. The population size = 1000, the Crossover Distribution indexes = 20. The penalty method is used to overcome the constraints generated by the NBI and NNC methods. The proposed algorithms are programmed in Matlab and run on Windows 7 with Intel Core Duo2 with 2.0 GHz and 2 GB of RAM. The results are presented in terms of CPU time (CPUT in seconds) and the number of evaluations of the objective function (NevF).

Results
The results of the comparison are presented in Figs. 9, 10, 11, 12, and Table 1.

It is observed that the NSGAII method is able to generate the Pareto front, except for $f_2 > 6.0$ (see Fig. 11) with CPU time and number of function evaluation higher than those for the NBI and NNC methods. However, the Pareto points are not uniformly distributed. The results also show that the NBI and NNC methods generate a Pareto front evenly distributed. The NBI method is slow and costly in terms of the function evaluation then the NNC method, this is due to the equality constraints that consume a lot of CPU time. The quality of the Pareto front in the region corresponding to f_1 located between 0.01 and 0.1 is not good enough, since 3 generated points are not sufficient to represent this part correctly (see Fig. 9 and 10).

We also observe that a very significant improvement in the quality of the Pareto curve is obtained by using VAMO. Indeed the 17 points that represent the Pareto front are distributed optimally to cover this curve with a faster convergence (The lowest CPU time and number of function evaluations) than the other methods.

*** Zitzler problem ZDT1**

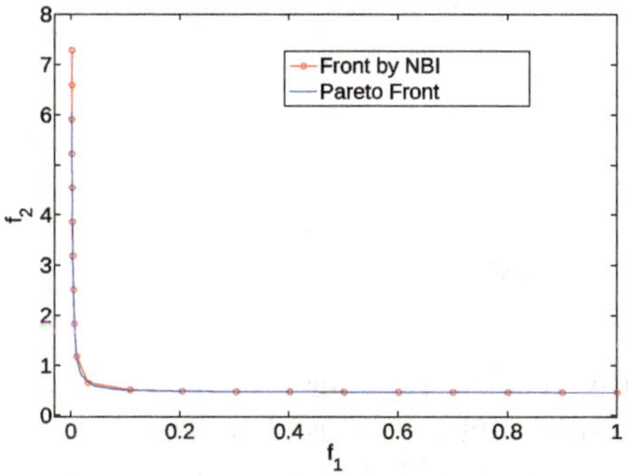

Fig. 9. Pareto Front generated by NBI method for Deb problem

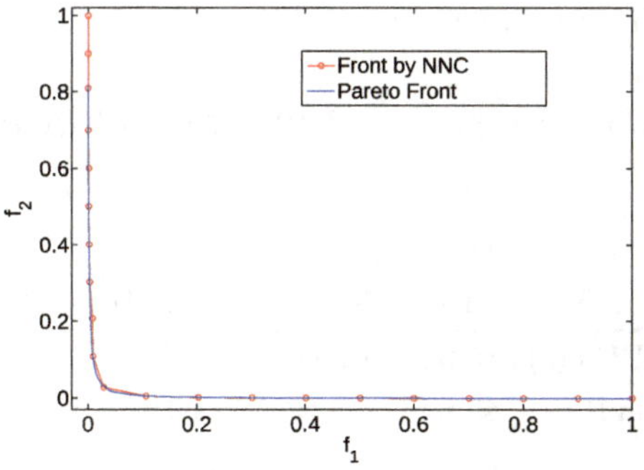

Fig. 10. Pareto Front generated by NNC method for Deb problem (normalized functions)

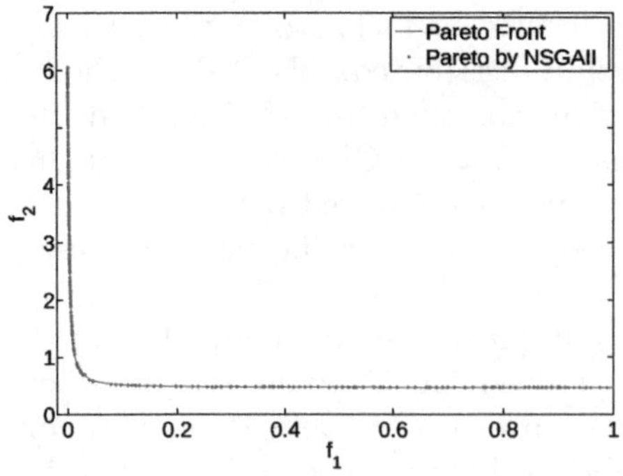

Fig. 11. Pareto Front generated by NSGAII method for Deb problem

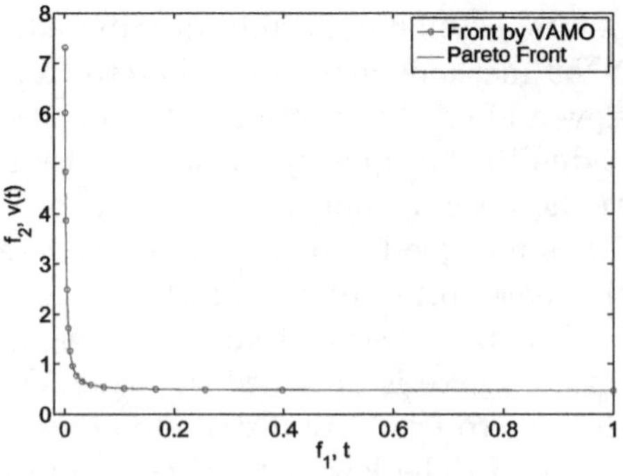

Fig. 12. Pareto Front generated by VAMO method for Deb problem

Its formulation is as follows [18]:

$$\begin{cases} \min \quad F(\mathbf{x}) = (f_1(\mathbf{x}), f_2(\mathbf{x}))^T, \text{ where} \\ \qquad f_1(\mathbf{x}) = \mathbf{x_1} \\ \qquad f_2(\mathbf{x}) = g(\mathbf{x}).(1 - \sqrt{\frac{\mathbf{x_1}}{g(\mathbf{x})}}) \\ \qquad g(\mathbf{x}) = 1 + 9.\frac{(\sum_{i=2}^{n} \mathbf{x_i})}{(n-1)} \\ \text{subject to} \quad 0 \le \mathbf{x_i} \le 1\, to\, 5. \end{cases} \tag{11}$$

<u>Parameters</u>
Concerning NBI and NNC, the number of points in the Pareto front is set to 21 and our RFNM hybrid method is used to solve the subproblems generated by these methods. Regarding NM, the standard parameters recommended by the authors are adopted. The used stopping criteria are: maximum number of function evaluations, $MAXFUN = 6000$, maximum number of iterations, $maxiter = 6000$; tolerance on improving the function $Tolf = 10^{-4}$, and tolerance on the relative variance of x is $Tolx = 10^{-4}$. Regarding the size of the sample for RFNM, it is set to a value less than 250. The NSGAII algorithm is used with the following parameters:

Table 1. Comparison between methods NBI, NNC, NSGA II, and VAMO for Deb problem

	Methods			
	NBI	NNC	NSGA II	VAMO
CPUT(s)	15.90	14.98	17.55	9.21
NevF	45 648	36 817	54 001	15 216

Probability of crossover = 0.8; Pareto fraction = 0.45. The population size = 500, the Crossover Distribution indexes = 20. The penalty method is used to overcome the constraints generated by the NBI and NNC methods. The proposed algorithms are programmed in Matlab and run on Windows 7 with Intel Core Duo2 with 2.0 GHz and 2 GB of RAM. The results are presented in terms of CPU time (CPUT in seconds) and the number of evaluations of the objective function (NevF).

Results

The results are presented in Figs. 13, 14, 15, and Table 2. We observe that the NSGAII method generates a Pareto solution located below the optimal Pareto front (see Fig. 14, with CPU time and number of function evaluation higher than those for NBI and NNC methods. The Pareto points are not uniformly distributed. The results show also that the NBI and NNC methods generate an evenly distributed Pareto front. The NBI method is slow and costly in terms of the function evaluation number compared to the NNC method. This is due to the equality constraints that consume a lot of CPU time.

VAMO method provided a Pareto curve with only 11 points with an optimum distribution and faster convergence (lowest CPU time and number of function evaluations) than other methods.

Table 2. Comparison between NBI, NNC, NSGAII and VAMO methods for ZDT1 problem

	Methods			
	NBI	NNC	NSGA II	VAMO
CPUT(s)	19.95	18.28	22.26	9.79
NevF	82 299	64 614	103 001	48 564

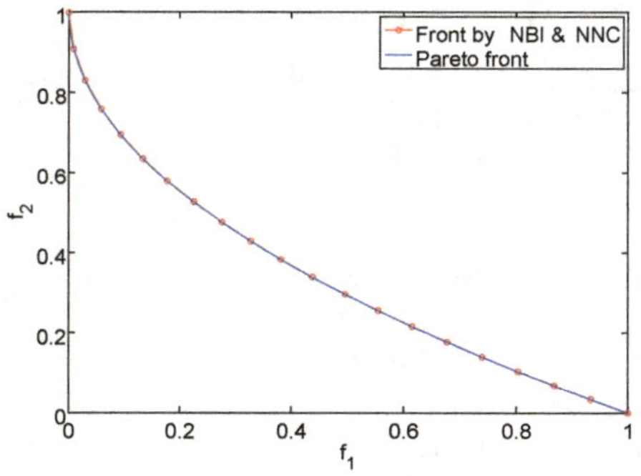

Fig. 13. Pareto Front generated by NBI and NNC methods for ZDT1 problem

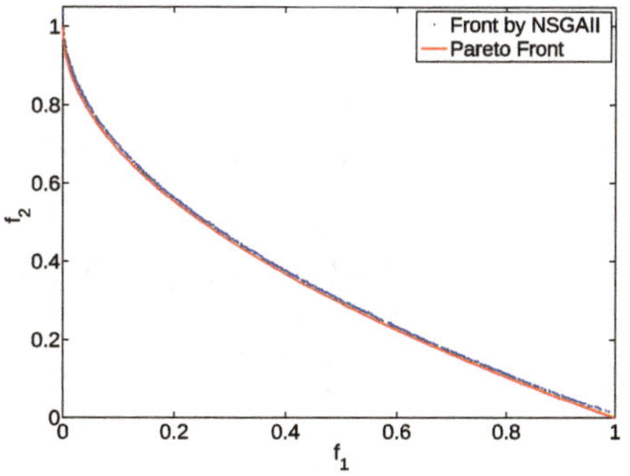

Fig. 14. Pareto Front generated by NSGAII method for ZDT1 problem

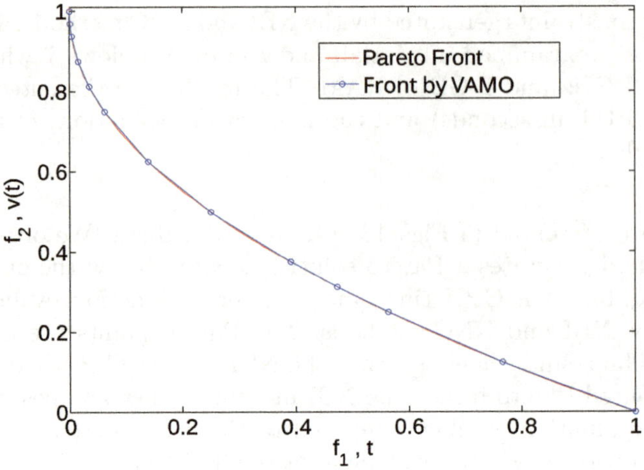

Fig. 15. Pareto Front generated by VAMO method for ZDT1 problem

5.2 Second Implementation

The second implementation consists in writing the decision variables x_i as a polynomial function of a variable $t \in [0, 1]$ $(x_i(t) = \sum_{i=1}^{n} a_i t^i)$ (see [16] for more details). It is valid for both convex and non convex problems. A bi-objective optimization problem can then be written as:

$$\min \int_{t=0}^{t=1} f_1(x(t)) df_2(x(t))$$

A convex problem: The problem is formulated as follows:

$$\begin{cases} \min \quad F(\mathbf{x}) = (f_1(\mathbf{x}), f_2(\mathbf{x}))^T, \quad \text{where} \\ \qquad f_1(\mathbf{x}) = \sum_0^1 x_i^2 \qquad i = 1 \ \ to \ \ 3 \\ \qquad f_2(\mathbf{x}) = \|(x-1)\|_2^2 \\ \text{subject to} \qquad -10 \leq \mathbf{x}_i \leq 10. \end{cases} \tag{12}$$

A nonconvex problem: The problem is formulated as follows:

$$\begin{cases} \min \quad F(\mathbf{x}) = (f_1(\mathbf{x}), f_2(\mathbf{x}))^T, \quad where: \\ \qquad f_1(\mathbf{x}) = |x_1| \\ \qquad g(\mathbf{x}) = 1 + 9 * |x_2| \\ \qquad f_2(\mathbf{x}) = g(x) * |1 - (\frac{f_1(x)}{g(x)})^2| \\ \text{subject to} \qquad -10 \leq \mathbf{x}_i \leq 10. \end{cases} \tag{13}$$

Solving the convex and the nonconvex problems with VAMO, based on the second approach, and using Matlab Nelder Mead toolbox with a relative error of 10^{-8} with a dimension of four leads to the Pareto front of Figs. 16 and 17.

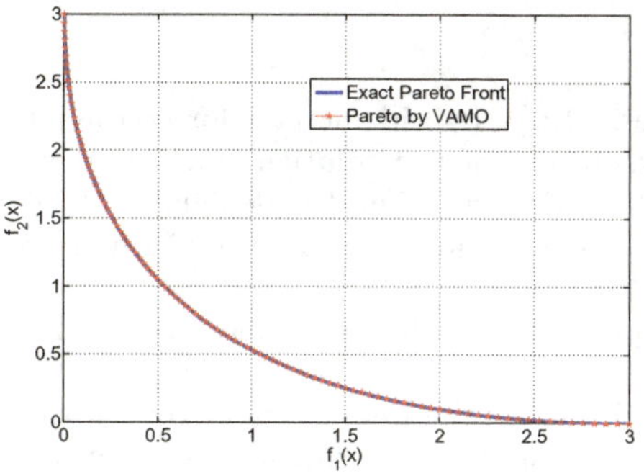

Fig. 16. Pareto Front generated by VAMO for convex problem

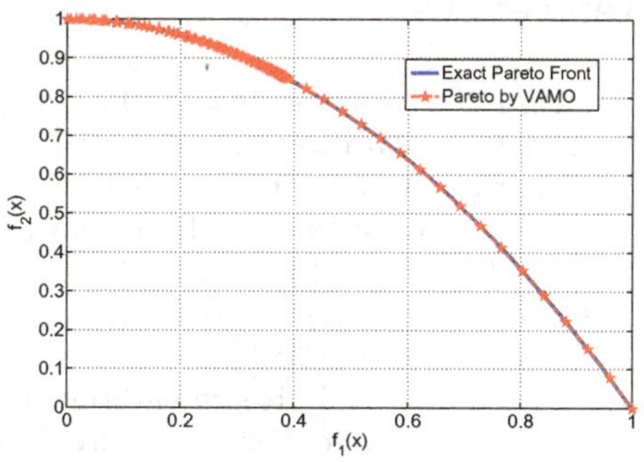

Fig. 17. Pareto Front generated by VAMO for nonconvex problem

For both problems the algorithm succeed in producing the whole extent of the Pareto Front. We observe that the points generated for the nonconvex problem are rather focused on the curvature of the Pareto front.

5.3 Discussion

The two proposed implementations give very satisfactory results and demonstrate the effectiveness of the approach adopted. The first is valid only for convex problems, but can be improved by introducing penalty constraints to be able to reach the points of the non-convex area. For the second implementation, valid for convex and non-convex problems, the choice of the polynomial basis can be not adequate for certain problems, a sinusoidal or finite element basis can also be considered. Other forms of implementation can be implemented, for example coupling with evolutionary multi-objective methods such as NSGA and SPEA2 which work with a population. These ideas, and others may arise in further studies.

6 Conclusion

A new method to solve multi-objective problems, called VAMO, is presented. It deals with convex and non-convex multi-objective optimization problems by transforming them into a problem of variational calculus, which may be solved in order to generate an approximation of the Pareto's front. Two implementations of the proposed approach were presented, concerning convex and non convex problems. Experiments on some test problems of the literature were performed and demonstrated the effectiveness and efficiency of this approach. Comparison with well-known methods in the field of multi-objective optimization have shown that VAMO is more effective and more efficient than other approaches. Applications to non-convex problems in the field of engineering science, and suggestions of new implementations of this method will be the subject of future works.

References

1. Bez, E.T., Souza de Cursi, J., Gonçalves, M.: A hybrid method for continuous global optimization involving the representation of the solution. In: 6th World Congress on Structural and Multidisciplinary Optimization, Rio de Janeiro (2005)
2. Charnes, A., Wolfe, M.: Extended pincus theorems and convergence of simulated annealing. Int J. Syst. Sci. **20**(8), 1521–1533 (1989)
3. Collette, Y., Siarry, P.: Multi-Objective Optimization : Principles and Case Studies. Springer, Berlin (2003)
4. Souza de Cursi, J.: Representation of solutions in variational calculus. In: Tarocco, E., de Souza Neto, E.A., Novotny, A.A. (eds.) Variational Formulations in Mechanics: Theory and Applications, pp. 87–106. CIMNE, Barcelona (2007)
5. Das, I., Dennis, J.: A closer look at drawbacks of minimizing weighted sums of objectives for pareto set generation in multicriteria optimization problems. Structural Optimization **14**, 63–69 (1997)
6. Das, I., Dennis, J.E.: Normal-boundary intersection: a new method for generating pareto optimal point in nonlinear multicriteria optimization problems. SIAM J. Optimizat. **8**(3), 631–657 (1998)
7. Deb, K.: Multi-Objective Optimization Using Evolutionary Algorithms. Wiley, Hoboken (2001)
8. Deb, K., Agrawal, S., Pratap, A.: A fast elitist non-dominated sorting genetic algorithm for multi-objective optimization:NSGA-II. In: Proceedings of the 6th International Conference on Parallel Problem Solving from Nature, pp. 849–858 (2000)
9. Falk, J.E.: Condition for global optimization in nonlinear programming. Oper. Res. **21**, 337–340 (1973)
10. Mavrotas, G.: Effective implementation of the e-constraint method in multi-objective mathematical programming problems. Appl. Math. Comput. **213**(2), 455–465 (2009). https://doi.org/10.1016/j.amc.2009.03.037. http://www.sciencedirect.com/science/article/pii/S0096300309002574
11. Messac, A., Ismail-Yahaya, A., Mattson, C.: The normalized normal constraint method for generating the pareto frontier. Struct. Multidisciplinary Optimizat. **25**(2), 86–98 (2003)
12. Miettinen, K.: Nonlinear Multiobjective Optimization. Kluwer Academic Publishers, Dordrecht (1999)
13. Pincus, M.: A closed formula solution of certain programming problems. Oper. Res. **16**(3), 690–694 (1968)
14. Stadler, W.: A survey of multicriteria optimization or the vector maximum problem, part I: 1776–1960. J. Optim. Theory Appl. **29**(1), 1–52 (1979)
15. Zhang, W.H., Gao, T.: A min-max method with adaptive weightings for uniformly spaced pareto optimum points. Comput. Struct. **84**(28), 1760–1769 (2006). Elsevier
16. Zidani, H., De Cursi, J.E.S., Ellaia, R.: Numerical approximation of the solution in infinite dimensional global optimization using a representation formula. J. Global Optimizati. **65**(2), 261–281 (2016). https://doi.org/10.1007/s10898-015-0357-5
17. Zidani, H., Pagnacco, E., Sampaio, R., Ellaia, R., de Cursi, J.E.S.: Multi-objective optimization by a new hybridized method: applications to random mechanical systems. Eng. Optimizat. **45**(8), 917–939 (2013). https://doi.org/10.1080/0305215X.2012.713355

18. Zitzler, E., Deb, K., Lothar, T.: Comparison of multiobjective evolutionary algorithms: empirical results. J. Evolutionary Comput. **8**(2), 173–195 (2000)
19. Zitzler, E.: Multiobjective evolutionary algorithms: a comparative case study and the strength pareto approach. IEEE Trans. Evolutionary Comput. **3**(4), 257–271 (1999)

Author Index

J. E. S. De Cursi (Ed.): Uncertainties 2020, LNME, pp. 471–472, 2021.
https://doi.org/10.1007/978-3-030-53669-5

CPSIA information can be obtained
at www.ICGtesting.com
Printed in the USA
BVHW010045030920
R11156500001B/R111565PG587325BVX18B/1

9 783030 536688